P9-AFZ-605

Methods in Cell Biology

VOLUME 47
Cilia and Flagella

Series Editors

Leslie Wilson
Department of Biological Sciences
University of California, Santa Barbara
Santa Barbara, California

Paul Matsudaira
Whitehead Institute for Biomedical Research and
Department of Biology
Massachusetts Institute of Technology
Cambridge, Massachusetts

Methods in Cell Biology

Prepared under the Auspices of the American Society for Cell Biology

VOLUME 47
Cilia and Flagella

Edited by

William Dentler
Department of Physiology and Cell Biology
University of Kansas
Lawrence, Kansas

and

George Witman
Worcester Foundation for Experimental Biology
Shrewsbury, Massachusetts

ACADEMIC PRESS
San Diego New York Boston London Sydney Tokyo Toronto

Cover photograph (paperback edition only): Immunofluorescence image of *Chlamydomonas reinhardtii* labeled with a monoclonal antibody against centrin. The autofluorescence of chlorophyll is red; centrin localization is shown in yellow/green. Both flagella and the nucleus-basal body connector stain with anticentrin antibodies. See chapter 24, figure 1B for further details.

This book is printed on acid-free paper. 

Academic Press, Inc.
A Division of Harcourt Brace & Company
525 B Street, Suite 1900, San Diego, California 92101-4495

United Kingdom Edition published by
Academic Press Limited
24-28 Oval Road, London NW1 7DX

International Standard Serial Number: 0091-679X

International Standard Book Number: 0-12-564148-6 (Hardcover)

International Standard Book Number: 0-12-209710-6 (Paperback)

PRINTED IN THE UNITED STATES OF AMERICA
95 96 97 98 99 00 EB 9 8 7 6 5 4 3 2 1

This volume is dedicated to Joel Rosenbaum, our mentor. He continues to be a major contributor of new knowledge, ideas, and spirit to all of us that study cilia and flagella.

William Dentler
George Witman

CONTENTS

PART II Culture and Isolation of Ciliated Epithelial Cells and Cilia

PART III General Methods

PART V Motors and Motion Analysis

PART IX Genetics and Molecular Tools for the Analysis of Cilia and Flagella

CONTRIBUTORS

Numbers in parentheses indicate the pages on which the authors' contributions begin.

David J. Asai (579) Department of Biological Sciences, Purdue University, West Lafayette, Indiana 47907

Andre T. Baron (341) Laboratory for Cell Biology, Department of Biochemistry and Molecular Biology, Mayo Clinic Foundation, Rochester, Minnesota 55905

Mitchell Bernstein (425) Department of Biology, Yale University, New Haven, Connecticut 06520

Robert A. Bloodgood (121, 273) Department of Cell Biology, University of Virginia School of Medicine, Charlottesville, Virginia 22908

Emmanuelle Boisvieux-Ulrich (75) Laboratoire de Cytophysiologie et de Toxicologie Cellulaire, Université Paris, 75251 Paris 05, France

G. Benjamin Bouck (25, 129, 355) Department of Biological Sciences, University of Illinois at Chicago, Chicago, Illinois 60607

Gerácimo E. Bracho (447) Department of Physiology, University of Kansas Medical Center, Kansas City, Kansas 66160

Angela Bremerich (315) Botanisches Institut der Universität zu Köln, D-50931 Köln, Germany

Monica Brito (385, 391) Instituto de Bioquímica, Universidad Austral de Chile, Valdivia, Chile

Charles J. Brokaw (231) Division of Biology, California Institute of Technology, Pasadena, California 91125

Beth Burnside (83) Department of Molecular and Cell Biology, Division of Cell and Developmental Biology, University of California at Berkeley, Berkeley, California 94720

Luis O. Burzio (385, 391) Instituto de Bioquímica, Universidad Austral de Chile, Valdivia, Chile

Gail L. Carlson (473) Department of Biochemistry, University of Wisconsin-Madison, Madison, Wisconsin 53706

Bernadette Chailley (75) Développement Normal et Pathologique des Fonctions Epithéliales, INSERM U319, Université Paris 75251 Paris 05, France

S.-J. Chen (129) Department of Biological Sciences, University of Illinois at Chicago, Chicago, Illinois 60607

Alan B. Clark (57) National Institute of Environmental Health Sciences, Laboratory of Pulmonary Pathobiology, Airway Cell Biology Group, Research Triangle Park, North Carolina 27709

Peggy S. Criswell (579) Department of Biological Sciences, Purdue University, West Lafayette, Indiana 47907

Lawrence J. Dangott (437) Department of Neurobiology, Harvard Medical School, Boston, Massachusetts 02115

William L. Dentler (13, 397, 407) Department of Physiology and Cell Biology, University of Kansas, Lawrence, Kansas 66045

Dennis R. Diener (545) Department of Biology, Yale University, New Haven, Connecticut 06511

Jacquelyn Dinusson (341) Laboratory for Cell Biology, Department of Biochemistry and Molecular Biology, Mayo Clinic Foundation, Rochester, Minnesota 55905

Ellen R. Dirksen (65, 289) Department of Anatomy and Cell Biology, University of California at Los Angeles, Los Angeles, California 90024

Susan K. Dutcher (323, 531) Department of Molecular, Cellular, and Developmental Biology, University of Colorado, Boulder, Colorado 80309

Ramesh Errabolu (341) Laboratory for Cell Biology, Department of Biochemistry and Molecular Biology, Mayo Clinic Foundation, Rochester, Minnesota 55905

Jennifer A. Felix (65) Department of Anatomy and Cell Biology, University of California at Los Angeles, Los Angeles, California 90024

Jacek Gaertig (559) Department of Zoology, The University of Georgia, Athens, Georgia 30602

Jean-Luc Gatti (47) Laboratoire de Physiologie de la Reproduction des Mammifères Domestiques, INRA Nouzilly, 37380 Monnaie, France

Martin A. Gorovsky (559, 571) Department of Biology, University of Rochester, Rochester, New York 14627

Thomas E. Gray (57) National Institute of Environmental Health Sciences, Laboratory of Pulmonary Pathobiology, Airway Cell Biology Group, Research Triangle Park, North Carolina 27709

Robert Hard (225) Department of Anatomy and Cell Biology, State University of New York at Buffalo, Buffalo, New York 14214

Annette T. Hastie (93) Department of Medicine, Jefferson Medical College, Philadelphia, Pennsylvania 19107

Lee Hon Cheung (43, 401) Department of Physiology, University of Minnesota, Minneapolis, Minnesota 55455

Jerry E. Honts (301) Department of Biochemistry, University of Arizona, Tucson, Arizona 85712

Harold J. Hoops (193) Department of Biology, State University of New York at Geneseo, Geneseo, New York 11454

Cynthia J. Horst (207) Biology Department, Carroll College, Waukesha, Wisconsin 53186

David R. Howard (257, 481) Department of Anatomy and Cell Biology, Emory University School of Medicine, Atlanta, Georgia 30322

Sumio Ishijima (239, 245) Biological Laboratory, Tokyo Institute of Technology, O-okayama, Meguro-ku, Tokyo 152, Japan

Jonathan W. Jarvik (307) Department of Biological Sciences, Carnegie Mellon University, Pittsburgh, Pennsylvania 15213

Karl A. Johnson (153) Department of Cell, Molecular, and Developmental Biology, Haverford College, Haverford, Pennsylvania 19041

Osamu Kagami (147, 487) National Institute of Bioscience and Human Technology, Tsukuba 305, Japan

Ritsu Kamiya (147, 487, 541) Zoological Institute, University of Tokyo, Tokyo 113, Japan

Pushpa Kathir (525) Department of Genetics and Cell Biology and Plant Biology, University of Minnesota, St. Paul, Minnesota 55108

Laura R. Keller (551) Department of Biological Science, Florida State University, Tallahassee, Florida 32306

Stephen M. King (9, 141, 503) Department of Biochemistry, University of Connecticut Health Center, Farmington, Connecticut 06030

Keith G. Kozminski (263) Department of Biology, Yale University, New Haven, Connecticut 06511

Ryoko Kuriyama (365) Department of Cell Biology and Neuroanatomy, University of Minnesota, Minneapolis, Minnesota 55455

Karl-Ferdinand Lechtreck (315, 335) Botanisches Institut, Universität zu Köln, D-50931 Köln, Germany

Paul A. Lefebvre (3, 513, 519, 525) Department of Genetics and Cell Biology and Plant Molecular Genetics Institute, University of Minnesota, St. Paul, Minnesota 55108

Richard W. Linck (365, 373) Department of Cell Biology and Neuroanatomy, University of Minnesota, Minneapolis, Minnesota 55455

Hans Machemer (419) Arbeitsgruppe Zelluläre Erregungsphysiologie, Ruhr-Universität, D-44780 Bochum, Germany

Silvio P. Marchese-Ragona (171, 177) TopoMetrix Corporation, Santa Clara, California 95054

David Mastronarde (183) Laboratory for Three-Dimensional Fine Structure, Department of Molecular, Cellular, and Developmental Biology, University of Colorado, Boulder, Colorado 80309

J. Richard McIntosh (183) Laboratory for Three-Dimensional Fine Structure, Department of Molecular, Cellular, and Developmental Biology, University of Colorado, Boulder, Colorado 80309

Michael Melkonian (315, 335) Botanisches Institut, Universität zu Köln, D-50931 Köln, Germany

Anthony G. Moss (281) Department of Zoology and Wildlife Sciences, Auburn University, Auburn, Alabama 36849

Yutaka Naitoh (211) Pacific Biomedical Research Center and Department of Microbiology, University of Hawaii at Manoa, Honolulu, Hawaii 96822

David L. Nelson (17, 473) Department of Biochemistry, University of Wisconsin-Madison, Madison, Wisconsin 53706

Julie A. Nelson (513) Department of Genetics and Cell Biology and Plant Molecular Genetics Institute, University of Minnesota, St. Paul, Minnesota 55455

Paul Nettesheim (57) National Institute of Environmental Health Sciences,

Laboratory of Pulmonary Pathobiology, Airway Cell Biology Group, Research Triangle Park, North Carolina 27709

Huân M. Ngô (25, 355) Department of Biological Sciences, University of Illinois at Chicago, Chicago, Illinois 60607

Eileen O'Toole (183) Laboratory for Three-Dimensional Fine Structure, Department of Molecular, Cellular, and Developmental Biology, University of Colorado, Boulder, Colorado 80309

Charlotte K. Omoto (507) Department of Genetics and Cell Biology, Washington State University, Pullman, Washington 99164

Lawrence E. Ostrowski (57) National Institute of Environmental Health Sciences, Laboratory of Pulmonary Pathobiology, Airway Cell Biology Group, Research Triangle Park, North Carolina 27709

Tim Otter (113) Department of Biology, Albertson College of Idaho, Caldwell, Idaho 83605

Kathryn Pagh-Roehl (83) Department of Molecular and Cell Biology, Division of Cell and Developmental Biology, University of California at Berkeley, Berkeley, California 94720

Gregory J. Pazour (281) Worcester Foundation for Experimental Biology, Shrewsbury, Massachusetts 01545

David G. Pennock (571) Department of Zoology, Miami University, Oxford, Ohio 45056

David M. Phillips (199) The Population Council, New York, New York 10021

Gianni Piperno (101, 107, 381) Department of Cell Biology and Anatomy, Mount Sinai School of Medicine, New York, New York 10029

Mark A. Pirner (373) Department of Cell Biology and Neuroanatomy, University of Minnesota, Minneapolis, Minnesota 55455

Mary E. Porter (183) Department of Cell Biology and Neuroanatomy, University of Minnesota School of Medicine, Minneapolis, Minnesota 55455

Scott H. Randell (57) Cystic Fibrosis/Pulmonary Research and Treatment Center, Department of Medicine, University of North Carolina at Chapel Hill, Chapel Hill, North Carolina 27599

T. K. Rosiere (129) Abbott Laboratories, Diagnostic Division, Cancer Business Unit, Abbott Park, Illinois 60064

Winfield S. Sale (257, 481) Department of Anatomy and Cell Biology, Emory University School of Medicine, Atlanta, Georgia 30322

J. L. Salisbury (163, 341) Department of Biochemistry and Molecular Biology, Laboratory for Cell Biology, Mayo Clinic Foundation, Rochester, Minnesota 55905

Nancy L. Salomonsky (121) Department of Cell Biology, University of Virginia School of Medicine, Charlottesville, Virginia 22908

Jovenal T. San Agustin (31, 135, 251) Male Fertility Program, Worcester Foundation for Experimental Biology, Shrewsbury, Massachusetts 01545

Mark A. Sanders (163) Department of Biochemistry and Molecular Biology, Laboratory for Cell Biology, Mayo Clinic Foundation, Rochester, Minnesota 55905

Michael J. Sanderson,[1] (65, 289) Department of Anatomy and Cell Biology, University of California at Los Angeles, Los Angeles, California 90024

Christèle Saudrais (47) Laboratoire de Physiologie des Poissons, INRA Campus de Rennes-Baulieu 35042 Rennes Cedex, France

Takashi Shimizu (497) National Institute of Bioscience and Human-Technology, Tsukuba, Ibaraki 305, Japan

Carolyn D. Silflow (525) Department of Genetics and Cell Biology and Department of Plant Biology, University of Minnesota, St. Paul, Minnesota 55108

Elizabeth F. Smith (491) Department of Genetics and Cell Biology, University of Minnesota, St. Paul, Minnesota 55108

William J. Snell (459) Department of Cell Biology and Neuroscience, University of Texas Southwestern Medical School, Dallas, Texas 75235

Raymond E. Stephens (37, 361, 431) Department of Physiology, Boston University School of Medicine, Boston, Massachusetts 03228

Bruce E. Taillon (307) Department of Biology, Washington University in St. Louis, St. Louis, Missouri 63130

Lai-Wa Tam (519) Department of Genetics and Cell Biology, University of Minnesota, St. Paul, Minnesota 55108

Joseph S. Tash (447) Department of Physiology, University of Kansas Medical Center, Kansas City, Kansas 66160

Robert M. Tombes (467) Division of Hematology and Oncology, Medical College of Virginia, Richmond, Virginia 23298

J. S. Wall (171) Department of Biology, Brookhaven National Laboratory, Upton, New York 11973

Norman E. Williams (301) Department of Biological Sciences, University of Iowa, Iowa City, Iowa 52242

George B. Witman (31, 135, 207, 251, 281) Male Fertility Program, Worcester Foundation for Experimental Biology, Shrewsbury, Massachusetts 01545

Robin Wright (413) Department of Zoology, University of Washington, Seattle, Washington 98195

Yuhua Zhang (459) Department of Cell Biology and Neuroscience, University of Texas Southwestern Medical School, Dallas, Texas 75235

[1] Present address: Department of Physiology, University of Massachusetts Medical Center, Worcester, Massachusetts 01655.

PREFACE

This book contains a collection of methods that we believe will be particularly useful to investigators studying cilia and flagella. These methods include both "classic" techniques that have stood the test of time and are likely to continue to be used with only minor modifications and more recently developed techniques that promise to revolutionize many aspects of research on cilia and flagella. Certain organisms have been emphasized more than others; this is due partly to the fact that some organisms have been more extensively studied than others and partly to our perception that certain model systems, particularly those amenable to genetic and biochemical analyses, will be most valuable for future investigations. However, it is emphasized that the vast majority of techniques described here should be generally applicable to other systems and thus serve as valuable starting points for the analysis of cilia and flagella from less well-studied organisms or tissues. We have aimed this book primarily at the researcher who is actively investigating cilia or flagella or who may have need for ciliary or flagellar components such as axonemes, tubulin, or dynein as biochemical reagents. Nevertheless, many of the techniques described could be readily adapted for use by teachers in a high school or undergraduate laboratory setting.

The first two sections contain methods for culturing ciliated and flagellated cells and for isolating cilia and flagella from these cells. We hope that the second section, focused on mammalian epithelial cilia, will stimulate increased investigation of these terminally differentiated cells. Sections 3–5 describe methods of general use for studying the composition, structure, and motor activity of cilia and flagella. Section 6 describes methods for purifying components of the cilia- or flagella-associated cytoskeleton; these components are physically connected to the flagellar axoneme and are important in ciliary morphogenesis and in establishing and maintaining the spatial relationships between cilia and other structures. Sections 7 and 8 describe methods for fractionating cilia and flagella and purifying or otherwise identifying components essential for ciliary and flagellar functioning. The final section describes techniques for the genetic and molecular genetic analysis of flagella. Many of the latter methods have only recently been developed but already are contributing immensely to our knowledge of flagellar structure and function at the molecular level.

While cilia and flagella are inherently interesting and are important objects of investigation unto themselves, they also provide opportunities for discovering and understanding proteins and processes of widespread distribution in cells. For example, dynein was first identified as a ciliary protein, and it was knowledge of its biochemical and structural properties that led to the realization that an isoform of this enzyme is responsible for many types of movement in the cytoplasm. Ciliary and flagellar dyneins are still the best understood of all dyneins, and, in contrast to cytoplasmic dynein, alterations to flagellar dynein are not lethal to the cell, so it is apt to be the superior model system for site-directed mutagenesis to investigate critical functional domains common to all dyneins. Other proteins, most readily studied in ciliary and flagellar preparations, are likely to have important roles far beyond these cell organelles. Thus, we anticipate that analysis of flagellar components, including many yet to be discovered, will continue to elucidate fundamental principles of cell function and architecture.

William Dentler
George Witman

INTRODUCTION

Cilia and eukaryotic flagella (these terms will be used interchangeably throughout this volume) have been studied for over 100 years. With each new study our appreciation for the complexity of these organelles increases. Best known for their role in propelling single cell organisms and sperm through water and for cleansing the surfaces of vertebrate respiratory and reproductive tracts, they serve a multitude of other organism-specific functions, including excretion and water balance in some excretory organs, circulation of fluids within the coelomic cavity, brain, and spinal cord, and selection and transport of food particles along the surfaces of gills. In addition to their roles in motility, cilia and flagella serve a variety of sensory functions, including cell–cell recognition during mating in protistans, mechanoreception in arthropods, geotaxis in mollusks, anchorage of parasitic protozoans to their hosts, and chemoreception in vertebrates. Modified cilia form photoreceptors in the vertebrate retina.

Most, but not all, cilia and flagella are composed of nine doublet microtubules that surround two single central microtubules to form a structure called the axoneme. Cilia and flagella move due to the highly coordinated sliding of the nine doublets, which, in at least some cases, is accompanied by rotation of the central microtubules. Sliding is driven by the dynein arms attached to the A-tubule of each of the doublet microtubules. The way in which the sliding movements are coordinated in a cilium (which can form effective and recovery strokes at rates of greater than 50 per second) is not fully understood. Indeed, we continue to discover new isoforms of dynein arms as well as new proteins involved in the control of dynein. In addition to the bending motions caused by microtubule sliding, several other types of motility are associated with the surrounding membrane, including the gliding of particles along flagellar surfaces and the rapid and directed movements of particles within or just beneath the flagellar membrane. Some of these movements can only be detected using computer-assisted high-resolution light microscopical techniques developed in the past few years. With the increased application of new morphological and molecular genetic techniques, we may discover additional structural specializations and types of motility and begin to understand how microtubule sliding is regulated to give rise to coordinated flagellar beating.

Cilia and flagella also provide some of the most interesting examples of biological assembly. The axoneme is a highly ordered and precisely assembled superstructure, in which inner and outer arms, radial spokes, kinesin-like proteins along the central microtubules, beaks within the proximal regions of individual doublets, caps at the microtubule tips, and scores of other components are brought together in an exact stoichiometry. Examination of a variety of organisms reveals that cells individually regulate the lengths of different sets of cilia on a single cell as well as the lengths of individual doublet and central microtubules within an individual organelle. Examination of the growth of ctenophore macrocilia revealed that the growth of individual axonemes occurs in a precisely orchestrated series of starts and stops. In some protozoans, particularly *Chlamydomonas,* flagella can be amputated and the cell will regrow new full-length flagella within about an hour, so we can examine the assembly process in some detail. This requires the coordinated (up and down) regulation of transcription and translation of flagellar genes as well as the packaging and transport of flagellar components to the tip of the growing organelle, where they are added to the microtubules. As we begin to understand the complexity of flagellar structure, the ability of a cell to carry out the assembly process becomes more impressive. Currently, we have little knowledge of the mechanisms by which cells select proteins or protein complexes to be shipped into the flagellum, although recent studies suggest a role for chaperones in the transport or folding processes. We also need a greater understanding of the ports through which flagellar proteins pass as they are transported into and out of the growing and disassembling flagellum. Possible avenues include the membrane, the space between the membrane and basal body in the transition region, and the core of the basal body. Another possibility is that some proteins are transported through the lumen of the microtubules, similar to the transport of flagellin proteins that pass through the lumen of bacterial flagella, which is less than half the diameter of a single microtubule. Clearly, eukaryotic flagella and cilia provide excellent systems for the study of a variety of biological assembly processes.

In addition to the elongation of the flagellar microtubules and associated structures, flagellar assembly initially requires the assembly and positioning of basal bodies, which specify the direction of ciliary beating. The movement and positioning of basal bodies is of great importance to ciliated epithelial tissues, which transport mucus, food particles, and other fluids in a directional manner over the surfaces of embryos, gills, and mammalian respiratory tissue. In these tissues, basal bodies move to the cell surface, begin to assemble cilia, and then gradually, but precisely, become aligned so that the cilia will beat in the correct direction. The mechanisms by which basal body movement and ciliary alignment are specified are poorly understood. However, failure to achieve proper alignment leads, in mammals, to a class of primary ciliary dyskinesia, one of the major genetically based respiratory diseases. Since the basal bodies are similar to mammalian cell centrioles, studies of the assembly of basal bodies and their transport to the cell surface also should contribute to our general understanding of the assembly and positioning of centrioles, which determines the orientation of the mitotic apparatus, cell polarity, and, possibly, the direction of movement of mammalian cells.

Cilia and eukaryotic flagella, therefore, provide us with valuable experimental systems with which to study the coordinated expression, synthesis, packaging, transport, and assembly of a cellular protein complex. In the past, these processes have been studied primarily by microscopic and biochemical techniques. However, the recent development of methods for transforming flagellated cells, which makes possible targeted gene disruption, gene tagging, gene replacement, and unequivocal identification of cloned genes by mutant rescue, should lead to a wealth of new information about the specific protein components involved in many aspects of flagellar assembly and growth control.

Methods for the culturing of respiratory epithelial cells and the ability to stimulate differentiation of ciliated tissue in culture should open up a new field to study the cytoskeletal organization within and among respiratory cells that specify the targeting of basal bodies and cilia and their orientation on the epithelial surface. This process is relevant to studies of tissue organization and development and for the investigation of respiratory diseases, including classes of primary ciliary dyskinesia, that result in defects in the ability of the respiratory tract to clear mucus. Some of these methods involve the growth and differentiation of these polarized epithelial cells on transwell culture dishes, which can be used to study drug and metabolite transport from one side of the epithelial layer to the other. In particular, these culture systems should facilitate studies of drug delivery across respiratory epithelium in all mammals.

We believe that this is an ideal time to assemble this collection of methods. For nearly 40 years investigators have been developing morphological, biochemical, and cell biological tools with which to investigate the structure, assembly, and composition of cilia and flagella. The result is a large body of precise and extremely effective methods for the isolation, fractionation, and biochemical analysis of cilia and flagella and their components. More recently, this body of methods has been greatly enhanced by the development of molecular genetic approaches for flagellated organisms, by new optical techniques for observing flagellar movements, and by methods for the culturing and differentiation of ciliated epithelial cells *in vitro*. Our goal has been to present here the most useful of the "classical" methods together with the most powerful of the new techniques that are likely to dominate the field for the foreseeable future.

PART I

Isolation of Cilia and Flagella

CHAPTER 1

Flagellar Amputation and Regeneration in *Chlamydomonas*

Paul A. Lefebvre

Department of Genetics and Cell Biology
University of Minnesota
St. Paul, Minnesota 55108

I. Introduction

Chlamydomonas cells shed their flagella in response to many stressful conditions, including extremes of temperature or pH and the presence of noxious agents such as detergents or alcohols in the medium. This process, called *flagellar autotomy* (Lewin and Burrascano, 1983), presumably serves a protective function for cells in the environment. The flagella present a large area of cell membrane to the medium, and under adverse conditions the flagella are shed to minimize exposure to environmental stress.

The flagella break at a defined site just above the transition region, in a Ca^{2+}-dependent process involving the Ca^{2+}-binding protein centrin (Sanders and Salisbury, 1994). Experiments using a variety of inhibitors and medium compositions have been used to predict the existence of two different signaling mechanisms for flagellar autotomy (Quarmby and Hartzell, 1994). One of these involves the influx of Ca^{2+} from the medium through specific channels; the other is proposed to be dependent on the release of intracellular Ca^{2+} stores. A rapid

METHODS IN CELL BIOLOGY, VOL. 47

increase in inositol 1,4,5-triphosphate (IP_3) is associated with flagellar autotomy (Quarmby *et al.*, 1992; Yueh and Crain, 1993), probably acting to release intracellular calcium.

The regeneration of cilia and flagella has been described in many unicellular eukaryotes, and in the multicellular embryos of sea urchins (reviewed in Lefebvre and Rosenbaum, 1986). In *Chlamydomonas* the regeneration of flagella has been studied to provide insight into the synthesis of proteins that constitute the flagella (Rosenbaum *et al.*, 1969; Lefebvre *et al.*, 1978) and to characterize mutants with defective flagellar assembly (Huang *et al.*, 1977; Barsel *et al.*, 1988).

II. Procedures for Flagellar Amputation

The procedures described here are used to amputate flagella under conditions that allow continued cell viability and flagellar regeneration. Efficient methods to remove flagella from large numbers of cells for biochemical purposes (when cell survival does not matter) are described elsewhere (Witman, 1986). Before starting any deflagellation procedure, check living cells to determine if flagella are present. To avoid mechanical deflagellation by the weight of the coverslip, place four small dabs of Vaseline on a glass slide, place a drop of cells on the slide, carefully place the coverslip over the Vaseline, and allow the coverslip to contact the droplet of cells gradually. The easiest way to apply Vaseline is to fill a small (1–10 ml) syringe with it and squeeze a droplet out the needle using the plunger. Ensure that the coverslip covers the entire Vaseline and be certain that none of the Vaseline contacts the microscope lenses.

A. pH Shock

The most widely used procedure and probably the easiest to adapt to any research or teaching laboratory is pH shock. Hartzell *et al.* (1993) discuss a number of different organic acids that can be used for flagellar amputation, but the mostly commonly used agent is acetic acid.

1. Grow cells in minimal medium (such as medium I or "M" of Sager and Granick, 1953; also see Chapter 45 in this volume). Cells can be deflagellated at any density or stage of growth, although in this medium cells will often become aflagellate within a day or two of reaching stationary phase (3 × 106/ml).
2. While stirring cells gently on a magnetic stirrer, add 0.5 *N* acetic acid dropwise until a pH of 4.5 is reached.
3. Let the mixture stir for 30 seconds, then raise the pH to 7.0 by the dropwise addition of 0.5 *N* KOH. *Be careful not to undershoot pH 4.5 or overshoot pH 7.0 to avoid killing cells.*

4. If desired, pellet cells with a clinical centrifuge and suspend in fresh medium for flagellar regeneration.

B. Mechanical Shear

Another simple deflagellation procedure commonly used is amputation of the flagella by mechanical shear. Cells can be grown in any medium for this procedure, whereas cells to be used for pH shock should not be grown in acetate-containing medium. On the other hand, cells to be deflagellated by mechanical shear must retain their cell walls to avoid cell rupture during the process.

1. Place cells in the fluted chamber of a VirTis homogenizer. As little as 10 ml in the 80-ml sample holder can be deflagellated.

2. Homogenize the cells at a setting near full speed for 1 minute. The speed of the blade should be calibrated for each homogenizer and for each size vessel. The appropriate speed is usually slightly faster than the setting needed to induce cavitation of the sample (i.e., the appearance of bubbles at the surface of the blade). At the proper speed, greater than 99% efficiency of flagellar amputation should be obtained, with no visible cell lysis.

3. If desired, pellet cells with a clinical centrifuge and suspend in fresh medium for flagellar regeneration.

C. Suction Deflagellation

Techniques have been developed to deflagellate small numbers of cells in each well of a 96-well culture dish (Lefebvre *et al.*, 1988). This procedure has been used to identify mutants with defective flagellar regeneration. Suction deflagellation presumably acts by pulling the flagella through the holes of a nitrocellulose filter while retaining the cells on the filter. Because there are a limited number of holes on such a filter, it is important to use a small number of cells in each well.

1. Grow the cells to a low density (early logarithmic phase, ca. 2×10^5 cells/ml) in the wells of a 96-well dish. For mutant screening, each well usually contains cells from a different colony of mutagenized cells.

2. Prepare a multiwell suction device (e.g., Minifold from Schleicher and Schuell) by placing a prewetted nitrocellulose filter (pore size 0.45 μm) between the top and bottom plates. *Note:* It is important to use nitrocellulose filters that have been manufactured in the absence of detergents (e.g., HATF filters, Schleicher and Schuell) or to boil the filters repeatedly in distilled water to remove any traces of the detergent.

3. Add 200 μl of medium to each well of the suction device, then add 200 μl of cells from the corresponding wells of the cell culture dish.

4. Apply gentle suction to the device such that the wells are emptied within 20 seconds. Turn off the suction, resuspend the cells in 200 μl of medium, and transfer to a new multiwell dish.

5. Within 60 minutes, wild-type cells regenerate their flagella and regain full motility. Mutants with defective regeneration can be identified as those that remain immotile after 60 minutes. It is simple to rapidly examine the surface of each well of the culture dish using a dissecting microscope to assay motility.

III. Assay of Flagellar Regeneration

Wild-type cells begin to regenerate flagella immediately after amputation, with regeneration being completed in about 45 to 60 minutes for vegetative cells and about 90 minutes for gametes. To prepare samples for determining a rate curve for regeneration, cells are fixed before and at different times after deflagellation by adding 2 drops of cells (from a Pasteur pipet) to capped microfuge tubes containing 1 drop of 10% glutaraldehyde. Small samples are then observed by phase-contrast or differential-interference-contrast microscopy. Flagellar length can easily be assayed by aligning the flagella on fixed cells with an ocular micrometer, which consists of a small glass disk that fits into the eyepiece of the microscope. The grid is calibrated by observing a field micrometer grid, which is a microscope slide etched with lines of a predetermined spacing.

References

Barsel, S. E., Wexler, D. E., and Lefebvre, P. A. (1988). Genetic analysis of long-flagella mutants of *Chlamydomonas reinhardtii*. *Genetics* **118,** 637–648.

Hartzell, L. B., Hartzell, H. C., and Quarmby, L. M. (1993). Mechanisms of flagellar excision: I. The role of intracellular acidification. *Exp. Cell Res.* **208,** 148–153.

Huang, B., Rifkin, M. R., and Luck, D. J. L. (1977). Temperature-sensitive mutations affecting flagellar assembly and function in *Chlamydomonas reinhardtii*. *J. Cell Biol.* **72,** 67–85.

Lefebvre, P. A., and Rosenbaum, J. L. (1986). Regulation of the synthesis and assembly of ciliary and flagellar proteins during regeneration. *Annu. Rev. Cell Biol.* **2,** 517–546.

Lefebvre, P. A., Barsel, S. E., and Wexler, D. E. (1988). Isolation and characterization of *Chlamydomonas reinhardtii* mutants with defects in the induction of flagellar protein synthesis after deflagellation. *J. Protozool.* **35,** 559–564.

Lefebvre, P. A., Nordstrom, S. A., Moulder, J. E., and Rosenbaum, J. L. (1978). Flagellar elongation and shortening in *Chlamydomonas*. IV. Effects of flagellar detachment, regeneration, and resorption on the induction of flagellar protein synthesis. *J. Cell Biol.* **78,** 827.

Lewin, R. A., and Burrascano, C. (1983). Another new kind of *Chlamydomonas* mutant with impaired flagellar autotomy. *Experientia* **39,** 1397–1398.

Quarmby, L. M., Yueh, Y. G., Cheshire, J. L., Keller, L. R., Snell, W. J., and Crain, R. C. (1992). Inositol phospholipid metabolism may trigger flagellar excision in *Chlamydomonas reinhardtii*. *J. Cell Biol.* **116,** 737–744.

Quarmby, L. M., and Hartzell, H. C. (1994). Two distinct, calcium-mediated, signal transduction pathways can trigger deflagellation in *Chlamydomonas reinhardtii*. *J. Cell Biol.* **124,** 807–815.

Rosenbaum, J. L., Moulder, J. E., and Ringo, D. L. (1969). Flagellar elongation and shortening in *Chlamydomonas*. The use of cycloheximide and colchicine to study synthesis and assembly of flagellar proteins. *J. Cell Biol.* **41,** 600–619.

Sager, R., and Granick, S. (1953). Nutritional studies with *Chlamydomonas reinhardi. Ann. N.Y. Acad. Sci.* **56,** 831–838.

Sanders, M. A., and Salisbury, J. L. (1994). Centrin plays an essential role in microtubule severing during flagellar excision in *Chlamydomonas reinhardtii. J. Cell Biol.* **124,** 795–805.

Witman, G. B. (1986). Isolation of *Chlamydomonas* flagella and flagellar axonemes. *Methods Enzymol.* **134,** 280–290.

Yueh, Y. G., and Crain, R. C. (1993). Deflagellation of *Chlamydomonas reinhardtii* follows a rapid transitory accumulation of inositol 1,4,5-triphosphate and requires calcium entry. *J. Cell Biol.* **123,** 869–875.

CHAPTER 2

Large-Scale Isolation of *Chlamydomonas* Flagella

Stephen M. King
Department of Biochemistry
University of Connecticut Health Center
Farmington, Connecticut 06030

I. Introduction

There are many advantages to the use of *Chlamydomonas* as a model system in which to examine flagellar function. This organism is easy to grow synchronously in large quantities in simple defined media and thus is readily amenable to biochemical analysis (for a detailed description of the media and culture facilities required for growing *Chlamydomonas,* see Harris, 1989 and Witman, 1986). Moreover, the cells can be labeled *in vivo* with a variety of radioactive tracers, i.e., ^{32}P (Piperno and Luck, 1981; King and Witman, 1994), ^{35}S (Lefebvre *et al.,* 1978; Remillard and Witman, 1982), and ^{3}H (Witman, 1975). A wealth of flagellar mutants with defined lesions have been isolated, e.g., mutations in 13 different loci that affect the outer dynein arms are currently available (Kamiya, 1988, reviewed in Witman *et al.,* 1994). Many of these genes have been located on the *Chlamydomonas* genetic map (see Harris, 1989). The development of methods for obtaining stable nuclear transformation in this organism (Kindle *et al.,* 1989, 1990) now allows the power of molecular genetics to be combined with biochemistry in the analysis of flagellar protein function. This already is providing significant insights into the function and regulation of several axonemal systems, e.g., radial spokes (Diener *et al.*, 1993) and dynein arms (Mitchell and Kang, 1993) and holds great promise for the future.

In this chapter, I describe how to obtain highly purified flagella from a large-scale culture of *Chlamydomonas*. The cultures routinely reach ~2 × 10^6 cells/ml and the volume required obviously will depend on the specific experiment. For example, 36 liters of cells provides sufficient flagella from which to purify ~1 nmole of outer-arm dynein (= ~2.5 mg). The most expedient way to harvest these large cultures is to use a tangential flow filtration system which allows for rapid concentration of cells. For smaller-scale experiments (using up to ~8 liters of culture), centrifugation in 250-ml bottles is adequate. Once harvested and washed, the cells routinely are deflagellated with dibucaine (Witman *et al.*, 1978; Pfister *et al.*, 1982), which is rapid and highly efficient; however, *Chlamydomonas* do not readily recover from this treatment. Therefore, if flagellar regeneration is required, e.g., for preparing mRNA enriched in flagellar sequences (see Chapter 79 in this volume), an alternative method such as pH shock should be employed (Witman *et al.*, 1972).

II. Methods

A. Solutions

10 m*M* 4-(2-hydroxyethyl)-1-piperazineethanesulfonic acid (Hepes), pH 7.5

10 m*M* Hepes, pH 7.5, 5 m*M* $MgSO_4$, 1 m*M* dithiothreitol, 4% sucrose (HMDS)

10 m*M* Hepes, pH 7.5, 5 m*M* $MgSO_4$, 1 m*M* dithiothreitol, 25% sucrose (HMD25%S)

25 m*M* dibucaine–HCl (Sigma; or Nupercaine–HCl, Ciba Pharmaceutical)

200 m*M* phenylmethylsulfonyl fluoride (PMSF) in isopropanol

100 m*M* ethylene glycol bis(ß-aminoethyl ether) *N,N'*-tetraacetic acid (EGTA)

0.5 *M* acetic acid (for deflagellation by pH shock only)

1 *M* $NaHCO_3$ (for deflagellation by pH shock only)

B. Procedure

1. Grow the required volume of *Chlamydomonas* culture as described by Witman (1986).

2. Check that the cells are flagellated and uncontaminated by phase-contrast microscopy.

3. For volumes less than 8 liters, harvest cells by centrifugation at 1100*g* and resuspend the pellets by gentle pipetting in 10 m*M* Hepes, pH 7.5 (to aid

resuspension and avoid shearing the flagella, it helps to use a 10-ml plastic pipet from which the end has been snapped off). Alternatively, a tangential flow filtration system (Pellicon system, Millipore) may be employed; this is especially useful if large volumes of cells (>8 liters) are required. In the latter case, concentrate cells until ~400 ml remains in the collection vessel. Pour this into a graduated 2-liter cylinder, and then flush the filter unit with 10 m*M* Hepes, pH 7.5, until a final volume of ~1100 ml is obtained (this removes nearly all of the cells from the filter unit). With this method, 36 liters of cells can be concentrated to <1200 ml in approximately 20 minutes. The filter unit should be washed extensively with water and subjected to mild bleach treatment prior to storage. *Note:* Some cell lines, especially those with flagellar defects, hatch poorly; often, they can be induced to hatch and become flagellated by vigorous stirring for 1–2 hours. Alternatively, treatment with gametic autolysin (the preparation of this solution from mating gametes is described by Harris, 1989) aids removal of the mother cell wall.

4. Dispense the concentrated cells (from a 36-liter culture) into 24 graduated 50-ml conical tubes (45 ml/tube) and spin at 1100*g* for 5 minutes at 20°C. At this stage, there should be ~4 ml of packed cells/tube. Remove the supernatant by aspiration and wash each cell pellet twice by resuspension in 45 ml 10 m*M* Hepes, pH 7.5, followed by centrifugation.

5. During the wash steps, set up, on ice, 24 tubes containing 28 ml HMDS and 100 μl 100 m*M* EGTA.

6. Following the last wash, add ice-cold HMDS to the 10-ml mark on the tube and resuspend the cells by pipetting. From this stage forward, all tubes and solutions should be kept on ice.

7. Deflagellate the cells by adding 2 ml 25 m*M* dibucaine to each tube and mix by pipetting up and down. Check that the cells are deflagellated by phase-contrast microscopy (almost 100% deflagellation occurs within 1–2 minutes). Add the cold HMDS/EGTA solution to each tube and spin at 1100*g* for 5 minutes at 4°C (turn the centrifuge brake off for this step to minimize contaminating the flagella with cell bodies). Alternatively, the cells can be deflagellated by pH shock (the advantage here is that the cells will regenerate their flagella). In this case, they should be resuspended in 0.1 vol of medium and deflagellated by quickly decreasing the pH to ~4.5 with 0.5 *M* acetic acid. After ~20 seconds, the pH should be rapidly returned to near neutrality by adding 1 *M* $NaHCO_3$. Throughout this procedure, the cells should be stirred and the pH monitored continually using a pH meter. The cells are pelleted by centrifugation and the supernatant is treated as described below.

8. Remove the flagella-containing supernatant by aspiration into an ice-cold flask. Once all tubes have been collected, add 0.005 vol of PMSF and mix rapidly by swirling the flask (the PMSF will precipitate but then go back into solution).

9. Dispense the solution of flagella into 50-ml round-bottomed tubes and harvest by centrifugation in a superspeed centrifuge (Sorvall SS34 rotor; 10,000 rpm, 10 minutes, 4°C).

10. Resuspend the flagella pellets by gentle pipetting in a small volume of HMDS (50–100 ml total) containing PMSF.

11. Place 25 ml of the concentrated flagella in a graduated conical tube and underlayer with 15 ml HMD25%S. (Although this is easily achieved using a 10-ml pipet with a rubber bulb to control the flow, it is probably worth practicing once or twice on "blank" tubes.) Spin at 1100*g* for 20 minutes at 4°C. Any cell bodies contaminating the flagella will pellet through the underlayer while flagella will remain in the upper layer and form a milky white band at the interface.

12. The interface and upper layer are removed by aspiration or with a 10-ml pipet. At this stage, the flagellar preparation should be almost completely devoid of cell bodies and is suitable for demembranation and the subsequent purification of flagellar components.

References

Diener, D. R., Ang, L. H., and Rosenbaum, J. L. (1993). Assembly of flagellar radial spoke proteins in *Chlamydomonas:* identification of the axoneme binding domain of radial spoke protein 3. *J. Cell Biol.* **123,** 183–190.

Harris, E. H. (1989). "The *Chlamydomonas* source book." Academic Press, New York.

Kindle, K. L. (1990). High-frequency nuclear transformation method for *Chlamydomonas reinhardtii. Proc. Natl. Acad. Sci. U.S.A.* **87,** 1228–1232.

Kindle, K. L., Schnell, R. A., Fernandez, E., and Lefebvre, P. A. (1989). Stable nuclear transformation of *Chlamydomonas* using a gene for nitrate reductase. *J. Cell Biol.* **109,** 2589–2601.

King, S. M., and Witman, G. B. (1994). Multiple sites of phosphorylation within the α-heavy chain of *Chlamydomonas* outer arm dynein. *J. Biol. Chem.* **269,** 5452–5457.

Lefebvre, P. A., Nordstrom, S. A., Moulder, J. E., and Rosenbaum, J. L. (1978). Flagellar elongation and shortening in *Chlamydomonas.* IV. Effects of flagellar detachment, regeneration, and resorption on the induction of flagellar protein synthesis. *J. Cell Biol.* **78,** 8–27.

Mitchell, D. R., and Kang, Y. (1993). Reversion analysis of dynein intermediate chain function. *J. Cell Sci.* **105,** 1069–1078.

Pfister, K. K., Fay, R. B., and Witman, G. B. (1982). Purification and polypeptide composition of dynein ATPases from *Chlamydomonas* flagella. *Cell Motil.* **2,** 525–547.

Piperno, G., and Luck, D. J. L. (1981). Inner arm dyneins from flagella of *Chlamydomonas reinhardtii. Cell* **27,** 331–340.

Remillard, S. P., and Witman, G. B. (1982). Synthesis, transport, and utilization of specific flagellar proteins during flagellar regeneration of *Chlamydomonas. J. Cell Biol.* **93,** 615–631.

Witman, G. B. (1975). The site of in vivo assembly of flagellar microtubules. *Ann. N.Y. Acad. Sci.* **253,** 178–191.

Witman, G. B. (1986). Isolation of *Chlamydomonas* flagella and flagellar axonemes. *Methods Enzymol.* **134,** 280–290.

Witman, G. B., Carlson, K., Berliner, J., and Rosenbaum, J. L. (1972). *Chlamydomonas* flagella. I. Isolation and electrophoretic analysis of microtubules, matrix, membranes and mastigonemes. *J. Cell Biol.* **54,** 507–539.

Witman, G. B., Plummer, J., and Sander, G. (1978). *Chlamydomonas* flagellar mutants lacking radial spokes and central tubules. Structure, composition and function of specific axonemal components. *J. Cell Biol.* **76,** 729–747.

CHAPTER 3

Isolation of Cilia from *Tetrahymena thermophila*

William L. Dentler
Department of Physiology and Cell Biology
University of Kansas
Lawrence, Kansas 66045

I. Introduction

Tetrahymena thermophila provides an excellent source of prokaryotic cilia. *Tetrahymena* cells can be grown at densities up to 5×10^6 cells/ml, and each cell contains 400–600 cilia. If treated gently, deciliated *Tetrahymena* cells regenerate cilia, although all cilia do not regenerate synchronously, and because regenerating cilia cannot be observed directly by light microscopy, *Tetrahymena* does not provide as good a system with which to study flagellar growth as does *Chlamydomonas*. Based on the ease of cell harvesting, cell culture, and the number of cilia per cell, *Tetrahymena* clearly provides a better source of ciliary protein than do most other organisms.

II. *Tetrahymena* Cell Culture

We use *T. thermophila* strain SB-715 because it does not secrete mucus on deciliation. This produces much cleaner preparations of cilia than can be accomplished with mucus-secreting strains. Cells are cultured in sterilized 2% proteose peptone (Difco Laboratories) supplemented with 0.1 m*M* $FeCl_3$ and 0.025% penicillin–streptomycin or in 1% powdered skim milk, 1% glucose, 0.5% yeast extract, and 0.003% ethylene diamine tetraacetic acid (EDTA) (Kiy and Tiedtke, 1992). Recently, we have changed to the latter medium, because it is much less expensive than proteose peptone and produces larger yields of cells and very clean cilia. We have also tried culturing cells on NZ-amine (Johnson, 1986), but find that it significantly changes the solubilization properties of the ciliary membranes and reduces the number of cilia with intact capping structures at their distal tips (Dentler, 1980; Miller *et al.*, 1990; Wang *et al.*, 1993, 1994).

For all media, cells are grown in 2.8-liter Fernbach flasks on a rotating shaker table at room temperature. The following procedure is for a typical preparation of 3 liters of cells (in medium), but can easily be scaled from 200 ml to 9 liters of cells without any significant change in cilia purity.

III. Isolation of *Tetrahymena* Cilia

These procedures were optimized to produce intact cilia with microtubule capping structures.

A. Solutions

HNMK (1 liter): 50 m*M* 4-(2-hydroxyethyl)-1-piperazineethanesulfonic acid (Hepes), pH 6.9, 36 m*M* NaCl, 0.1 m*M* $MgSO_4$, 1 m*M* KCl

HNMKS (1 liter): 50 m*M* Hepes, pH 6.9, 36 m*M* NaCl, 0.1 m*M* $MgSO_4$, 1 m*M* KCl, 250 m*M* sucrose

HEEMS (1 liter): 50 m*M* Hepes, pH 7.4, 1 m*M* EDTA, 1 m*M* ethylene glycol bis(β-aminoethyl ether)-*N,N′*-tetraacetic acid (EGTA), 3 m*M* Mg^{2+}-acetate, 250 m*M* sucrose, 0.1 m*M* dithiothreitol (DTT)

1000× Leupeptin stock: 40 m*M* leupeptin in water, stored at −20°C

1000× PMSF stock: 29 m*M* phenylmethylsulfonyl fluoride (PMSF) in propanol, stored at −20°C

B. Procedure

1. Harvest 1500 ml of cells by centrifugation at 4000*g* for 5 minutes at 20°C.
2. Suspend cells in approximately 900 ml of HNMK by swirling cells in centrifuge bottles. Pellet cells as in step 1. After each wash, carefully resuspend cells by swirling buffer and cells.
3. Suspend cells in 200 ml of HNMKS. Add 200 mg of dibucaine dissolved

in a small amount of HNMKS and swirl for 4 minutes. Check cells for deciliation with a phase microscope.

4. Immediately dilute the suspension with ice-cold HNMKS and add PMSF from the 1000X stock: pour approximately 50 ml of cells into each of four centrifuge bottles containing ~200 ml of HNMKS and swirl to dilute cells.
5. Pellet cell bodies in a cold rotor, 4000*g* for 5 minutes at 20°C.
6. Pour off supernatant into a clean cold centrifuge bottle. Check suspension with a phase microscope; if there are too many cell bodies, repeat step 5 to obtain a cleaner supernatant.
7. Pellet cilia from the supernatant by centrifugation at 17,000g_{max}, 4°C, for 30–40 minutes.
8. Gently suspend cilia in 30–40 ml of HEEMS. Add PMSF and leupeptin. Be careful with suspension to avoid breakage of the cilia. Examine cilia with the phase microscope.
9. Pellet cilia by centrifuging the cilia suspension (if clean) at 17,400*g*, 20 min, 4°C.

References

Dentler, W. L. (1980a). Microtubule–membrane interactions in cilia. I. Isolation and characterization of ciliary membranes from *Tetrahymena pyriformis*. *J. Cell Biol.* **84,** 364–380.

Dentler, W. L. (1980b). Structures linking the tips of ciliary and flagellar microtubules to the membrane. *J. Cell Sci.* **42,** 207–220.

Johnson, K. A. (1986). Preparation and properties of dynein from *Tetrahymena* cilia. *Methods Enzymol.* **134,** 306–317.

Kiy, T., and Tiedtke, A. (1992). Mass cultivation of *Tetrahymena thermophila* yielding high cell densities and short generation times. *Appl. Microbiol. Biotech.* **37,** 576–579.

Miller, J. M., Wang, W., Balczon, R., and Dentler, W. L. (1990). Ciliary microtubule capping structures contain a mammalian kinetochore antigen. *J. Cell Biol.* **110,** 703–714.

Wang, W., Suprenant, K. A., and Dentler, W. L. (1993). Reversible association of a 97kD protein complex found at the tips of ciliary microtubules with in vitro assembled microtubules *J. Biol. Chem.* **268,** 24796–24807.

Wang, W., Himes, R. H., and Dentler, W. L. (1994). The interaction of a plus-end binding ciliary microtubule protein complex with microtubules is regulated by endogenous protein kinase and phosphatase activities. *J. Biol. Chem.* **269,** 21,460–21,466.

CHAPTER 4

Preparation of Cilia and Subciliary Fractions from *Paramecium*

David L. Nelson

Department of Biochemistry
College of Agricultural and Life Sciences
University of Wisconsin—Madison
Madison, Wisconsin 53706

I. Introduction

For biochemical studies of ciliary motility in *Paramecium* it is useful, and sometimes essential, to obtain quantities of pure cilia, uncontaminated by other subcellular fractions of similar size (such as mitochondria or trichocysts) or by bacteria from the culture medium. An ideal preparation would include the entire cilium including the basal body and associated components, all axonemal proteins in their normal state of association, the soluble components of the ciliary matrix, and the ciliary membrane, all unaltered by proteolysis during the fractionation, and in good yield. In practice, one can reproducibly obtain quite pure preparations of the distal portion of the cilium, lacking only the basal structures, still containing functional axonemes and at least some of the soluble proteins of the matrix, and still surrounded by the ciliary membrane. When stored frozen, these ciliary preparations retain enzymatic activities for weeks or months.

With purified cilia as the starting material, it is also possible to obtain three distinct subciliary fractions: the axoneme including its dynein components, the ciliary membrane, and the soluble proteins of the matrix. This chapter describes in detail a commonly used deciliation procedure (Ca^{2+} or Ba^{2+} shock deciliation) and a protocol for subciliary fractionation on sucrose gradients and, in less detail, alternate deciliation procedures employing Mn^{2+} or dibucaine as deciliating agents. For electrophysiological studies, it is sometimes useful to remove cilia (and their membrane) from cells without compromising the electrical integrity of the plasma membrane, which can be accomplished by deciliation with ethanol or chloral hydrate. Protocols for doing this are also provided here.

The mechanisms underlying Ca^{2+}-induced deflagellation in *Chlamydomonas reinhardtii* appears to involve the Ca^{2+}-binding protein centrin, which is part of the stellate fibers of the transition zone, the contraction of which causes severing of microtubules in the transition zone (Sanders and Salisbury, 1994). As centrin is found in all ciliated and flagellated cells (including *Paramecium*), this may be a general mechanism for Ca^{2+}- or Ba^{2+}-induced deflagellation and deciliation. The point of breakage during deciliation appears to be in the transition zone near the proximal end of the cilium, just distal to the axosome and between the plaques and the necklace (Kennedy and Brittingham, 1968; Blum, 1971). This means that isolated cilia lack the axosome and associated structures, as well as the necklace. The stub of the broken cilium apparently seals over quickly and completely, allowing very little protein leakage out of the cell body and leaving the plasma membrane of carefully deciliated cells electrically intact. None of the methods detailed below is effective at removing the cilia lining the oral groove.

II. Solutions

Dryl's solution: 1 m*M* NaH_2PO_4, 1 m*M* Na_2HPO_4, 2 m*M* sodium citrate, 1.5 m*M* $CaCl_2$, pH 6.8

STEN: 0.5 *M* sucrose, 20 m*M* Tris–HCl, 2 m*M* ethylenediaminetetraacetic acid (EDTA), 6 m*M* NaCl, pH 7.5

SMEN: 0.5 *M* sucrose, 20 m*M* 4-morpholinepropanesulfonic acid (Mops), 2 m*M* EDTA, 6 m*M* NaCl, pH 7.5

Storage buffer: 0.5 m*M* EDTA, 5 m*M* Mops, pH 7.5, 0.3 m*M* phenylmethylsulfonyl fluoride (PMSF); for storage at −20°C, include 50% glycerol

MMKED: 20 m*M* Mops, 1 m*M* $MgCl_2$, 100 m*M* KCl, 0.1 m*M* EDTA, 1 m*M* dithiothreitol (DTT), pH 7.5

Tris–EDTA buffer: 1 m*M* Tris–HCl, 0.1 m*M* EDTA, pH 8.3

HMS buffer: 5 m*M* $MgSO_4$, 4% sucrose, 25 m*M* 4-(2-hydroxyethyl)-1-piperazineethanesulfonic acid (Hepes), pH 7.0

Deciliation of *Paramecium* releases a potent sulfhydryl-activated protease (Davis and Steers, 1978). We therefore routinely include protease inhibitors in the solutions used in ciliary isolations. The minimal protease inhibitor is PMSF at 0.3 m*M*. The most complex inhibitor cocktail contains 2 m*M* EDTA, 0.3 m*M* PMSF, 0.3 m*M* tosylarginine methyl ester (TAME), 10 μg/ml leupeptin, 2 μg/ml pepstatin A, and 0.2 unit/ml aprotinin (all final concentrations), a mixture found empirically to inhibit protease action effectively even in whole-cell extracts of *Paramecium*. A useful compromise between cost and effectiveness is a mixture of EDTA, TAME, and PMSF.

III. Methods of Deciliation

A. Deciliation by Ba^{2+} or Ca^{2+} Shock

This procedure was originally developed with Ca^{2+} as the deciliating agent (Adoutte *et al.*, 1980; modification of the method of Hansma and Kung, 1975). We have found that Ba^{2+} is at least as effective as Ca^{2+}, and is less likely to cause damage to the cell body; we therefore routinely use Ba^{2+}.

1. Culture cells of *Paramecium tetraurelia* strain 51s either in wheat grass extract bacterized with *Aerobacter aerogenes* (prepared as for the Cerophyll medium described by Sonneborn, 1970) or in the crude axenic medium of Soldo, modified as described by Van Wagtendonk (1974).

2. Harvest cells in late exponential growth phase by centrifugation at $400g_{max}$ in pear-shaped oil-testing centrifuge tubes at room temperature. Centrifuge tubes of this shape prevent cells from swimming back up from the pellet and make it unnecessary to centrifuge so hard as to form a tight pellet.

3. Gently resuspend the cells in Dryl's solution at room temperature, and wash twice in this solution at 4°C. All subsequent steps are at 4°C. The combination of the cold shock and the mechanical stimulation from centrifugation and resuspension causes massive trichocyst discharge. The discharged trichocysts form a fluffy pellet above the cell pellet, which can be removed by gentle aspiration. If trichocyst contamination of the ciliary preparation proves to be a serious problem, consider using one of the mutants of *P. tetraurelia* that cannot discharge its trichocysts (Beisson *et al.*, 1980).

4. After the last wash, suspend cells in a 1 : 1 mixture of Dryl's solution and either STEN or SMEN solution; each milliliter of packed cells is suspended in

about 10 ml of the mixture. Within 10–20 min at 4°C, all cells are immobilized but most still retain their cilia. Add PMSF to bring the final concentration to 0.3 m*M*.

5. Detach cilia by adding concentrated $BaCl_2$ (or $CaCl_2$, see above) and KCl to bring the final concentrations to 10 m*M* Ba^{2+} (or Ca^{2+}) and 30 m*M* K^+. Monitor deciliation in the phase-contrast microscope; it usually takes from 3 to 10 minutes and is at least 90% complete. A good preparation should show no blistering or blebbing of the deciliated cell bodies; this is crucial to the purity and integrity of the cilia. Cell lysis can also be monitored by measuring the release of the cytosolic enzyme catalase, which is not present in cilia and is not released during a good deciliation (Riddle *et al.*, 1982).

6. Remove cell bodies by centrifugation twice for 2–3 minutes at $850g_{max}$; then recover cilia by transferring the low-speed supernatant to centrifuge bottles and centrifuging at $28{,}000g_{max}$ for 20 minutes. From 1 ml of packed cells (about 100 mg of protein) one obtains 3–5 mg of ciliary protein.

7. Suspend ciliary pellet in storage buffer at 5–20 mg protein per milliliter. At this stage cilia can be used directly, subjected to subciliary fractionation, or stored in storage buffer plus 50% glycerol at −20°C. Frozen cilia retain several enzymatic activities (dynein, cAMP- and cGMP-dependent protein kinases) for weeks or months.

8. The high-speed supernatant fluid (deciliation supernatant) from step 7, containing a few hundred micrograms per milliliter protein derived at least in part from the ciliary matrix, can be concentrated by ultrafiltration (with Amicon filters with molecular weight cutoff of 10,000). A Ca^{2+}-ATPase in this fraction (Riddle *et al.*, 1982) is stable for weeks at 4°C, provided that 3 m*M* NaN_3 is added to inhibit bacterial growth.

B. Deciliation with Mn^{2+}

Because Mn^{2+} does not support trichocyst exocytosis (as Ca^{2+} does), this protocol minimizes the possibility that trichocysts released during deciliation will contaminate the isolated cilia (Fukushi and Hiwatashi, 1970; Brugerolle, 1980).

1. Suspend washed cells in 50 ml of a 50 m*M* $MnCl_2$ solution in 10 m*M* Tris–HCl, pH 7.2, containing 0.02 m*M* PMSF. After 2 minutes at 4°C, remove cells by centrifugation and resuspend them in the same solution. After 20 minutes of gentle shaking, 90–95% of cells are deciliated, as judged by phase-contrast microscopy.

2. Remove deciliated bodies by centrifugation at $400g_{max}$ for 5 minutes.

3. Recover cilia by centrifugation at $12{,}000g_{max}$ for 15 minutes at 4°C.

C. Deciliation with Dibucaine

This procedure yields cilia that continue to "wriggle" for many seconds after falling off cells and that still show cAMP regulation of (dynein) ATPase (N. M. Bonini, Ph.D. thesis, University of Wisconsin—Madison, 1987; adapted from Witman *et al.*, 1978).

1. Wash axenically grown cells twice at room temperature with 10 m*M* Hepes, pH 7.0.
2. Suspend cells in 40 ml HMS buffer.
3. Add 8 ml of 25 m*M* dibucaine–HCl. Gently swirl cell suspension for 1–2 minutes.
4. Add 40 ml HMSE (HMS plus 1 m*M* EDTA).
5. Remove cell bodies by low-speed centrifugation ($850g_{max}$ for 1–2 minutes).
6. Harvest cilia by centrifugation at $28{,}000g_{max}$ for 20 minutes.
7. Suspend cilia in 300 μl 50 m*M* potassium acetate, 10 m*M* ethylene glycol bis(ß-aminoethyl ether) *N,N*′ tetraacetic acid (EGTA), 10 m*M* potassium phosphate, pH 6.4

D. Deciliation with Ethanol

See Ogura (1977) and Machemer and Ogura (1979).

1. All procedures are at room temperature. Wash *Paramecium caudatum*, cultured in hay infusion, in 1 m*M* $CaCl_2$, 1 m*M* KCl, 1 m*M* Tris–HCl, pH 7.3, and suspend cells in the same buffer.
2. After 1 hour for equilibration, transfer 1.9 ml of the cell suspension to a glass tube. Add 0.1 ml absolute ethanol, then immediately shake gently by hand for 1–2 minutes.
3. Dilute the cell suspension with buffer and transfer to a glass dish. Nonswimmers are easily recognized and can be picked for electrophysiological study.
4. Cells treated with ethanol but not shaken retain most of their cilia, and are appropriate controls for the effects of ethanol alone, without deciliation.
5. This same technique has been applied successfully to *P. tetraurelia;* complete deciliation may require slightly more vigorous conditions: 6% ethanol, vortexing.

E. Deciliation with Chloral Hydrate

See Grebecki and Kuznicki (1961), Kuznicki (1963), and Dunlap (1977).

1. Wash *P. caudatum* grown in hay infusion with control solution (5 m*M*

$MgCl_2$, 4 m*M* KCl, 1 m*M* $CaCl_2$, 1 m*M* Tris–HCl, 0.1 m*M* EDTA, pH 7.2) at room temperature.

2. Incubate washed cells in control solution plus 4 m*M* chloral hydrate at 20°C for 20 hours. Monitor deciliation by microscopy. It may be helpful to agitate the cell suspension gently in chloral hydrate with a stream of fine bubbles.
3. When returned to control solution, these deciliated cells begin immediately to regenerate cilia over a period of about 10 hours.

IV. Preparation of Subciliary Fractions

A. Preparation of Subciliary Fractions from Isolated Cilia

See Adoutte *et al.* (1980).

1. Suspend freshly prepared or frozen cilia in Tris–EDTA buffer. Gently vortex for about 2 minutes at 4°C, then centrifuge at 48,000g_{max} for 30 minutes to recover all particulate material. The supernatant is the Tris–EDTA extract, which contains soluble proteins released from the ciliary matrix. The ciliary cGMP- and cAMP-dependent protein kinases and two casein kinases are present in this extract, but dynein remains in the insoluble fraction.

2. Suspend the pellet in 10 m*M* Tris, pH 8.0, to a concentration of 3–10 mg protein/ml, and layer this suspension atop a step sucrose gradient consisting of 0.7 ml of 66% (w/w) sucrose under 1.7 ml of 55% sucrose under 1.7 ml 45% sucrose under 0.7 ml 20% sucrose, all containing 10 m*M* Tris–HCl, pH 8.0.

3. Centrifuge the step sucrose gradient in an SW50.1 rotor at 45,000 rpm for 1.5 hours at 4°C. The material that accumulates near the interfaces can be recovered by careful aspiration with a Pasteur pipet. Near the top of the 45% sucrose layer are ciliary membrane vesicles (CMVs, density 1.15–1.20 g/ml, protein/lipid mass ratio about 2.5), which represent 30–40% of the total protein recovered from the gradient. On sodium dodecyl sulfate–polyacrylamide gel electrophoresis (SDS–PAGE), this fraction is enriched for a protein of M_r 200,000–250,000, the immobilization antigen, and tubulin is only a minor component. At the 45–55% interface are incompletely demembranated cilia (IDCs), which in the electron microscope are seen to consist of broken segments of axoneme still surrounded by membranes (density 1.28 g/ml, protein/lipid ratio 5), representing 40–50% of recovered protein. Both tubulin and immobilization antigen are prominent. At the 55–66% interface are demembranated axonemes (density 1.29 g/ml, protein/lipid >20). CMVs, IDCs, and axonemes can be recovered from the sucrose solutions by dilution into appropriate buffers and high-speed centrifugation.

B. Preparation of Ciliary Membranes

This procedure is an alternative to protocol A (Brugerolle *et al.*, 1980). It employs Mn^{2+}-induced deciliation of *P. tetraurelia*, demembranation by dialysis of isolated cilia (a procedure modified from Witman *et al.*, 1972), and fractionation of ciliary components on a continuous sucrose gradient. It yields two subfractions of ciliary membranes as well as demembranated axonemes.

1. Suspend isolated cilia obtained by Mn^{2+} deciliation (Section III,B) in the dialysis medium of Witman *et al.* (1972) containing 0.01% (v/v) mercaptoethanol, 0.02 m*M* PMSF, and 1 m*M* NaN_3.

2. Dialyze this suspension against the same buffer for 7 hours at 4°C.

3. Layer the dialysate onto a 20–60% (w/v) continuous sucrose gradient in 10 m*M* Tris–HCl, pH 7.4, containing 0.01% mercaptoethanol, 5 m*M* $MgCl_2$, and 0.02 m*M* PMSF (10 ml of dialysate on 20 ml of gradient).

4. Centrifuge the gradient tube for 14 hours at 65,000g_{max} in an SW25 rotor.

5. Two distinct bands of ciliary membranes are separated on the gradient, with densities of 1.146 and 1.214 g/ml. The less dense fraction of membranes is primarily small vesicles; the more dense fraction is flat sheets of membrane. Axonemes can be recovered from the bottom of the gradient. Membrane fractions are recovered from the gradient by aspiration, dilution into 20 m*M* Tris–HCl, pH 7.5, and centrifugation at 165,000g_{max} in a 50 Ti rotor.

C. Demembranation of Axonemes with Triton X-100

This procedure is an alternative to protocols A and B that is useful when the goal is not to recover ciliary membranes, but to obtain pure axonemes or proteins (such as dynein) derived from axonemes.

1. All manipulations are done at 4°C, and all solutions should contain protease inhibitors (0.5 μg/ml leupeptin, 0.3 m*M* PMSF, 0.1 trypsin-inhibiting unit of aprotinin/ml). Suspend isolated cilia in Tris–EDTA buffer and vortex for about 2 minutes. Centrifugation at 48,000g_{max} for 30 minutes yields the Tris–EDTA extract (containing soluble matrix proteins) and a pellet containing axonemes and ciliary membrane.

2. Suspend cilia in MMKED buffer at about 1–2 mg/ml protein. Add Triton X-100 to a final concentration of 1%. After 20–30 minutes, collect axonemes by centrifugation for 20 minutes at 30,000 g_{max}; wash pellet twice with MMKED and resuspend in MMKED plus 40% glycerol at 1–2 mg/ml for storage at −20°C.

D. High-Salt Extraction of Dynein and Protein Kinases

1. Beginning with fresh or frozen cilia demembranated as in Section IV,C and suspended in MMKED, add concentrated KCl to bring final concentration to 0.6 *M*. Incubate 15 minutes at 4°C.

2. Centrifuge for 60 minutes at 120,000g_{max}. The pellet contains extracted axonemes. The supernatant is the high-salt extract, which contains 12 S and 22 S dyneins (Travis and Nelson, 1988) and a number of other proteins, including a casein kinase (Walczak *et al.,* 1993).

References

Adoutte, A., Ramanathan, R., Lewis, R. M., Dute, R. R., Ling, K.-Y., Kung, C., and Nelson, D. L. (1980). Biochemical studies of the excitable membrane of *Paramecium tetraurelia*. III. Proteins of cilia and ciliary membranes. *J. Cell Biol.* **84,** 717–738.

Beisson, J., Cohen, J., Lefort-Tran, M., Pouphile, M., and Rossignol, M. (1980). Control of membrane fusion in exocytosis. Physiological studies on a *Paramecium* mutant blocked in the final step of the trichocyst extrusion process. *J. Cell Biol.* **85,** 213–227.

Blum, J. J. (1971). Existence of a breaking point in cilia and flagella. *J. Theor. Biol.* **33,** 257–263.

Brugerolle, G., Andrivon, C., and Bohatier, J. (1980). Isolation, protein pattern and enzymatic characterization of the ciliary membrane of *Paramecium tetraurelia*. *Biol. Cellulaire* 37, 251–260.

Davis, R. H., Jr., and Steers, E., Jr. (1978). Purification of the i-antigen 51A from *Paramecium tetraurelia* by immunoaffinity chromatography. *Immunochemistry* **15,** 371–378.

Dunlap, K. (1977). Localization of calcium channels in *Paramecium caudatum*. *J. Physiol.* **271,** 119–133.

Fukushi, J., and Hiwatashi, K. (1970). Preparation of mating reaction cilia from Paramecium caudatum by $MnCl_2$. *J. Protozool.* **17,** Suppl., 21.

Grebecki, A., and Kuznicki, L. (1961). Immobilization of *P. caudatum* in the chloral hydrate solutions. *Biol. Acad. Pol. Sci.* (Cl. II Ser. Sci. Biol.) **9,** 459–462.

Hansma, H., and Kung, C. (1975). Studies of the cell surface of *Paramecium*. Ciliary membrane proteins and immobilization antigens. *Biochem. J.* **152,** 523–528.

Kennedy, J. R., Jr., and Brittingham, E. (1968). Fine structure changes during chloral hydrate deciliation of *Paramecium caudatum*. *J. Ultrastruct. Res.* **22,** 530–545.

Kuznicki, L. (1963). Recovery in *P. caudatum* immobilized by chloral hydrate treatment. *Acta Protozool.* **1,** 177–185.

Machemer, H., and Ogura, A. (1979). Ionic conductances of membranes in ciliated and deciliated *Paramecium*. *J. Physiol.* **296,** 49–60.

Ogura, A. (1977). Non-lethal deciliation of *Paramecium* with ethanol. M. S. thesis. University of Tokyo.

Riddle, L. M., Rauh, J. J., and Nelson, D. L. (1982). A Ca-activated ATPase specifically released by Ca shock from *Paramecium tetraurelia*. *Biochim. Biophys. Acta* **688,** 525–540.

Sanders, M. A., and Salisbury, J. L. (1994). Centrin plays an essential role in microtubule severing during flagellar excision in *Chlamydomonas reinhardtii*. *J. Cell Biol.* **124,** 795–805.

Sonneborn, T. M. (1970). Methods in *Paramecium* research. *Methods Cell Physiol.* **4,** 241–339.

Travis, S. M., and Nelson, D. L. (1988). Purification and properties of dyneins from *Paramecium* cilia. *Biochim. Biophys. Acta* **966,** 73–83.

Van Wagtendonk, W. J. (1974). Nutrition of *Paramecium*. *In* "*Paramecium*' A Current Survey" (W. J. Van Wagtendonk, ed.), pp. 339–376. Elsevier Scientific Publishing, New York.

Walczak, C., Anderson, R. A., and Nelson, D. L. (1993). Identification of a family of casein kinases in *Paramecium:* biochemical characterization and localization. *Biochem. J.* **296,** 729–735.

Witman, G. B., Carlson, K., Berliner, J., and Rosenbaum, J. L. (1972). *Chlamydomonas* flagella. I. Isolation and electrophoretic analysis of microtubules, matrix, membranes and mastigonemes. *J. Cell Biol.* **54,** 507–539.

Witman, G. B., Plummer, J., and Sander, G. (1978). *Chlamydomonas* flagellar mutants lacking radial spokes and central tubules. *J. Cell Biol.* **76,** 729–747.

CHAPTER 5

Isolation of *Euglena* Flagella

Huân M. Ngô and G. Benjamin Bouck

Department of Biological Sciences
University of Illinois at Chicago
Chicago, Illinois 60607

I. Introduction

The *Euglena* flagellum is distinct from most other eukaryotic flagella with respect to architectural design and biochemical composition. Its surface is uniformly coated with a glycoprotein layer termed xyloglycorein and the flagellum is partially enveloped with a stable array of filamentous mastigonemes (Bouck *et al.,* 1978, 1990). Internally, the conventional 9 + 2 microtubular axoneme is paralleled along most of its length with a distinctive coiled-sheet paraxonemal rod (PR) (Bouck *et al.,* 1990). In the proximal region, there is a crystalline paraxonemal body (PB) which is presumed to be a photoreceptor (Häder and Brodhun, 1991; Nebenführ *et al.*, 1991). The *Euglena* locomotory flagellum extends from a basal body at the base of the reservoir through a narrow canal and finally to the outside of the cell. Flagellar beat is planar in the reservoir–canal region and progressively more helical outside of the cell.

Methods for deflagellating *Euglena* have been refined somewhat from the

METHODS IN CELL BIOLOGY, VOL. 47

original procedure of Rosenbaum and Child (1967). Mechanical shearing, and cold or pH shock consistently detach the *Euglena* flagellum at the transitional region at the canal opening, whereas a protocol using a combined Ca^{2+} and cold shock treatment developed by Gualtieri *et al.* (1986) can reportedly induce flagellar excision deeper within the reservoir. Interestingly, structures or fibers resembling the contractile stellate structure found in the transitional zone of *Chlamydomonas* cannot be detected, suggesting a microtubule-severing mechanism in *Euglena* operates differently from that proposed during flagellar amputation in *Chlamydomonas* (Sanders and Salisbury, (1989).

Summarized here are procedures for the cultivation of *Euglena* and protocols for flagellar isolation. See Chapter 48 for additional procedures on purification of the paraxonemal rod proteins, Chapter 20 for the enrichment of flagellar glycosyltransferase, and Bouck *et al.* (1978) for the isolation of mastigonemes.

II. Cell Culture

Because of its relatively simple nutrient requirements and short generation time of 10–24 hours, *Euglena gracilis* is easily maintained and grown in culture. All media and culturing glassware should be autoclaved and standard sterile techniques should be used when culturing *Euglena,* as it can be easily overgrown by bacteria and fungi. Axenic stock cultures of *Euglena* can be obtained from the University of Texas Algae Culture Collection (Department of Botany, University of Texas, Austin) or the Culture Centre of Algae and Protozoa (36 Storey's Way, Cambridge, England). Alternative methods for long-term maintenance and growth of *Euglena* cultures as well as additional sources for stock cultures are described in Cook (1968) and Lyman and Traverse (1980).

Cells of *E. gracilis* strain Z are routinely grown in acetate-containing medium of Cramer and Myers (1952) under continuous fluorescent illumination at room temperature (~22°C). For biochemical and molecular assays, larger quantities of flagella are needed and these can be isolated from a 15-liter batch culture, whereas 500-ml cultures are adequate for smaller-scale experiments, e.g., light and electron microscopy.

1. For 500-ml cultures, inoculate 10 ml of cells from a continous liquid culture into 500 ml of medium in a 1-liter Erlenmeyer flask. Grow cells for 3 days under continuous illumination with no shaking.

2. For batch 15-liter cultures, transfer a 3-day 500-ml culture into 15 liters of sterile medium in a 20-liter Pyrex carboy and culture for 5 days. Batch cultures are stirred continuously with a magnetic stirrer to allow aeration and also to keep the cells from attaching to the glass surface.

III. Isolation of Flagella

Harvesting 15 liters of a 5-day *Euglena* culture at a density of approximately 1×10^6 cells/ml should yield ~15 g of cells from which 3–6 mg of flagellar proteins can be collected after deflagellation.

A. Harvesting of Cells

1. Examine a sample of culture by light microscopy to be sure that it is not contaminated with bacteria or fungi.

2. Harvest intact cells from the 15 liters of culture by repetitive rounds of centrifugation of 400-ml batches at 1000–2000 rpm for 5 minutes at room temperature. Gentle handling and low-speed centrifugation at room temperature are critical to avoid premature deflagellation due to excessive packing or temperature shock.

B. Deflagellation by Cold Shock

1. Solutions

Culture medium: ice cold

PET buffer: 0.1 *M* piperazine-*N*,*N*′-bis(2-ethanesulfonic acid), 1 m*M* ethylene glycol bis(β-aminoethyl ether)-*N*,*N*′-tetraacetic acid (EGTA), 10 m*M* *p*-tosyl-L-arginine methyl ester–HCl (TAME), pH 7.0

2. Procedure

1. Resuspend the cell pellet in 30 ml of ice-cold fresh medium and incubate on ice for 1–2 hours. Evaluate the extent of deflagellation by light microscopy.
2. Pellet cell bodies by centrifugation at 1000*g* for 5 minutes in a clinical centrifuge and collect the supernatant containing the flagella.
3. Resuspend the cell pellet with 20 ml of fresh cold medium, spin, and add the supernatant to the first spin supernatant. To increase the flagellar yield, repeat this step four or five times to collect detached flagella that may have been sedimented along with cell bodies.
4. Spin supernatant at 15,000 rpm in a Sorvall SS34 rotor for 30 minutes to yield a white pellet of flagella. Contamination with cell bodies are easily detected by a green hue in the pellet.
5. Resuspend the pellet in 3 ml of PET buffer; if necessary, cell bodies can be removed by an additional centrifugation at 1000*g* in a clinical centrifuge for 5 minutes.
6. Remove supernatant and sediment flagella in a microcentrifuge at 12,000 rpm for 10 minutes to collect the highly purified flagellar pellet.

C. Deflagellation by pH Shock

1. Resuspend the pellet of intact cells in 20 ml of fresh medium and lower the pH of solution to pH 4.6 with 1 *N* HOAc while gently stirring.
2. After 2 minutes, increase the pH to 6.8 with 1 *N* KOH and follow steps 2–5 of procedure B.

D. Deflagellation by Mechanical Shearing

1. Gently pellet 1.5 liters of culture, resuspend in 3 ml of fresh medium, and transfer to a fluted glass tube (Rosenbaum and Child, 1967).
2. Agitate the tube and contents on a Vortex homogenizer for 2–3 minutes.
3. Isolate flagella as in steps 2–5 of procedure B.

E. Deflagellation by Ca^{2+}/Cold Shock

This procedure is taken from Gualtieri *et al.* (1986).

1. Gently pellet 1.5 liters of culture, resuspend in 375 ml of 25 m*M* Na acetate, pH 7.0, and cool to 4°C on a stirrer in an ice bath.
2. Slowly add 1875 ml of a solution containing 25 m*M* Na acetate, 12% ethanol, 0.12% ethylenediaminetetraacetic acid (EDTA), 7% sucrose at pH 7.0. Then add 75 ml of 1 *M* $CaCl_2$ and stir for 30 minutes at 4°C.
3. Isolate flagella as in steps 2–5 of procedure B.

F. Comments

Although flagellar detachment using mechanical shearing is rapid and is useful for regeneration studies (Levasseur *et al.*, 1994), it is not the preferred method for critical biochemical analysis because of potential contamination of other organelles and of cytoplasmic protein complexes from cell breakage. Similarly, the pH shock method is also rapid, but isolated flagella appear swollen and are not optimally preserved for critical morphological observation. The protocol using a combined Ca^{2+} and cold shock has the advantage of yielding a fraction of flagella containing the putative photoreceptor (PB); nevertheless, it produces an inconsistent yield of PB-containing flagella (10–40%). To date, the most consistent and reliable method is cold shock, which generates intact *Euglena* flagella containing most of the axoneme, paraxonemal rod, and flagellar surface.

Acknowledgments

G. Benjamin Bouck is funded by National Science Foundation Grant MCD9105226, and Huân M. Ngô is partially funded by a fellowship from the Laboratory for Molecular Biology of the University of Illinois at Chicago.

References

Bouck, G. B., Rogalski, A., and Valaitis, A. (1978). Surface organization and composition of Euglena. II. Flagellar mastigonemes. *J Cell Biol.* **77,** 805–826.

Bouck, G. B., Rosiere, T. K., and Levasseur, P. J. (1990). *Euglena gracilis:* A model for flagellar surface assembly, with reference to other cells that bear flagellar mastigonemes and scales. *In* "Ciliary and Flagellar Membranes" (R. A. Bloodgood, ed.), pp. 65–90. Plenum Press, New York.

Cook, J. R. (1968). The cultivation and growth of *Euglena. In* "The Biology of *Euglena*" (D. E. Buetow, ed.) Vol. I. pp. 243–314. Academic Press, New York.

Cramer, M. L., and Myers, J. (1952). Growth and photosynthesis characteristics of *Euglena gracilis. Arch. Microbiol.* **17,** 384–402.

Gualtieri, P., Barsanti, L., and Rosati, G. (1986). Isolation of the photoreceptor (paraflagellar body) of the phototactic flagellate *Euglena gracilis. Arch. Microbiol.* **145,** 303–305.

Häder, D.-P., and Brodhun, B. (1991). Effects of ultraviolet radiation on the photoreceptor proteins and pigments in the paraflagellar body of the flagellate, *Euglena gracilis. J. Plant Physiol.* **137,** 641–646.

Levasseur, P. J., Meng, Q., and Bouck, G. B. (1994). Tubulin genes in the algal protist *Euglena gracilis. J. Eukaryotic Microbiol.* **41,** 468–477.

Lyman, H., and Traverse, K. (1980). *Euglena:* mutations, chloroplast "bleaching," and differentiation. *In* "Handbook of Phycological Methods. Developmental and Cytological Methods" (E. Gantt, ed.), pp. 107–141. Cambridge University Press, London.

Nebenführ, A., Schäfer, A., Galland, P., Senger, H., and Hertel, R. (1991). Riboflavin-binding sites associated with flagella of *Euglena:* a candidate for blue-light photoreceptor? *Planta* **185,** 65–71.

Rosenbaum, J. L., and Child, F. M. (1967). Flagellar regeneration in protozoan flagellates. *J. Cell Biol.* **34,** 345–364.

Sanders, M. A., and Salisbury, J. L. (1989). Centrin-mediated microtubule severing during flagellar excision in *Chlamydomonas reinhardtii. J. Cell Biol.* **108,** 1751–1760.

CHAPTER 6

Isolation of Ram Sperm Flagella

Jovenal T. San Agustin and George B. Witman
Male Fertility Program
Worcester Foundation for Experimental Biology
Shrewsbury, Massachusetts 01545

I. Introduction

Methods designed to isolate mammalian sperm substructures broadly fall into two categories: chemical and physical.

Chemical methods include such procedures as protease treatment (Edelman and Millette, 1971; Millette *et al.*, 1973) and dithiothreitol exposure followed by treatment with protease (Millette *et al.*, 1973) or detergent (Hernandez-Montes *et al.*, 1973; Fisher and Bartoov, 1980). Dithiothreitol was shown to cause removal of large areas of plasma membrane, and depending on the detergent used, selective solubilization of heads or tails was achieved.

Physical methods have been more widely used than chemical methods. Nitrogen cavitation has been employed to fractionate bovine and rabbit sperm (Pihlaja *et al.*, 1973; Wang *et al.*, 1986). Under the high pressures used (950 psi), practically all of the plasma membranes over the head region and about 70% of the membranes over the tail were removed (Peterson *et al.*, 1980). Sonication in its early use yielded mammalian sperm tail fragments (Zittle and O'Dell, 1941). Calvin (1976) described a sonication procedure for the separation of heads and tails of rat sperm that was applied to rabbit sperm (Bouchard *et al.*, 1980). Adjustment of the energy input and time of sonication enabled the fractionation of bovine, porcine, and murine sperm into substructures such as heads and tails (Feinberg *et al.*, 1981; Ji *et al.*, 1981; Little *et al.*, 1983; Thakkar *et al.*, 1983; de Curtis *et al.*, 1986; Yagi and Parenko, 1992) and sperm tails without midpieces (Schoff *et al.*, 1989). Depending on the conditions used, plasma membranes may remain intact (Ji *et al.*, 1981) or be largely lost (Feinberg *et al.*, 1981; Yagi and Parenko, 1992).

We describe in this chapter a method for obtaining ram sperm tails by sonication that is based largely on that developed by Calvin (1976). The method also yields nearly pure sperm heads. We include a demembranation procedure for sperm tails that may be employed in studies, for example, where direct access to the sperm axoneme is desired. We have used it to study the cAMP-dependent phosphorylation of ram sperm axonemal proteins and the effects of various inhibitors of sperm motility on the phosphorylation of these proteins. We expect that these methods could be applied, with minor modifications, to the sperm of many other mammalian species.

II. Preparation of Ram Sperm Flagella

A. Solutions and Materials

Phosphate-buffered saline with protease inhibitors (PBSI): 140 m*M* NaCl, 20 m*M* potassium phosphate, pH 6.5, 1 m*M* phenylmethylsulfonyl fluoride (PMSF), 10 m*M* 4-aminobenzamidine, 4 m*M* ethylenediaminetetraacetic acid (EDTA), 4 m*M* dithiothreitol (DTT).

2.2 *M* sucrose solution: 753.1 g sucrose + 528.8 g buffer consisting of 20 ml 1 *M* potassium phosphate, pH 6.5, 20 ml 50 m*M* EDTA, and water. Make 2.5 m*M* in 4-aminobenzamidine and 1 m*M* in DTT before use.

2.05 *M* sucrose solution: 701.7 g sucrose + 561.6 g buffer consisting of 20 ml 1 *M* potassium phosphate, pH 6.5, 20 ml 50 m*M* EDTA, and water. Make 2.5 m*M* in 4-aminobenzamidine and 1 m*M* in DTT before use.

Buffer B: 200 m*M* sucrose, 25 m*M* monopotassium glutamate, 1 m*M* DTT, 40 m*M* 4-(2-hydroxyethyl)-1-piperazineethanesulfonic acid) (Hepes), pH 7.9, 0.25 m*M* PMSF, 2.5 m*M* 4-aminobenzamidine, 1 m*M* EDTA.

B. Washing of Sperm

See Chapter 36 in this volume.

C. Separation of Sperm Flagella from Sperm Heads

1. Mix the washed sperm with 3 vol of PBSI (total volume of 8 ml from 2 ml of washed sperm).

2. Prepare four sucrose step gradients as follows: Into each tube (Beckman Ultra-Clear centrifuge tube, 25 × 89 mm), pipet 13 ml 2.05 *M* sucrose solution. Fill a 20-ml syringe with 2.2 *M* sucrose solution and fit it with a long needle. With the needle tip touching the bottom of the centrifuge tube, push out 13 ml of the 2.2 *M* sucrose solution, which then settles at the bottom of the tube. Keep the tubes on ice.

3. Sonicate the sperm suspension on ice for 45 seconds using a Branson Sonifier Cell Disruptor 200 equipped with a 3-mm microtip, set at 25% watt scale (or 37.5 W).

4. Mix 18 ml of 2.2 *M* sucrose with 8 ml of sonicated sperm. The resulting suspension is about 1.5 *M* sucrose with a density of about 1×10^8 sonicated sperm/ml.

5. Layer half (13 ml) of the sperm suspension on top of the 2.05 *M* sucrose solution in each of two of the sucrose gradient tubes. Keep the remaining two tubes for step 9.

6. Centrifuge at 91,000*g* (23,000 rpm, SW28 rotor, Beckman L5-75 ultracentrifuge) at 4°C for 1.5 hours.

7. Collect the sperm flagella at the 1.5 *M*–2.05 *M* sucrose interface and dilute with cold buffer B in a 12-ml tube. Undecapitated sperm collect at the 2.05 *M*–2.20 *M* interface, whereas the bottom of the tube contains mostly heads with a small amount of whole sperm. If heads are also to be collected, be sure to remove the sucrose layers above the pelleted heads carefully to prevent undue contamination with undecapitated sperm. Then gently pipet 1 ml of cold buffer B over the head pellet and swirl slowly. Decant and repeat, and then disperse heads in 10 ml of cold buffer B. The sonication conditions used here are relatively mild, producing a large proportion of whole flagella, although leaving a significant number of undecapitated sperm. Longer sonication times or higher wattages were found to result in preparations containing fragmented flagella.

8. Centrifuge at 5000*g* (for tails) or 2500*g* (for heads) at 4°C for 10 minutes (6500 and 4500 rpm, respectively, SS-34 rotor, RC-5 Sorvall Superspeed centrifuge, DuPont Instruments). For the first centrifugation, a higher speed may be needed to pellet the tails or heads because of the presence of residual sucrose. Resuspend the pellet in ~10 ml cold buffer B using gentle vortexing.

9. Centrifuge at 1000g (tails) or 300g (heads) at 4°C for 10 minutes (3000 and 1500 rpm, respectively, Sorvall SS-34 rotor). A loose pellet collects at the bottom of the tube. Remove the supernatant carefully and then add 4 ml buffer B. Vortex gently to disperse the pellet.

10. Using the remaining tubes from step 5, perform a second sucrose gradient centrifugation to remove traces of heads and whole sperm from the flagellar preparation and whole sperm from the head preparation. If the density of the washed sperm is too high, a significant amount of heads and undecapitated sperm may be trapped in the flagella layer at the 1.5 *M*–2.05 *M* sucrose interface.

11. Resuspend the isolated flagella (Fig. 1A) and heads (Fig. 1B) in buffer B at a density of 1.5×10^9/ml.

12. Divide into 100-ml aliquots and store in liquid nitrogen.

III. Demembranation of Ram Sperm Flagella

Triton X-100 is employed to demembranate the sperm flagella. We use centrifugation steps to move the sperm flagella into and out of the demembranating solution. This treatment ensures that the flagella are exposed to the detergent for only a short period, and is virtually identical to that which we use to prepare cytosol-free sperm models for reactivation (see Chapter 36). For some studies, the demembranated tails undoubtedly could be centrifuged into a pellet from the demembranating solution.

A. Solutions

10× Solution A, Solution A, 90% Percoll, Solution B: see Chapter 36 for compositions.

Demembranation medium (DM): 0.2% Triton X-100, 200 m*M* sucrose, 25 m*M* monopotassium glutamate, 1 m*M* EDTA, 2.5 m*M* 4-aminobenzamidine, 0.2 m*M* PMSF, 5% w/v Ficoll type 70 (Sigma), 10 m*M* Hepes, pH 7.9.

40% Percoll: 4.44 vol 90% Percoll + 5.56 vol solution A.

B. Procedure

1. Pipet 800 μl of 40% Percoll, 200 μl 90% Percoll, and 200 μl DM into a 1.5-ml microfuge tube as described for ram sperm demembranation (see Chapter 36).

2. Measure out 50 μl of the sperm flagella (1.5×10^9/ml) and carefully pipet it on top of DM.

3. Demembranate for 35 seconds and then spin for 55 seconds. The demembranated flagella collect on top of the 90% Percoll layer. Smaller fragments remain suspended in the 40% Percoll layer.

4. Collect the demembranated tails with a disposable transfer pipet as described for ram sperm (see Chapter 36 in this volume). Make up the volume to about 300 μl using solution A. The final density is about $1\text{–}2 \times 10^8$ tails/ml.

References

Bouchard, P., Gagnon, C., Phillips, D. M., and Bardin, C. W. (1980). The localization of protein carboxyl-methylase in sperm tails. *J. Cell Biol.* **86,** 417–423.

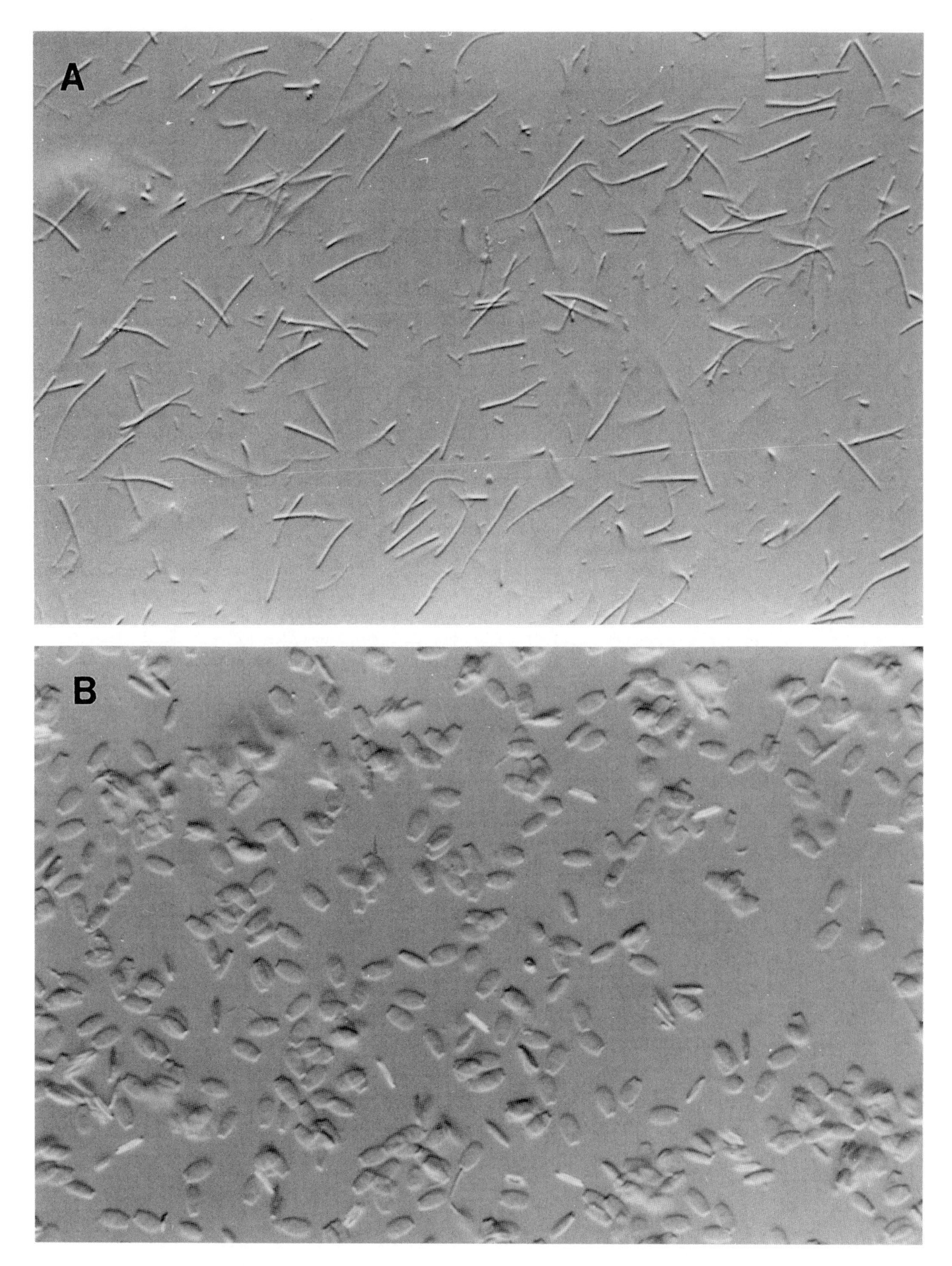

Fig. 1 Light micrographs of isolated ram sperm tails (A) and heads (B) as viewed by differential-interference-contrast optics (× 550).

Calvin, H. I. (1976). Isolation and subfractionation of mammalian sperm heads and tails. *In* "Methods in Cell Biology" (Prescott D. M., ed.) Vol 13, pp. 85–104. Academic Press, New York.
de Curtis, I., Fumagalli, G., and Borgese, N. (1986). Purification and characterization of two plasma membrane domains from ejaculated bull spermatozoa. *J. Cell Biol.* **102,** 1813–1825.
Edelman, G. M., and Millette, C. F. (1971). Molecular probes of spermatozoan structures. *Proc. Natl. Acad. Sci. U.S.A.* **68,** 2436–2440.
Feinberg, J., Weinman, J., Weinman, S., Walsh, M. P., Harricane, M. C., Gabrion, J., and Demaille, J. G. (1981). Immunocytochemical and biological evidence for the presence of calmodulin in bull sperm flagellum. Isolation and characterization of sperm calmodulin. *Biochim. Biophys. Acta* **673,** 303–311.
Fisher, J., and Bartoov, B. (1980). DNase II in bull and ram sperm tail and mitochondria. *Arch. Androl.* **4,** 157–170.
Hernandez-Montes, H., Iglesias, G., and Mujica, A. (1973). Selective solubilization of mammalian spermatozoa structures. *Exp. Cell Res.* **76,** 437–440.
Ji, I., Yoo, B. Y., and Ji, T. H. (1981). Surface proteins and glycoproteins of ejaculated bovine spermatozoa. II. Molecular composition of the midpiece and mainpiece. *Biol. Reprod.* **24,** 627–636.
Little, M., Rohricht, C., and Schroeter, D. (1983). Pig sperm tail tubulin. Its extraction and characterization. *Exp. Cell Res.* **147,** 15–22.
Millette, C. F., Spear, P. G., Gall, W. E., and Edelman, G. M. (1973). Chemical dissection of mammalian spermatozoa. *J. Cell Biol.* **58,** 662–675.
Morton, B. E. (1968). A disruption and fractionation of bovine epididymal spermatozoa. *J. Reprod. Fertil.* **15,** 113–119.
Peterson, R., Russell, L., Bundman, D., and Freund, M. (1980). Evaluation of the purity of boar sperm plasma membranes prepared by nitrogen cavitation. *Biol. Reprod.* **23,** 637–645.
Pihlaja, D. J., Roth, L. E., and Consigli, R. A. (1973). Bovine sperm fractionation. I. Selective degradation and segment separation. *Biol. Reprod.* **8,** 311–320.
Schoff, P. K., Cheetham, J., and Lardy, H. A. (1989). Adenylate kinase activity in ejaculated bovine sperm flagella. *J. Biol. Chem.* **264,** 6086–6091.
Thakkar, J. K., East, J., Seyler, D., and Franson, R. C. (1983). Surface-active phospholipase A_2 in mouse spermatozoa. *Biochim. Biophys. Acta* **754,** 44–50.
Wang, L. F., Miao, S. Y., Cao, S. L., Wu, B. Y., and Koide, S. S. (1986). Isolation and characterization of a rabbit sperm tail protein. *Arch. Androl.* **16,** 55–66.
Yagi, A., and Parenko, J. (1992). Localization of actin, a-actinin, and tropomyosin in bovine spermatozoa and epididymal epithelium. *Anat. Rec.* **233,** 61–74.
Zittle, C. A., and O'Dell, R. A. (1941). Chemical studies of bull spermatozoa. Lipid, sulfur, cystine, nitrogen, phosphorus, and nucleic acid content of whole spermatozoa and of the parts obtained by physical means. *J. Biol. Chem.* **140,** 899–907.

CHAPTER 7

Isolation of Molluscan Gill Cilia, Sperm Flagella, and Axonemes

Raymond E. Stephens
Department of Physiology
Boston University School of Medicine
Boston, Massachusetts 02118

I. Introduction

The isolation of cilia from marine organisms is easily accomplished by brief exposure of actively beating cilia to hypertonic treatment, an approach first introduced by Auclair and Siegel (1966) for sea urchin embryos. This hypertonic salt method was applied to scallop gill cilia by Linck (1973a), the only substantive change being an increased length of treatment that was required for this far larger and more complex tissue.

In other bivalves, however, where most of the cilia may not be beating actively, the yields can be quite low unless extensive agitation is used, in which case the preparation may become heavily contaminated with mucus and exfoliated epithelial tissue. Taking advantage of the fact that active cilia are removed by hypertonic salt, Stommel (1984) isolated one specific type of cilia from mussel gills by first briefly exposing the tissue to hypertonic salt, then specifically activating and releasing only the neuronally stimulated lateral cilia by transferring the tissue to isotonic seawater containing serotonin.

The isolation of molluscan sperm flagella is also derived from methods first used with sea urchins (Stephens, 1970, 1986), the basis of which is simple mechanical shear to sever the head from the tail, followed by differential centrifugation to selectively sediment the heads. This approach was used by Linck (1973a,b) to obtain scallop sperm flagella for structural and biochemical comparison with gill cilia. In most of the isolation procedures for marine cilia and flagella, 9 + 2 axonemes are produced by solubilization of the membrane with detergents such as Triton X-100 and Nonidet P-40 (Linck, 1973a; Stephens, 1977).

The detailed methods presented here are based on those outlined by Stephens and Prior (1992) for the comparison of activation mechanisms of mussel or clam gill cilia with those of sperm flagella. These procedures, in turn, were derived primarily from those of Stephens (1977), Linck (1973a), and Stommel (1984), references that should be consulted for additional background.

II. Biological Material

If one does not have direct access to the seacoast for purposes of collecting, blue mussels and several clam species can be obtained at markets in most larger American cities. These animals are generally quite suitable for gill cilia preparations, but one must be sure that the animals are alive and healthy. This is easily determined by the fact that dead or even marginal bivalves cannot keep their shells closed. In the case of sperm preparations, animals shipped during their breeding season will retain a good deal of sperm within their gonads during cold shipment; only when later placed in seawater will they spawn, especially if that water is warm.

III. Methods

A. Preparation of Gill Cilia

1. Solutions

Tris-buffered seawater (TBSW): Natural or artificial seawater, filtered through Whatman No. 1 paper, containing 10 m*M* Tris–HCl, pH 8

3× TBSW: 64 g of NaCl in 1 liter of TBSW

2× TBSW: 32 g of NaCl in 1 liter of TBSW

Calcium-free seawater (CFSW): 423 m*M* NaCl, 10 m*M* KCl, 58 m*M* $MgCl_2$, and 10 m*M* Tris–HCl, pH 8.0

1 m*M* serotonin: 0.4 mg/ml serotonin–creatinine complex (Calbiochem, No. 5659) in water, prepared fresh (oxygen and light sensitive)

100 m*M* PMSF stock: 17 mg/ml phenylmethylsulfonyl fluoride (PMSF) in isopropanol (sensitive to moisture)

2. Procedure

1. Dissect sufficient mussels (about 100) or surf clams (about 10) to obtain 100 g wet wt of tissue, collecting the gills in 1 liter of cold TBSW.

2. Remove the gills with plastic forks and place in 1 liter fresh, cold TBSW; stir occasionally for about 10 minutes. This step removes silt and mucus.

3. Transfer the gills to 1 liter of cold 3× TBSW and stir occasionally and very gently for 3–4 minutes. The gills will eventually sink, usually after 2–3 minutes, an indication that they have equilibrated with this very hypertonic medium.

4. Decant most of the solution and gently pour the gills over a 15 × 15-cm piece of vinyl window screen to quickly filter out any remaining hypertonic seawater. Invert the screen, shaking the gills loose into 1 liter of cold, isotonic TBSW containing 2.5×10^{-7} *M* serotonin (immediately diluted from a freshly prepared stock). Separate the clumped gills quickly with plastic forks, dispersing them uniformly into the solution. Serotonin activates the lateral cilia while they are still attached to the hypertonically shocked tissue. The beating cilia pinch off above the basal plate and are released (cf. Fig. 2A in Stephens and Prior, 1992).

5. Stir occasionally and moderately over 12–13 minutes to prevent the gills from settling and to wash fluid through the gills, expelling the now-free cilia. After about 8 minutes, the solution should become cloudy or silky with free cilia.

6. Filter through a single layer of Miracloth (Calbiochem, No. 75855) wet with seawater to remove the gills, mucus strands, and small clumps of cells. Do not squeeze the retained material to increase the yield of filtrate; this will only contaminate the preparation with mucus and free cells.

7. Distribute the filtrate into 250-ml polycarbonate centrifuge bottles and spin at 10,000*g* for 10 minutes to harvest the cilia. Pour off quickly in a direction opposite the white or yellow-white pellet, as a "streak" of cilia will accumulate and adhere to the side of the bottle above the pellet.

8. Using a large 5-ml Pipetman or equivalent, wash the "streak" of material down from the side of the bottle into the pellet, resuspending each pellet in about 35 ml of cold CFSW containing 0.25 m*M* PMSF. The calcium-free medium helps to disperse contaminating mucus, allowing a much cleaner separation in the next step.

9. Transfer to 50-ml centrifuge tubes and spin at 1000*g* for 5 minutes. This step removes any remaining cell debris or mucus strands.

10. Carefully decant into fresh tubes and spin at 10,000*g* for 10 minutes to recover the cilia. Thoroughly pour off the supernatant, wiping any excess from the walls of the tubes with Kimwipes. Cap and store on ice until use. Axonemes may be prepared by following the detergent extraction procedure given below.

Note: To prepare scallop cilia by the original method of Linck (1973a), start with the same amount of material and relative solution volumes as above but skip steps 3 and 4. Carry out step 5 using 2× TBSW at room temperature, stirring for 15 minutes. Chill the preparation on ice after filtering through Miracloth in step 6. The cilia on these gills, although mostly laterals, continue to beat for many hours after excision, even without neuronal input. A related procedure, involving 3-minute exposure of gills to 3× TBSW, has been devised for isolating scallop cilia (Sizov and Zhadan, 1989). This is said to give purer cilia, and although this is true, the yield may be somewhat less.

B. Preparation of Sperm Flagella

1. Solutions

Tris-buffered EDTA–seawater (TESW): 10 m*M* Tris, pH 8, and 0.1 m*M* EDTA in filtered natural or artificial seawater

100 m*M* PMSF stock: see above

2. Procedure

The preparation must be kept cold at all times.

1. Collect 10 g (ml) of testicular tissue in a glass beaker. At peak season, this can be obtained from 5–10 mussels or 1 surf clam. Bring the volume to 20 ml with TESW.

2. Finely mince the tissue with scissors. Let the preparation stand on ice for 10 minutes, with occasional stirring. Add TESW to 40 ml, mix well, and filter through either a single layer of Miracloth or four layers of cheesecloth into a 50-ml centrifuge tube. This maceration step releases the sperm and the filtration removes spent tissue.

3. Recover the sperm by spinning at 5000*g* for 5 minutes. After careful decantation, sperm may be stored on ice for many hours or even overnight.

4. Suspend the pellet in 40 ml TESW, transfer to a Teflon/glass homogenizer (A. H. Thomas, Size C), and homogenize with 20 vigorous strokes. Avoid foaming or cavitation. Check in the phase microscope to see that most of the heads are sheared. If not, continue the homogenization with 5 additional strokes and check again.

5. Transfer the broken sperm to a 50-ml centrifuge tube and spin at 1000*g* for 5 minutes to sediment the heads. Remove the tubes from the centrifuge carefully to avoid disturbing the pellets.

6. Carefully decant the suspension of tails into another 50-ml centrifuge tube, avoiding pellet interface material, and spin at 10,000*g* for 5 minutes to sediment the tails. Discard the supernatant.

7. Resuspend the pellet thoroughly in 40 ml of TESW. While stirring or pipetting vigorously, add 0.1 ml 100 m*M* PMSF stock to inhibit proteases.

8. Spin at 1000*g* for 5 minutes to sediment any remaining heads or debris.

9. Again decant carefully into a clean tube and spin at 10,000*g* for 5 minutes to recover the tails. Pour off as much of the supernatant as possible, wiping the inside of the tube dry with a Kimwipe. Cap with Parafilm, storing on ice, or continue with demembranation to produce axonemes.

C. Preparation of 9 + 2 Axonemes

1. Solutions

Low-salt solution (LSS): 0.1 *M* NaCl, 4 m*M* $MgSO_4$, 0.1 m*M* EDTA, 1 m*M* dithiothreitol, 5 m*M* imidazole, pH 7; add 1 μl/ml of 2-mercaptoethanol and 2.5 μl/ml PMSF (above) just before use.

10% Nonidet P-40 (NP-40): "Surfact-Amps P-40" (Pierce, No. 28323), for addition to LSS, immediately before use, to prepare 0.25% NP-40 in LSS.

2. Procedure

1. Thoroughly and rapidly, but without foaming, resuspend each pellet in more than 25 pellet volumes of 0.25% NP-40 in low-salt solution (LSS). This is typically about 10 ml for sperm flagella or 5 ml for gill cilia, assuming the above quantities and steps. The volume can be greater if one has no desire to produce a concentrated membrane extract for use in other procedures.

2. Extract on ice for 10 minutes, with occasional pipetting. This step dissolves the membrane and also disperses any contaminating cellular debris.

3. Spin at 10,000*g* for 10 minutes. Higher speeds may be used but the pellet becomes difficult to resuspend after demembranation. Save the supernatant if the membrane/matrix fraction is desired (see Chapter 61).

4. Thoroughly resuspend the pellet in the same volume (step 1) of fresh 0.25% NP-40 in LSS and immediately spin at 10,000*g* for 10 minutes. This step washes away residual, interstitial extract. Discard the supernatant and wipe the tube dry with a Kimwipe.

5. Resuspend the pellet in the same volume (step 1) of LSS (without detergent) and spin at 10,000*g* for 10 minutes. This step washes away residual, interstitial detergent.

The pellets of detergent-free axonemes are now ready for preparation of remnants (see Chapter 49) or for dynein extraction by high salt (cf. Stephens and Prior, 1992). For storage at −20°C, the pellets should be resuspended in LSS and an equal volume of pure glycerol then added, with thorough mixing. Do not freeze directly. Similarly, resuspend in LSS first for gel sample preparation, then add concentrated sodium dodecylsulfate gel buffer. Do not boil until the axonemes are thoroughly dispersed.

References

Auclair, W., and Siegel, B. W. (1966). Cilia regeneration in sea urchin embryos: evidence for a pool of ciliary proteins. *Science* **154,** 913–915.

Linck, R. W. (1973a). Comparative isolation of cilia and flagella from the mollusc, *Aequipecten irradians. J. Cell Sci.* **12,** 345–367.

Linck, R. W. (1973b). Chemical and structural differences between cilia and flagella from the lamellibranch mollusc, *Aequipecten irradians. J. Cell Sci.* **12,** 951–981.

Sizov, A. V., and Zhadan, P. M. (1989). Protein differences in cilia of mechanoreceptor hair and gill cells of the scallop *Mizuhopecten yessoensis. Comp. Biochem. Physiol.* **94B,** 277–284.

Stephens, R. E. (1970). Thermal fractionation of outer doublet microtubules into A- and B-subfiber components: A- and B-tubulin. *J. Mol. Biol.* **47,** 353–363.

Stephens, R. E. (1977). Major membrane protein differences in cilia and flagella: evidence for a membrane-associated tubulin. *Biochemistry* **16,** 2047–2058.

Stephens, R. E. (1986). Isolation of embryonic cilia and sperm flagella. *Methods Cell Biol.* **27,** 217–227.

Stephens, R. E., and Prior, G. (1992). Dynein from serotonin-activated cilia and flagella: extraction characteristics and distinct sites for cAMP-dependent protein phosphorylation. *J. Cell Sci.* **103,** 999–1012.

Stommel, E. W. (1984). Calcium activation of mussel gill abfrontal cilia. *J. Comp. Physiol.* **155A,** 457–469.

CHAPTER 8

Isolation of Flagella and Their Membranes from Sea Urchin Spermatozoa

Lee Hon Cheung
Department of Physiology
University of Minnesota
Minneapolis, Minnesota 55455

I. Introduction

The initiation of sea urchin sperm motility is a Na^+-dependent process. On dilution of semen into seawater, motility is activated and is accompanied by acid extrusion. Both of these processes do not occur in the absence of external Na^+ (Nishioka and Cross, 1978). Artificial elevation of the internal pH of the sperm with nigericin and/or NH_4Cl can overcome the requirement of external Na^+ (Lee *et al.*, 1982, 1983). Conversely, lowering the pH of regular seawater reduces the internal pH of the sperm and inhibits motility, which can be reversed by monesin, a Na^+/H^+ exchange ionophore (Hansbrough and Garbers, 1981; Christen *et al.*, 1982; Repaske and Garbers, 1983). These results are consistent with the causal trigger of motility being elevation of internal pH of the sperm mediated by a Na^+/H^+ exchanger. To facilitate investigation of the ionic mechanisms involved in sperm motility initiation, procedures are described in this chapter to isolate sea urchin sperm flagella and their membranes that are suitable for measurements of ion transport properties (Lee, 1984, 1985). The advantage of these preparations is that they are single-compartment systems, thus allowing unambiguous analysis and interpretation of ion transport measurements. An added advantage of the flagellar membrane preparation is that the ionic conditions in both the internal and external media can be controlled, therefore allowing the determination of which ion(s) is required for the operation of the flagellar Na^+/H^+ exchanger. In Chapter 56, methods for measuring various ion transport properties such as membrane potential and Na^+ and H^+ movement are described.

II. Experimental Procedures

A. Sperm Collection

Strongylocentrotus purpuratus are purchased from Pacific Bio-Marine (Venice, CA). The animals are maintained at 9°C in an aquarium with artificial seawater made up with Instant Ocean Synthetic Sea Salts and fed with dried seaweed. Gamete shedding is induced by injection of 0.5 *M* KCl into the coelomic cavity. The sperm are stored undiluted at 4°C and used within 1–2 hours of collection.

B. Media

Na^+-free seawater (NaFSW) contains 460 m*M* choline chloride, 27 m*M* $MgCl_2$, 28 m*M* $MgSO_4$, 8 m*M* KCl, 10 m*M* $CaCl_2$, 2 m*M* $KHCO_3$, 5 m*M* Tris, 5 m*M* 4-(2-hydroxyethyl)-1-piperazineethanesulfonic acid (Hepes), pH 7.9.

Regular seawater (NaSW) has the same composition as NaFSW except choline chloride is replaced with 460 m*M* NaCl.

The internal medium (IM) used for isolation of flagellar membranes contains 20 m*M* K_2SO_4, 2.5 m*M* 1,4-piperazinediethanesulfonic acid (Pipes), 50 μM 8-hydroxypyrene-1,2,6-trisulfonic acid (pyranine), and its pH is adjusted to 6.5 with methylglucamine. The composition of this medium was designed with K^+ being the only permeant ion. This simplifies the calculation of membrane potential across the flagellar vesicles and the interpretation of the ion transport measurements. The pyranine is an impermeant pH indicator which is used as both a marker for the intravesicular space and a probe for measuring internal pH of the flagellar vesicles.

C. Isolation of Sperm Flagella

A 0.2-ml aliquot of semen is diluted with 50 ml of cold NaFSW. The large dilution of the semen with NaFSW is necessary to prevent activation of motility by the residual Na^+ in the seminal fluid. The sperm suspension is filtered through a 20-μm Nitex screen (Tetko, Elinsford, NY) to remove debris, and homogenized on ice with a 50-ml Potter–Elvehjem glass homogenizer fitted with a Teflon pestle. About 15–20 strokes are enough to detach most of the flagella as monitored by phase-contrast microscopy. The homogenate (45 ml) is centrifuged (0–5°C) for 30 minutes at 4000 rpm with a Beckman JS7.5 rotor (2000g at r_{ave}) to pellet the sperm heads and intact sperm. The supernatant fluid, containing detached flagella, is centrifuged at 6000g (at r_{ave}) for 30 minutes, and the flagellar pellet is resuspended with 1–2 ml of ice-cold NaFSW. Normally, the contamination of the sperm heads or intact sperm in the final flagellar suspension is no more than 0.05% of the starting number of sperm as determined by direct count with phase-contrast microscopy or by cytochrome c oxidase assay (Lee, 1984).

If larger quantities of flagella are needed, 1–2 ml of semen can be washed with 250 vol of cold NaFSW by centrifugation at 2000g (at r_{ave}) for 5 minutes. The sperm pellets are resuspended with 45 ml of NaFSW (0–5°C) and homogenized as described above. Freshly prepared flagella should be used in experiments within 1–2 hours.

D. Isolation of Flagellar Membranes

Because sea urchin sperm flagella contain no internal membranous organelles, a simple lysis procedure can be used to release the plasma membrane from the axonemes. Detached flagella from 1–2 ml of semen are prepared as described above and pelleted by centrifugation at 6000g for 30 minutes. Next, they are osmotically swollen to disk-shaped structures by resuspension in 0.8 ml of IM. This step is used to break possible linkages between the plasma membrane and axoneme. The swollen flagella are finally lysed by dilution to 8 ml of IM. Further homogenization with a 10-ml glass homogenizer fitted with a Teflon pestle ensures complete detachment of the membranes. The composition of the IM can be modified appropriately if it is desirable to introduce other substances inside the flagellar membrane vesicles. For example, Cs_2SO_4 or KCl can be used instead of K_2SO_4. The total osmolarity of the IM should, however, be kept low to ensure efficient lysis. The axonemes are removed by centrifugation at 13,000g for 4 minutes using a microfuge (Fisher) in a cold room. The flagellar membrane vesicles in the supernatant can be collected by centrifugation (5°C) at 45,000 rpm for 15 minutes using a Beckman Ti50 rotor (130,000g at r_{ave}). The membrane pellet is rinsed gently with an external medium (EM), which has the same composition as IM except without pyranine and with the addition of 10 mM $MgSO_4$. The pellet is resuspended by homogenization with 9 ml of EM and centrifuged at 130,000g for 15 minutes at 5°C. The washed vesicles, free of extracellular pyranine, are finally resuspended into 1 ml of EM. Inclusion of Mg^{2+} in the EM is not necessary for preparation of the flagellar membranes, but it does have important effects on the voltage sensitivity of the flagellar Na^+/H^+ exchanger (Lee, 1985).

The flagellar membrane fraction contains about 4% of the protein in the total flagellar lysate which has 9.1 ± 0.8 mg protein/ml of semen. The intravesicular space is 16.3 ± 2.8 μl/mg of membrane protein as determined by the entrapment of pyranine. Vesicle sizes are heterogeneous, with most having a diameter of about 0.1 μm, but vesicles with diameters as large as 1 μm can be found (Lee, 1985).

Acknowledgments

I thank Richard Graeff for suggestions and critical reading of the manuscript. The research in my laboratory is supported by a grant from the National Institutes of Health.

References

Christen, R., Schackman, R. W., and Shapiro, B. M. (1982). Elevation of the intracellular pH activates respiration and motility of sperm of the sea urchin, *Strongylocentrotus purpuratus*. *J. Biol. Chem.* **257,** 14881–14890.

Hansbrough, J. R., and Garbers, D. L. (1981). Sodium-dependent activation of sea urchin spermatozoa by speract and monensin. *J. Biol. Chem.* **256,** 2235–2241.

Lee, H. C., Forte, J. G., and Epel, D. (1982). The use of fluorescent amines for the measurement of pH_i: Applications in liposomes, gastric microsomes, and sea urchin gametes. *In* "Intracellular pH: Its Measurement, Regulation and Utilization in Cellular Functions" (R. Nuccitelli and D. Deamer, eds.), pp. 135–160. Alan R. Liss, New York.

Lee, H. C., Johnson, C., and Epel, D. (1983). Changes in internal pH associated with initiation of motility and acrosome reaction of sea urchin sperm. *Dev. Biol.* **95,** 31–45.

Lee, H. C. (1984). Sodium and proton transport in flagella isolated from sea urchin spermatozoa. *J. Biol. Chem.* **259,** 4957–4963.

Lee, H. C. (1985). The voltage-sensitive Na^+/H^+ exchange in sea urchin spermatozoa flagellar membrane vesicles studied with an entrapped pH probe. *J. Biol. Chem.* **260,** 10794–10799.

Nishioka, D., and Cross, N. (1978). The role of external sodium in sea urchin fertilization. *In* "Cell Reproduction" (E. R. Dirsken, D. Prescott, and D. F. Fox, eds.), pp. 403–413. Academic Press, New York.

Repaske, D. R., and Garbers, D. L. (1983). A hydrogen ion flux mediates stimulation of respiratory activity by speract in sea urchin spermatozoa. *J. Biol. Chem.* **258,** 6025–6029.

CHAPTER 9

Isolation of Fish Sperm Flagella[1]

Christèle Saudrais* **and Jean-Luc Gatti**[†,2]

*Laboratoire de Physiologie des Poissons
INRA Campus de Rennes-Beaulieu
35042 Rennes cedex, France
†URA 1291 INRA-CNRS
Laboratoire de Physiologie de la Reproduction des Mammifères Domestiques
INRA Nouzilly
37380 Monnaie, France

I. Introduction

Flagella and cilia have been obtained from various sources but only from a small number of vertebrates and, mainly, from mammals. Fish spermatozoa offer a good source from which to isolate and purify large amounts of vertebrate flagella and axonemes. Sperm from two teleost species, trout and carp, have been used extensively to study the mechanisms involved in the initiation of sperm motility. Each of these species has advantages as well as disadvantages. During the spawning season, an adult male trout produces a large number of sperm and can give up to 100 ml of semen at 10^{10} sperm/ml. Thus, the trout is a good choice for biochemical study; however, the spawning season is limited

[1] This paper is dedicated to the memory of Dr. M.-P. Cosson.
[2] To whom correspondence should be addressed.

and it is necessary to use different strains (summer, autumn, and spring spawners) to extend the workable season. Individual carp give fewer sperm than trout (5 to 10 ml at 0.5 to 4 $\times$ 10^{10} sperm/ml). Although carp naturally spawn in autumn, sperm production can be induced throughout the year by repetitive injection of pituitary extracts (Billard *et al.*, 1986). Therefore, the carp is an excellent choice for physiological studies of flagellar movement.

Trout and carp spermatozoa are quite simple and their flagella (total length 50–60 μm) contain a classical 9 + 2 axoneme without periaxonemal structures. Moreover, the absence of an acrosomal vesicle decreases the risk of contamination of the preparation with proteases. The preparation of large quantities of fish sperm flagella was described by Pautard (1962), who used trout and perch flagella for X-ray diffraction and biophysical studies. The same preparation was used by Tibbs (1962) to study some properties of the flagellar ATPase and acetylcholinesterase. Subsequently, very few studies on fish sperm flagellar proteins were reported until the isolation and purification of trout sperm outer-arm dynein ATPase (Gatti *et al.*, 1989; King *et al.*, 1990; Moss *et al.*, 1991). Here we briefly describe methods for purifying axonemes (see Moss *et al.*, 1991, for a more detailed protocol) and focus more on how to prepare intact flagella to analyze their membrane proteins and to reactivate them to study axonemal motility mechanisms.

II. Media

Initiation of trout sperm motility can be inhibited by media containing large amounts of potassium or having a low pH (for reviews, see Stoss, 1983; Billard and Cosson, 1992). Therefore, a simple trout sperm inhibitory medium (modified from Billard, 1983) contains 28 m*M* KCl, 110 m*M* NaCl, and 20 m*M* Tris–HCl, pH 8.0. Carp sperm activation is controlled by a decrease in the external osmotic pressure and is not dependent on the external ions (Morisawa and Suzuki, 1980; Redondo-Müller *et al.*, 1991). Therefore, a solution containing 200 m*M* KCl and 30 m*M* Tris–HCl at pH 8.0 (osmotic pressure 380 mOsm) is sufficient to inhibit carp sperm motility.

The extraction buffer (IMEN) used to isolate the axonemes contains: 5 m*M* imidazole, 5 m*M* 2-mercaptoethanol, 0.5 m*M* ethylenediaminetetraacetic acid (EDTA), 100 m*M* NaCl, 4 m*M* $MgCl_2$, 0.5 m*M* phenylmethylsulfonyl fluoride (PMSF), and 1% Triton X-100 at pH 7.0. Although we did not notice proteolysis of the axonemal proteins during the preparation (Gatti *et al.*, 1989; King *et al.*, 1990), it may be advisable to include different protease inhibitors (e.g., 2 m*M* *para*-aminobenzamidine) in this medium, as proteasomes are present on trout sperm plasma membrane (Inaba *et al.*, 1993). Imidazole can be replaced with Tris–HCl without a noticable change in the results.

The sperm demembranation buffer contains 150 m*M* K-acetate, 0.5 m*M* $CaCl_2$, 0.5 m*M* EDTA, 1 m*M* dithiothreitol, 20 m*M* Tris–HCl, pH 8.2, and 0.01% Triton X-100. The reactivation buffer has the same composition but with

2 m*M* $MgCl_2$ and 2 m*M* ATP (vandium-free from Sigma) instead of Triton X-100.

III. Collection of Sperm

Both fish species are available from fish farms and can be kept in tanks with running tap water. For trout, the water should be aerated with bubbling and frequently renewed. The quality of the water is very important and chlorine must be avoided. The temperature for carp should be about 20°C and not less than 15°C, whereas trout need a lower temperature (5–14°C). For injection or sperm collection, fish must be anesthetized by immersion for a few minutes in a 0.25 to 0.5% solution of 2-phenoxyethanol or a 2% solution of aminobenzoic acid ethyl ester (MS-222). For carp, spermiation can be induced throughout the year by intraabdominal injection of carp pituitary extract (2 mg/kg body wt; Argent Chemical Laboratory, Redmond, WA). Spermiation occurs within 12 to 20 hours and reasonable amounts (about 2–5 ml of semen) are collected each day for several days (Billard *et al.,* 1986). For both species, semen is obtained by applying gentle pressure along the abdomen, but care should be taken not to contaminate the semen with feces or urine. An alternative method is to cannulate the genital papilla, but this carries the risk of inducing blood contamination. When possible, or when necesssary if using male trout obtained by sex reversal induced by steroid administration, testes can be removed and cut in pieces in motility inhibitory medium, and the solution filtered through 150-μm nylon mesh. Only a fraction of the sperm obtained from a single trout can be used directly for axoneme isolation. Therefore, if only biochemical studies are to be carried out, some of the sperm may be frozen in liquid nitrogen for later use (Billard, 1990; Moss *et al.,* 1991). Otherwise, the semen is kept on ice until use.

IV. Isolation of Axonemes

Trout sperm axonemes are isolated as described by Moss *et al.* (1991). Briefly, 5 ml of trout sperm at up to 5×10^{10} sperm/ml are diluted with 10 ml ice-cold IMEN in a 15-ml Dounce homogenizer (glass–glass, A pestle). After four to five gentle strokes, 10 ml additional ice-cold IMEN is introduced and the solution centrifuged at 750*g* for 10 minutes at 4°C. At this step bubbling of the axonemal suspension should be avoided, and the sample kept on ice. The supernatant must be very carefully removed to avoid head contamination. A second homogenization–centrifugation of the resuspended pellet can be done to improve the yield of axonemes but this is not always necessary. The axonemes are then collected by centrifugation at 12,000*g* for 10 minutes at 4°C. To remove the residual cytoplasmic proteins and detergent, axonemes are gently resuspended in ice-cold IMEN without Triton X-100 and then centrifuged at 12,000*g* for 10 minutes at 4°C. This washing step is repeated twice. Plastic pipets and tubes are always used to prevent axonemes from sticking to the glass surface. Routinely 10 mg of pure axonemes can be obtained by this procedure; this is sufficient to yield about 1 mg of outer-arm dynein (Gatti *et al.*, 1989). When more axonemes are needed, this procedure can be scaled up by doubling all the volumes and using a 40-ml Dounce homogenizer.

V. Isolation of Flagella

This procedure is quite straightforward. For trout, sperm are first concentrated by centrifugation at 250*g* for 15 minutes; then the speed is increased to 500*g* for 10 minutes (4°C). The seminal plasma is removed and the pellet is resuspended to the initial volume in immobilizing medium and centrifuged again (250*g* for 15 minutes followed by 500*g* for 10 minutes). The final pellet is diluted with 2 vol of an immobilizing medium containing 190 m*M* instead of 110 m*M* NaCl. This increase in osmotic pressure from 285 m to 475 mOsm seems necessary to maintain good integrity of the plasma membrane during the following step. A syringe with a 23-gauge needle (0.6 × 25 mm) is fitted with a polyethylene catheter (70-cm length, 1-mm o.d., 0.5-mm i.d.). The sperm suspension is drawn in and out of the syringe seven times to detach the heads from the tails. The suspension is then centrifuged at 750*g* for 10 minutes (4°C). If necessary, the sperm/head pellet can be gently resuspended and the operation repeated to increase the yield of flagella. The supernatants (Fig. 1A) are pooled and centrifuged at 2000*g* for 20 minutes (4°C) to pellet the flagella (lower speed may be used to reduce the risk of breakage). Very little fragmentation occurs and very few membrane vesicles are visible (Fig. 1B). The presence of membrane on the flagella is indicated by the observation that some flagella become highly coiled as a result of an osmotic effect (Fig. 1B, arrow). The low level of flagella plasma membrane loss was also checked by sodium dodecylsulfate–polyacrylamide gel electrophoresis, using the 42-kDa major plasma membrane protein (Labbé and Loir, 1991) as a marker (Fig. 1B, inset).

For carp sperm the same procedure is used except that sperm are diluted in the immobilizing medium at a high osmotic pressure (see Figs. 2A and B). In this case, plasma membrane integrity can be checked by the absence of reactivation when Triton X-100 is omitted from the demembranation medium (see below).

VI. Demembranation and ATP Reactivation of Isolated Carp Sperm Flagella

This procedure is adapted from that described for sea urchin sperm (Gibbons and Gibbons, 1972) and used to reactivate trout and carp sperm (Morisawa and

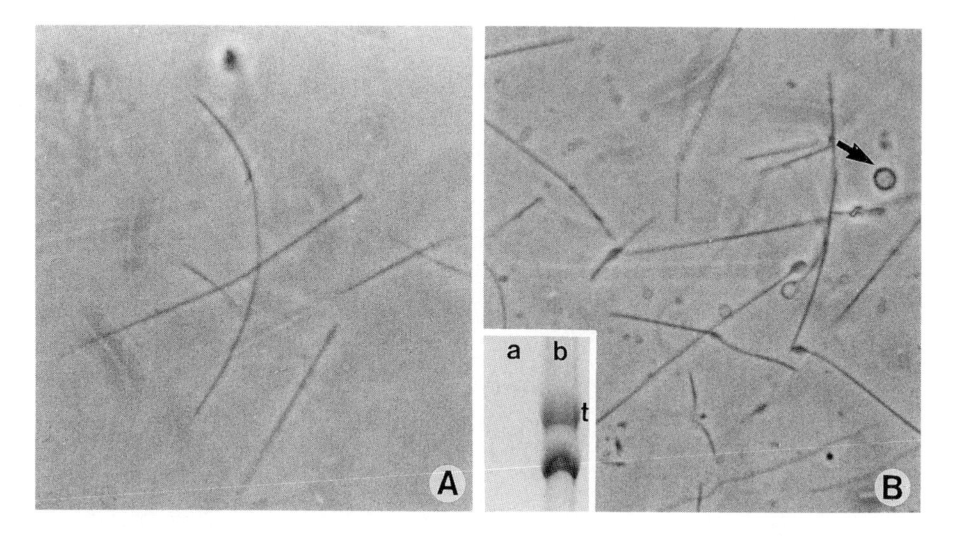

Fig. 1 Isolation of trout sperm flagella. Flagella were obtained after the first low-speed centrifugation (A) and after the final high-speed centrifugation (B). Note the presence of highly coiled flagella (arrow in B), indicating the presence of the plasma membrane. Plasma membrane integrity was also assessed on sodium dodecyl sulfate–polyacrylamide gel electrophoresis by the complete absence of the plasma membrane 42-kDa protein on the high-speed supernatant (a) and its presence on the flagella fraction (b). t, tubulins. Coomassie brillant blue-stained gel.

Okuno, 1982; Cosson and Gagnon, 1988). Flagella are first diluted 1/20 in a test tube containing cold demembranation solution and gently shaken on ice for 30 seconds. Then 1 μl is mixed with 20 μl of the reactivation solution directly on a microscope slide coated with bovine serum albumin. Typically, 50 to 70% of isolated carp flagella are reactivated. A majority of these reactivated flagella show a planar beat (Fig. 2C) similar to what is observed with live spermatozoa. No inclusion of cAMP is necessary to obtain the reactivation (Cosson *et al.*, 1986).

VII. Conclusion

Teleost sperm can provide large amounts of material for the isolation of axonemes or intact flagella. The large amounts of material available make possible the extraction and analysis of specific components of the axoneme and the flagella plasma membrane. The ability to isolate intact carp sperm flagella that can be reactivated permits study of the role of axonemal proteins in flagellar beating and allows direct investigation of the effect of intracellular effectors (Ca^{2+}, phosphorylations, etc.) on the reactivated movement.

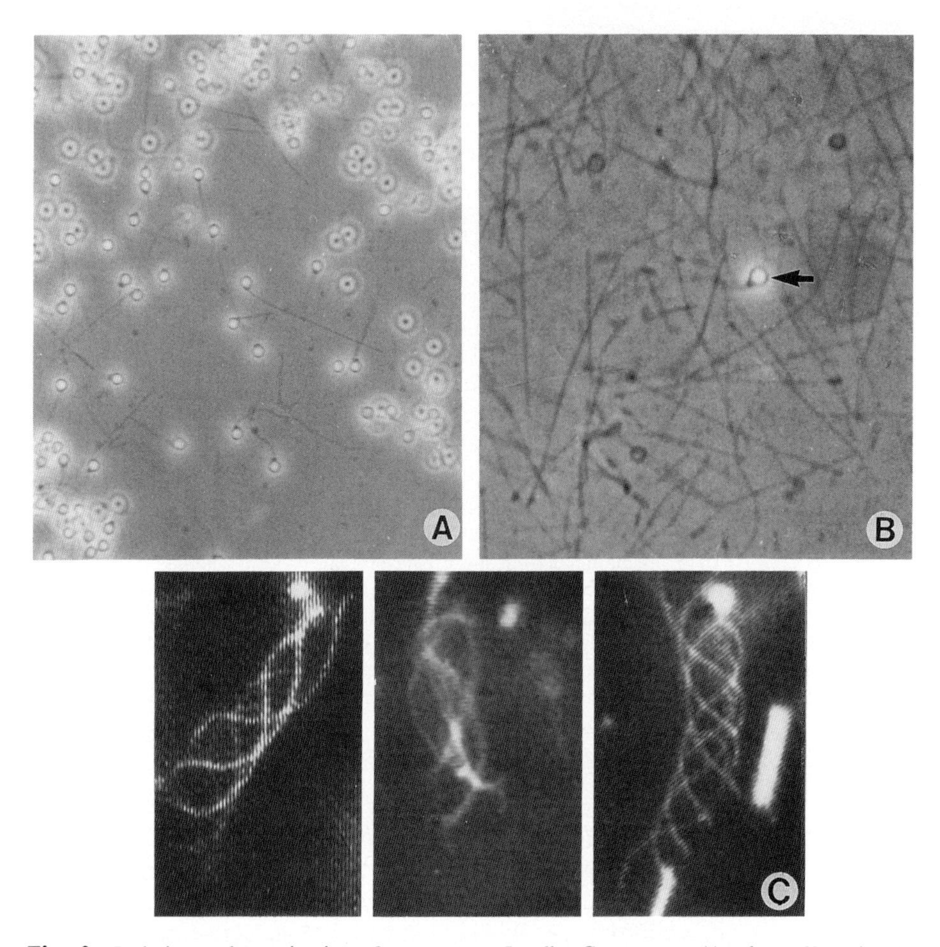

Fig. 2 Isolation and reactivation of carp sperm flagella. Carp sperm (A, about 60 μm) were deflagellated and their flagella concentrated (B). Very few heads are still present (arrow). The flagella were reactivated as described and micro-videotaped under stroboscopic illumination. The monitor was photographed with a 1-second continuous exposure. A majority of flagella had a symmetrical planar beating pattern (C, left and right); a few exhibited a more complex movement (C, center).

Acknowledgments

The authors thank Dr. G. B. Witman for corrections and comments on the manuscript. Dr. M.Loir, Dr. J. Cosson, and Prof. R. Billard are thanked for helpful discussions and A. Beguey for photographic assistance. Christèle Saudrais is supported by a grant from Ministère de l'Enseignement superieur et de la Recherche.

References

Billard, R., (1983). Effect of coelomic and seminal fluids and various saline diluents on the fertilizing ability of spermatozoa in the rainbow trout. *J. Reprod. Fertil.* **68,** 77–84.

Billard, R., Christen, R., Cosson, M.-P., Gatti, J.-L. Letellier, L., Renard, P., and Saad., A. (1986). Biology of the gametes of some teleost species. *Fish Phys. Biochem.* **2,** 115–120.

Billard, R. (1990). Artificial insemination in fish. *In* "Marshall's Physiology of Reproduction" (G. E. Lamming, ed.), p. 870. Churchill Livingston, Edinburgh.

Billard, R., and Cosson, M.-P. (1992). Some problems related to the assessment of sperm motility. *J. Exp. Zool.* **261,** 122–131.

Cosson, M.-P., Gatti, J.-L., Christen, R., and Billard, R. (1986). Carp sperm, a new model for studying sperm motility regulation. *In* "Aquaculture of Cyprinids" (R. Billard and J. Marcel, eds.). p. 165, INRA, Paris.

Cosson, M.-P., and Gagnon, C. (1988). Protease inhibitor and substrates block motility and microtubule sliding of sea urchin and carp spermatozoa. *Cell Motil. Cytoskel.* **10,** 518–527.

Gatti, J.-L., King, S. M., Moss, A. G., and Witman, G. B. (1989). Outer arm dynein from trout spermatozoa. *J. Biol. Chem.* **264,** 11450–11457.

Gibbons, B. H., and Gibbons, I. R. (1972). Flagellar movement and adenosine triphosphatase activity in sea urchin sperm extracted with Triton X 100. *J. Cell. Biol.* **54,** 75–97.

Inaba, K., Akazome, Y., and Morisawa, M. (1993). Purification of proteasomes from salmonid fish sperm and their localization along sperm flagella. *J. Cell. Sci.* **104,** 907–915.

King, S. M., Gatti, J.-L., Moss, A. G., and Witman, G. B. (1990). Outer arm dynein from trout spermatozoa: substructural organization. *Cell Motil. Cytosk.* **16,** 266–278.

Labbé, C., and Loir, M. (1991). Plasma membrane of trout spermatozoa: I. Isolation and partial characterization. *Fish Phys. Biochem.* **9,** 325–338.

Moss, A. G., Gatti, J.-L., King, S. M., and Witman, G. B. (1991). Purification and characterization of *Salmo gairdneri* outer arm dynein. *Methods Enzymol.* **196,** 201–222.

Morisawa, M., and Suzuki, K. (1980). Osmolarity and Potassium ion: Their role in the initiation of sperm motility in Teleosts. *Science* **210,** 1145–1147.

Morisawa, M., and Okuno, M. (1982). cAMP induces maturation of trout sperm axoneme to initiate motility. *Nature* **295,** 703–704.

Pautard, F. G. E. (1962). Biomolecular aspect of spermatozoan motility. *In* "Spermatozoan motility" (D. W. Bishop, ed.), p. 189. American Association for the Advancement of Science, Washington, D.C.

Redondo-Müller, C., Cosson, M.-P., Cosson, J., and Billard, R. (1991). *In vitro* maturation of the potential for movement of carp spermatozoa. *Mol. Reprod. Dev.* **29,** 259–270.

Stoss, J. (1983). Fish gamete preservation and spermatozoan physiology. *In* "Fish Physiology Vol. IXB" (W. S. Hoar, D. Randall, and E. M. Donaldson, eds.), Vol. IXB, p. 305. Academic Press, New York.

Tibbs, J. (1962). Adenosine triphosphatase and acetylcholinesterase in relation to sperm motility. *In* "Spermatozoan motility" (D. W. Bishop, ed.), p. 233. American Association for the Advancement of Science, Washington, D.C.

PART II

Culture and Isolation of Ciliated Epithelial Cells and Cilia

CHAPTER 10

Ciliogenesis of Rat Tracheal Epithelial Cells *in Vitro*

Lawrence E. Ostrowski,* Scott H. Randell,† Alan B. Clark,* Thomas E. Gray,* and Paul Nettesheim*

* Airway Cell Biology Group
Laboratory of Pulmonary Pathobiology
National Institute of Environmental Health Sciences
Research Triangle Park, North Carolina 27709
† Cystic Fibrosis/Pulmonary Research and Treatment Center
Department of Medicine
University of North Carolina at Chapel Hill
Chapel Hill, North Carolina 27599

I. Introduction

As early as 1984, the differentiation of ciliated cells in primary cultures of hamster tracheal epithelial cells was reported (Lee *et al.*, 1984). When airway epithelial cells were cultured on plastic and cells were submerged in medium, however, the level of ciliated cell differentiation was generally poor, with most cells assuming a squamous morphology. The use of explant cultures or very high seeding densities allowed for the maintenance of ciliated cells present in the original tissue (e.g., Baeza-Squiban *et al.*, 1994; van Scott *et al.*, 1988), but these conditions were unsuitable for studies of the process of ciliogenesis. Recently, methods for culturing dissociated airway epithelial cells that promote *de novo* ciliogenesis *in vitro* have been developed for hamster, rat, dog, guinea pig, and human respiratory cells (Kondo *et al.*, 1991; Kaartinen *et al.*, 1993;

Adler *et al.*, 1990; Whitcutt *et al.*, 1988; De Jong *et al.*, 1994; Yamaya *et al.*, 1992; Moller *et al.*, 1989; Jorissen *et al.*, 1990). Several of these studies have demonstrated the importance of a preformed extracellular matrix and/or culture on permeable supports at an air–liquid interface for the development of ciliated cells (Kaartinen *et al.*, 1993; Kondo *et al.*, 1991; De Jong *et al.*, 1994; Whitcutt *et al.*, 1988; Lee *et al.*, 1984). In some species or under certain conditions, however, ciliogenesis occurs in the submerged state and in the absence of supplied extracellular matrix (i.e., Lee *et al.*, 1984; Jorissen *et al.*, 1990; Moller *et al.*, 1989). To facilitate studies of the regulation of airway cell differentiation, an *in vitro* system was developed that supports proliferation and then mucociliary differentiation of dissociated primary rat tracheal epithelial (RTE) cells (Kaartinen *et al.*, 1993). A detailed description of the procedures used has been previously published (Kaartinen *et al.*, 1993).

II. Isolation and Culture of Cells

All reagents are obtained from Sigma (St. Louis, MO) unless otherwise noted. Transwell-Col tissue culture inserts (24.5 mm, 0.4-μm pore size) are obtained from Costar (Cambridge, MA). Recently, we have also used Transwell-Clear tissue culture inserts (also from Costar) with similar results. Cannulas, used for the isolation of RTE cells, are made from 5-cm pieces of polyethylene tubing (PE-190, Becton Dickinson, Parsippany, NJ). One end of the tubing is pressed against a heated scalpel blade to form a slight flange. These are stored in 70% ethanol.

A. Solutions and Materials

Pronase solution contains 1% Pronase E, type XIV, in Ham's F-12. DNase solution contains 0.5 mg/ml crude pancreatic DNase I and 10 mg/ml bovine serum albumin (BSA), in Ham's F-12.

Basic Medium. Supplement Dulbecco's modified Eagle's medium/Ham's F-12 (DME/F12, Sigma No. D-9785) with 0.45 m*M* L-leucine, 0.50 m*M* L-lysine, 6.5 m*M* L-glutamine, 0.12 m*M* L-methionine, 0.30 m*M* $MgCl_2$, 0.40 m*M* $MgSO_4$, 1.05 m*M* $CaCl_2$, 8.6 mg/liter phenol red, and 1.2 g/liter $NaHCO_3$. Store at 4°C.

Complete Medium (CM). Supplement basic medium with 10 μg/ml insulin, 0.1 μg/ml hydrocortisone, 0.1 μg/ml cholera toxin (CT), 5 μg/ml transferrin (Collaborative Research, Bedford, MA), 25 ng/ml epidermal growth factor (EGF, Collaborative Research), 50μM phosphoethanolamine, 80 μM ethanolamine, 30 m*M* 4-(2-hydroxyethyl)-1-piperazineethanesulfonic acid (Hepes), 0.5 mg/ml BSA, 50 U/ml–50 μg/ml penicillin–streptomycin (pen–strep, Gibco, Grand Island, NY) and 1% bovine pituitary extract (BPE).

Bovine Pituitary Extract. To prepare, thaw pituitaries (Pel-Freeze, Rogers, AK), drain excess blood, and add 2 ml phosphate buffered saline per gram of tissue. Mince pituitaries in a Waring blender at low speed for 10 minutes at 4°C and centrifuge at 1500*g* for 10 minutes at 4°C. Store supernatant frozen in small aliquots and centrifuge again (1500*g* for 10 minutes) before adding the supernatant to the medium. Sterile-filter CM using a 0.45-mm filter and store for up to 1 week at 4°C. To prevent clogging, use a prefilter and add the BPE supernatant *after* all the medium has passed through the filter. Add retinoic acid to the CM to a final concentration of 5×10^{-8} *M* and use within 2 days.

Tissue Culture Insert. Twenty-four hours before obtaining the RTE cells, form a collagen gel on top of the culture insert. Spread 0.4 ml of 0.3% type I collagen (rat tail collagen, Collaborative Research) over the surface of each membrane. Saturate a gauze pad with 30% ammonium hydroxide and place in the bottom of a covered 160-mm petri dish. Place the membranes on the lid of a 100-mm petri dish and place the dish on top of the gauze pad; cover. After 3 minutes the collagen should be a firm gel. Place the membranes back in the culture wells in which the lower compartments are filled with 3 ml of sterile water. After 4–8 hours, replace the water with sterile DME/F-12 medium and incubate overnight at room temperature. Before preparing the RTE cells for plating (see below), replace the DME/F-12 in the lower chamber with CM (2.5 ml) containing 10% fetal bovine serum and 3 mg/ml BSA (final concentration).

B. Procedures

1. Isolation of Rat Tracheal Epithelial Cells

1. Euthanize male Fisher 344 rats (10–16 weeks old) by CO_2 asphyxiation.

2. Expose the trachea and separate it from the esophagus and underlying tissue using sterile instruments. Make a small incision below the larynx, insert the flanged end of a cannula (drained of ethanol), and tie it in place with silk thread.

3. Cut the trachea just above its bifurcation and place it in a tube of sterile Ham's F-12/pen–strep on ice. All subsequent operations are performed in a sterile tissue culture hood.

4. Tie the trachea loosely at the lower end and flush with a few drops of 1% Pronase solution using a syringe fitted with an 18-gauge needle. After tightening the lower knot, fill the trachea with the Pronase solution until gently distended. Double over the cannula and pinch shut with a bulldog clamp.

5. Place the tracheas in tubes containing enough Ham's F-12/pen–strep to cover them and incubate at 4°C overnight.

6. Flush RTE cells from the trachea in Ham's F-12/pen–strep containing 5% fetal bovine serum (about 5 ml/trachea) using a syringe and 18-gauge needle. Collect cells into tubes with a funnel and pellet by centrifugation (500*g*, 4°C, 10 minutes).

7. Resuspend in DNase solution (1 ml/trachea). Incubate cells in an ice bath for 5 minutes and collect by centrifugation as above. Then resuspend cells in CM containing 3 mg/ml BSA (1 ml/trachea, about 1×10^6 cells/ml) and count with a hemacytometer. The viability of the cells should be 90% or greater.

2. Plating and Growth of Cells

1. Dilute the cell suspension with CM containing 3 mg/ml BSA to 1.2×10^5 cells/ml.

2. Add 1 ml of the suspension to the top of the collagen gel, prepared as above.

3. Gently rotate the culture insert with forceps to evenly spread the cell solution over the surface.

4. Place in a humidified incubator at 35°C with 3% CO_2 (day 0 of culture).

5. On the following day (day 1), remove medium from the upper (apical) and lower (basal) compartments using a Pasteur pipet and gentle suction. Slightly tilt the inserts to aid removal of the apical media. Do not puncture the gel. Then add 2.5 ml of CM to the basal compartment and 0.5 ml to the apical compartment. Medium should be changed as described on days 3, 5, and 6 of culture.

3. Differentiation of Rat Tracheal Epithelial Cells

1. On day 7 of culture, create an air–liquid interface. Remove medium as described above and add 2.5 ml of CM to the lower compartment only.

2. Change medium daily until the end of the experiment, usually day 14. Daily medium change is important to prevent acidification of the medium. Any fluid on the apical surface should be removed at the time of medium change.

At day 7, the cultures should be mostly confluent to the edges. If the cultures are not confluent by day 7 or 8, leakage of medium through the membrane results in partially submerged rather than air–liquid interface cultures. Submerged cultures do not differentiate fully into a mucociliary epithelium. By days 9–10 of culture, the apical fluid will be very viscous, the result of mucus production by secretory cells. Between days 10 and 12, ciliated cells become apparent by phase microscopy and, by day 14, cover 5–20% of the surface area of the dish, depending on the culture. Cells are difficult to remove from the culture membranes. Currently, we add a 0.15% trypsin/1.6 m*M* ethylenediaminetetraacetic acid (EDTA, Gibco)/0.1% Pronase E solution to the cells followed by 30–40 minutes of incubation at 35°C. Cultures should contain between 2 and 8×10^6 cells at day 14.

4. Quantitation of Ciliated Cell Differentiation

Ciliated cells are visible in the viable cultures by phase microscopy. Estimates of the percentage of ciliated cells can be obtained by standard light and scanning electron microscopy techniques. To estimate the extent of ciliated cell differentiation in large numbers of cultures, an immunostaining procedure was developed using a monoclonal antibody, RTE-3, which reacts specifically with ciliated cells (Ostrowski *et al.*, 1994; Shimizu *et al.*, 1992). Culture inserts are fixed *in situ* and reacted with RTE-3 using standard immunostaining techniques. Binding of RTE-3 is detected using peroxidase-conjugated goat anti-mouse antibody and diaminobenzidine. The culture insert is examined by low-power light microscopy, and an image analysis program is used to determine the percentage of the culture surface area stained (ciliated) in a series of equally spaced fields. RTE-3 can also be used to identify ciliated cells in cytospins prepared from dissociated cultures.

III. Comments and Conclusions

Under the described conditions, ciliated cells present in the initial cell preparation are slowly lost from the culture, so that by day 7, the cultures contain less than 1% ciliated cells. If the cultures are maintained in the submerged state, the number of ciliated cells does not increase significantly. Creation of an air–liquid interface increases ciliated cell differentiation so that by day 14, about 10% if the culture surface is covered by ciliated cells. However, we have observed (Clark *et al.*, 1995) that, if some of the growth-stimulating compounds (i.e., BPE, EGF, CT) are removed from the medium after the cells have reached confluence (day 7), the number of ciliated cells present at day 14 can be two- to fourfold higher than that obtained when the cells are cultured with CM. Removal of EGF and CT from the medium at the same time as the air–liquid interface is created is the condition we have explored in most detail. Cytospin preparations of dissociated cells from day 14 cultures demonstrated more than 30% ciliated cells in the cultures lacking EGF and CT, compared with 8% for the CM cultures. By comparison, the normal rat trachea has been estimated to contain between 17 and 33% ciliated cells (Jeffery and Reid, 1975).

This procedure reproducibly results in the production of a mucociliary epithelium *in vitro* that is similar to the *in vivo* tissue (Fig. 1). Ciliated cell differentiation occurs in a significant percentage of the cell population over a short period, and can be modulated by changing the culture conditions (i.e., submersion) or medium (i.e., removal of EGF and CT). This model will be useful for studies of the factors and processes regulating the differentiation and function of ciliated cells of the airways.

Fig. 1 Comparison of rat tracheal epithelial culture to *in vivo* rat trachea. (a) A Day 16 culture of RTE cells grown for 2 days in the absence of BPE was photographed by scanning electron microscopy. A heavily ciliated section of the membrane is shown. (b) Portion of an adult rat trachea showing mature ciliated cells. Magnification = 1000×.

Acknowledgments

The authors thank Dr. Bob Bagnell and Vicky Madden for their excellent scanning electron microscopy services.

References

Adler, K. B., Cheng, P.-W., and Kim, K. C. (1990). Characterization of guinea pig tracheal epithelial cells maintained in biphasic organotypic culture: cellular composition and biochemical analysis of released glycoconjugates. *Am. J. Respir. Cell Mol. Biol.* **2,** 145–154.

Baeza-Squiban, A., Boisvieux-Ulrich, E., Guilianelli, C., Houcine, O., Geraud, G., Guennou, C., and Marano, F. (1994). Extracellular matrix-dependent differentiation of rabbit tracheal epithelial cells in primary culture. *In Vitro Cell. Dev. Biol.* **30A,** 56–67.

Clark, A. B., Randall, S. H., Nettesheim, P., Gray, T. E., Bagnell, B., and Ostrowsk, L. E. (1995). Regulation of ciliated cell differentiation in cultures of rat tracheal epithelial cells. *Am. J. Respir. Cell Mol. Biol.* **12,** 329–338.

De Jong, P. M., Van Sterkenburg, M. A. J. A., Hesseling, S. C., Kempenaar, J. A., Mulder, A. A., Mommaas, A. M., Dijkman, J. H., and Ponec, M. (1994). Ciliogenesis in human bronchial epithelial cells cultured at the air-liquid interface. *Am. J. Respir. Cell Mol. Biol.* **10,** 271–277.

Jeffery, P. K., and Reid, L. (1975). New observations of rat airway epithelium: a quantitative and electron microscopic study. *J. Anat.* **120,** 295–320.

Jorissen, M., Van der Schueren, B., Van den Berghe, H., and Cassiman, J.-J. (1990). Ciliogenesis in cultured human nasal epithelium. *ORL J. Otorhinolaryngol. Relat. Spec.* **52,** 368–374.

Kaartinen, L., Nettesheim, P., Adler, K. B., and Randell, S. H. (1993). Rat tracheal epithelial cell differentiation *in vitro*. *In Vitro Cell. Dev. Biol.* **29A,** 481–492.

Kondo, M., Finkbeiner, W. E., and Widdicombe, J. H. (1991). A simple technique for culture of highly differentiated cells from dog tracheal epithelium. *Am. J. Physiol.* (*Lung Cell. Mol. Physiol.* 5) **261,** L106–L117.

Lee, T.-C., Wu, R., Brody, A. R., Barrett, J. C., and Nettesheim, P. (1984). Growth and differentiation of hamster tracheal epithelial cells in culture. *Exp. Lung Res.* **6,** 27–45.

Moller, P. C., Partridge, L. R., Cox, R. A., Pellegrini, V., and Ritchie, D. G. (1989). The development of ciliated and mucus cells from basal cells in hamster tracheal epithelial cell cultures. *Tissue Cell* **21,** 195–198.

Ostrowski, L. E., Randell, S. H., Clark, A. B., Gray, T. E., and Nettesheim, P. (1994). Rapid quantitation of ciliated cell differentiation in *in vitro* cultures of rat tracheal epithelial cells *Am. J. Respir. Crit. Care Med.* **149,** A995.

Shimizu, T., Nettesheim, P., Eddy, E. M., and Randell, S. H. (1992). Monoclonal antibody (Mab) markers for subpopulations of rat tracheal epithelial (RTE) cells. *Exp. Lung Res.* **18,** 323–343.

van Scott, M. R., Lee, N. P., Yankaskas, J. R., and Boucher, R. C. (1988). Effect of hormones on growth and function of cultured canine tracheal epithelial cells. *Am. J. Physiol.* **255,** C237–C245.

Whitcutt, M. J., Adler, K. B., and Wu, R. (1988). A biphasic chamber system for maintaining polarity of differentiation of cultured respiratory tract epithelial cells. *In Vitro Cell. Dev. Biol.* **24,** 420–428.

Yamaya, M., Finkbeiner, W. E., Chun, S. Y., and Widdicombe, J. H. (1992). Differentiated structure and function of cultures from human tracheal epithelium. *Am. J. Physiol.* **262,** L713–L724.

CHAPTER 11

Preparation of Explant and Organ Cultures and Single Cells from Airway Epithelium

Ellen R. Dirksen, Jennifer A. Felix, and Michael J. Sanderson[1]

Department of Anatomy and Cell Biology
UCLA School of Medicine
University of California
Los Angeles, California 90024

I. Introduction

The inaccessibility of the ciliated epithelium of the mammalian respiratory tract has hampered *in vivo* studies of its structure and function, including those regarding the regulatory mechanisms of mucociliary transport. Consequently, several *in vitro* approaches have been developed to maintain and study airway tissues in culture (see Chapter 10; Jorissen *et al.,* 1991). These include (1) explant cultures, where pieces of tissue are seeded on a substrate to encour-

[1] Present address: Department of Physiology, University of Massachusetts Medical Center, Worcester, Massachusetts 01655.

age the spreading and outgrowth of the epithelial cells (Sanderson and Dirksen, 1985; Van Scott *et al.*, 1986); (2) organ cultures, in which the tissue is maintained to preserve its original structural and functional form (Sanderson and Sleigh, 1981); (3) monolayer cell cultures, where enzymatically dissociated cells are plated onto a noncellular substrate (Welsh, 1985); and (4) feeder layer cultures, where isolated epithelial cells are grown on a cellular substrate such as treated fibroblasts (Gray *et al.*, 1983).

In this chapter, we describe our methods for preparing explant and organ cultures as well as the enzymatic isolation of cells. Explant cultures are advantageous because they express epithelial characteristics for several weeks, form monolayer outgrowths suitable for light microscopy, and retain large numbers of ciliated cells (Fig. 1A). We have used explant cultures for the measurement of ciliary beat frequency, including the timings of the phases of the beat cycle, of single (Sanderson and Dirksen, 1985) and multiple cells (Sanderson *et al.*, 1988), as well as for the analysis of intercellular Ca^{2+} signaling (Sanderson *et al.*, 1990; Boitano *et al.*, 1992). We maintain the tracheal mucosa in organ culture for up to a week to provide a source of tissue for the enzymatic isolation of cells (Dirksen and Zeira, 1981) or intact sheets of epithelial cells (Sanderson and Sleigh, 1981). Isolated epithelial cells (Fig. 1B) have been used in patch-clamp studies (Kim *et al.*, 1993) and in assays to measure inositol trisphosphate levels. Although we use primarily rabbits, these methods could apply to other mammals.

II. Methods

Epithelial cells are cultured on glass coverslips that have been coated with a thin, flat, translucent gel of collagen for compatibility with light microscopy and the promotion of cell attachment and growth. We recommend the use of collagen extracted from rat tails, as this promotes rapid cell growth and has excellent optical properties (Bornstein, 1958; Sanderson and Dirksen, 1985).

A. Preparation of Rat Tail Collagen

USE STERILE TECHNIQUE THROUGHOUT

1. A fresh rat tail or one that has been previously frozen may be used. Pour ~100 ml of 95% ethanol into one-half of a 150 × 15-mm Petri dish and ~100 ml of sterile double-distilled H_2O (ddH_2O) into the other. Soak the rat tail in the ethanol for 5 minutes to sterilize.

2. Hold the tail ~2 cm from the tip with a large hemostat. Fracture the tail bones below the hemostat with bone cutters, cutting through skin without severing the tail. Firmly hold the lower part of the tail, just below the cut, with

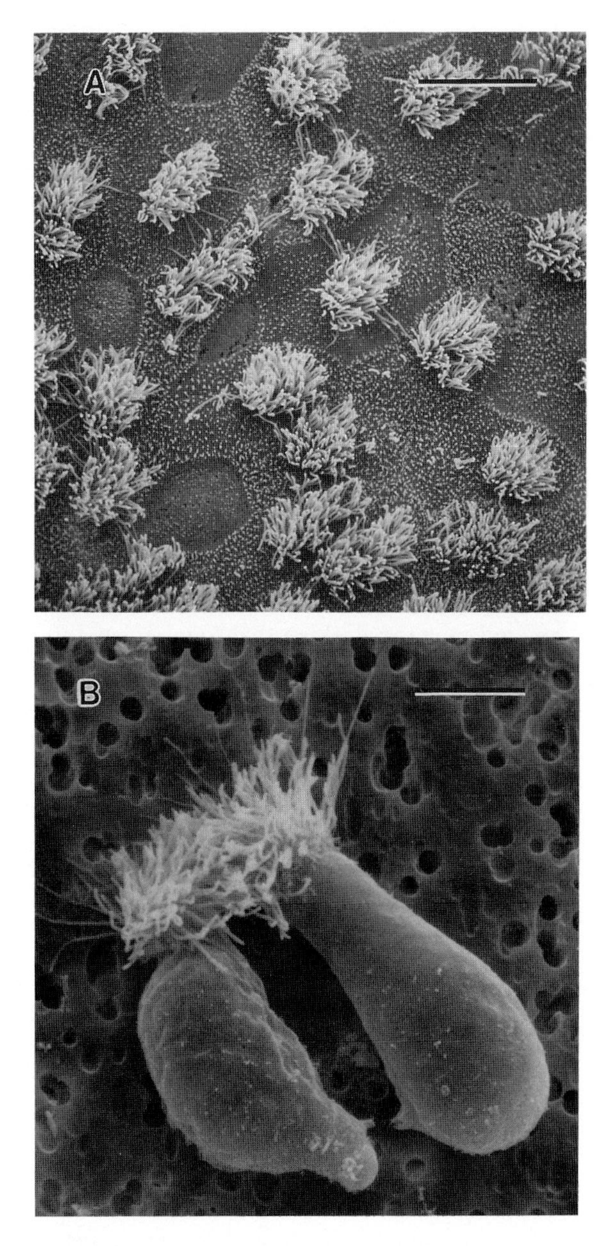

Fig. 1 Scanning electron micrographs of (A) a typical area of an outgrowth from a tracheal epithelial explant culture (bar = 10 μm) and (B) two ciliated epithelial cells enzymatically isolated from tracheal mucosal tissue (bar = 5 μm).

the bone cutters or another hemostat, and pull the tip of the tail away until the attached silvery tendons are free of the tail.

3. With fine scissors cut the tendons free from the tip and allow them to drop into the water in the large Petri dish. Discard the tip.

4. Repeat steps 2 and 3, successively removing small pieces from the end of the tail and recovering attached tendons until only half of the rat tail remains.

5. Using two pairs of fine forceps, tease the tendons apart into finer strands to increase their surface area and transfer to a 250-ml centrifuge bottle containing 150 ml 0.1% (v/v) acetic acid. Allow 48–72 hours (at 4°C) for the acid to dissolve the tendons and form a thick collagen suspension.

6. Centrifuge the collagen solution at 10,000*g* for 2 hours at 4°C and transfer 40 ml of the supernatant to a sterile container. Seal and store at 4°C for up to 6 months.

7. Prepare a "dialysis flask" in the following manner. Seal and weigh down one end of a 6-in. piece of dialysis tubing (Spectro/Por 3 dialysis tubing, molecular weight cutoff of 3.5 kDa). With a rubber band or string attach the open end to a wide-bore pipet inserted through a two-holed stopper of an 1000-ml Erlenmeyer flask. Fill the flask with ~1000 ml of ddH_2O. Wrap the stopper in foil and plug the flask so that the dialysis tubing is immersed. Cover the pipet opening with an inverted 100-ml beaker and autoclave.

8. Prior to use, the stored acetic acid–collagen solution must be dialyzed. Transfer 5 ml of collagen into the tubing in the dialysis flask with a syringe and 18-gauge spinal needle. Cover the flask with the beaker and equilibrate at 4°C for 6–9 hours. The collagen solution will become more viscous with a longer dialysis time.

9. Remove dialyzed collagen with a fresh syringe and store in a 15-ml centrifuge tube at 4°C for up to 2 months.

B. Preparation of Collagen-Coated Coverslips

USE STERILE TECHNIQUE THROUGHOUT

1. Materials

Fluorescent illuminator, daylight spectrum

Three four-well culture multidishes (Fisher Scientific, Catalog No. 12-565-72)

150 × 15-mm glass Petri dish containing a 75 × 45-mm glass slide placed on filter paper

12 acid-washed 15-mm No. 1 circular coverslips (Carolina Biological, Catalog No. 633030)

100 × 15-mm Petri dish containing two pieces of Whatman No. 1 filter paper

~75 × 40-mm glass dish containing a coverslip holder (Thomas Scientific, Catalog No. 8542-E40)

100 ml Hanks' balanced salt solution supplemented with 25 m*M* 4-(2-hydroxyethyl)-1-piperazineethanesulfonic acid (HEPES), pH 7.4 (sHBSS)

0.05 ml 0.2% (w/v) sterile flavin mononucleotide (FMN, Sigma, Catalog No. F-6750) solution

0.5 ml dialyzed rat tail collagen (Section II,A)

5 ml Dulbecco's modified Eagle medium without $NaHCO_3$ (Gibco-BRL, Catalog No. 31600-034) supplemented with 25 m*M* HEPES (pH 7.4), 10% fetal bovine serum, 100 U/ml penicillin, 100 μg/ml streptomycin, and 0.25 μg/ml amphotericin B (sDMEM)

2. Procedure

1. With a cotton-plugged pipet, transfer 3 ml of ddH_2O to the filter paper in the 150-mm Petri dish for humidification and place 12 small, evenly spaced droplets of water at the periphery of the glass slide.

2. With fine tweezers, place one coverslip on each drop of water such that one-third of the coverslip protrudes from the edge of the slide. With the correct amount of water, surface tension will firmly hold the coverslip in place.

3. To prevent the premature gelling of the collagen solution by light activation, reduce the amount of light in the work area.

4. Mix 0.5 ml (~10 drops) of dialyzed collagen with 0.05 ml (~1 drop) of FMN. The ratio of collagen to FMN may be modified, depending on the viscosity of the collagen. Mix FMN and collagen by drawing the solution in and out of a pipet.

5. Place one drop of the collagen mixture on each coverslip and spread by manipulating an air bubble blown into the collagen. Withdraw excess collagen to leave a thin layer. Work quickly because collagen will gel after adding FMN. The gels will shrink unless humidified.

6. Cover the dish and illuminate with fluorescent light for 20–40 minutes. Surplus collagen can be used to test when collagen has solidified.

7. Fill the dish containing a coverslip holder with 90 ml sterile ddH_2O and fill a beaker with 80 ml 5% glutaraldehyde. After collagen has set, immerse each coverslip in glutaraldehyde for ~10 seconds to crosslink the collagen. To remove excess fixative, blot the coverslip edge on sterile filter paper inside the petri dish and place in coverslip holder in glass dish.

8. Wash coverslips three times with ddH_2O for 20 minutes each and do a final 20-minute wash in sHBSS.

9. Add 1.5 ml H_2O to the middle well of each four-well multidish to humidify. Transfer a coverslip, collagen side uppermost, to each well. Add sufficient

sDMEM to each well to completely cover the collagen. These can be stored at room temperature for weeks before use.

C. Dissection of Rabbit Trachea

USE ASEPTIC TECHNIQUE THROUGHOUT

1. Prepare two 50-ml centrifuge tubes with 20 ml sHBSS in each.
2. Restrain a New Zealand White rabbit in a cage and cover its eyes.
3. Inject 100 mg/kg body wt Nembutal (sodium pentobarbital) into the large vein on the dorsal side of the ear using a 21-gauge syringe needle. Ensure the animal is dead before proceeding.
4. Place the rabbit on its back to expose the neck. Shave the neck area with animal clippers and remove fur with a vacuum cleaner hose. Douse the shaved area with ethanol to sterilize.
5. Make a single incision through the skin from the collar bone to above the larynx with a scalpel (No. 3 handle, No. 11 blade). Attach hemostats to the edges of the incision and use their weight to hold the incision open and expose the neck muscles.
6. Cut through the muscle and connective tissue overlying the trachea. Remove the trachea by severing it between the larynx and the collar bone.
7. Transfer the trachea to a tube of sHBSS to wash off blood and mucus and then transfer to a second tube of sHBSS. If the trachea is not used immediately, place in sDMEM and store at 4°C for up to 24 hours.

D. Preparation of Tracheal Epithelial Explant Culture

USE STERILE TECHNIQUE THROUGHOUT

1. Materials

Rabbit trachea (Section II, C)

100 × 15-mm glass Petri dish half-filled with polymerized Sylgard 184 silicone elastomer (K. R. Anderson, Santa Clara, CA)

12 collagen-coated coverslips (Section II, B)

25 ml sHBSS, 5 ml sDMEM

5 ml sDMEM with 0.37% (w/v) $NaHCO_3$ equilibrated to pH 7.4 in 10% CO_2 atmosphere (sDMEM with HCO_3^-).

2. Procedure

1. Transfer 1 ml of sDMEM to a small Petri dish (60 × 15 mm) and set aside.
2. Place 4 ml and 6 ml of sHBSS in a small and Sylgard-lined Petri dish, respectively.

3. Place the trachea in sHBSS in the small Petri dish and cut in half, between cartilage rings, with a No. 22 scalpel blade (No. 4 handle) and work with half the tissue at a time.

4. With fine dissecting scissors make a lengthwise cut along the trachealis muscle. Transfer a tracheal half to the Sylgard-filled dish containing sHBSS.

5. To stretch the tissue, pin the four corners of the trachea, with its luminal ciliated side uppermost, to the Sylgard base with needles bent at ~150°.

6. Use a No. 15 scalpel blade (No. 3 handle) to make four shallow cuts through only the mucosa, near the edges of the trachea, creating the outline of the mucosal section to be removed.

7. While viewing through a dissecting microscope, hold and lift one corner of the mucosa with fine tweezers. Use a No. 15 scalpel blade to separate the mucosa from the cartilaginous backing by cutting through the connective tissue.

8. Transfer the mucosa to sDMEM in the small Petri dish and spread out with the ciliated surface uppermost. The isolated mucosa tends to curl up with the cilia on the convex surface.

9. Cut the mucosa into 0.5- to 1.0-mm^2 explants by making rolling cuts with a No. 22 blade and add an additional 2 ml of sDMEM to the dish.

10. Aspirate the sDMEM from the multidishes without disturbing the collagen on the coverslips.

11. Draw one to three mucosal explants, together with sDMEM, into a pipet and transfer to a collagen-coated coverslip without touching the collagen. Quickly add at least one explant with sDMEM to each coverslip to prevent desiccation of the collagen. Place two to four, well-spaced, additional explants on each coverslip.

12. Remove the excess sDMEM to prevent explants from floating; cover and allow ~5 minutes for the explants to settle and stick to the collagen.

13. Very slowly, add sufficient sDMEM with HCO_3^- to cover the explants to the side of each well of the multiwell dish. If the explants float, repeat steps 10–12.

14. Incubate the cultures in a humidified atmosphere of 10% CO_2 at 37°C. Outgrowths should be evident after 24–36 hours. Beating cilia indicate cell viability. Groups of ciliated cells are more abundant near the explant. For use, remove the coverslip from the culture dish with the aid of a fine metal dental probe.

E. Preparation of Organ Culture

In organ culture, the tracheal mucosa is supported at a liquid–air interface to mimic *in vivo* conditions of nutrient and gas exchange. Organ cultures require some materials additional to those needed for preparing explant cultures. Collagen-coated coverslips are not required.

1. Materials

Stainless-steel wire grids (60 mesh), ~2-cm sided triangle, with the corners bent to form a low "table"
Sterile lens paper (~2 × 2-cm pieces)

2. Procedure

1. Isolate the tracheal mucosa as described in Section II, D, steps 1–8, and ensure that the ciliated surface is uppermost.
2. Slice the mucosa into ~5 × 10-mm strips with a rolling action, with a No. 22 scalpel blade.
3. Place a piece of lens paper beside a strip of mucosa in the small Petri dish and drag the mucosal strip onto the lens paper, preserving its orientation.
4. Place a wire grid inside a small Petri dish (30 × 10 mm) and add sufficient sDMEM with HCO_3^- to wet the grid and align the fluid meniscus with the level of the grid.
5. With fine tweezers, transfer the lens paper together with the tissue atop the wire grid. Ensure that the tissue is protruding slightly above the surface of the medium, but not floating.
6. To incubate, cover the dish and immediately place in a humidified atmosphere of 10% CO_2 at 37°C to prevent the pH of the sDMEM with HCO_3^- from becoming basic.
7. Repeat the preceding steps for the remaining slices. Mucosal slices can be maintained in this manner for up to 1 week.

F. Preparation of Enzymatically Isolated Cells

Isolated cell preparations are obtained either from freshly dissected tissues or from tissues that had previously been placed in organ culture. If the isolated cells are to be used immediately, it is not necessary to maintain aseptic technique; however, sterile technique is required if cell cultures are to be initiated by plating isolated cells onto collagen-coated coverslips or Petri dishes. Digestion of mucosal tissue from a single trachea with a nonspecific protease produces a suspension of about 2.5×10^6 single cells with a few multicell clusters and a viability of 85–95%, as determined by hemocytometry and trypan blue exclusion (Fig. 1B; Dirksen and Zeira, 1981).

1. Materials

Mucosa from one trachea (Section II, D, steps 1–8, and Section II, E)
3 ml of 0.5% (w/v) Pronase solution (type XIV bacterial protease, Sigma

Catalog No. P-5147) in Dulbecco's modified Eagle medium supplemented only with 25 m*M* Hepes (pH 7.4) in a 15-ml centrifuge tube

3 ml sDMEM, 40 ml sHBSS

2. Procedure

1. Use a freshly dissected mucosa or mucosal slices maintained in organ culture.
2. Cut mucosa into thin strips with a No. 22 scalpel blade and place mucosal strips in Pronase solution.
3. Incubate at 37°C for 50 minutes with constant gentle agitation (e.g., on a slow rotational mixer). Once or twice during the incubation vigorously shake the tube to increase cell yield.
4. After 50 minutes, discard tissue and add 3 ml sDMEM to inhibit further digestion.
5. Centrifuge at 100*g* for 5 minutes, aspirate the supernatant, add 10 ml sHBSS to the cell pellet, and pipet to resuspend the cells.
6. Repeat step 5 twice and in the final resuspension add the desired volume of sHBSS.

Acknowledgments

This work was supported by the Smokeless Tobacco Research Council, Inc., and the Cigarette and Tobacco Surtax Fund of the State of California through the Tobacco-Related Disease Research Program of the University of California. Michael J. Sanderson is a recipient of National Institutes of Health Grant HL49288.

References

Boitano, S., Dirksen, E. R., and Sanderson, M. J. (1992). Intercellular propagation of calcium waves mediated by inositol trisphosphate. *Science* **258,** 292–295.

Bornstein, M. B. (1958). Reconstituted rat-tail collagen used as a substrate for tissue cultures on coverslips in Maximow slides and roller tubes. *Lab. Invest.* **7,** 134–137.

Dirksen, E. R., and Zeira, M. (1981). Microtubule sliding in cilia of the rabbit trachea and oviduct. *Cell Motil.* **1,** 247–260.

Gray, T. E., Thomasson, D. G., Mass, M. J., and Barrett, J. C. (1983). Quantification of cell proliferation, colony formation, and carcinogen induced cytotoxicity of rat tracheal epithelial cells grown in culture on 3T3 feeder layers. *In Vitro* **19,** 559–570.

Jorissen, M., Van der Schueren, B., Van den Berghe, H., and Cassiman, J. J. (1991). Contribution of *in vitro* culture methods for respiratory epithelial cells to the study of the physiology of the respiratory tract. *Eur. Respir. J.* **4,** 210–217.

Kim, Y. K., Dirksen, E. R., and Sanderson, M. J. (1993). Stretch-activated channels in airway epithelial cells. *Am. J. Physiol.* **265**(Cell Physiol. 34), C1306–C1318.

Sanderson, M. J., and Sleigh, M. (1981). Ciliary activity of cultured rabbit tracheal epithelium: beat pattern and metachrony. *J. Cell Sci.* **47,** 331–347.

Sanderson, M. J., and Dirksen, E. R. (1985). A versatile and quantitative computer-assisted photoelectric technique used for the analysis of ciliary beat cycles. *Cell Motil.* **5,** 267–292.

Sanderson, M. J., Chow, I., and Dirksen, E. R. (1988). Intercellular communication between ciliated cells in culture. *Am. J. Physiol.* **254**(Cell Physiol. 23), C63–C74.

Sanderson, M. J., Charles, A. C., and Dirksen, E. R. (1990). Mechanical stimulation and intercellular communication increases intracellular Ca^{2+} in epithelial cells. *Cell Regul.* **1,** 585–596.

Van Scott, M. R., Yankaskas, J. R., and Boucher, R. C. (1986). Culture of airway epithelial cells: Research techniques. *Exp. Lung Res.* **11,** 75–94.

Welsh, M. J. (1985). Ion transport by primary cultures of canine tracheal epithelium: Methodology, morphology, and electrophysiology. *J. Membr. Biol.* **88,** 149–163.

CHAPTER 12

Induction of Ciliogenesis in Oviduct Epithelium

Bernadette Chailley*[*] and Emmanuelle Boisvieux-Ulrich[†]

[*]Développement normal et pathologique des Fonctions Épithéliales
INSERM U319, Université Paris 7
75251 Paris 05, France
[†]Laboratoire de Cytophysiologie et de Toxicologie Cellulaire
Université Paris 7
75251 Paris 05, France

I. Introduction

In contrast with a number of organs, the oviduct differentiates during postnatal sexual maturation. In the developing oviduct, estrogens influence both epithelial proliferation and differentiation. During the proliferative phase, oviduct length and wall height increase. During the differentiation phase, the cellular composition of the luminal epithelium is modified in response to hormonal influence. Initially, the epithelium is characterized by a predominance of ciliated cells which are found in the female genital tract of all vertebrates. In a second step, the characteristic oviduct epithelium consists of ciliated and secretory cells, forming, in most cases, a mucociliary epithelium as in airways of the respiratory tract. The purpose of this chapter is to describe a method for inducing ciliogenesis in a model of arrested avian oviduct development.

A. Natural Development of the Bird Oviduct

In birds, only the left ovary and oviduct develop. The right gonad and oviduct remain rudimentary. The immature quail oviduct consists of a thin, short tube (40 mm long, 2 mm in diameter). Immature oviduct weight is about 10 mg. The luminal epithelium is composed of a single layer of cuboidal cells which are undifferentiated with a primary cilium.

Oviduct growth initiates at the fourth week posthatching. High cell proliferation occurs and the oviduct increases to 200 mm in length and 10 mm in diameter at the mature stage. The oviduct is nearly 1000-fold heavier (9000 mg) in the laying bird. The oviduct of the laying bird is subdivided into five functionally distinct segments for egg formation: the fimbria and the ampulla forming the infundibulum, the magnum, the isthmus, and the shell gland or so-called uterus. All these segments are ciliated (Pageaux *et al.,* 1986). Ciliary beat is oriented from the infundibulum to the uterus, and the determination of ciliary polarity is established before differentiation (Boisvieux-Ulrich and Sandoz, 1991).

From the sixth week, the luminal epithelium comprises two cell types, with ciliated and mucous cells regularly interspersed. In the middle part of the oviduct, tubular serous glands, which invaginate from the luminal epithelium in the underlying stroma, are responsible for the secretion of egg white proteins. The magnum weighs nearly 75 mg when the tubular glands begin to form and 250 mg when glandular cells begin to elaborate serous secretory granules.

B. Ultrastructural Events during Ciliogenesis

Ciliogenesis occurs in the same basic pattern in oviduct cells of mammals and birds (see Dirksen, 1991). In brief, in the quail oviduct, the first step is centriologenesis, during which approximately 200 centrioles are formed from fibrogranular masses (generative complexes) present near the Golgi apparatus (Lemullois *et al.,* 1988). Centrioles move under the cellular surface and attach to the apical plasma membrane which has previously developed microvilli (Chailley and Boisvieux-Ulrich, 1985). Centrioles are then called basal bodies, from which the cilia grow at the cell surface by lengthening of the axonemal microtubules. This corresponds to the second step, cilium formation proper (Chailley *et al.,* 1982, 1990).

C. Hormonal Induction of Ciliogenesis

Natural maturation of the quail oviduct is prevented by ovariectomy of the left ovary. Young quails recover easily from ovariectomy and their oviduct then remains in an immature state. The maturation process can be restored by hormonal treatment, which can then be applied within a broad period. Immature 3-week-old quails also can be stimulated for oviduct maturation; in this case, development of the oviduct is accelerated and occurs within 6 days instead of 2 weeks.

According to the nature and dose of hormones and the duration of treatment, development is oriented toward either the ciliogenic or the secretory path of epithelial differentiation. Estrogen treatment induces ciliogenesis; undifferentiated epithelium turns into both undifferentiated cells and cells undergoing ciliogenesis (Sandoz *et al.,* 1975). Progesterone hyperstimulation in association with estrogen treatment strongly induces the secretory process and inhibits the differentiation of ciliated cells (Perche *et al.,* 1989). The balance between estrogen and progesterone is critical for harmonious development of the oviduct (Laugier *et al.,* 1975). Estrogenic treatment has also been shown to control ciliogenesis in the oviduct epithelium of other species, such as the cat (Verhage and Brenner, 1975) and chick (Anderson and Hein, 1976).

II. Methods

A. Animals

Two-week-old female immature quails (*Coturnix coturnix japonica*) are used. The breeding conditions: light–dark cycle (14 h/10 h), temperature of 23 ± 2°C, and standard feed and water *ad libitum*.

B. Ovariectomy

To avoid interference with endogenous sexual hormone secretion, the immature quails are ovariectomized between days 15 and 18.

1. Immobilize the quail lying on its back, on a dissection board with rubber bands linking wings and legs. Avoid the use of anesthetics that are hazardous for birds. Only the quail's skin is sensitive to pain; internal viscera are not sensitive.
2. Remove the feathers on the left side, at the level of the rib cage, and disinfect with 70% alcohol. Incise skin and muscle successively with a scalpel between the two last left ribs. Separate the two ribs using a retractor to reach the ovary, which lies on the dorsal floor of abdominal cavity. Preserve blood vessels close to the ovary from injury to avoid hemorrhage.
3. Completely remove the ovary with jaw forceps, without leaving any fragment that could regenerate and develop mature oocytes.
4. Stitch the muscle, then the skin with sterile thread (American black silk with curved needle).

5. Return the birds to breeding conditions. They can be stimulated from 3 days to 3 months later.

C. Hormonal Treatment

1. Further dilute estradiol benzoate (Benzogynestryl, 5 mg/ml, Roussel-Uclaf, Romainville, France) in olive oil to obtain a concentration of 1 mg/ml, 1:5 (v/v) (stock solution kept at 4°C).

2. Prepare the hormone solution at a final concentration of 200 μg/ml by a second dilution in olive oil (amount sufficient for the expected injections). Use a 1-ml syringe with a 19- to 23-gauge needle.

3. Make daily 0.1-ml injections (containing 20 μg estradiol benzoate) into the pectoral muscle of ovariectomized or immature ~3-week-old quails. Control birds receive only the oil vehicle.

4. After the last injection, wait 1 day before beginning further experimentation or observations.

Ciliogenesis is not synchronous (Fig. 1) but it is possible to observe one major determined step of ciliogenesis according to the number of daily injections and hormonal dosage (Fig. 2) as follows.

1. First Protocol: Daily Injections of 20 μg Estradiol Benzoate

After 3 injections	Centriologenesis is induced in a few cells.
After 4 injections	Centriologenesis is induced in numerous cells and centrioles migrate to the apical surface.
After 5 injections	Centrioles are numerous, migrating to the surface at which they link. Cilia begin to emerge (5% mature ciliated cells, 25% cells in ciliogenesis, and 70% unciliated cells, as observed in scanning electron microscope).
After 6 injections	About 45% of epithelial cells are ciliated.

At the end of the *in vivo* treatment inducing ciliogenesis, complete ciliary differentiation can be carried on in *in vitro* organotypic culture within 48 hours (Boisvieux-Ulrich *et al.*, 1987).

2. Second Protocol: Varying Doses on a 6-Day Treatment

It is possible to vary estrogen doses to obtain ciliogenesis plus the induction of secretory differentiation (Laugier *et al.*, 1975):

10 μg estradiol benzoate	Tubular glands are generated underneath epithelium.
20 μg estradiol benzoate	Secretory process is induced in the glands.
50 μg estradiol benzoate	Secretory process is strongly stimulated in the glands, and at the epithelium level, mucous secretory granules appear in nonciliated cells.

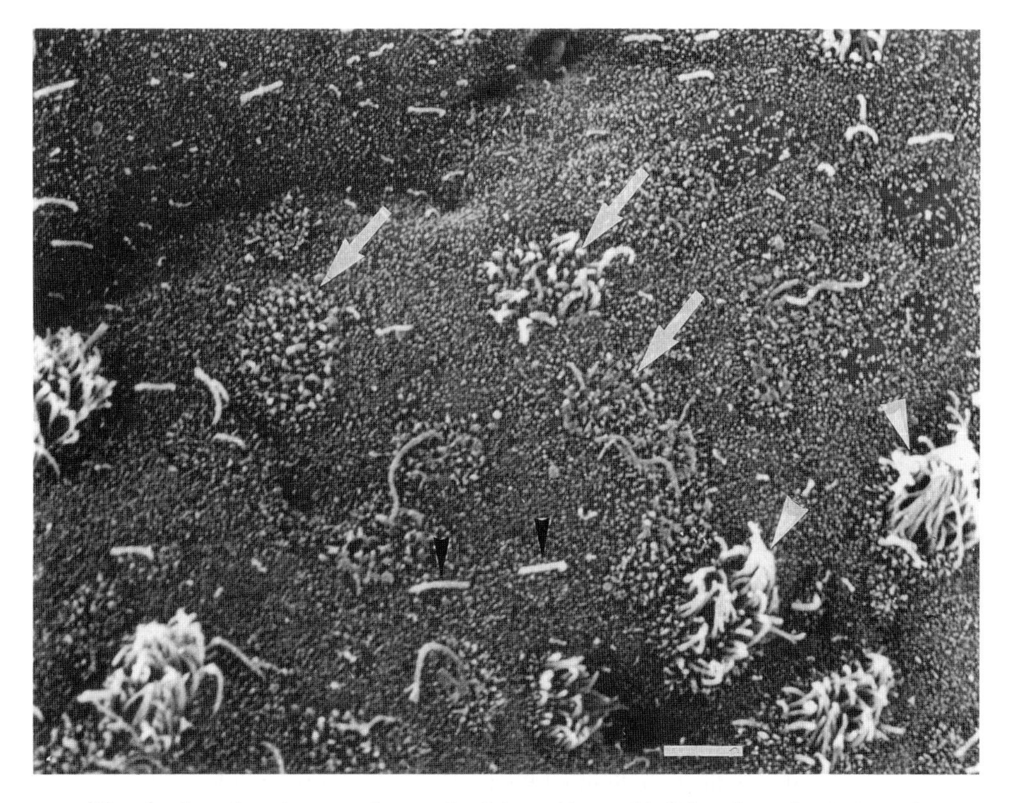

Fig. 1 Scanning electron micrograph of the oviduct epithelial surface after 3 days of 20 μg estradiol benzoate treatment and fixation 1 day later. Numerous undifferentiated cells are recognizable by a single primary cilium (black arrowheads). Cells in ciliogenesis (arrows) are characterized by development of microvilli and by emergence of short cilia. A few cells have more elongated cilia (arrowheads). Bar = 5 μm.

Ciliogenesis is nearly inhibited by six daily injections of a combination of estradiol benzoate (50 μg/day) and progesterone (4 μg/day) diluted together in olive oil; however, mucous cell differentiation is stimulated in quail (Laugier *et al.*, 1975; Sandoz *et al.*, 1975) and in chick (Oka and Schimke, 1969) oviduct. In all instances, withdrawal of the estradiol/progesterone treatment allows ciliogenesis to occur in few days. Mucous cells transdifferentiate into ciliated cells (Sandoz *et al.*, 1976).

III. Troubleshooting

Steps of ciliogenesis are not induced as expected. (1) The injections have not been successful; this can result from the loss of solution at the time of

CHAPTER 13

Preparation of Teleost Rod Inner and Outer Segments

Kathy Pagh-Roehl and Beth Burnside
Division of Cell and Developmental Biology
Department of Molecular and Cell Biology
University of California
Berkeley, California 94720

I. Introduction

In the vertebrate rod photoreceptor, light detection is mediated by the outer segment, a highly modified, nonmotile cilium consisting of a "9 + 0" axoneme and a stack of photopigment-packed, membranous disks, derived from and surrounded by the ciliary membrane. The outer segment is attached to the cell by a structure analogous to the transition zone of motile cilia, the connecting cilium, which arises from a basal body housed within a more proximal part of the rod, called the inner segment (Besharse and Horst, 1990) (Figs. 1a and c). The inner segment is connected by a slender myoid to the rod perinuclear region, which forms junctions with adjacent cells and extends an axon proximally to synapse with inner retinal neurons. As the distal portions of the rod project from the surface of the retina into the subretinal space and lack adhesive contacts to other cells or extracellular structures, it is easy to isolate rod outer segments (ROS) or rod inner segments with attached outer segments (RIS–ROS) from the more proximal part of the cell (nucleus and axon) and from other retinal cells (Fig. 1b).

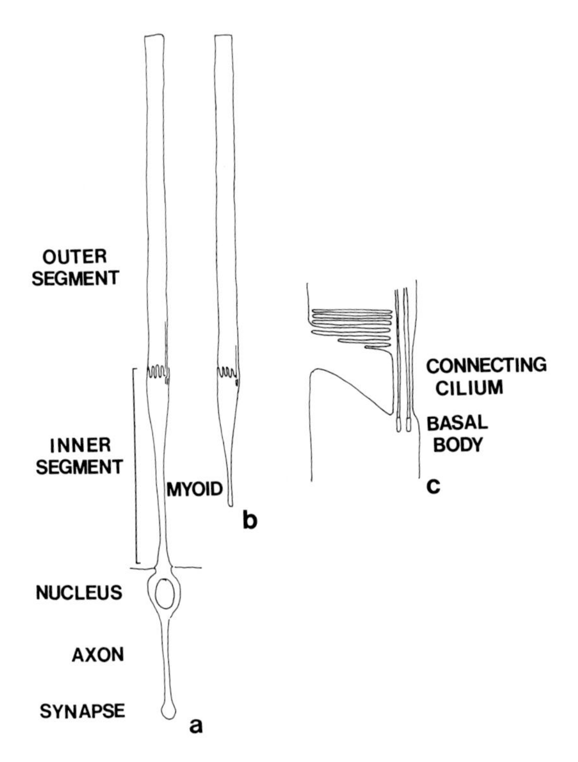

Fig. 1 Diagram of (a) a rod photoreceptor, (b) a RIS–ROS cell fragment, and (c) the connecting cilium region of a rod. (a) Rod photoreceptors of teleost retinas consist of outer and inner segments that project into the subretinal space. A myoid attaches these distalmost segments to the perinuclear region of the cell. (b) RIS–ROS are obtained by shaking the retina in culture medium. The rods break off from the retina at their slender myoids. (c) The outer segment is a modified cilium that is attached to the rod cell via a connecting cilium. The axonemal microtubules of the outer segment arise from a basal body situated in the distalmost portion of the inner segment.

Teleost retinas provide an excellent source of photoreceptor cilia. When retinas are shaken in culture medium, rods break off just distal to the perinuclear region, at the myoid, releasing RIS–ROS into suspension with minimal cell damage. The suspension of teleost RIS–ROS differs from rod preparations obtainable from other vertebrate species. Teleost preparations are highly enriched in RIS–ROS, with few other cell types and only 5% ROS fragments. The plasma membrane reseals after detachment so that less than 5% of the RIS–ROS are leaky (Liepe and Burnside, 1993). Preparations of RIS–ROS can be obtained quickly and reproducibly. Isolated RIS–ROS are morphologically similar to cells observed *in situ* and are ideal for immunocytochemistry (Pagh-Roehl *et al.*, 1991).

In contrast, bovine, rat, and chicken retinas require homogenization or vortexing to detach the distalmost rod fragments (Hamm and Bownds, 1986), and the rods usually break at the connecting cilium. The preparations contain ROS with basal bodies and variable amounts of ROS fragments and RIS–ROS, and must be further purified to remove significant amounts of retinal contaminants (Papermaster and Dreyer, 1974; Uhl *et al.,* 1987). Another advantage of the teleost retina is that it is avascular; most vertebrate retinas contain blood vessels, adding contaminants to retinal cilia preparations. For purification of mammalian retinal axonemes, see Fleischman *et al.* (1980) and Horst *et al.* (1987). To prepare ROS and RIS–ROS from frog retina, see Hamm and Bownds (1986) and Biernbaum and Bownds (1985).

We obtain 5×10^6 RIS–ROS ($\sim$100 μg of protein) from the 4-in.-long green sunfish (*Lepomis cyanellus*) and three to four times more RIS–ROS per retina using a 6-in.-long striped bass (*Morone saxatilis*). These yields are insufficient for preparative biochemistry, but are adequate for many applications, including protein synthesis and transport studies, protein phosphorylation, and Western blotting (Kirsch and Burnside, 1992; Pagh-Roehl *et al.,* 1993). For preparative biochemistry, 40–200 bovine retinas, purchased from local slaughterhouses, provide $\sim$0.5 mg ROS protein per retina (Molday and Molday, 1993).

The major contaminants in RIS–ROS preparations are cone inner–outer segments (CIS–COS). In the intact retina, approximately 10% of the sunfish photoreceptors are cones, which are larger than rods. When fish are dark-adapted for several hours, the cones elongate, and CIS–COS detach from the retina during dissection and remain associated with the retinal pigmented epithelium (RPE), from which they can be purified (Burnside *et al.,* 1993). The RIS–ROS suspension contains 1–2% CIS–COS (Pagh-Roehl *et al.,* 1993). After density gradient purification on "isosmotic" Percoll (Hamm and Bownds, 1986; Biernbaum and Bownds, 1985; Shuster and Farber, 1984), CIS–COS and nuclear contamination can be reduced to less than 0.4% (Pagh-Roehl *et al.,* 1993). Preparations of RIS–ROS are highly enriched for rod ciliary axonemes, but contain other rod outer- and inner-segment components. To enrich for axonemes, RIS–ROS cytoskeletons can be prepared by extraction with Triton X-100 (Pagh-Roehl *et al.,* 1992) (Fig. 2). RIS–ROS cytoskeletons also contain a robust inner-segment actin cytoskeleton and other Triton-insoluble material associated with the basal body. With most buffers, the outer-segment axoneme with attached basal body dissociates from the actin cytoskeleton (Fig. 3), making the separation of outer- and inner-segment cytoskeletal components a possibility.

II. Method

A. Solutions

2× HERT (Hepes-Based Earle's Ringer with Taurine). Prepare on the day of use 0.0336 g $NaHCO_3$, 0.3603 g glucose, 0.5005 g *N*-2-hydroxyethylpiper-

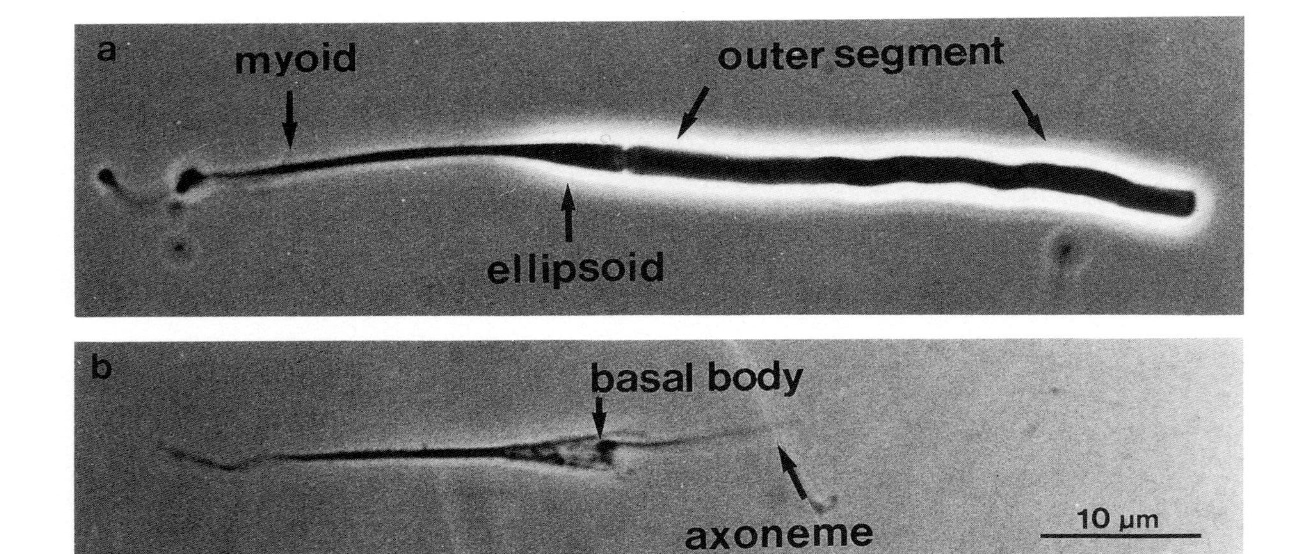

Fig. 2 Rod inner segments with attached outer segments (a) before and (b) after extraction with 1% Triton X-100 in a cytoskeleton stabilizing buffer. Detergent solubilizes the outer segment, revealing the axoneme connected to its basal body. Actin microfilaments and microtubules make up an inner-segment cytoskeleton, including a longer slender myoid. Reprinted with permission from Pagh-Roehl *et al.* (1992).

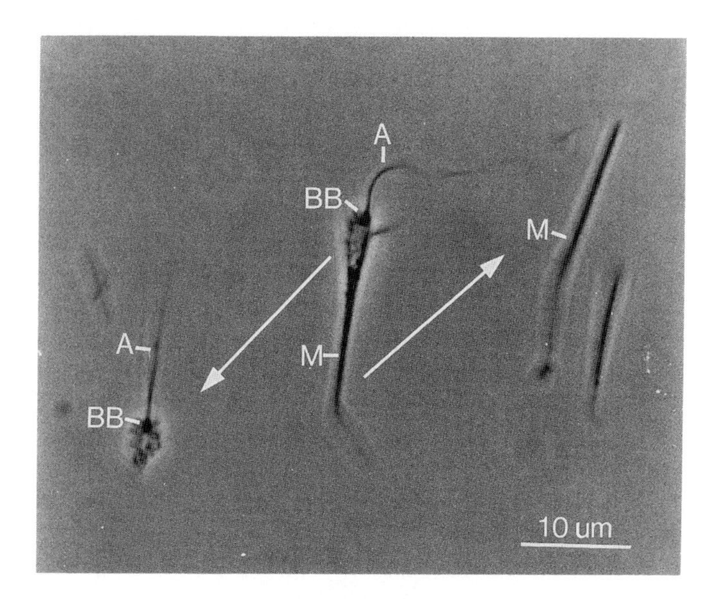

Fig. 3 Dissociation of RIS–ROS into two components after detergent extraction: (1) an axoneme (A) with attached basal body (BB) and associated detergent-insoluble material, and (2) bundles of actin filaments and microtubules which make up the inner-segment cytoskeleton, including the long myoid (M). Cells were extracted with a cytoskeleton stabilization buffer containing phalloidin and taxol, fixed with 1% glutaraldehyde, applied to polylysine-coated coverslips, and then photographed with a 63× phase objective.

azine-*N*-2-ethanesulfonic acid (Hepes), 0.0176 g ascorbic acid, 0.0626 g taurine, 40 ml of deionized, distilled water (ddH_2O), and 10 ml of 10× Earle's balanced salt solution (EBSS, without sodium bicarbonate and phenol red, Gibco). Adjust the pH to 7.2 with NaOH. Dilute with an equal volume of ddH_2O for cultures.

Percoll (Sigma) Solutions. For two discontinuous gradients, used to purify RIS–ROS from 10–12 retinas, make 8 ml of 33% Percoll/HERT (2.7 ml Percoll, 4 ml 2×HERT, 1.3 ml ddH_2O) and 8 ml of 50% Percoll/HERT (4 ml Percoll, 4 ml 2×HERT). To a 15-ml Corex centrifuge tube, add 3.8 ml of 50% Percoll/HERT, then overlay with 3.8 ml of 33% Percoll/HERT. Mark the boundary with a marking pen. Store tubes upright in the cold until use. (Despite the density stated on the bottle, we find that the actual densities of Percoll vary. For each new batch of Percoll, check that the density of a 50% Percoll solution is not less than 1.07 g/ml. We have not succeeded in purifying RIS–ROS from nuclei using similar discontinuous gradients of Nycodenz.)

Cytoskeleton Extraction Buffer: For 1 ml, mix 50 μl 1 *M* Hepes, pH 7.5, 50 μl 100 m*M* $MgSO_4$, 50 μl 100 m*M* ethylene glycol bis(B-aminoethyl ether)-*N*-*N'*-tetraacetic acid (EGTA), pH 7.5, 50 μl 1 *M* KCl, 100 μl 10% dimethyl sulfoxide, 10μl 1 *M* dithiothreitol, 20 μl 5 m*M* GTP, 531 μl ddH_2O, 25 μl 1 mg/ml phalloidin, 2 μl 10 m*M* taxol (Calbiochem), 100 μl 10% Triton X-100 (membrane research grade, Boehringer-Mannheim), 2 μl protease inhibitor cocktail (0.0783 g benzamidine, 0.005 g leupeptin, 0.005 g aprotinin, 0.005 *N* α-*p*-tosyl-1-arginine methyl ester; dissolve in 1 ml of ddH_2O and store at -20°C).

Add 10 μl of phenylmethylsulfonyl fluoride immediately before use.

B. Equipment

Dissection tools should include small iridectomy scissors (curved, with spring handle), bent-tip tissue forceps; and fine-tip (e.g., 3c Dumont) forceps. Assemble a glass Petri dish lined with dental wax; a small paring knife; 35-mm Petri dishes; 24 multiwell culture plates; nylon mesh (100 μm, Tektco, Elmsford, NY); Corex or other clear, conical, 15-ml centrifuge tubes; and blue pipet tips, with 5 mm cut off from the end, and then flamed.

For dissections a dark room with dim incandescent lighting (~1 footcandle) is required. Never use fluorescent lights for working with isolated retinas and RIS–ROS.

Place a bench-top orbital shaker and a refrigerated, swinging-bucket clinical centrifuge in the dissection room.

A lighttight dark-adaptation box is also kept in the dissection room. Ours is made from wood (12 × 16 in. inner dimensions, 7.75 in. high) with an overhanging lid. For aeration, a small hole is drilled into the side through which recurved, metal tubing is inserted to prevent light transmission. This tubing is connected on the outside either to an aquarium aerator or to "house" air, and on the inside to aeration stones. Plastic containers (we use standard rodent cages) are placed into the dark box and filled with water.

C. Fish

We purchase green sunfish, 3–5 in. long, from Fender Fish Farms (Baltik, Ohio), but RIS–ROS can probably be obtained from most teleost species. We maintain fish on a light cycle of 14 hours light :10 hours darkness, in 150-gal tanks (up to 150 fish per tank) with a continuous flow of filtered, fresh water, held at approximately 20°C.

D. Procedure

1. *Dark-adapting the fish:* Transfer fish from the aquarium into plastic containers, with no more than four fish per container. As fish are sensitive to abrupt changes in temperature, transport fish to the dissection room in containers filled with aquarium water. Cover plastic containers with plastic lattice lids to prevent the fish from jumping out, and close the main lid to the dark-adaptation box. Dark-adapt the fish for 2–3 hours. We usually make RIS–ROS in the afternoon, when cone elongation is optimal.

2. *Catching the fish:* Turn out all room lights before opening the dark box. Fish must remain dark-adapted just prior to dissection to permit removal of the retina from the RPE. Scoop up the fish from its front end, flattening the dorsal fins against a cupped hand. Close the dark-adaptation box before turning on the lights.

3. *Killing the fish:* With a small paring knife, sever the spinal cord just behind the operculum (the flap of skin covering the gill slit). Pith the brain by inserting the knife tip into the spinal chord and pushing it toward the head until the eye rotates. Avoid cutting the dorsal aorta as this immediately stops circulation.

4. *Removing the eye:* Gently push down on one side of the eye using the side of your forceps to expose the sclera, which indicates the boundary between eyeball and socket. Make a small incision in the connective tissue between the eye and socket; this creates a flap of tissue that can be held with forceps. With iridectomy scissors directed at a 45° angle, cut the connective tissue around the eye, pulling the eye up gently with forceps at the same time, to reveal where the eye is still attached. Position the scissors behind the eye, and cut the optic nerve flush with the surface of the eyeball. Avoid puncturing the eye during enucleation as intraocular pressure aids in surgically opening the eye. Remove both eyes before proceeding.

5. *Making an eyecup:* Place the eye on a small piece of paper towel, lens side up. Holding the eye with thumb and forefinger of one hand, insert the point of the iridectomy scissors near the equator of the eyeball at the transition from light to dark pigmentation. Push the scissor blade all the way through the eye until it touches the finger holding the other side of the eye; then make a long first cut while pushing against the finger. Hold the anterior segment (top half) with forceps and cut around the remainder of the equator. Discard the anterior segment including the lens. The remaining tissue is the eyecup or posterior segment, which contains the retina, RPE, and vascular choroid. The level of the cut at the equator is important. If the cut is too high, it may be above the ora serrata where the retina is attached to the RPE, making it difficult to remove the retina. In general, it is better to err by cutting too low; the retina falls out more easily with a shallow eyecup.

6. *Removing the retina:* Place the retina in a wax-lined Petri dish. The retina appears in the eyecup as a clear tissue lying on top of a black background, consisting of the RPE and underlying choroid. Locate a small white dot, which is the remnant of the optic nerve, and note that it is closer to one side of the eyecup than the other. Hold the eyecup with forceps on the side furthest away from the optic nerve, avoiding the edge of the retina. Position the eyecup vertically, place a small drop of HERT on the wax, slightly evert the eyecup by touching it to the surface of the wax near the drop, and either grasp the retina with another pair of fine forceps and pull it down until it resists or push the retina along its sides to help it fall free of the eyecup. The surface tension of the drop should help draw out the retina. When roughly two-thirds of the

retina is hanging free of the eyecup, position the scissors behind the retina and snip the optic nerve. The retina should fall out, appearing as a uniformly pink hemisphere. If small pieces of RPE, which have the appearance of black specks, adhere to the surface, remove them carefully with fine-tipped forceps. With some experience, you can do a dry dissection, without the drop of buffer.

7. *Collecting the retinas:* Place each retina in the well of a 24-well culture dish, containing 0.5 ml of HERT, and incubate on an orbital shaker at 45 rpm until all of the retinas have been collected.

8. *Preparing retinal shake-offs:* Add 0.75 ml of HERT to a 35-mm Petri dish, one for each retina. With tissue forceps, place the retina, convex surface down (this is the pinkish surface containing the photoreceptors), into the buffer. Shake the retina back and forth 20–30 times, running the surface of the retina along the bottom of the dish. Shake vigorously but stop if the retina starts to tear. The buffer should become turbid.

9. *Pooling the shake-offs:* Transfer the cell suspension using the cut and flamed, blue plastic pipet tip. Filter the suspension through the 100-μm nylon mesh, prewet with HERT and secured with a rubber band to a conical, calibrated centrifuge tube (the mesh is rinsed with water, scrubbed clean with a brush, and reused many times before being discarded). Pool the retinal shake-offs. The solution should be turbid but free of large particles. At this point, the crude RIS–ROS suspension can be used for immunocytochemistry or sodium dodecyl sulfate–polyacrylamide gel electrophoresis, or it can be purified on a density gradient as follows:

10. *Performing Percoll purification:* Chill the tube of RIS–ROS suspension on ice. Layer on no more than 4 ml of crude RIS–ROS carefully over the 33% Percoll step of the gradient. Too many cells will overload the gradient. Centrifuge at 1000*g* for 20 minutes at 4°C. As RIS–ROS are very sensitive to ambient lighting, only dim, incandescent light should be used.

11. *Recovering RIS–ROS:* After centrifugation, the HERT solution above the 33% Percoll should appear clear. Some particles and turbid material may appear at the buffer/33% Percoll interface. Remove this together with the HERT layer and most of the 33% Percoll with a Pasteur pipet. Make sure that the 33%/50% interface is turbid and that most of the 50% Percoll is clear. (If the turbid area extends into the 50% Percoll, the gradient may have been overloaded with cells, the Percoll density was miscalculated, or the original 33%/50% interface was disrupted.) Using a fresh cut and flamed blue tip, remove the purified RIS–ROS from the 33%/50% interface. Recover up to 3.5 ml per gradient, and add to a fresh centrifuge tube. Avoid taking Percoll near the pellet; this contains CIS–COS, RPE, and choroidal blood cells.

12. *Concentrating RIS–ROS:* Dilute the recovered RIS–ROS fraction 1 : 1 with HERT, gently mix, and centrifuge again at 1000*g* for 20 minutes. RIS–ROS will appear as a pink pellet. Remove the supernatant with a Pasteur pipet. At any point in the procedure, RIS–ROS can be fixed with 1% glutaraldehyde

(in 0.1 *M* $NaPO_4$ buffer, pH 7) and viewed with a 40× phase objective or counted in a hemocytometer.

13. *Preparing RIS–ROS cytoskeletons:* Aspirate off as much of the culture medium as possible from the RIS–ROS pellet. Resuspend the pellet in 100 μl per retina of extraction buffer. We have observed the best morphology when the buffer is ice cold. Triturate the lysate several times to ensure that the pellet is completely resuspended, and transfer the suspension to an Eppendorf tube. Cells are extracted on ice for 10 minutes. We examine cell morphology by fixing the entire lysate and applying it to coverslips. To prepare cytoskeletons for gel electrophoresis, we centrifuge the suspension at 8000*g* for 10 minutes at 4°C. The cytoskeleton is clear with some brownish material (presumably mitochondrial DNA and residual pigment granules) toward the margins.

Acknowledgments

We thank Dr. Christina King Smith, Mr. Homero Rey, Mr. David Hillman, and Mr. Leon Su for reviewing the methods. This work was supported by National Institutes of Health Grant EY03575 to Beth Burnside.

References

Besharse, J. C., and Horst, C. J. (1990). The photoreceptor connecting cilium. A model for the transition zone. *In* "Ciliary and Flagellar Membranes" (R. A. Bloodgood, ed.), pp 389. Plenum Publishing Corporation, New York.

Biernbaum, M. S., and Bownds, M. D. (1985). Frog rod outer segments with attached inner segment ellipsoids as an *in vitro* model for photoreceptors on the retina. *J. Gen. Physiol.* **85,** 83.

Burnside, B., Wang, E., Pagh-Roehl, K., and Rey, H. (1993). Retinomotor movements in isolated teleost retinal cone inner-outer segment preparations (CIS–COS): Effects of light, dark and dopamine. *Exp. Eye Res.* **57,** 709.

Fleischman, D., Denisevich, M., Raveed, D., and Pannbacker, R. G. (1980). Association of guanylate cyclase with the axoneme of retinal rods. *Biochim. Biophys. Acta* **630,** 176.

Hamm, H. E., and Bownds, M. D. (1986). Protein complement of rod outer segments of frog retina. *Biochemistry* **25,** 4512.

Horst, C. J., Forestner, D. M., and Besharse, J. C. (1987). Cytoskeletal-membrane interactions: A stable interaction between cell surface glycoconjugates and doublet microtubules of the photoreceptor connecting cilium. *J. Cell Biol.* **105,** 2973.

Kirsch, M., and Burnside, B. (1992). Opsin transport in fish rod photoreceptors. *Invest. Opthalmol. Vis. Sci.* **33**(Suppl), 738.

Liepe, B. A., and Burnside, B. (1993). Cyclic nucleotide regulation of teleost rod photoreceptor inner segment length. *J. Gen. Physiol.* **102,** 75.

Molday, R. S., and Molday, L. L. (1993). Isolation and characterization of rod outer segment disk and plasma membranes. *In* "Photoreceptor Cells" (P. Hargrave, ed.), *Methods in Neuroscience* **15,** 131.

Pagh-Roehl, K., Wang, E., and Burnside, B. (1991). Posttranslational modifications of tubulin in teleost photoreceptor cytoskeletons. *Cell. Mol. Neurobiol.* **11** 593.

Pagh-Roehl, K., Brandenburger, J., Wang, E., and Burnside, B. (1992). Actin-dependent myoid elongation in teleost rod inner/outer segments occurs in the absence of net actin polymerization. *Cell. Motil. Cytoskel* **21,** 235.

Pagh-Roehl, K., Han, E., and Burnside, B. (1993). Identification of cyclic nucleotide-regulated phosphoproteins, including phosducin, in motile rod inner-outer segments of teleosts. *Exp. Eye Res.* **57,** 679.

Papermaster, D. S., and Dreyer, W. J. (1974). Rhodopsin content in the outer segment membranes of bovine and frog retinal rods. *Biochemistry* **13,** 2438.

Shuster, T. A., and Farber, D. B. (1984). Phosphorylation in sealed rod outer segments: Effects of cyclic nucleotides, *Biochemistry* **23,** 515.

Uhl, R., Desel, H., Ryba, N., and Wagner, R. (1987). A simple and rapid procedure for the isolation of intact bovine rod outer segments (ROS). *J. Biochem. Biophys. Methods* **14,** 127.

CHAPTER 14

Isolation of Respiratory Cilia

Annette T. Hastie
Department of Medicine
Jefferson Medical College
Philadelphia, Pennsylvania 19107

I. Introduction

The membrane surrounding the ciliary axoneme of mammals, although not required for beating movement, interacts with the external environment which may alter ciliary function. Successful methods for isolating cilia from mammalian airway epithelium employ the nonionic detergent Triton X-100, Ca^{2+} influx, and a shear force (vortexing or shaking) (Dirksen and Zeira, 1981; Hastie *et al.,* 1986), but produce axonemes that are demembranated. Methods avoiding detergent use for *Tetrahymena* cilia, e.g., dibucaine (Thompson *et al.,* 1974) or sucrose/ethanol/$CaCl_2$ (Gibbons, 1965), have not worked well with mammalian cells (Hastie *et al.,* 1990). Zwitterionic 3-[(3-cholamidopropyl)dimethylammonio]-1-propanesulfonate (CHAPS) detergent at 0.1% is effective combined with brushing the epithelium (Hastie *et al.,* 1990). This injury multiplies the yield of cilia approximately 10-fold compared with nontraumatic extraction, but also increases cellular debris. CHAPS at 0.5% without brushing appears to release ciliary membranes into solution during axoneme isolation (Salathe *et al.,* 1993). A different approach homogenizes epithelial cells scraped off the trachea, followed by $MgCl_2$ aggregation and ouabain affinity chromatography to remove intracellular and basolateral membranes from the apical membrane preparation; however, this method results in a mixture of epithelial cells other than just ciliated ones (Shen *et al.,* 1991). The method given below specifically yields isolated intact cilia, derives membranes from the isolated cilia, and identifies apical membrane surface components by prior labeling of the epithelium with biotin, as similarly applied to *Chlamydomonas* (Reinhart and Bloodgood, 1988). Solubilizing ciliary membranes releases the dense matrix proteins from within the axoneme (Gibbons, 1965). Reconstitution of membrane vesicles after detergent removal separates integral membrane components from soluble matrix proteins (Hastie *et al.*, 1990). Use of Triton X-114 in a study of *Tetrahymena* ciliary membrane proteins facilitated further partitioning into aqueous and detergent phases (Dentler, 1992).

II. Materials

Bovine tracheas are obtained from animals of about 250 kg at a local abattoir under permit from the U.S. Department of Agriculture (USDA). The length of the trachea extends from just below the larynx to the carini, generally 20–30 cm. Each individual trachea is put into a separate plastic bag for transport (15–20 minutes), on ice if necessary. These animals, healthy by USDA standards, often have a bacterial infection of the respiratory airway epithelium which reduces yield of cilia (Hastie *et al.*, 1993). It is therefore best to obtain twice the number of tracheas expected to be needed if all were uninfected.

Chemicals are obtained from Fisher or Sigma, with the exception of *N*-hydroxysuccinimido-LC-biotin reagent and ExtractiGel D (Pierce), Bio-Rad Protein Assay Reagent (Bio-Rad), and streptavidin colloidal gold suspension (E-Y Laboratories, San Mateo, CA).

Solutions are prepared and chilled in an ice-water bath just prior to use:

0.9% NaCl solution

1 *M* $CaCl_2$ stock solution

PBS: 0.12 *M* NaCl, 2.7 m*M* KCl, 5 m*M* potassium phosphate, pH 7.4

NHS-LC-biotin solution: 7.5 mg of extended spacer arm analog of *N*-hydroxysuccinimido-LC-biotin in 30 ml PBS (freshly prepared)

Extraction buffer: 50 m*M* NaCl, 20 m*M* Tris–HCl, pH 7.5, 1 m*M* ethylenediaminetetraacetic acid (EDTA), 7 m*M* β-mercaptoethanol, with 0.1% CHAPS detergent (An equal volume of this solution without CHAPS also is prepared.)

Resuspension buffer: 50 m*M* potassium acetate, 20 m*M* Tris–HCl, pH 8.0, 4 m*M* $MgSO_4$, 1 m*M* dithiothreitol, 0.5 m*M* EDTA, 0.1 mg/ml soybean trypsin inhibitor (A small aliquot, 2–5 ml, of this solution is prepared with 0.5% Triton X-100 just prior to use for removal of ciliary membranes.)

Reactivation buffer: 10 m*M* ATP, 20 m*M* Tris–HCl, pH 8.0, 0.4 m*M* potassium acetate, 3 m*M* $MgSO_4$, 0.5 m*M* EDTA, 1 m*M* dithiothreitol

HCMNT buffer: 10 m*M* 4-(2-hydroxyethyl)-1-piperazineethanesulfonic acid (Hepes), pH 8.0, 0.1 m*M* $CaCl_2$, 0.1 m*M* $MnCl_2$, 0.15 *M* NaCl, 0.1% Tween 20, 0.1 m*M* phenylmethylsulfonyl fluoride, 2 μg/ml leupeptin, 0.1 mg/ml pepstatin

Streptavidin colloidal gold suspension buffer: 1/10 dilution of streptavidin-coated colloidal gold (10 nm size) and 1% gelatin in HCMNT buffer

III. Procedure

1. Each trachea is trimmed of excess tissue, cut into quarter lengths, 6–8 cm, and rinsed with 2 × 50 ml of 0.9% NaCl to remove any blood clots, excess mucus, or other debris. The trachea is handled only by the outer surface to avoid premature damage to the epithelium.

2. To identify apical surface-accessible components of ciliated epithelium, the tracheal quarter is treated with 30 ml NHS-LC-biotin solution for 30 minutes at room temperature with gentle rotation in a tightly capped specimen jar placed on its side on a roller or rocking platform. The biotin solution is discarded and the tracheal quarter rinsed with 2 × 50 ml PBS or extraction buffer without CHAPS. Tris buffer inactivates unreacted NHS-LC-biotin, although less than 5% remains unreacted according to the manufacturer.

3. Each tracheal quarter is lightly brushed with a nylon bristle brush (2 cm diameter × 8 cm long) inserted into one end of the lumen, placed against the epithelium, and pushed through to the opposite end while rotating the brush 180°. The brush is pulled back through the lumen again, employing a similar twisting movement while maintaining contact with the epithelial surface. The in/out brushing procedure is repeated once more. Material clinging to the bristles is rinsed off into a specimen jar using 40 ml of extraction buffer. The tracheal quarter is placed into this jar and 0.4 ml of 1 *M* $CaCl_2$ is added. The jar is immediately capped tightly and shaken vigorously for 30 seconds.

4. Suspensions from two quarters are decanted into a 250-ml centrifuge bottle. Each tracheal quarter is rinsed with 35 ml of extraction buffer without CHAPS which is combined with the first suspension. The first centrifugation step at 2000*g* for 2 minutes pellets cell debris and some cilia. Washing the pellet does not retrieve enough cilia to be worthwhile. All centrifugation is at 4°C.

5. The supernatant material is carefully poured into clean centrifuge bottles and spun at 12,000*g* for 5 minutes. The supernatant fluid is discarded. Pellets in each centrifuge bottle are gently dispersed in 2 ml of resuspension buffer and transferred in two equal aliquots to two 2-ml centrifuge tubes. Each centrifuge bottle is rinsed with an additional 2 ml of resuspension buffer which is added to the first two portions.

6. The suspension is pelleted at 12,000*g* for 1.5 minutes. The supernatant fluid is discarded and each pellet is gently dispersed into 2 ml of resuspension buffer. This suspension is recycled through centrifugation at 2000*g* for 1.5 minutes (pellets discarded) and subsequently at 12,000*g* for 1.5 minutes (supernates discarded).

7. Cilia pellets are resuspended in 2 ml total of resuspension buffer. Ten microliters each of cilia preparation and of reactivation buffer are mixed on a microscope slide, overlaid by a coverslip, and viewed by oil immersion, phase-contrast optics at 1000–1500× magnification to assess yield, contamination, and reactivation of ciliary beating. A preparation of good yield will have a large number of single and aggregated cilia. Retention of an intact membrane prevents access to ATP, and thus isolated cilia should not reactivate. Axonemes without membranes, but otherwise structurally intact, will beat. Reactivation can occur at 20–23°C, although at a slower beat frequency than at 37°C. Presence of membranes is ascertained by transmission electron microscopy (TEM) as shown in Fig. 1. TEM also detects a well-camouflaged Gram-negative bacterium, the cilia-associated respiratory bacillus (see inset to Fig. 1). This microbe is morphologically indistinguishable from cilia by light microscopy unless immunostained with specific antisera (Hastie *et al.*, 1993). Biotin-labeled cilia may be pretreated with streptavidin-coated colloidal gold to identify labeled sites. Fifty microliters of the cilia suspension is pelleted at 12,000*g* for 3 minutes, resuspended, and incubated for 1 hour in 200 μl of streptavidin-coated colloidal gold particle solution, pelleted as before, washed with HCMNT buffer, and repelleted prior to fixation in 2.5% glutaraldehyde in HCMNT buffer. Standard TEM techniques are employed in embedment, sectioning, staining, and examination.

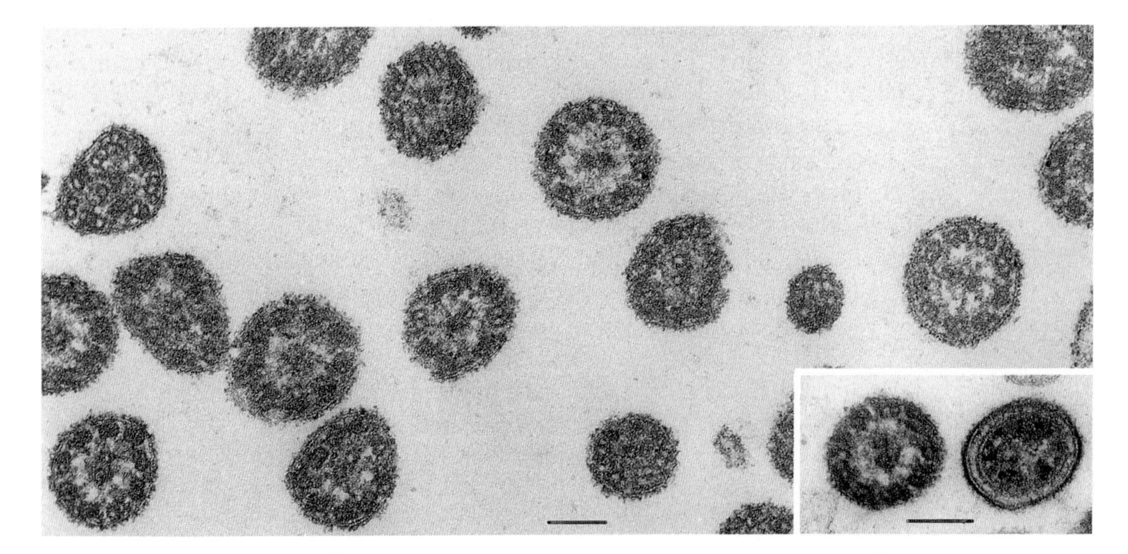

Fig. 1 Transmission electron micrograph of isolated cilia shows bilayer membranes surrounding axonemes in cross section. Matrix proteins within the internal axoneme structure obscure components other than the microtubules. The inset contains cross sections of a single isolated cilium and a cilia-associated respiratory bacillus (lower right), a Gram-negative bacterium that frequently infects bovine tracheal epithelium (Hastie *et al.*, 1993). Bar = 100 nm.

8. The membranes and internal matrix material are released from the ciliary axonemes by Triton X-100 treatment. One-half of the cilia preparation, in 1 ml (for full preparation, adjust volumes twice), is pelleted at 12,000*g* for 2–3 minutes and the supernate discarded. The pellet is gently resuspended in 0.4 ml of resuspension buffer containing 0.5% Triton X-100 for 15 minutes and repelleted at 12,000*g* for 5 minutes. The supernate, which is the membrane + matrix fraction, is carefully removed and retained for further processing (step 10). The pellet of ciliary axonemes is washed with resuspension buffer, pelleted again, and gently dispersed into 0.5 ml of resuspension buffer.

9. The demembranated axonemes from step 8 are reexamined for reactivation as described (step 7). A much greater proportion of these should now beat. The average yield of axoneme protein from one-half of a trachea is 0.6 mg (Bio-Rad Protein Assay Reagent micromethod). This material is sufficient for other studies, including TEM, sodium dodecyl sulfate–polyacrylamide gel electrophoresis (SDS–PAGE), and subfractionation to obtain other axoneme components, e.g., tektins.

10. Triton X-100 is removed from the 0.4 ml of membranes + matrix fraction on a 1-ml column of ExtractiGel D, eluted with resuspension buffer. One-half-milliliter fractions are collected and assessed for protein (Bio-Rad Protein Assay Reagent micromethod). The two fractions containing peak protein are combined, frozen at −85°C, and thawed. Reconstituted membrane vesicles are pelleted at 48,700*g* for 15 minutes. The supernate is retained as the soluble matrix protein fraction. The pelleted membrane vesicles are resuspended in 0.1 ml of resuspension buffer. Membrane vesicles and matrix protein fractions may be further analyzed by TEM and SDS–PAGE to identify biotin-labeled, i.e., surface-accessible, components.

References

Dentler, W. L. (1992). Identification of *Tetrahymena* ciliary surface proteins labeled with sulfosuccinimidyl 6-(biotinamido) hexanoate and Concanavalin A and fractionated with Triton X-114. *J. Protozool.* **39,** 368–378.

Dirksen, E. R., and Zeira, M. (1981). Microtubule sliding in cilia of the rabbit trachea and oviduct. *Cell Motil.* **1,** 247–260.

Gibbons, I. R. (1965). Chemical dissection of cilia. *Arch. Biol.* (Liege) **76,** 317–352.

Hastie, A. T., Dicker, D. T., Hingley, S. T., Kueppers, F., Higgins, M. L., and Weinbaum, G. (1986). Isolation of cilia from porcine tracheal epithelium and extraction of dynein arms. *Cell Motil.* **6,** 25–34.

Hastie, A. T., Krantz, M. J., and Colizzo, F. P. (1990). Identification of surface components of mammalian respiratory tract cilia. *Cell Motil. Cytoskel.* **17,** 317–328.

Hastie, A. T., Evans, L. P., and Allen, A. M. (1993). Two types of bacteria adherent to bovine respiratory tract ciliated epithelium. *Vet. Pathol.* **30,** 12–19.

Reinhart, F. D., and Bloodgood, R. A. (1988). Membrane-cytoskeleton interactions in the flagellum: A 240,000 Mr surface-exposed glycoprotein is tightly associated with the axoneme in *Chlamydomonas moewusii. J. Cell Sci.* **89,** 521–531.

Salathe, M., Pratt, M. M., and Wanner A. (1993). Cyclic AMP-dependent phosphorylation of a

26 kD axonemal protein in ovine cilia isolated from small tissue pieces. *Am. J. Respir. Cell Mol. Biol.* **9,** 306–314.

Shen, B.-Q., Yang, C.-M., and Widdicombe, J. H. (1991). Rapid procedure for obtaining tracheal apical membranes. *Am. J. Physiol.* **261,** L102–L105.

Thompson, G. A. Jr., Baugh, L. C., and Walker, L. F. (1974). Nonlethal deciliation of *Tetrahymena* by a local anesthetic and its utility as a tool for studying cilia regeneration. *J. Cell Biol.* **61,** 253–257.

PART III

General Methods

CHAPTER 15

Electrophoretic Separation of Dynein Heavy Chains

Gianni Piperno

Department of Cell Biology and Anatomy
Mount Sinai School of Medicine
New York, New York 10029

I. Introduction

It is probable that each kind of motile axoneme contains multiple dynein heavy chains (DHCs). Studies of molecular properties of axonemal dyneins have demonstrated that each outer and inner dynein arm contains at least two DHCs (Gibbons, 1989) and that inner dynein arms may be composed of different heavy chains along the length of the axoneme (Piperno and Ramanis, 1991).

All DHCs characterized until now have an apparent molecular weight above 380,000 (Gibbons, 1989). They cannot be resolved by two-dimensional polyacrylamide gel electrophoresis and are resolved by one-dimensional polyacrylamide gel electrophoresis only under specific conditions.

The following electrophoretic procedure is a modified version of the procedure of Neville (1971). It can be applied to resolve at least some of the DHCs present in cilia or eukaryotic flagella. It was developed to obtain optimal resolution of the 9 (Piperno and Ramanis, 1991) or 11 (Kagami and Kamiya, 1992) DHCs that are assembled in the axoneme of *Chlamydomonas* flagella. The procedure is capable of resolving six bands, each comprising one or two or three of the DHCs (Piperno *et al.,* 1990). Although changes in the composition of the gel or buffers used for the electrophoresis cause changes in the electrophoretic patterns formed by the DHCs (Piperno, 1988; Kagami and Kamiya, 1992), all DHCs cannot be separated by gel electrophoresis alone because they are so similar in structure and present in the axonemes at different concentrations (Piperno *et al.*, 1990; Piperno and Ramanis, 1991). Complete separation of DHC subsets can be achieved through gel electrophoresis if the number of DHCs under analysis is reduced by mutation (Huang *et al.*, 1979) or chromatographic fractionation (Piperno and Luck, 1981; Goodenough *et al.*, 1987; Kagami and Kamiya, 1992).

II. Chemicals, Solutions, and Apparatus

The following chemicals are used: sodium dodecyl sulfate (AnalaR grade, BDH Laboratory Supplies); acrylamide (molecular biology certified, International Biotechnologies, Subsidiary of Eastman Kodak); *N,N'*-methylenebisacrylamide (electrophoresis grade, Eastman Kodak); *N,N,N',N'*-tetramethylethylenediamine (TEMED, electrophoresis grade, Eastman Kodak); ammonium persulfate [reagent ACS (crystals), Eastman Kodak]; tris(hydroxymethyl)aminomethane, Trizma base (reagent grade, Sigma); 2-mercaptoethanol, 98% (2-Met), (Aldrich Chemical); glycine (electrophoresis grade) and sucrose (ultrapure) (Schwarz/Mann Biotech, Division of ICN Biomedicals). All other reagents are certified ACS from Fisher Chemical.

Nine stock solutions are prepared and kept at room temperature (if not indicated otherwise):

(1) 30% acrylamide, 1.2% bisacrylamide (kept at 4°C and made monthly)
(2) 50 m*M* Tris, 0.384 *M* glycine, 0.2% sodium dodecyl sulfate
(3) 2.12 *M* Tris–chloride, pH 9.4
(4) 0.216 *M* Tris–sulfate, pH 6.1
(5) 0.82 *M* Tris–borate, pH 8.64, 2% sodium dodecyl sulfate
(6) 0.2 *M* ethylenediaminetetraacetic acid (EDTA), pH 7.5
(7) 60% sucrose
(8) 10% sodium dodecyl sulfate
(9) 0.2% bromphenol blue in 60% sucrose

The electrophoresis is performed in a homemade, vertical apparatus that holds 35.5-cm-wide, 26.5-cm-long, 0.4-cm-thick glass plates. Thinner glass plates are not adequate for optimal resolution of dynein heavy chains because they are flexible and do not provide a gel slab that has homogeneous thickness. Glass plates should be cleaned with Alconox (Alconox, distributed by Fisher Scientific) before use.

III. Procedure

Table I lists the volumes of solutions used to prepare (1) the 9.4% polyacrylamide used to seal the mold of the gel; (2) a discontinuous gel slab, 30 cm wide, 22.5 cm long, 0.1 cm thick, composed of a 3.5-cm-long 3.2% polyacrylamide layer (stacking gel) and a 19-cm-long 3.6–5% polyacrylamide gradient (resolving gel); and (3) cathodic and anodic buffers.

A mold for the gel slab is created by 0.1-cm-thick, 0.7-cm-wide plastic spacers that are held in place by paper clips. Moreover, 9.4% acrylamide is injected and polymerized in the space left between the spacers and the edges of the plates while the mold is kept horizontal. During this operation the 9.4% acrylamide is stirred in a beaker to prevent premature polymerization. Any acrylamide that leaks inside the mold before polymerization should be absorbed by paper strips.

To form a 3.6–5% polyacrylamide gradient resolving gel, 35 ml of 3.6% and 29 ml of 5% acrylamide are mixed in a conical gradient maker and poured into the mold. The mold should be filled to a level that is 3 mm higher than the level desired for the upper limit of the resolving gel. An overlay solution of 0.424 *M* Tris–chloride, pH 9.4, 0.1% sodium dodecyl sulfate is layered over the acrylamide solution before the polymerization occurs (polymerization occurs in less than 1 hour).

Following polymerization, the overlay solution is removed and the 3.2% stacking gel is polymerized over the resolving gel. Sixteen slots, each 1.3 cm

Table I
Volumes of Solutions Required for the Electrophoresis of Dynein Heavy Chains

Solution	Sealing gel 9.4%	Resolving gel 3.6%	Resolving gel 5.0%	Stacking gel 3.2%	Cathodic buffer	Anodic buffer
1	15.0 ml	5.4 ml	7.5 ml	2.0 ml		
2	5.0 ml					
3		9.0 ml	9.0 ml			200 ml
4				5.0 ml		
5					50 ml	
6					5 ml	
7		3.8 ml	13.0 ml			
8		0.45 ml	0.45 ml	0.2 ml		
H_2O		26.1 ml	15.0 ml	12.4 ml	944 ml	800 ml
2-Met					1 ml	
TEMED	15 μl	20 μl	10 μl	20 μl		
AP[a]	0.3 ml	0.14 ml	0.14 ml	0.10 ml		

[a] AP is a 12.5% ammonium peroxydisulfate solution that is prepared the same day of the experiment. AP and TEMED must be added to the pool of other solutions while they are stirred and just before the final product is poured into the mold.

wide by 1.5 cm long, are created in the stacking gel by inserting a plastic comb that has a tight fit. The distance of the lower end of the slot and the interface between stacking and resolving gels should be 2 cm.

Polymerizations are performed at room temperatures and gels are aged at least 24 hours before use. Aging is performed at room temperature inside a closed plastic bag containing some water at the bottom.

Unlabeled or ^{35}S-labeled axonemal proteins are dissolved in 2% sodium dodecyl sulfate, 1% 2-mercaptoethanol at a concentration of 2–5 mg/ml and diluted in 1- to 2-μg aliquots by 100 μl of sample buffer formed by 9 parts of cathodic buffer and 1 part of 0.2% bromphenol blue in 60% sucrose. Each of these aliquots is injected into a slot through the cathodic buffer. Cathodic and anodic buffers respectively are poured into the upper and lower chambers of the apparatus. After that the electrophoretic apparatus is assembled.

Electrophoresis is performed at 20 mA, constant current, for 26 hours and at room temperature. Detection of polypeptides is performed by either silver staining of wet gels or autoradiography of dried gels. Samples of *Chlamydomonas* axonemes also can be analyzed in amounts larger than 3 μg, to detect DHCs by Coomassie blue staining; however, the electrophoretic resolution of DHCs decreases as the amount of protein loaded is increased.

IV. Pitfalls

The reagents used in this procedure must be as pure as possible. Batches of acrylamide and sodium dodecyl sulfate that are not of the highest quality contain impurities that affect optimal resolution of DHCs or change their electrophoretic mobility.

The glass plates used to assemble the gel mold must have constant thickness, be inflexible, and be cleaned from any fingerprints. Traces of silicone lubricant that may be used to seal the glass plates against the electrophoresis apparatus should not contaminate the glass surface that is in contact with the polyacrylamide. Traces of silicone on the internal sides of the mold decrease the sharpness of electrophoretic bands formed by the DHCs.

The plastic comb used to create slots in the polyacrylamide must have a tight fit in the mold formed by the glass plates. If a thin layer of polyacrylamide is formed inside the slots it affects the stacking of the polypeptides and, consequently, the sharpness of the electrophoretic bands.

Removal of the plastic comb from the polyacrylamide should be performed by pushing the comb 1–2 mm toward a side and subsequently pulling it out of the mold. Therefore, air enters the slots during removal of the comb. The air–stacking gel interface should not be deformed. Any alteration of that surface or detachment of the stacking gel from the glass plate has an adverse effect on the resolution of the DHCs.

References

Gibbons, I. R. (1989). Microtubule-based motility: an overview of a fast-moving field. *In* "Cell Movement" (F. D. Warner, P. Satir, and I. R. Gibbons, eds.), Vol 1, pp. 3–22. Alan R. Liss, New York.

Goodenough, U. W., Gebhart, B., Mermall, V., Mitchell, D. R., and Heuser, J. E. (1987). High-pressure liquid chromatography fractionation of *Chlamydomonas* dynein extracts and characterization of inner-arm dynein subunits. *J. Mol. Biol.* **194,** 481–494.

Huang, B., Piperno, G., and Luck, D. J. (1979). Paralyzed flagella mutants of *Chlamydomonas reinhardtii.* Defective for axonemal doublet microtubule arms. *J. Biol. Chem.* **254,** 3091–3099.

Kagami, O., and Kamiya, R. (1992). Translocation and rotation of microtubules caused by multiple species of *Chlamydomonas* inner-arm dynein. *J. Cell Sci.* **103,** 653–664.

Neville, D. M. (1971). Molecular weight determination of protein-dodecyl sulfate complexes by gel electrophoresis in a discontinuous buffer system. *J. Biol. Chem.* **246,** 6328–6334.

Piperno, G. (1988). Isolation of a sixth dynein subunit adenosine triphosphatase of *Chlamydomonas* axonemes. *J. Cell Biol.* **106,** 133–140.

Piperno, G., and Luck, D. J. (1981). Inner arm dyneins from flagella of *Chlamydomonas reinhardtii. Cell* **27,** 331–340.

Piperno, G., and Ramanis, Z. (1991). The proximal portion of *Chlamydomonas* flagella contains a distinct set of inner dynein arms. *J. Cell Biol.* **112,** 701–709.

Piperno, G., Ramanis, Z., Smith, E. F., and Sale, W. S. (1990). Three distinct inner dynein arms in *Chlamydomonas* flagella: molecular composition and location in the axoneme. *J. Cell Biol.* **110,** 379–389.

CHAPTER 16

Two-Dimensional Separation of Axonemal Proteins

Gianni Piperno

Department of Cell Biology and Anatomy
Mount Sinai School of Medicine
New York, New York 10029

I. Introduction

The development of two-dimensional electrophoretic procedures capable of resolving nearly all axonemal proteins provided the opportunity to identify the molecular components of all major axonemal substructures (Piperno *et al.*, 1977, 1981, 1992; Huang *et al.*, 1979; Adams *et al.*, 1981; Segal *et al.*, 1984; Piperno and Luck, 1981). It was evident from the beginning of this analysis that the axoneme is composed of hundreds of distinct polypeptides that have different concentrations (Piperno *et al.*, 1977) and are post-translationally modified (Piperno *et al.*, 1981).

The procedure described here was developed to resolve the molecular components of *Chlamydomonas* axonemes but it can be applied to all kinds of axonemes for various purposes, namely, the analysis of axonemal fractions, the identification of defective gene products (Luck *et al.*, 1977), the recognition of cytoskeletal proteins (Piperno and Luck, 1979; Piperno *et al.*, 1992), and, finally, the definition of "molecular signatures" that are formed by specific substructures. These signatures consist of characteristic two-dimensional patterns of spots.

The following procedure consists of two successive gel electrophoreses of polypeptides that, first, are resolved in a pH gradient and, then, are separated by conventional sodium dodecyl sulfate–polyacrylamide gel electrophoresis (SDS–PAGE). During the first step the polypeptides retain their own electric charge but do not reach their isoelectric point. The procedure works very well for separating proteins that are smaller than 200,000 Da, but larger proteins are poorly resolved and should be analyzed by one-dimensional SDS–PAGE (see Chapter 15).

II. Chemicals, Solutions, and Apparatus

The following chemicals are used: urea (ultrapure, Schwarz/Mann Biotech Division of ICN Biomedicals); Nonidet P-40 (BDH Laboratory Supplies); ampholines pH 3.5–10, 9–11, 4–6, 5–7 (ampholines are 40% w/v with the exception of the ampholine pH 9–11, which is 20% w/v) (Pharmacia-LKB); riboflavin 5′-phosphate (electrophoresis purity reagent, Bio-Rad Laboratories); Agarose SeaKem LE (FMC BioProducts). All other reagents are listed in Chapter 15.

Eight stock solutions (numbered sequentially with those described in Chapter 15) are prepared:

(10) 28.38% acrylamide, 1.62% bisacrylamide (kept at 4°C and made monthly)
(11) 10% (w/v) Nonidet P-40
(12) 14 ml ampholine pH 3.5–10, 3 ml ampholine pH 9–11, 1 ml ampholine pH 4-6, 1 ml ampholine pH 5–7 (the mixture is kept sterile at 4°C)
(13) 0.27 m*M* riboflavin 5′-phosphate, 1% TEMED (kept at 4°C and made monthly)
(14) 9.5 *M* urea, 0.8% ampholine pH 3.5–10, 8% w/v Nonidet P-40, 5% 2-mercaptoethanol (kept at − 20°C in 1-ml aliquots)
(15) 5 *M* urea, 0.8% ampholine pH 5–7, 0.2% ampholine pH 3.5–10 (kept at −20°C in 2-ml aliquots)
(16) 0.01 *M* H_3PO_4
(17) 0.02 *M* NaOH

Solution 14 is prepared by first dissolving 57.06 g of urea in 35 ml H_2O with stirring and heating at 60°C, then adding this solution to 8 g of Nonidet P-40 and 5 ml of 2-mercaptoethanol. After thorough mixing of this solution, the ampholine and H_2O are added to a final volume of 100 ml. Aliquots of 1 ml are prepared while the solution has a temperature above 25°C. Stock solutions 1–9, which are described in Chapter 15, also are used in the following procedure.

The electrophoreses are performed in two homemade, vertical apparatus that hold 16-cm-wide, 19-cm-long, 0.3-cm-thick glass plates and 35.5-cm-wide,

26.5-cm-long, 0.4-cm-thick glass plates, respectively. Glass plates are cleaned as indicated in Chapter 15. The dimensions of the apparatus for the second electrophoresis are designed so that two samples can be run side-by-side in an identical acrylamide gradient.

III. Procedure

Table I shows the volumes of solutions used to prepare (1) a discontinuous gel slab, 30 cm wide, 23.5 cm long, 0.1 cm thick, composed of a 2-cm-long 3.2% polyacrylamide layer (stacking gel) and a 21.5-cm-long 4–11% polyacrylamide gradient (resolving gel); and (2) a continuous gel slab 13.5 cm wide, 14 cm long, 0.075 cm thick, for the electrophoresis in a pH gradient. The 9.4% polyacrylamide used to seal the mold of the gel, the stacking gel, and the cathodic and anodic buffers for the second electrophoresis are as described in Chapter 15.

The discontinuous gel is prepared first and aged 24 hours before use as described in Chapter 15. The mold of the discontinuous gel slab is identical to that previously described. The plates are marked at 2.7 and 0.7 cm from the upper edge. These are the levels respectively reached by resolving and stacking gels after polymerization. To reach these levels, acrylamide solutions are poured to levels that are 0.3 cm above the marks. The 0.6 to 0.7-cm space, left

Table I
Volumes of Solutions Required for Two-Dimensional Analysis of Axonemal Proteins

Solution	Resolving gel 4.0%	Resolving gel 11.0%	pH gradient gel
1	6.0 ml	16.6 ml	
3	9.0 ml	9.0 ml	
7	3.8 ml	13.0 ml	
8	0.45 ml	0.45 ml	
10			4.0 ml
11			6.0 ml
12			1.9 ml
13			60 μl
H_2O	25.5 ml	6.0 ml	5.8 ml
Urea			16.5 g
TEMED	20 μl	10 μl	
AP[a]	0.14 ml	0.14 ml	

[a] AP is a 12.5% ammonium peroxydisulfate solution that is prepared the same day of the experiment. AP and TEMED must be added to the pool of other solutions while that is stirred and just before the final product is poured into the mold.

between the upper layer of the stacking gel and the upper edge of the plate, is used to adapt two polyacrylamide strips that are cut from the gel used for the electrophoresis in a pH gradient.

To form a resolving gel, made of a 4–11% polyacrylamide gradient, 37 ml of 4% and 31 ml of 11% acrylamide are mixed in a conical gradient maker and poured into the mold. A solution of 0.424 *M* Tris–chloride, pH 9.4, 0.1% sodium dodecyl sulfate is layered over the acrylamide solution before the polymerization occurs (polymerization occurs in less than 1 hour). The overlay solution is removed after that time and the 3.2% acrylamide for the stacking gel is poured over the resolving gel. Isobutanol is overlaid before the polymerization of the stacking gel occurs.

The mold for the continuous gel used for the electrophoresis in a pH gradient is formed by 0.7-cm-wide, 0.075-cm-thick plastic spacers. These spacers are kept in place by paper clips and 9.4% polyacrylamide, as described in Chapter 15.

The pH gradient gel is prepared a few hours before use and polymerized by a fluorescent light for 2 hours, at room temperature (illumination is performed by two neon tubes, cool white, 20 W, General Electric, located 2 cm from the plate and 6 cm from each other). Six slots, each 1.3 cm wide × 1.5 cm long, are created by inserting a plastic comb in the mold.

Unlabeled or ^{35}S-labeled axonemal proteins are dissolved in 2% sodium dodecyl sulfate, 1% 2-mercaptoethanol at a concentration of 2–5 mg/ml. Aliquots of 30 μl, containing 10–100 μg of proteins, are mixed with 10 μl of Nonidet P-40 and 60 μl of stock solution 14. These aliquots are introduced in a slot and overlaid with stock solution 15. Solutions of 0.01 *M* H_3PO_4 and 0.02 *M* NaOH are poured respectively into the upper and lower chambers of the apparatus. Electrophoresis is performed from the positive (anode, upper chamber) to the negative (cathode, lower chamber) pole at 1.4-mA constant current for 14 hours at room temperature.

Following this step the mold is opened and 0.7-cm-wide, 14-cm-long, 0.075-cm-thick polyacrylamide strips are cut from the middle of each lane with the edge of a plastic ruler. Care must be taken to create polyacrylamide strips that are rectangular. Pairs of these strips are lifted by Parafilm (America National Can) and subsequently fitted end-to-end into the space left in the upper part of the mold used for the second electrophoresis. That space first is washed with 0.216 *M* Tris–sulfate, pH 6.1, 1% sodium dodecyl sulfate. The two strips are sealed in place by a solution of melted 1% agarose in 0.216 *M* Tris–sulfate, pH 6.1, 1% sodium dodecyl sulfate. Before starting the electrophoresis a drop of 0.2% bromphenol blue, 60% sucrose is injected into the agarose holding the polyacrylamide strips. Electrophoresis is performed at 20-mA constant current until the bromphenol blue runs off the gel.

Following the second electrophoresis, detection of polypeptides is performed either by silver staining of wet gels or autoradiography of dried gels.

IV. Advantages

Nonequilibrium electrophoresis of polypeptides in a gradient of pH offers two advantages. First, all polypeptides under analysis remain in the gel. Second, polypeptides that could irreversibly precipitate at their isoelectric point in a focusing gel remain soluble and can be separated in the gel used for the second electrophoresis.

The use of a gel slab for the electrophoresis in a gradient of pH and the specific dimensions of both gel slabs create the opportunity to compare two distinct samples of axonemal proteins that are analyzed under identical conditions.

V. Pitfalls

The polymerization of the gel used for the separation in the first dimension occurs after approximately 1 hour of illumination. During that time the volume of the polyacrylamide gel decreases significantly. It is important to compensate this volume reduction by multiple additions of the original acrylamide solution. Otherwise, the creation of slots in the upper part of the gel may be impossible.

The first electrophoresis is performed from the positive to the negative pole to avoid the migration of sodium dodecyl sulfate in the polyacrylamide gel. The presence of sodium dodecyl sulfate in the gradient of pH affects the migration of polypeptides.

The polyacrylamide slab used for the separation in the first dimension is elastic, is difficult to cut, and may stick to the edge used for the cutting or to the Parafilm used for transferring the gel. The addition of a few drops of 0.216 *M* Tris–sulfate, pH 6.1, 1% sodium dodecyl sulfate over the line to be cut will facilitate severing, lifting, and transferring operations.

References

Adams, G. M., Huang, B., Piperno, G., and Luck, D. J. (1981). Central-pair microtubular complex of *Chlamydomonas* flagella: polypeptide composition as revealed by analysis of mutants. *J. Cell Biol.* **91,** 69–76.

Huang, B., Piperno, G., and Luck, D. J. (1979). Paralyzed flagella mutants of *Chlamydomonas reinhardtii.* Defective for axonemal doublet microtubule arms. *J. Biol. Chem.* **254,** 3091–3099.

Luck, D., Piperno, G., Ramanis, Z., and Huang, B. (1977). Flagellar mutants of *Chlamydomonas:* studies of radial spoke-defective strains by dikaryon and revertant analysis. *Proc. Natl. Acad. Sci. U.S.A.* **74,** 3456–3460.

Piperno, G., Huang, B., and Luck, D. J. (1977). Two-dimensional analysis of flagellar proteins from wild-type and paralyzed mutants of *Chlamydomonas reinhardtii. Proc. Natl. Acad. Sci. U.S.A.* **74,** 1600–1604.

Piperno, G., Huang, B., Ramanis, Z., and Luck, D. J. (1981). Radial spokes of *Chlamydomonas* flagella: polypeptide composition and phosphorylation of stalk components. *J. Cell Biol.* **88,** 73–79.

Piperno, G., and Luck, D. J. (1979). An actin-like protein is a component of axonemes from *Chlamydomonas* flagella. *J. Biol. Chem.* **254,** 2187–2190.

Piperno, G., and Luck, D. J. (1981). Inner arm dyneins from flagella of *Chlamydomonas reinhardtii*. *Cell* **27,** 331–340.

Piperno, G., Mead, K., and Shestak, W. (1992). The inner dynein arms I2 interact with a "dynein regulatory complex" in *Chlamydomonas* flagella. *J. Cell Biol.* **118,** 1455–1463.

Segal, R. A., Huang, B., Ramanis, Z., and Luck, D. J. (1984). Mutant strains of *Chlamydomonas reinhardtii* that move backwards only. *J. Cell Biol.* **98,** 2026–2034.

CHAPTER 17

Analysis of Flagellar Calcium-Binding Proteins

Tim Otter
Department of Biology
Albertson College of Idaho
Caldwell, Idaho 83605

I. Background and Rationale

Calcium is an important regulator of flagellar function. In addition to controlling movement, Ca^{2+} is involved in flagellar growth, regeneration, abscission, flagellar surface motility, and adhesion (reviewed by Otter, 1989, and Bloodgood, 1991). Furthermore, flagella offer distinct advantages over many other types of organelles for the study of Ca^{2+}-regulated processes. For example, mutant cells with structural defects in their axonemes may be used for mapping the locations of calcium-binding proteins (CaBPs). The purpose of this chapter is to describe methods for identifying CaBPs that may control these processes in flagella.

One such method that has been developed in my laboratory is called *diagonal electrophoresis* (Fig. 1). The name derives from the patterns that result from this simple strategy: first, flagellar proteins are separated in the presence of ethylenediamine tetraacetic acid (EDTA), and then a second electrophoresis is carried out in the presence of Ca^{2+}, orthogonal to the first dimension. Most proteins do not exhibit Ca^{2+}-sensitive migration in gels (Burgess *et al.*, 1980) and these form the diagonal. CaBPs (those with Ca^{2+}-dependent electrophoretic mobility) lie conspicuously off the diagonal in a region of very low background staining (Fig. 1B). Diagonal electrophoresis is highly sensitive and it greatly reduces the problems of comigrating bands. For example, diagonal gels may be used to identify CaBPs in mammalian sperm, which are notoriously difficult to study by conventional electrophoresis.

Despite its advantages, diagonal electrophoresis is limited to the detection of proteins that bind Ca^{2+} under the conditions of electrophoresis and that migrate differently in the presence or absence of Ca^{2+}. Consequently, as discussed by Garrigos *et al.* (1991), it should be used in combination with other techniques such as $^{45}Ca^{2+}$ overlays of protein blots. In addition, preparing samples for the diagonal technique described here involves boiling in the presence of mercaptoethanol and sodium dodecyl sulfate (SDS), followed by electrophoresis in the presence of both these chemicals plus urea. Thus, it may seem surprising that any protein could bind Ca^{2+} under these conditions, but several do, including calmodulin, troponin C, and parvalbumin, all members of the helix-turn-helix superfamily. In principle, diagonal electrophoresis could be performed with "native" gel systems to identify CaBPs that lose their Ca^{2+} binding properties in the presence of mercaptoethanol and SDS.

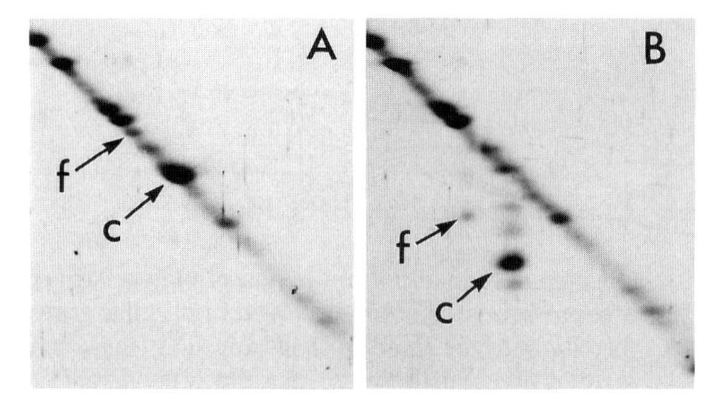

Fig. 1 Coomassie blue-stained diagonal gels of *Chlamydomonas* axonemal proteins, containing calf brain calmodulin (c) as a positive control. In both panels, the region of the gels corresponding to ca. 22,000 < M_r < 12,000 is shown. (A) Both dimensions of electrophoresis contained EDTA. (B) Dimension 2 contained 0.1 mM Ca^{2+}; both calmodulin and an axonemal CaBP (f) lie off the diagonal in a region of low background staining.

II. Materials

A. Chemicals

Acrylamide (Bio-Rad); bisacrylamide (Bio-Rad); piperazine diacrylamide (PDA, Bio-Rad); SDS (Bio-Rad); Tris base (Sigma No. T-1503 or Bio-Rad); β-mercaptoethanol (β-MeSH, Sigma No. M-6250); *N,N,N',N'*-tetramethylethylenediamine (TEMED, Kodak No. 8178 or Bio-Rad); ammonium persulfate; glycine (Sigma No. G-7126); $CaCl_2$ (Mallinckrodt); EDTA (ICN No. 800682); methanol (Mallinckrodt); acetic acid, glacial (Mallinckrodt); glutaraldehyde (EM Sciences, 50%, biological grade); Coomassie blue R-250 (Sigma); silver nitrate, ultrapure (Kodak No. X491); ammonium hydroxide (Mallinckrodt); agarose, low gel temperature (ICN No. 800257); and HCl (Mallinckrodt) are used.

B. Apparatus

I use a minigel apparatus, such as Model No. 1001X from Idea Scientific (Corvallis, OR), with cooling fan assembly (No. 2008), 0.8-, 1.0-, and 1.5-mm-thick spacers, and 10 × 10-cm gel plates (No. 1017). Other minislab gel units may be suitable, but some of them use shorter gel plates and so produce inferior results. Also, not all minislab units accommodate 1.5-mm-thick spacers. The power supply should be capable of producing ca. 500 V, 100 mA at constant current (e.g., Hoefer PS500X).

C. Stock Solutions

In the following list, explicit instructions are provided only for solutions that are not routine or that may require special care. Otherwise follow standard procedures, particularly the method of Laemmli (1970), for making solutions. Except as noted, solutions may be stored at room temperature, with a shelf life of at least 1 month.

1. Acrylamide/PDA stock solution: 50 g acrylamide, 1.33 g PDA, 48.67 ml deionized H_2O (dH_2O). Stir over low heat to dissolve. For Coomassie blue staining, PDA may be replaced on a per gram basis with bisacrylamide.
2. Electrode buffer, after Laemmli (1970): The electrode buffer for the first dimension should contain the same concentration of EDTA as in the gel slab (usually 0.1 m*M*), and the electrode buffer for the second dimension should contain $CaCl_2$, as in the second-dimension slab (usually 0.2 m*M*). Adding Ca^{2+} will cause SDS to precipitate, but this does not seem to affect the results adversely.
3. Low-gel-temperature agarose, 1%, containing 0.1 m*M* $CaCl_2$: Dissolve over moderate heat; store as 4-ml aliquots in refrigerator.

4. "2×" Sample buffer according to Laemmli (1970): Prepare without β-MeSH and store frozen; on day of use, thaw, add β-MeSH, and mix thoroughly.
5. SDS reducing buffer (for soak between first and second dimensions): 62.5 m*M* Tris–HCl, pH 6.8, 10% glycerol, 2% SDS, 5% β-MeSH, 0.0025% bromphenol blue.
6. Ammonium persulfate (AP), 0.025%: Prepare *fresh,* just before use!
7. TEMED, 50 μl per 2 ml H_2O: Prepare *fresh,* just before use!
8. Tris–HCl, 2.2 *M,* pH 8.8 (separating gel buffer).
9. Tris–HCl, 1.25 *M,* pH 6.8 (stacking gel buffer).
10. 20% SDS.
11. EDTA, 100 m*M*, pH 7.0.
12. $CaCl_2$, 100 m*M*.

III. Diagonal Gel Procedures

A. Overview

The standard configuration for diagonal gels that I find most convenient is the minislab, where dimension 1 (0.8 mm thick) contains 0.1 m*M* EDTA and dimension 2 (1.0 mm thick) contains 0.2 m*M* Ca^{2+}. Several variations on this scheme are, however, possible. Dimension 1 may be run in thin tubes. Other spacer thicknesses may be used, provided that dimension 2 is slightly thicker than dimension 1. If dimension 1 is a slab, lanes of interest may be excised with a pizza cutter and stored frozen. Other Ca^{2+} concentrations, as well as metals other than Ca^{2+}, may be used in dimension 2 (Cd^{2+} or Ni^{2+} gives interesting results). If the sequence of electrophoresis is reversed (Ca^{2+} first, EDTA second) the proteins of interest usually lie above the diagonal. Other buffer systems, gel strengths, or gels without urea may be substituted. In my experience, gradient gels produce smaller mobility shifts than uniform-strength gels, and therefore gradients are less desirable. Finally, in our early trials with diagonal gels calmodulin purified from calf brain was added to the flagellar samples as a positive control (Fig. 1). This is useful, especially when learning the diagonal technique or when adjusting the gel format, buffer system, or other parameters. Under optimal conditions, the entire procedure can be completed within 24 hours if the gels are stained with Coomassie blue. Silver staining requires an additional 8 hours.

B. Pouring the Gels

Use 0.8- and 1.0-mm spacers for the first- and second-dimension gels, respectively. The recipe below is for 16% gels, appropriate for identifying CaBPs of ca. 10 to 25 kDa. Allow about 3 cm for a stacking gel (dimension 1 only) to be added later. Carefully overlayer each gel with about 1 cm of 0.1% SDS solution containing 0.1 m*M* EDTA for the first-dimension gel and 0.1% SDS containing 0.1 m*M* $CaCl_2$ for the second-dimension gel. Place a piece of Scotch tape over the top of each gel sandwich to prevent the SDS solution from evaporating. Polymerization is evident within 1 hour, but complete polymerization takes several hours. Gels that have been allowed to polymerize overnight (8–12 hours) give the best results. Before pouring the stacking gel, discard the overlayer solution using filter paper to remove any solution trapped between the gel plates.

16% Separating Gel (10 ml). Acrylamide, 3.2 ml; Tris–HCl, pH 8.8, 2.5 ml; urea (solid), 2.4 g; EDTA (0.1 m*M* final concentration, dimension 1), 10 μl, *or* $CaCl_2$ (0.2 Ca^{2+} final concentration, dimension 2), 20 μl; dH_2O, to 9.5 ml; SDS, 0.1 ml; ammonium persulfate, 0.2 ml; TEMED, 0.2 ml. Warm gently to dissolve the urea.

C. Running the Gels

1. Dimension 1: 0.1 m*M* EDTA in Gel and Electrode Buffer

Select a narrow-toothed comb, pour a 3% stacking gel (see below), and allow 25–30 minutes for it to polymerize. While the stacking gel polymerizes, prepare the protein samples (mix 1 : 1 with 2× Laemmli sample buffer, place in boiling water for 90 seconds). Load 100–200 μg of protein sample into each well, separating it from other samples by wells containing 15 μl of 1× Laemmli sample buffer. Heavier than normal loading of protein are desirable, especially if one is trying to detect minor CaPBs. The electrophoresis in the first dimension should be as narrow as possible, so highly concentrated (5–10 μg protein/μl of whole flagella or axonemes) samples are necessary. Keep this in mind when samples are being prepared during isolatioin and fractionation of flagella.

Stacking Gel (5 ml). Acrylamide, 0.3 ml; Tris–HCl, pH 6.8, 0.75 ml; dH_2O, 3.65 ml; ammonium persulfate, 0.2 ml; TEMED, 0.1 ml.

Begin dimension 1 electrophoresis at a constant current of 15 mA. When proteins enter the resolving gel, increase the current to 20 mA. Although it is not absolutely necessary, some type of cooling is recommended.

2. Dimension 2: 0.1–0.2 m*M* Ca^{2+} in Gel and Electrode Buffer

When the dye front (bromphenol blue) has reached the bottom of the resolving gel (2.5–3 hours later), disconnect and disassemble the gel unit. Slice out a lane(s) for the second dimension, using a standard kitchen rotary knife or pizza cutter. Cut as narrow a slice as possible without removing parts of gel bands. Quickly staining a companion lane may help to guide your gel slicing. Soak the gel slice in reducing buffer (solution 5, above) for 15–30 minutes while preparing the second-dimension gel. First, mount the gel sandwich in a clean electrophoresis unit. Using a piece of Parafilm to support the gel slice, carefully slide it into the second-dimension gel sandwich on top of the second-dimension gel slab. Wear gloves and do not touch the gel slice with your fingers. A small volume of reducing buffer carried over with the gel slice provides lubrication. A spacer from the first dimension is handy for pushing the gel slice into place. Seal the gel slice in place with melted 1% low-gel-temperature agarose that has been precooled to just above the gel point. This step is critical because very hot agarose may denature some CaBPs. When the agarose has solidified, fill the top and bottom buffer reservoirs with electrode buffer containing 0.1–0.2 m*M* $CaCl_2$. Run at 20-mA constant current until the dye front fully enters the resolving gel; then increase current to 25 mA. When the tracking dye reaches the bottom, turn off the unit, disassemble it, remove the gel from the plates, and begin staining procedures.

D. Staining

1. Coomassie Blue

Use standard procedures for staining with Coomassie blue (0.0025% Coomassie blue in 45% methanol, 10% acetic acid). After 2–10 hours in stain, transfer the stained gel into destain (20% methanol, 7.5% acetic acid) with several changes, until the background is clear. Adding a small piece of foam plug (Baxter Scientific No. T1385) greatly speeds up the destaining process. This staining procedure routinely detects 25 ng of calmodulin. Such high sensitivity is possible because the CaBPs are located off of the diagonal in a region of extremely low background staining.

2. Silver Staining

Calmodulin and related CaBPs do not stain well, if at all, with conventional silver stains or silver stain "kits." As noted by Schleicher and Watterson (1983), deposition of silver on CaBPs requires pretreatment with glutaraldehyde. In addition, color of CaBPs develops more slowly than most other proteins, but this is not a problem if the CaBP is well resolved from the diagonal. CaBPs often appear blue-gray, in contrast to the brown color of proteins along the diagonal. To minimize background, use the thinnest gel possible for silver staining. Glass dishes are used throughout the procedure. Plastic containers give inferior results. The procedure below, modified from Poehling and Neuhoff (1981), is an "ammoniacal" silver stain that involves some caustic, potentially explosive reagents. Refer to Oakley *et al.* (1980) for instructions on making the ammoniacal silver solution, the silver stain developer, and safety precautions. Wash volumes are 100–150 ml.

Treat gel with 50% methanol/10% acetic acid (60 minutes); 15% methanol (three washes, 10 minutes each); 10% glutaraldehyde (unbuffered, in deionized H_2O, 30 minutes); 15% methanol (5 minutes); 15% methanol (overnight); 15% methanol (10 minutes); deionized water (10 minutes); ammoniacal silver solution (ca. 12 minutes); deionized water (two washes, 3 minutes each); developer (3–5 minutes); photographic fixer (1 minute, optional); deionized water (several changes).

Gels polymerized in the presence of Ca^{2+} are unusually brittle and they tend to crack extensively during drying. Therefore, it is best to photograph diagonal gels before you attempt to dry them.

Acknowledgments

I thank Bernie Galgoci, Ev Hatch, and Kristin Morrison for their contributions to the development of diagonal electrophoresis as a method for identifying calcium-binding proteins in flagella.

References

Bloodgood, R. A. (1991). Transmembrane signalling in cilia and flagella. *Protoplasma* **164,** 12–22.

Burgess, W. H., Jemiolo, D. K., and Kretsinger, R. H. (1980). Interaction of calcium and calmodulin in the presence of sodium dodecylsulfate. *Biochim. Biophys. Acta* **623,** 257–270.

Garrigos, M., Deschamps, S., Viel, A., Lund, S., Champeil, P., Moller, J. V., and leMaire, M. (1991). Detection of Ca^{2+}-binding proteins by electrophoretic migration in the presence of Ca^{2+} combined with $^{45}Ca^{2+}$ overlay of protein blots. *Anal. Biochem.* **194,** 82–88.

Laemmli, U. K. (1970). Cleavage of structural proteins during assembly of the head of bacteriophage T4. *Nature (London)* **227,** 680–685.

Oakley, B. R., Kirsh, D. R., and Morris, N. R. (1980). A simplified ultrasensitive silver stain for detecting proteins in polyacrylamide gels. *Anal. Biochem.* **105,** 361–363.

Otter, T. (1989). Calmodulin and the control of flagellar movement. *In* "Cell Movement, Volume 1: The Dynein ATPases" (F. D. Warner, P. Satir, I. R. Gibbons, eds.), pp. 281–298. Alan R. Liss, New York.

Poehling, H-M., and Neuhoff, V. (1981). Visualization of proteins with a silver "stain": a critical analysis. *Electrophoresis* **2,** 141–147.

Schleicher, M., and Watterson, D. M. (1983). Analysis of differences between Coomassie Blue stain and silver stain procedures in polyacrylamide gels: conditions for the detection of calmodulin and troponin C. *Anal. Biochem.* **131,** 312–317.

CHAPTER 18

Phosphorylation of *Chlamydomonas* Flagellar Proteins

Robert A. Bloodgood and Nancy L. Salomonsky
Department of Cell Biology
University of Virginia School of Medicine
Charlottesville, Virginia 22908

I. Introduction

Protein phosphorylation and dephosphorylation are widely used as switches for regulating a wide variety of biological processes. Phosphorylation can be studied *in vivo* using [^{32}P]orthophosphoric acid, but steady-state labeling of phosphoproteins *in vivo* depends on the coordinated actions of endogenous protein kinases and phosphatases; the extent of radioactive labeling depends on the rate at which particular phosphate groups turn over and the specific activity of the cellular ATP pool (Garrison, 1993). Protein phosphorylation also can be studied *in vitro* using [γ-^{32}P]ATP (Hasegawa *et al.*, 1987) or [*S*-γ-^{32}P]ATP and [γ-^{32}P]ATP (Segal and Luck, 1985) and endogenous or exogenous protein kinases.

Cilia and flagella possess a large number of phosphoproteins, and most studies of them have focused on axonemal proteins and the role of phosphorylation in regulating axonemal motility (Chapter 63 of this volume; Satir *et al.*, 1993). In

Chlamydomonas, protein phosphorylation had been suggested to be involved in the regulation of axonemal motility (Hasegawa *et al.,* 1987), the photophobic response (Segal and Luck, 1985), flagellar assembly (May and Rosenbaum, 1983; May, 1984), flagellar signaling during mating (Pasquale and Goodenough, 1987; Zhang and Snell, 1994), and flagellar glycoprotein dynamics (Bloodgood and Salomonsky, 1991; Bloodgood, 1992b). *In vivo* labeling with [^{32}P]orthophosphoric acid coupled with two-dimensional polyacrylamide gel electrophoresis (PAGE) has revealed greater than 80 phosphorylated *Chlamydomonas* axonemal polypeptides, including components of dynein arms, radial spokes, the central pair complex, and beak projections within selected doublet microtubules (Adams *et al.,* 1981; Huang *et al.,* 1981; King and Witman, 1994; Piperno and Luck, 1976, 1981; Piperno *et al.,* 1981; Segal *et al.,* 1984). Extraction of purified flagella with nonionic detergents such as Nonidet P-40 and Triton X-100 solubilizes the flagellar membrane, resulting in an axonemal fraction, which can be pelleted, and a membrane–matrix fraction, which remains in the supernatant. Phosphorylated polypeptides in the latter fraction have been identified by *in vivo* and *in vitro* labeling (May, 1984; Bloodgood, 1992a,b; Bloodgood and Salomonsky, 1994). *In vivo* studies demonstrated the presence of greater than 30 phosphoproteins in the membrane–matrix compartment, including a low-abundance surface-exposed flagellar membrane phosphoglycoprotein, FMG-3C, and a 60-kDa phosphoprotein associated with the 350-kDa major membrane glycoprotein (Bloodgood and Salomonsky, 1994).

Important questions to be addressed before carrying out a labeling experiment include the following: (1) Do *in vivo* and *in vitro* labeling with ^{32}P produce similar patterns of flagellar protein phosphorylation? (2) Does the labeling vary according to different solution conditions? We find significant differences among the patterns of flagellar membrane–matrix polypeptides phosphorylated *in vitro* as a function of the concentration of free Ca^{2+} (Bloodgood, 1992b), as well as major differences between the pattern of the membrane–matrix polypeptides phosphorylated *in vivo* and the pattern of those phosphorylated *in vitro* by endogenous protein kinases and phosphatases at high or low free Ca^{2+} concentrations (Fig. 1). A pattern intermediate between *in vivo* and *in vitro* phosphorylation patterns was obtained when [γ-^{32}P]ATP was added to purified whole flagella (Bloodgood, unpublished data); presumably, [γ-^{32}P]ATP permeated the flagella because of partially disrupted membranes. In other studies, axonemal and membrane–matrix tubulin is phosphorylated *in vitro* but not *in vivo* (Segal and Luck, 1985; Bloodgood, 1992b). Axonemal and membrane–matrix proteins are labeled primarily on serine residues both *in vivo* and *in vitro* (Segal and Luck, 1985; Bloodgood and Salomonsky, 1994). If one does label flagella or axonemes *in vitro,* it is important to include sodium vanadate to inhibit the high ATPase activity produced by axonemal dynein and thereby maintain the amount of [γ-^{32}P]ATP available for phosphorylation (Wang *et al.,* 1994).

Based on the presence of polypeptides phosphorylated *in vitro* that do not appear to be labeled *in vivo* and vice versa, it is preferable to perform *in vivo*

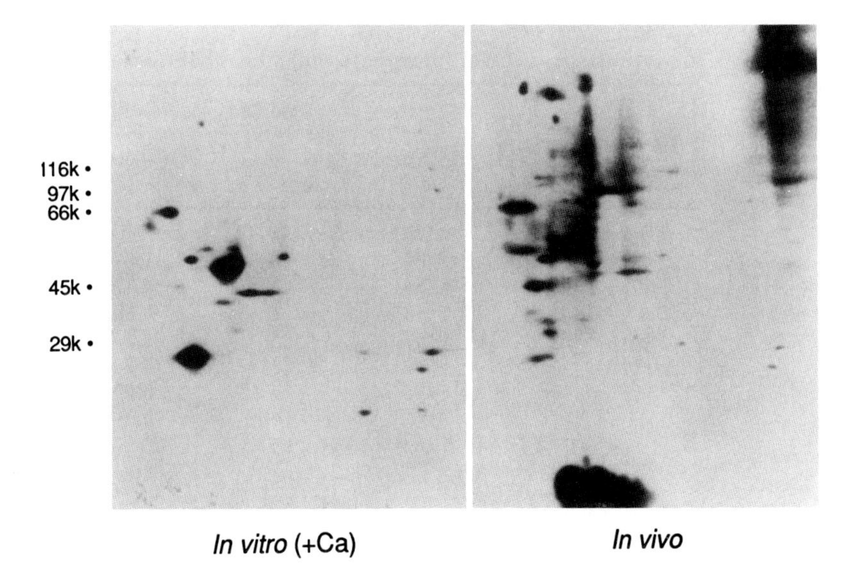

Fig. 1 Autoradiograms of two-dimensional polyacrylamide gels comparing *Chlamydomonas reinhardtii* flagellar membrane–matrix proteins phosphorylated (left) *in vitro* (in the presence of micromolar Ca^{2+} and [^{32}P]ATP) and (right) *in vivo* (using [^{32}P]orthophorphoric acid). There is a significant difference in the patterns of polypeptides phosphorylated *in vivo* and *in vitro*. The large spot of label at the bottom of the *in vivo* autoradiogram represents labeled phospholipids. Reprinted with permission from Bloodgood (1992b).

phosphorylation (with [^{32}P]orthophosphoric acid). Caution still must be taken, however, as for most *in vivo* labeling studies, cells are grown in medium containing 10% or less of the normal levels of phosphate used in normal culture conditions. Total phosphate starvation of *Chlamydomonas reinhardtii* results in the appearance of new phosphatases, some of which are secreted into the culture medium (Matagne *et al.*, 1976; Dumont *et al.*, 1990), so one must be vigilant to prevent dephosphorylation of flagellar proteins during deflagellation and subsequent purification and fractionation of the flagella.

II. *In Vivo* Labeling of *Chlamydomonas* Flagellar Proteins with [^{32}P]Orthophosphoric Acid

A. Solutions

Medium I of Sager and Granick containing 10% of the normal potassium phosphate plus 10 m*M* 4-(2-hydroxyethyl)-1-piperazineethanesulfonic acid (Hepes), pH 7.2–7.4 (Table I)

Medium I containing 20 m*M* Hepes, pH 7.2, and none of the potassium phosphate

Table I
Composition of Low Phosphate and Phosphate-Free Media

		10% Phosphate medium		Phosphate-free medium	
Component	Stock concn (w/v)	ml/8 liters	Final concn (m*M*)	ml/8 liters	Final concn (m*M*)
1. Trace metals	(see below)	8	(see below)	8	(see below)
2. Na citrate · $2H_2O$	10%	40	1.7	40	1.7
3. $FeCl_3 \cdot 6H_2O$	17%	8	0.37	8	0.37
4. $CaCl_2 \cdot 2H_2O$	5.3%	8	0.36	8	0.36
5. $MgSO_4 \cdot 7H_2O$	10%	24	1.2	24	1.2
6. NH_4NO_3	10%	24	3.7	24	3.7
7. KH_2PO_4	10%	0.8	0.074	0	0
8. K_2HPO_4	10%	0.8	0.057	0	0
9. Hepes, pH 7.4	1 *M*	80	10	160	20

Trace metal stock solution

Component	mg/liter
H_3BO_3	1000
$ZnSO_4 \cdot 7H_2O$	1000
$MnSO_4 \cdot H_2O$	303
$CoCl_2 \cdot 6H_2O$	200
$Na_2MoO_4 \cdot 2H_2O$	200
$CuSO_4$	40

5mCi [^{32}P]orthophosphoric acid (New England Nuclear NEX-053, 8500–9100 Ci/mmole in water)

HMDS: 10 m*M* Hepes, pH 7.4, 5 m*M* $MgSO_4$, 1 m*M* dithiothreitol, 4% sucrose (Witman, 1986)

STOP solution: 20 m*M* Hepes, pH 7.2, 20 m*M* ethylene glycol bis(β-aminoethyl ether)-*N,N'*-tetraacetic acid (EGTA), 200 m*M* sodium fluoride, 4% sucrose, 200 U/ml Trasylol (aprotinin, Sigma), 1 μM Microcystin-LR (Calbiochem), 0.1 μM okadaic acid (Gibco-BL, No. 31665A)

0.05% Nonidet P-40, 20 m*M* Hepes, pH 7.2 (for electrophoresis)

100 m*M* NaCl, 5 m*M* EGTA, 1 m*M* $MgCl_2$, 100 U/ml aprotinin, 20 m*M* Tris–HCl, pH 8.3 (for immunoprecipitation)

B. Cell Culture

1. Grow *C. reinhardtii* for 3 days at 21°C (light/dark cycle of 14 : 10 hours) in 200-ml cultures in medium I containing only 10% of the normal level of phosphate (Table I). Inoculate 8-liter bottles containing the same medium from the small culture and grow for an additional 3 days on the same light/dark cycle.

2. Collect cells by centrifugation and wash into medium I containing 20 m*M* Hepes, pH 7.2, and potassium phosphate. Bubble cells with air in front of bright fluorescent lights for ~1 hour. Adjust cell concentration to 4–4.5 × 10^7 cells/ml.

C. Labeling

1. Add 5 mCi of [^{32}P]orthophosphoric acid to 200 ml cells, for a final concentration of 25 μCi/ml as follows: (a) Divide cells into two 100-ml aliquots (control and experimental). (b) Dilute 1 ml containing 5 mCi of [^{32}P]orthophosphoric acid with 9 ml of culture medium. (c) Add dropwise to cells while they are being mixed on a platform shaker (5 ml for each aliquot).

2. Incubate cells for up to 60 minutes at room temperature in 250-ml round-bottomed polycarbonate centrifuge bottles on a platform shaker illuminated with fluorescent cool-white lights. Adequate labeling for resolution on one- or two-dimension polyacrylamide gels is obtained with a 10-minute labeling period. Labeling for up to 70 minutes produces no significant difference in the pattern of labeled polypeptides. If cells are to be treated with drugs, antibodies, or a biotinylation reagent, label first with the [^{32}P]orthophosphoric acid.

3. At the end of the labeling period, dilute cells with an equal volume of ice-cold phosphate-free medium.

4. Centrifuge at 1500*g* for 7 minutes 4°C, aspirate the supernatant, and suspend the pellet in 10 ml of HMDS.

D. Flagellar Fractionation

1. Deflagellate cells by adding 2 ml of ice-cold 25 m*M* dibucaine in water to the cell suspension in 10 ml of HMDS. Pull up and down in a 10-ml plastic disposable pipet 10 times to deflagellate the cells.

2. Check deflagellation using a phase microscope. As soon as cells are deflagellated, add an equal volume of ice-cold STOP solution. Keep flagella at 4°C and add 1 μM Microcystin-LR for all remaining steps.

3. For membrane–matrix fractions, extract flagella on ice for 15 minutes in either 0.05% Nonidet P-40, 20 m*M* Hepes, pH 7.2 (for electrophoresis), or in 100 m*M* NaCl, 5 m*M* EGTA, 1 m*M* $MgCl_2$, 100 U/ml aprotinin, 20 m*M* Tris–Cl, pH 8.3 (for immunoprecipitation). Pellet axonemes by centrifugation at 89,000*g* for 20 minutes. The supernatant is defined as the "membrane–matrix fraction." Overnight incubation of the fraction in the immunoprecipitation buffer at 4°C resulted in minimal loss of protein-associated ^{32}P.

A typical experiment involving a 60-minute labeling period and 4.5 × 10^9 cells yields approximately 300 μl of flagellar membrane–matrix extract containing approximately 50,000 cpm/μl extract. With typical protein concentrations of

0.5–1.0 mg/ml, the specific activity of this extract is 25,000–50,000 cpm/μg of protein. We typically load 100,000–500,000 cpm of ^{32}P on each isoelectric focusing tube gel, which requires a typical exposure time for the second-dimension slab gel of ~12–48 hours using Kodak XAR film and a DuPont Cronex enhancing screen. Spots on the second-dimension gels have sufficient label to allow phosphoamino acid analysis with exposure times of 1–2 weeks (Hemmings *et al.*, 1984).

References

Adams, G. M. W., Huang, B., Piperno, G., and Luck, D. J. L. (1981). Central pair microtubular complex of *Chlamydomonas* flagella: polypeptide composition as revealed by analysis of mutants. *J. Cell Biol.* **91,** 69–76.

Bloodgood, R. A. (1992a). Calcium-regulated phosphorylation of proteins in the membrane-matrix compartment of the *Chlamydomonas* flagellum. *Exp. Cell Res.* **198,** 228–236.

Bloodgood, R. A. (1992b). Directed movements of ciliary and flagellar membrane components. *Biol. Cell* **76,** 291–301.

Bloodgood, R. A., and Salomonsky, N. L. (1991). Regulation of flagellar glycoprotein movements by protein phosphorylation. *Eur. J. Cell Biol.* **54,** 85–89.

Bloodgood, R. A., and Salomonsky, N. L. (1994). The transmembrane signaling pathway involved in directed movements of *Chlamydomonas* flagellar membrane glycoproteins involves the dephosphorylation of a 60 kDa phosphoprotein that binds to the major flagellar membrane glycoprotein. *J. Cell Biol.* **127,** 803–811.

Dumont, F., Loppes, R., and Kremers, P. (1990). New polypeptides and *in-vitro*-translatable mRNAs are produced by phosphate-starved cells of the unicellular alga *Chlamydomonas reinhardtii*. *Planta* **182,** 610–616.

Garrison, J. C. (1993). Study of protein phosphorylation in intact cells. *In* "Protein Phosphorylation. A Practical Approach" (D. G. Hardie, ed.), pp. 1–29. IRL Press, Oxford.

Hasegawa, E., Hayashi, H., Asakura, S., and Kamiya, R. (1987). Stimulation of *in vitro* motility of *Chlamydomonas* axonemes by inhibition of cAMP-dependent phosphorylation. *Cell Motil. Cytoskel.* **8,** 302–311.

Hemmings, H. C., Nairn, A. C., and Greengard, P. (1984). DARPP-32, a dopamine- and adenosine 3′:5′-monophosphate-regulated neuronal phosphoprotein. *J. Biol. Chem.* **259,** 14491–14497.

Huang, B., Piperno, G., Ramanis, Z., and Luck, D. J. L. (1981). Radial spokes of *Chlamydomonas* flagella: genetic analysis of assembly and function. *J. Cell Biol.* **88,** 80–88.

King, S. M., and Witman, G. B. (1994). Multiple sites of phosphorylation within the γ heavy chain of *Chlamydomonas* outer arm dynein. *J. Biol. Chem.* **269,** 5452–5457.

Matagne, R. F., Loppes, R., and Deltour, R. (1976). Phosphatases of *Chlamydomonas:* Biochemical and cytochemical approach with specific mutants. *J. Bacteriol.* **125,** 937–950.

May, G. S. (1984). Flagellar protein kinases and protein phosphorylation during flagellar regeneration and resorption in *Chlamydomonas reinhardtii*. Ph.D. Dissertation. Yale University. University Microfilms No. 8509723.

May, G. S., and Rosenbaum, J. L. (1983). Flagellar protein phosphorylation during flagellar regeneration and resorption in *Chlamydomonas reinhardtii*. *J. Cell Biol.* **97,** 195a.

Pasquale, S. M., and Goodenough, U. W. (1987). Cyclic AMP functions as a primary sexual signal in gametes of *Chlamydomonas reinhardtii*. *J. Cell Biol.* **105,** 2279–2292.

Piperno, G., and Luck, D. J. (1976). Phosphorylation of axonemal proteins in *Chlamydomonas reinhardtii*. *J. Biol. Chem.* **251,** 2161–2167.

Piperno, G., and Luck, D. J. (1981). Inner arm dyneins from flagella of *Chlamydomonas reinhardtii*. *Cell* **27,** 331–340.

Piperno, G., Huang, B., Ramanis, Z., and Luck, D. J. L. (1981). Radial spokes of *Chlamydomonas* flagella: polypeptide composition and phosphorylation of stalk components. *J. Cell Biol.* **88,** 73–79.

Sager, R., and Granick, S. (1953). Nutritional studies with *Chlamydomonas reinhardtii. Ann. N.Y. Acad. Sci.* **56,** 831–838.

Satir, P., Barkalow, K., and Hamasaki, T. (1993). The control of ciliary beat frequency. *Trends Cell Biol.* **3,** 409–412.

Segal, R. A., and Luck, D. J. (1985). Phosphorylation in isolated *Chlamydomonas* axonemes: a phosphoprotein may mediate the Ca^{2+}-dependent photophobic response. *J. Cell Biol.* **101,** 1701–1712.

Segal, R. A., Huang, B., Ramanis, Z., and Luck, D. J. L. (1984). Mutant strains of *Chlamydomonas reinhardtii* that move backwards only. *J. Cell Biol.* **98,** 2026–2034.

Tash, J. S. (1989). Protein phosphorylation: the second messenger signal transducer of flagellar motility. *Cell Motil. Cytoskel.* **14,** 332–339.

Wang, W., Himes, R. H., and Dentler, W. L. (1994). The binding of a ciliary microtubule plus-end binding protein complex to microtubules is regulated by ciliary protein kinase and phosphatase activities. *J. Biol. Chem.* **269,** 21,460–21,466.

Witman, G. B. (1986). Isolation of *Chlamydomonas* flagella and flagellar axonemes. *Methods Enzymol.* **134,** 280–290.

Zhang, Y. H., and Snell, W. J. (1994). Flagellar adhesion dependent regulation of *Chlamydomonas* adenylyl cyclase *in vitro:* A possible role for protein kinases in sexual signaling. *J. Cell Biol.* **125,** 617–624.

CHAPTER 19

In Situ Glycosylation of Flagellar Lipids

S.-J. Chen,[†] T. K. Rosiere,[*] and G. Benjamin Bouck[†]

[*]Cancer Business Unit
Diagnostic Division
Abbott Laboratories
Abbott Park, Illinois 60064
[†]Department of Biological Sciences (m/c 066)
University of Illinois at Chicago
Chicago, Illinois 60607

I. Introduction

Proteins and lipids are often significantly modified by the enzymatic addition of sugars (van den Eijnden and Joziasse, 1993; Shaper and Shaper, 1992). The specific glycosyltransferases that carry out this process are important for many functions including glycosylation of proteins for sorting (e.g., lysosomal glycoproteins, Gieselmann *et al.*, 1992), modification of the proteins and lipids that determine antigenic properties of cell surfaces (e.g., ABO blood groups, Ginsburg and Robbins, 1984), and mediation of a variety of cell–cell interactions (e.g., Needham and Schnaar, 1993). Glycosyltransferases can covalently bind sugars to a variety of lipids, including sphingolipids (cerebrosides, van Echten and Sandhoff, 1993), sterols (steryl glucosides, Zimowski, 1992), and glycerides (galactosylglycerides). As many of these glycolipids are associated with the plasma membrane, it is not surprising that glycosylated lipids are also present in the flagellar membrane. In *Euglena,* steryl glucosides have been identified in lipid extracts of flagella, and some of these appear to be distinct from the steryl glucosides present in the adjacent cell surface membrane (Chen, 1983). Interestingly, at least some of these lipids are glycosylated *in situ* as isolated flagella, and detergent extracts of flagella can glycosylate a number of endogenous and exogenous lipid substrates, indicating that functional glycosyltransferases are residents of the flagellar membrane (Chen, 1983; Chen and Bouck, 1984; Rosiere, 1989). The function of glycosylation of flagellar lipids is not yet known. Flagellar-specific glycosyltransferases that use UDP-glucose (e.g., glucosyltransferases) may, however, provide unique markers for identifying flagellar membranes and for following the ontogeny and development of flagellar membranes. Here we summarize procedures for the detection and characterization of lipid substrates and lipid glucosyltransferases in *Euglena* flagella.

Specific reactions mediated by flagellar GTase(s) are illustrated in the following diagram:

glucosyltransferase (GTase) and cofactors
↓
UDP-glucose + membrane lipids → lipid-glucose + UDP
(donor) (substrate acceptor) (product)

II. Protocol for Demonstrating Glucosyltransferase Activity

A. Solutions and Materials

10 mM 4-(2-hydroxyethyl)-1-piperazineethanesulfonic acid (Hepes), pH 7.0
10 mM Hepes, 10 mM $MgCl_2$
Chloroform/methanol (1/2, v/v)
Chloroform
water
0.5 μCi of UDP-[^{3}H]glucose (specific activity = 3.26–4.7 Ci/mmole), 10 mM $MgCl_2$

B. Donor, Acceptor, and Product

1. Harvest 1.5 liters of *Euglena gracilis* strain Z cells by centrifugation and deflagellate by incubation on ice for 1.5 hours.
2. Collect flagella by centrifugation and resuspend in 1 ml of 10 mM Hepes.

3. Transfer 70–80 μl of flagellar suspension to a glass vial containing dried flagellar lipid extract (see Section II,C) or cholesterol, and incubate in 10 m*M* Hepes, 0.5 μCi of UDP-[^{3}H]glucose (specific activity = 3.26–4.7 Ci/mmole).

4. After 1 hour at room temperature, add 1.5 ml of chloroform/methanol to the reaction mixture and vortex for 15 seconds. Add an additional 0.5 ml of chloroform and vortex; then add 0.5 ml H_2O with vortexing, and finally centrifuge the mixture at 1000*g* in a clinical centrifuge.

5. Remove the aqueous (upper) phase, reextract the organic phase with 0.5 ml of H_2O, and centrifuge again. Discard the aqueous phase. Transfer the organic phase to a glass vial, dry it, and resuspend in scintillation fluid for tritium counting.

C. Comments and Controls

1. Donors

One requisite for sugar transfer is a nucleotide diphosphate-activated sugar donor. The sugar donor is generally specific for both nucleotide and saccharide. In *Euglena* flagella, enzymatic incorporation of [^{3}H]glucose occurs only when flagella are incubated with UDP-[^{3}H]glucose as the sugar donor. Neither UDP-[^{14}C]xylose or -[^{3}H]glucose is incorporated significantly into the acceptor (Chen, 1983). Controls should include the addition of unlabeled nucleotide sugars and unlabeled sugars to assess their competitive effects on the incorporation of UDP-[^{3}H]glucose (Chen, 1983; Chen and Bouck, 1984).

2. Acceptor Substrates

Endogenous acceptors can be extracted from flagella with chloroform/methanol/H_2O (1/1/0.5, v/v/v) which, after emulsification by vortexing and centrifugation, produces a biphasic mixture. The upper phase is discarded and the lower organic phase is dried under a stream of nitrogen gas in a glass vial. The vials with dried lipids can be stored desiccated and frozen until use. Various assays indicate that the native acceptors have properties in common with sterols, and that one of these acceptors is unique to the flagellum whereas the other major acceptor can also be extracted from the cell surface membrane (Chen and Bouck, 1984).

The use of an exogenous sterol acceptor can greatly expedite the assay for and facilitate the identification of the flagellar glycosyltransferase(s). Cholesterol (100 μg/assay, Sigma) dissolved in 100% ethanol, then emulsified in buffer with 10 m*M* 3-[(3-cholamidopropyl)-dimethylammonio)]-1-propanesulfonate (CHAPS, Calbiochem–Behring) and 10% glycerol, is an excellent substitute for the endogenous acceptor (Rosiere, 1989). The choice of sterol appears to be important as lanosterol is not glycosylated. Indeed lanosterol inhibits incorporation into the endogenous acceptor of *Euglena* flagella. Butylated hydroxyl toluene (0.1%, Sigma) should be added to both endogenous and exogenous acceptors to reduce lipid oxidation.

3. Cofactors

Maximum glucosylation of acceptor requires divalent cations. Magnesium and calcium at 5–20 m*M* are the most effective, whereas copper is inhibitory and cobalt has no effect. Manganese stimulates at 5–10 m*M*, but is inhibitory at 20 m*M*. Zinc is inhibitory at the lowest concentration tested (5 m*M*, Chen 1983; Bouck and Chen, 1984).

4. Products

Flagellar lipids radiolabeled after UDP-[^{3}H]glucose incorporation can be solubilized in chloroform/methanol and separated by column chromatography using Sephadex LH-60 (Pharmacia) equilibrated with chloroform. Two well-defined peaks (II and IV) which contain nearly all the radiolabel can be eluted with chloroform; one peak (II) is unique to the flagellum, whereas peak IV is also glycosylated in cell surface fractions. When applied to silica gel G plates and developed with chloroform/methanol/H_2O (50/21/3, v/v/v), peak II migrates as a nonpolar lipid just below a cholesterol standard. Peak IV migrates as a more polar lipid on silica gel G plates. In two different solvent systems peak IV migrates to the same position as a steryl glucoside standard (Chen and Bouck, 1984).

It is important to confirm that the incorporated radiolabel is glucose. Peak fractions from the LH-60 column can be collected and hydrolyzed with 1 *N* HCl in a sealed vial at 100°C for 1 hour. After neutralization and deproteination the sample is applied to a cellulose thin-layer chromatography plate (Fisher) and developed in butanol/pyridine/water (6/4/3, v/v/v). Radiolabeled spots should be compared with migration of a glucose standard (Chen, 1983).

III. Glucosyltransferase(s): Enrichment and Identification on Polyacrylamide Gels

A. Solutions and Materials

Hepes/Mg/CHAPS: 10 m*M* Hepes, pH 7.0, 10 m*M* $MgCl_2$, 10 m*M* Chaps
Glycerol

B. Flagellar Preparation

1. Harvest 16 liters of *Euglena* by low-speed centrifugation and amputate flagella by cold shock. After removing cell bodies by centrifugation, pellet the flagella in the supernatant at 10,000 rpm in a Sorvall SS34 rotor.

2. Extract flagellar pellets with a solution containing Hepes/Mg/CHAPS at room temperature for 1 hour.

3. Centrifuge extracted flagella for 30 minutes in the Sorvall SS34 rotor at 17,500 rpm. Remove the supernatant and make 10% in glycerol. These soluble preparations can be frozen at −78°C without appreciable loss of enzyme activity (Rosiere, 1989).

C. Two Methods for Enrichment of Enzyme Activity

1. Hydrophobic Interaction Chromatography

1. Equilibrate phenyl-Sepharose (Sigma) in a solution containing 10 m*M* Hepes, pH 7.0, with 10 m*M* $MgCl_2$, 10 m*M* CHAPS, and 10% glycerol. Pour the slurry into a column with 2–5 ml total bed volume. Load the flagellar extract from Section III,B, step 3, on the column and wash with Hepes/$MgCl_2$/CHAPS/glycerol. Elute enzyme activity with the same solution with the addition of 0.32% Nonidet P-40 (Polysciences). Collect 0.75- to 1.0-ml fractions; assay 70- to 80-μl aliquots from each fraction for enzyme activity. Emulsified cholesterol or flagellar lipids can be used as acceptor substrates; UDP-[^{3}H]glucose is the sugar donor.

2. Highly reproducible preparations are generated with Pharmacia's FPLC system equipped with a Superose-12 HR 10/30 column (Pharmacia). Equilibrating, loading, eluting, and assaying are carried out exactly as with the phenyl-Sepharose columns, but each of these steps is automated with the FPLC system.

D. CHAPS Extraction of Flagellar Vesicles

Resuspend flagella from Section III,B, step 1, in 50 m*M* NaOH on ice for 15 minutes. After centrifugation for 10 minutes at 12,000*g* in a microfuge, resuspend the pellet in 10 m*M* Hepes buffer at pH 7.0 and recentrifuge. The resulting pellet contains flagellar membrane vesicles and mastigonemes (flagellar hairs). Extraction of these pellets with Hepes/$MgCl_2$/CHAPS/glycerol yields soluble glucosyltransferase activity when incubated with emulsified cholesterol and UDP-[^{3}H]glucose.

E. Comments

The zwitterionic detergent CHAPS is far more effective in extracting glucosyltransferase(s) activity from flagellar membranes than Triton X-100, Nonidet P-40, or 12-*o*-tetradecanoyl-phorbol-13-acetate (Chen 1983). Thus although more than 50% of the activity remains in the flagellar pellet after CHAPS treatment, CHAPS is still the detergent of choice. When eluting the phenyl-Sepharose columns, it is important not to exceed 0.32% Nonidet P-40, as this detergent significantly reduces enzyme activity at higher concentrations (Rosiere, 1989).

Proteins in fractions enriched in glucosyltransferase(s) prepared by these methods can be electrophoresed through sodium dodecyl sulfate–polyacrylamide gels to determine the molecular weight of the monomeric glucosyltransferase, and to check for contamination with other proteins. In all of the preparations (phenyl-Sepharose column, FPLC, or vesicle extracts) there are several polypeptides. Only one polypeptide (ca. 64 kDa), however, is common to all three of these preparations which are enriched in glucosyltransferase activity (Rosiere, 1989; Bouck *et al.*, 1990). Antibodies against the 64-kDa polypeptide band will provide further identification and characterization of the presumptive glucosyltransferase, and can be used, for example, for selection of the presumptive glucosyltransferase from a *Euglena* cDNA library.

References

Bouck, G. B., and Chen, S. J. (1984). Synthesis and assembly of the flagellar surface. *J. Protozool.* **31,** 21–24.

Bouck, G. B., Rosiere, T. K., and Levasseur, P. J. (1990). *Euglena gracilis:* A model for flagellar surface assembly, with reference to other cells that bear flagellar mastigonemes and scales. *In* "Ciliary and Flagellar Membranes" (R. A. Bloodgood, ed.), pp. 65–90. Plenum Press, New York.

Chen, S. J. (1983). Surface Glycolipids and Their Glycosyltransferases in *Euglena gracilis*. Ph.D thesis. University of Illinois at Chicago.

Chen, S. J., and Bouck, G. B. (1984). Endogenous glycosyltransferases glucosylate lipids in flagella of *Euglena*. *J. Cell Biol.* **98,** 1825–1835.

Gieselmann, V., Schmidt, B., and von Figura, K. (1992). *In vitro* mutagenesis of potential N-glycosylation sites of arylsulfatase A, effects on glycosylation, phosphorylation and intracellular sorting. *J. Biol. Chem.* **267,** 13262–13266.

Ginsburg, V., and Robbins, P. W. (1984). "Biology of Carbohydrates", Vol. 2, John Wiley & Sons, New York.

Needham, L. K., and Schnaar, R. L. (1993). Carbohydrate recognition in the peripheral nervous system: A calcium-dependent membrane binding site for HNK-1 reactive glycolipids potentially involved in Schwann cell adhesion. *J. Cell. Biol.* **121,** 397–408.

Rosiere, T. (1989). Studies of Cell Surface and Flagellar Integral Membrane Proteins and the Paraxial Rod Proteins in *Euglena gracilis*. Ph.D thesis. University of Illinois at Chicago.

Shaper, J. H., and Shaper, N. L. (1992). Enzymes associated with glycosylation. *Curr. Opin. Struct. Biol.* **2,** 701–709.

van den Eijnden, D. H., and Joziasse, D. H. (1993). Enzymes associated with glycosylation. *Curr. Opin. Struct. Biol.* **3,** 711–721.

van Echten, G., and Sandhoff, K. (1993). Ganglioside metabolism, enzymology, topology and regulation. *J. Biol. Chem.* **268,** 5341–5344.

Zimowski, J. (1992). Specificity and some other properties of cytosolic and membranous UDPGlc: 3ß-hydroxysteroid glucosyltransferases from *Solanum tuberosum* leaves. *Phytochemistry* **31,** 2977–2981.

CHAPTER 20

Detection of Flagellar Protein Kinases on Polyvinylidene Difluoride Membranes Following Sodium Dodecyl Sulfate–Polyacrylamide Gel Electrophoresis

Jovenal T. San Agustin and George B. Witman
Male Fertility Program
Worcester Foundation for Experimental Biology
Shrewsbury, Massachusetts 01545

I. Introduction

In situ renaturation of protein kinases in polyacrylamide gels after electrophoretic separation was first demonstrated by Geahlen *et al.* (1986). Their procedure was later modified by Kameshita and Fujisawa (1989) to include a denaturation step before renaturation of the kinases. Most subsequent work involving renaturation of specific kinases in polyacrylamide gels (Gotoh *et al.*, 1990; Heider *et al.*, 1994; Wang and Erikson, 1992; Durocher *et al.*, 1992; Ding and Badwey, 1993) used modified versions of the method of Kameshita and Fujisawa. An extension of the method was introduced by Celenza and Carlson (1986) wherein proteins resolved by sodium dodecyl sulfate–polyacrylamide gel electrophoresis (SDS–PAGE) were transferred to nitrocellulose before renaturation. This method was subsequently improved by Ferrell and Martin (1989, 1991) who used polyvinylidene difluoride (PVDF) membranes instead of nitrocellulose. Renaturation on PVDF membranes enabled the detection of up to 20 protein kinases in fibroblast culture cells (Ferrell and Martin, 1989), and has been used to identify nuclear protein kinases (Rachie *et al.,* 1993) and to detect CaM kinase II in purified samples or in crude tissue homogenates (Shackelford and Zivin, 1993).

Protein kinases are known to be present in the cilia and flagella of many organisms (see, e.g., Chapter 66 in this volume) and are involved in the control of motility, mating, nondynein surface motility, and dynein force generation (Brokaw, 1987; San Agustin and Witman, 1994; Zhang *et al.,* 1991; Bloodgood, 1992; Walczak and Nelson, 1994; Hamasaki *et al.,* 1991). We describe below a procedure for the detection of protein kinases in flagella based on the method of Ferrell and Martin. It is demonstrated here using demembranated ram sperm flagella and *Chlamydomonas* axonemes, but may be applied to the cilia and flagella of other organisms.

II. Methods

A. Solutions

SDS–PAGE sample buffer: 10% glycerol, 3% SDS, 0.03% bromphenol blue, 50 m*M* dithiothreitol (DTT), 62.5 m*M* Tris, pH 6.8

Transfer buffer: 50 m*M* Tris base, 192 m*M* glycine, 20% methanol, 0.01% SDS

Denaturation buffer: 7 *M* guanidine hydrochloride (ICN), 2 m*M* ethylenediaminetetraacetic acid (EDTA), 50 m*M* DTT, 50 m*M* Tris–HCl, pH 8.3

TBS: 140 m*M* NaCl, 30 m*M* Tris-HCl, pH 7.4

Renaturation buffer: 140 m*M* NaCl, 10 m*M* Tris–HCl, pH 7.4, 2 m*M* EDTA, 1% w/v bovine serum albumin (BSA, Calbiochem, fatty acid free), 0.1% v/v Tween 20 (Bio-Rad), 2 *M* DTT (Other nonionic detergents may be used.)

Blocking solution: 5% BSA (Calbiochem, fatty acid free), 30 m*M* Tris–HCl, pH 7.4 (If a particular protein kinase is being investigated, BSA may be replaced with a specific substrate of that kinase.)

Phosphorylation buffer: 100 μCi/ml [γ-^{32}P]ATP (3000 Ci/mmole), 10 m*M* $MgCl_2$, 2 m*M* $MnCl_2$, 30 m*M* Tris–HCl, pH 7.4

Wash buffer A: 30 m*M* Tris–HCl, pH 7.4

Wash buffer B: 30 m*M* Tris–HCl, pH 7.4, 0.05% Tween 20

1 *N* potassium hydroxide

B. Preparation and Demembranation of Flagella

Unless the retention of plasma membrane is specifically desired, it is best to work with demembranated flagella. From our experience, the lanes in the PVDF blot show less background when demembranated flagella are used, and the kinase bands appear sharper. Several procedures for demembranation of flagella and cilia are described in this volume. In our laboratory, we are particularly interested in ram sperm axonemal kinases which are active under conditions that promote reactivation. We therefore treat the demembranated sperm tails with our standard reactivation medium (see Chapter 36 in this volume) prior to dissolving them in SDS–PAGE sample buffer (see below).

C. SDS–PAGE Gradient Gel Electrophoresis

This procedure was adapted from that of King *et al.* (1986).

1. The day before the experiment, pour a 5–15% gradient gel, 12 × 14 cm, 1.5 mm thick. Top with a thin layer of *n*-butanol. When the gel line becomes visible below the *n*-butanol layer, remove the *n*-butanol and replace it with about 1 ml of the gradient gel buffer (with 0.1% SDS). Let the gel stand overnight.
2. (Optional) Incubate the demembranated flagella 3–4 minutes in the chosen medium, 0.5–1.5 × 10^8 flagellum equivalents per milliliter. Typical final volume in our experiments was 200 μl.
3. Centrifuge 40 seconds in the microfuge.
4. Remove the supernatant and add boiling SDS–PAGE sample buffer to the pelleted flagella (about 50 μl sample buffer for a 200-μl incubation volume). Vortex to dissolve and use immediately or store at −20°C until use.
5. Pour the stacking gel.
6. While the stacking gel is polymerizing, boil frozen samples 4–5 minutes. Cool, and then centrifuge 4 minutes in the microfuge.
7. Introduce the samples to the stacking wells using a Hamilton syringe, 4–6 × 10^6 flagellum equivalents per well. Fill empty wells with about the same volume of sample buffer as the samples. Top off each well with stacking gel buffer.
8. Run at 60 V while stacking, then at 80 V when the dye front reaches the gradient gel.

D. Transfer of Proteins to PVDF Membrane and Subsequent Renaturation

1. Equilibrate the gel with the transfer buffer for 20 minutes.
2. Wet a sheet of Immobilon P (Millipore, 0.45 μm, 15 × 15 cm for a 12 × 14-cm gel) with methanol, then rinse it with distilled water. Immerse it in transfer buffer until use.

3. Do the transfer under the following conditions (adapted from Otter *et al.*, 1987) using a model TE transfer apparatus (Hofer Scientific): 1 hour at 28 V to allow transfer of low-molecular-weight proteins, then 14–17 hours at 84 V for the transfer of larger proteins. Employ a cooling manifold or perform the transfer at 4°C to prevent overheating of the buffer.

4. After the transfer, rinse the blot in wash buffer B for 15 minutes with gentle shaking.

5. Cut the blot to the desired dimensions and then place it in the denaturation buffer. A 7.5 × 15-cm blot will require at least 50 ml of denaturation buffer. Incubate 2–3 hours at room temperature.

6. Drain off the denaturation buffer and rinse the blot with 250 ml TBS until it turns opaque again (about 1 minute).

7. Incubate with 250 ml renaturation buffer overnight (16–20 hours) at 4°C with gentle shaking (10 rpm, Orbit Shaker, Lab Line). The choice of the detergent component in the renaturation buffer is very important. We have tried other nonionic detergents like Triton X-100 and Nonidet P-40 procured from several sources. Tween 20 (Bio-Rad, EIA grade) works best in our hands. It is important that the detergent be of the highest purity available.

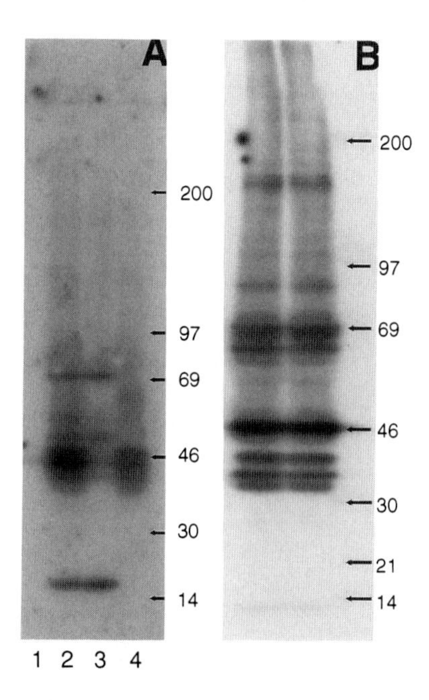

Fig. 1 (A) Protein kinase blot of catalytic subunit of porcine heart cAMP-dependent protein kinase (lane 1), demembranated ram sperm (lane 2), demembranated ram sperm tails (lane 3), and demembranated ram sperm heads (lane 4). (B) Protein kinase blot of isolated axonemes of *Chlamydomonas*. Molecular weight markers are indicated to the right of each blot.

8. Remove the renaturation buffer and treat the blot with 250 ml blocking solution for 1 hour at room temperature.

9. Transfer the blot to the phosphorylation buffer. For a 7.5 × 15-cm blot, prepare about 15 ml of phosphorylation buffer. Incubate 30 minutes at room temperature. Use an orbit shaker to ensure that the blot is immersed in the phosphorylation buffer at all times.

10. Transfer the blot to 100 ml wash buffer A for a brief rinse, then wash it gently in an orbit shaker in the following sequence:

Wash buffer A: 250 ml, 10 minutes, two times
Wash buffer B: 10 minutes
Wash buffer A: 10 minutes, two times
1 *N* KOH, 10 minutes
Wash buffer A, 10 minutes, two times

11. Let the blot dry (at least 1 hour).

12. Expose the blot to film (Kodak X-OMAT) for 12–36 hours at −70°C with intensifying screen or scan in a phosphorimager (Molecular Dynamics).

Figure 1 shows kinases in ram sperm and *Chlamydomonas* axonemes detected according to the above protocol.

References

Bloodgood, R. A. (1992). Directed movements of ciliary and flagellar membrane components: A review. *Biol. Cell* **76,** 291–301.

Brokaw, C. J. (1987). Regulation of sperm flagellar motility by calcium and cAMP-dependent phosphorylation. *J. Cell. Biochem.* **35,** 175–184.

Celenza, J. L., and Carlson, M. (1986). A yeast gene that is essential for release from glucose repression encodes a protein kinase. *Science* **233,** 1175–1180.

Ding, J., and Badwey, J. A. (1993). Stimulation of neutrophils with a chemoattractant activates several novel protein kinases that can catalyze the phosphorylation of peptides derived from the 47-kDa protein component of the phagocyte oxidase and myristoylated alanine-rich C kinase substrate. *J. Biol. Chem.* **268,** 17326–17333.

Durocher, Y., Chapdelaine, A., and Chevalier, S. (1992). Identification of cytosolic protein tyrosine kinases of human prostate by renaturation after SDS/PAGE. *Biochem. J.* **284,** 653–658.

Ferrell, J. E., Jr., and Martin, G. S. (1989). Thrombin stimulates the activities of multiple previously unidentified protein kinases in platelets. *J. Biol. Chem.* **264,** 20723–20729.

Ferrell, J. E., Jr., and Martin, G. S. (1991). Assessing activities of blotted protein kinases. *Methods Enzymol.* **200,** 430–435.

Geahlen, M., Anostario, M., Jr., Low, P. S., and Harrison, M. L. (1986). Detection of protein kinase activity in sodium dodecyl sulfate-polyacrylamide gels. *Anal. Biochem.* **153,** 151–158.

Gotoh, Y., Nishida, E., Yameshita, T., Hoshi, M., Kawakami, M., and Sakai, H. (1990). Microtubule-associated-protein (MAP) kinase activated by nerve growth factor and epidermal growth factor in PC 12 cells. *Eur. J. Biochem.* **193,** 661–669.

Hamasaki, T., Barkalow, K., Richmond, J., and Satir, P. (1991). cAMP-stimulated phosphorylation of an axonemal polypeptide that copurifies with the 22S dynein arm regulates microtubule translocation velocity and swimming speed in *Paramecium. Proc. Natl. Acad. Sci. U.S.A.* **88,** 7918–7922.

Heider, H., Hug, C., and Lucocq, J. M. (1994). A 40-kDa myeline basic protein kinase, distinct from erk1 and erk2, is activated in mitotic HeLa cells. *Eur. J. Biochem.* **219,** 513–520.

Kameshita, I., and Fujisawa, H. (1989). A sensitive method for detection of calmodulin-dependent protein kinase II activity in sodium dodecyl sulfate-polyacrylamide gel. *Anal. Biochem.* **183,** 139–143.

King, S. M., Otter, T., and Witman, G. B. (1986). Purification and characterization of *Chlamydomonas* flagellar dyneins. *Methods Enzymol.* **134,** 291–306.

Otter, T., King, S. M., and Witman, G. B. (1987). A two-step procedure for efficient electrotransfer of both high-molecular-weight (>400,000) and low-molecular-weight (<20,000) proteins. *Anal. Biochem.* **162,** 370–377.

Rachie, N. A., Seger, R., Valentine, M. A., Ostrowski, J., and Bomsztyk, K. (1993). Identification of an inducible 85-kDa nuclear protein kinase. *J. Biol. Chem.* **268,** 22143–22149.

San Agustin, J. T., and Witman, G. B. (1994). Role of cAMP in the reactivation of demembranated ram spermatozoa. *Cell Motil. Cytoskel.* **27,** 206–218.

Shackelford, D. A., and Zivin, J. A. (1993). Renaturation of calcium/calmodulin-dependent protein kinase activity after electrophoretic transfer from sodium dodecyl sulfate-polyacrylamide gels to membranes. *Anal. Biochem.* **211,** 131–138.

Walczak, C. E., and Nelson, D. L. (1994). Regulation of dynein-driven motility in cilia and flagella. *Cell Motil. Cytoskel.* **27,** 101–107.

Wang, H. C. R., and Erikson, R. L. (1992). Activation of protein serine/threonine kinases p42, p63, and p87 in Rous sarcoma virus-transformed cells: signal transduction/transformation-dependent MBP kinases. *Mol. Biol. Cell* **3,** 1329–1337.

Zhang, Y. H., Ross, E. M., and Snell, W. J. (1991). ATP-dependent regulation of flageller adenylylcyclase in gametes of *Chlamydomonas reinhardtii. J. Biol. Chem.* **266,** 22954–22959.

CHAPTER 21

Measurement of ATPase Activity Using [γ-^{32}P]ATP

Stephen M. King
Department of Biochemistry
University of Connecticut Health Center
Farmington, Connecticut 06030

I. Introduction

Measurement of ATPase activity is an important aspect of the biochemical characterization of ciliary and flagellar enzymes such as dynein. The technique provides a useful criterion for assessing the purity of an enzyme preparation (e.g., Piperno and Luck, 1979; Pfister *et al.*, 1982) as well as for investigating the mechanisms involved in coupling the ATPase and translocation activities of molecular motors and in regulating these motors (Johnson, 1985; Paschal *et al.*, 1987; Pfister and Witman, 1984).

The two methods described here employ radiolabeled nucleotide and depend on measuring the release of inorganic phosphate. The first involves the formation of a phosphomolybdate complex that subsequently is extracted into an organic phase away from the unhydrolyzed nucleotide (Conway and Lipmann, 1964). This assay is relatively rapid and very sensitive. The detection limit is approximately 0.04 nmole phosphate released per minute and the assay is linear over the range 0.04–6.7 nmole phosphate released per minute (Pfister *et al.*, 1982). This method has two disadvantages: (1) there are multiple pipetting steps, some of which involve organic solvents, and thus significant errors can be introduced

METHODS IN CELL BIOLOGY, VOL. 47

if care is not taken; and (2) the organic extraction requires the use of benzene, which is a carcinogen. The second method is very simple and quantitation is achieved either by liquid scintillation counting or, preferably, using a phosphorimager. Here a 1-μl sample of the enzyme reaction is spotted onto a polyethyleneimine (PEI)-cellulose thin-layer chromatography (TLC) plate. The plate is developed in LiCl/HCOOH and the ATP and phosphate spots are quantified directly in the phosphorimager or identified by autoradiography prior to counting. In both cases, the sensitivity is excellent and the small volume required enables measurements to be made in many different situations (e.g., enzymatic activity inside a motility chamber with a final volume of ~10 μl can be measured; see Moss *et al.,* 1992). The only potential disadvantage to the TLC-based method is that, when using autoradiography, it may be necessary to expose the film overnight if the specific activity of the ATP solution is low.

Both of the methods detailed here are considerably more sensitive than nonradioactive methods for measuring phosphate production, e.g., the malachite green-based colorimetric procedure (see Chapter 22 in this volume). It is however, necessary to use colorimetric methods to follow the hydrolysis of nucleotides for which radiolabeled versions are not commercially available (e.g., Gatti *et al.,* 1989).

II. Methods

A. Phosphomolybdate Solution Assay

1. Reagents and Solutions

Assay buffer (e.g., 30 m*M* Tris–Cl, pH 7.5, 5 m*M* $MgSO_4$, 1 m*M* dithiothreitol, 0.5 m*M* ethylenediaminetetraacetic acid (EDTA), 25 m*M* KCl, which is used for the assay of *Chlamydomonas* flagellar dyneins)

20 m*M* silicotungstic acid in 10 m*M* H_2SO_4 (STA)

5% Ammonium molybdate in 4 *N* sulfuric acid (AM)

2 m*M* KH_2PO_4

Isobutanol : benzene (1 : 1)

Liquid scintillation cocktail

[γ-^{32}P]ATP in 20 m*M* ATP (need at least 1000 cpm/nmole)

2. Procedure

1. Place the enzyme fraction to be assayed (10 μl) and the assay buffer (180 μl) in a 13 × 100-mm glass tube. (*Note:* it is important to wash the tubes as some batches contain a residue which can inhibit ATPase activity.) The buffer should contain the required cation (Mg^{2+}, Ca^{2+}, Mn^{2+}, etc.). Also prepare three "blank" tubes that contain buffer but no enzyme.

2. Start the reaction by adding 10 μl of the [γ-^{32}P]ATP solution (final concentration = 1 mM) and mix briefly by vortexing. To allow sufficient time for manipulating the samples, start the next tube 0.2 minute later.

3. After the desired period, stop the reaction by adding 100 μl STA solution. Routinely, the assay time varies between 5 and 60 minutes, depending on the enzyme activity and the number of tubes being processed.

4. Once all reactions have been stopped, add 100 μl AM solution and 200 μl 2 mM KH_2PO_4 as cold carrier phosphate. A phosphomolybdate complex will form and the solution should turn a pale yellow color (the color intensity increases with the age of the AM solution).

5. The next step is to extract the phosphomolybdate complex into an organic layer. To do this, add 500 μl isobutanol : benzene and vortex vigorously for 30 seconds. To save time this can be done in batches of eight tubes. Allow the two phases to separate; this may be accomplished simply by letting the tubes sit on the lab bench for about 5 minutes. As benzene is carcinogenic, *all* these manipulations must be performed in a fume hood. Remove 200 μl of the organic upper layer, place in a scintillation vial, add scintillant, and count.

6. Determine the specific activity of the ATP stock solution by counting three 10-μl aliquots.

7. The following formula may be used to convert the data obtained by liquid scintillation counting into ATPase activity, with units of nmole phosphate released/min/ml enzyme solution:

$$\text{activity} = 50{,}000\,[(C - B)/AT]$$

where A = average cpm of the 10-μl aliquots of the ATP stock solution, B = average cpm found in the blank tubes, C = cpm found in the sample tube, and T = time of assay (minutes).

8. For samples that contain nonionic detergents such as Nonidet P-40, the reaction should be stopped with 25 μl 20% sodium dodecyl sulfate prior to addition of 100 μl STA solution. For fractions that contain large amounts of phosphate (such as those from a hydroxyapatite column) the assay should be altered as follows: use 0.5 ml of both the STA and AM solutions, substitute 1.25 ml water for the 2 mM KH_2PO_4, extract with 2.5 ml isobutanol : benzene, and use 1 ml of the organic layer for liquid scintillation counting.

B. Phosphate Assay Using Thin-Layer Chromatography

1. Reagents and Solutions

Assay buffer (see Section II,A,1)

20 mM [γ-^{32}P]ATP stock solution

PEI-cellulose thin-layer chromatography plate (plastic-backed)

0.5 M LiCl/1 M HCOOH

2. Procedure

1. Place 180 μl of assay buffer (which should include the appropriate cation) in a microfuge tube and add 10 μl of the enzyme fraction to be assayed. Also prepare "blank" tubes which contain no enzyme.

2. Start the reaction by adding 10 μl [γ-^{32}P]ATP stock solution (1 m*M* final concentration) and mix briefly by vortexing. Note that the reaction volume can be scaled down as necessary without affecting subsequent steps.

3. On a PEI-cellulose TLC plate draw a faint horizontal pencil line approximately 1 cm from the bottom. Put a mark every 1 cm along this line.

4. Stop the reaction at the required time by pipetting 1 μl of each reaction solution directly onto a mark on the TLC plate.

5. Once all spots are dry, carefully stand the PEI-cellulose sheet in a chromatography tank containing 0.5 *M* LiCl/1 *M* HCOOH to a depth of approximately 0.5 cm. Allow the solvent front to migrate approximately 9 cm. This allows the [^{32}P]phosphate spot to become well separated from unhydrolyzed ATP, which migrates very slowly and remains close to the origin. (*Note:* If very low ATPase activities are expected, it is helpful to hydrolyze an aliquot of the [γ-^{32}P]ATP stock with apyrase so that the migration of free phosphate may be readily monitored.)

6. Following chromatography, allow the TLC plate to dry in the fume hood, wrap it in plastic film, and subject it to autoradiography. The exposure time required obviously will depend on the specific activity of the ATP stock solution and the enzymatic activity of the sample (several hours to overnight with a screen at -70°C is routine).

7. The developed film serves as a template to accurately locate the radioactive ATP and phosphate spots. These are cut from the TLC plate and the activity in each measured by liquid scintillation counting.

8. Alternatively, the activity present in each spot may be determined using a phosphorimager.

9. This technique also can be used to follow the hydrolysis of radiolabeled nucleotide analogs such as [γ-^{35}S]ATPγS. In this particular case, the chromatogram should be developed for a longer time, as thiophosphate migrates more slowly than phosphate in this solvent system.

References

Conway, T. W., and Lippman, F. (1964). Characterization of a ribosome-linked guanosine triphosphatase in *Escherichia coli* extracts. *Proc. Natl. Acad. Sci. U.S.A.* **52,** 1462–1469.

Gatti, J.-L., King, S. M., Moss, A. G., and Witman, G. B. (1989). Outer arm dynein from trout spermatozoa: purification, polypeptide composition and enzymatic properties. *J. Biol. Chem.* **264,** 11450–11457.

Johnson, K. A. (1985). Pathway of the microtubule-dynein ATPase and the structure of dynein. *Annu. Rev. Biophys. Biophys. Chem.* **14,** 161–188.

King, S. M., Otter, T., and Witman, G. B. (1986). Purification and characterization of *Chlamydomonas* flagellar dyneins. *Methods Enzymol.* **134,** 291–306.

Lanzetta, P. A., Alvarez, L. J., Reinach, P. S., and Candia, O. A. (1979). An improved assay for nanomole amounts of inorganic phosphate. *Anal. Biochem.* **100,** 95–97.

Moss, A. G., Gatti, J.-L., and Witman, G. B. (1992). The motile β/IC1 subunit of sea urchin sperm outer arm dynein does not form a rigor bond. *J. Cell Biol.* **118,** 1177–1188.

Paschal, B. M., King, S. M., Moss, A. G., Collins, C. A., Vallee, R. B., and Witman, G. B. (1987). Isolated flagellar outer arm dynein translocates brain microtubules *in vitro*. *Nature* **330,** 672–674.

Pfister, K. K., Fay, R. B., and Witman, G. B. (1982). Purification and polypeptide composition of dynein ATPases from *Chlamydomonas* flagella. *Cell Motil.* **2,** 525–547.

Pfister, K. K., and Witman, G. B. (1984). Subfractionation of *Chlamydomonas* 18S dynein into two unique subunits containing ATPase activity. *J. Biol. Chem.* **259,** 12072–12080.

Piperno, G., and Luck, D. J. L. (1979). Axonemal adenosine triphosphatases from flagella of *Chlamydomonas reinhardtii:* purification of two dyneins. *J. Biol. Chem.* **254,** 3084–3090.

CHAPTER 22

Nonradioactive Method for ATPase Assays

Osamu Kagami and Ritsu Kamiya

Zoological Institute
Graduate School of Science
University of Tokyo
Tokyo 113, Japan

I. Introduction

Conventional nonradioactive ATPase assays follow the change in orthophosphate concentration by colorimetry, after quenching an ATP-hydrolyzing reaction at regular time intervals and converting phosphate into a phosphomolybdate complex (Fiske and Subbarow, 1925; Taussky and Shorr, 1953). A very sensitive method of this kind determines the quantity of phosphomolybdate as a complex with malachite green (Itaya and Ui, 1966; Anner and Moosmayer, 1975). The method described below, developed by Kodama and colleagues (Kodama *et al.*, 1986; Ohno and Kodama, 1991), is sensitive enough to measure 0.2–15 μM phosphate, a concentration range that other colorimetric methods cannot measure. The method is simple and reliable; the citric acid added after initiation of the coloring reaction removes excess molybdate and thereby greatly reduces the nonenzymatic hydrolysis of ATP catalyzed by molybdate (Lanzetta *et al.*, 1979). Because of its high sensitivity, this assay was used for measuring the ATPase activities of myosin and kinesin on a single microscope coverslip in *in*

METHODS IN CELL BIOLOGY, VOL. 47

vitro motility assays (Harada *et al.,* 1990; Vale *et al.,* 1989). We use it to measure the ATPase activities in *Chlamydomonas* inner-arm dynein fractions in which only small amounts of proteins are available (Kagami and Kamiya, 1992; also see Chapter 68 in this volume).

II. Method

A. Solutions

To prepare the malachite green reagent (MG), dissolve 6.3 g of Na_2-$MoO_4 \cdot 2H_2O$, 0.15 g of malachite green oxalate, and 0.25 g of Sterox SE (polyethyleneglycol mono-*p*-nonylphenyl ether, Monsanto, St. Louis, MO) in 500 ml of 0.7 *M* HCl. If Sterox SE is not available, Triton X-100 can be used instead, although the resulting solution will form precipitates more easily. Add to this solution 13 ml (15.3 g) of 35% (w/w) HCl; leave the mixture standing for 24 hours, and filter it through a sintered-glass filter. This solution can be stored in a refrigerator for at least a month. Precipitates may form during storage. After prolonged storage, the solution should be filtered again before use.

The other solutions required are 0.6 *M* perchloric acid (PCA) and 34% (w/v) sodium citrate.

B. Procedure

1. Suspend the sample to be measured in 1 ml of appropriate buffer.
2. Place the sample in a water bath set at the desired temperature (usually 25°C).
3. Start ATP hydrolysis by quickly adding 0.005 vol of 100 m*M* ATP to the sample with stirring. Also prepare a reference sample to which no ATP is added.
4. After enough time for sufficient hydrolysis to have occurred (usually 20 minutes in our experiments), add 1 ml of ice-cold 0.6 *M* PCA, vortex the mixture, and store on ice. Add ATP to the reference sample after adding PCA.
5. Centrifuge the mixture at 1500*g* for 1 minute and save the supernate.
6. Mix 1 ml of the supernate with 1 ml of MG, vortex 10–15 seconds, add 100 μl of 34% sodium citrate, and vortex.
7. Incubate the mixture at 25°C for 10–15 minutes.
8. Measure absorbance at 650 nm of each sample against that of the reference sample.

C. Notes

1. Because commercial sodium citrate is often contaminated with trace amounts of phosphate, it is advisable to prepare it from citric acid and NaOH, rather than purchase it. Dissolve 22.2 g of citric acid in a small amount of distilled and deionized water (DDW), add 50 ml of 6 *M* NaOH while cooling the solution, adjust the pH to 8.0 with NaOH, and bring the total volume to 100 ml with DDW.

2. Use disposable test tubes. Use of tubes that have been washed with detergent and distilled water often results in a large scatter in the data. When disposable test tubes cannot be used, use test tubes that have been rinsed with 0.5 *N* HCl and DDW.

3. As long as the ATP concentration is below 1 m*M*, nonenzymatic degradation of ATP in the coloring mixture is negligible and thus the optical density may be read any time between 10 and 15 minutes after the onset of the reaction. With higher concentrations of ATP, keep the time for the coloring reaction strictly constant (e.g., 12 minutes).

4. Trichloroacetate (TCA) cannot be used in place of PCA because TCA will form precipitates when mixed with malachite green.

5. Glass cuvettes used in this assay should be washed thoroughly with ethanol.

6. The calibration curve for phosphate determination should be drawn using 2–30 μM KH_2PO_4. Absorbance at 650 nm should be about 0.55 for 10 μM phosphate. Because the calibration curve is not linear beyond 10 μM phosphate, optical density should be measured at various concentrations of phosphate.

References

Anner, B., and Moosmayer, M. (1975). Rapid determination of inorganic phosphate in biological systems by a highly sensitive photometric method. *Anal. Biochem.* **65,** 305–309.

Fiske, C. H., and Subbarow, Y. (1925). The colorimetric determination of phosphorus. *J. Biol. Chem.* **66,** 375–400.

Harada, Y., Sakurada, K., Aoki, T., Thomas, D. D., and Yanagida, T. (1990). Mechanochemical coupling in actomyosin energy transduction studied by *in vitro* movement assay. *J. Mol. Biol.* **216,** 49–68.

Itaya, K., and Ui, M. (1966). A new micromethod for the colorimetric determination of inorganic phosphate. *Clin. Chim. Acta* **14,** 361–366.

Kagami, O., and Kamiya, R. (1992). Translocation and rotation of microtubules caused by mutliple species of *Chlamydomonas* inner-arm dynein. *J. Cell Sci.* **103,** 653–664.

Kodama, T., Fukui, K., and Kometani, K. (1986). The initial phosphate burst in ATP hydrolysis by myosin and subfragment-1 as studied by a modified malachite green method for determination of inorganic phosphate. *J. Biochem.* (*Tokyo*) **99,** 1465–1472.

Lanzetta, P. A., Alvarez, L. J., Reinac, P. S., and Candia, O. A. (1979). An improved assay for nanomole amounts of inorganic phosphate. *Anal. Biochem.* **100,** 95–97.

Ohno, T., and Kodama, T. (1991). Kinetic of adenosine triphosphate hydrolysis by shortening myofibrils from rabbit psoas muscle. *J. Physiol.* **441,** 685–702.

Taussky, H. H., and Shorr, E. (1953). A microcolorimetric method for the determination of inorganic phosphorus. *J. Biol. Chem.* **202,** 675–685.

Vale, R. D., Soll, D. R., and Gibbons, I. R. (1989). One-dimensional diffusion of microtubules bound to flagellar dynein. *Cell* **59,** 915–925.

PART IV

Structure

CHAPTER 23

Immunoelectron Microscopy

Karl A. Johnson
Department of Molecular, Cell, and Developmental Biology
Haverford College
Haverford, Pennsylvania 19041

I. Introduction

Immunoelectron microscopy is a powerful tool for the ultrastructural localization of flagellar/ciliary components. The purpose of this chapter is to provide a practical introduction to the technique. For additional background, there are several excellent reviews on immunolocalization (Bendayan, 1993; Roth, 1989; Stirling, 1990).

II. Primary Antibodies

As polyclonal antibodies generally recognize multiple antigenic determinants per target molecule, they generally produce better signals than monoclonal antibodies. Antibody specificity is critical and should be evaluated by immunoblot analysis; because immunoblot analyses are often performed at 10- to 1000-fold greater dilutions than the corresponding immunolocalization experiments, nonspecific cross-reactions, due to high antibody concentration, may be significant. Affinity purification of polyclonal antisera, while not essential, is recommended (see Pringle *et al.*, 1989; Smith and Fisher, 1984). Affinity purifications from protein blots are relatively straightforward, but these procedures may yield antibodies that bind well to denatured antigen but poorly recognize the protein in its folded state. In this respect, affinity purification from native protein bound to columns may be superior (see Pringle *et al.*, 1989; Williams *et al.*, 1989). Prior to investing time in electron microscopic immunolocalization, it is important to test the antibodies using immunofluorescence microscopy (see Chapters 24 and 64).

III. Specimen Preparation

A. Embedding and Thin Sectioning

A traditional approach in immunoelectron microscopy is to localize material on thin sections of fixed, epoxy resin-embedded material. Fixation is usually kept to the absolute minimum required to preserve satisfactory cellular ultrastructure. Often, paraformaldehyde alone or in combination with a low concentration (0.1–0.5%) of glutaraldehyde is used, as extensive glutaraldehyde fixation is usually associated with loss of antigenicity (Bendayan *et al.*, 1987). Osmication is avoided as it completely masks most antigens. Material is embedded in a hydrophilic resin such as L. R. White (London Resin Co., Ltd., Hampshire England) or Lowicryl K4M (Chemische Werke Lowi GmbH, Waldkraiburg, FRG) and, following polymerization, is thin-sectioned and processed for labeling.

B. Embedment of *Chlamydomonas* in L.R. White

1. Solutions and Materials

Chlamydomonas cells
Medium MI (Sager and Granick, 1953)
2% Glutaraldehyde in medium MI (adjusted to pH 7.0)
2% Low-gelling-temperature agar (FMC, Rockland, ME; previously heated into solution and cooled to 37°C)
75% ethanol (in water)
L. R. White resin
Ice bucket with ice
Gelatin capsules
37°C water bath
50°C oven

2. Procedure

1. Collect cells from an actively growing culture by centrifugation.
2. Gently resuspend in 10–20 ml growth medium MI using a large-bore pipet.
3. Fix by adding an equal volume of 2% glutaraldehyde in MI; place on ice for 30 minutes.
4. Collect the fixed cells by low-speed centrifugation and resuspend in a small volume of water; transfer into a microfuge tube.
5. Briefly warm the cells in a 37°C water bath and mix with an equal volume of 2% low-gelling-temperature agar. Quickly centrifuge to concentrate the cells and place the tube on ice to harden the agar.
6. Cut the pellet free from the tube with a fresh razor blade or scalpel and section it into a number of smaller pieces. Inclusion of cells in agar is optional; however, it facilitates the handling of material by eliminating the need to subject cells to repeated rounds of centrifugation/resuspension during dehydration and embedding.
7. Transfer chunks of agarose-encapsulated cells directly into 75% ethanol for 30 minutes at 4°C; repeat using fresh 75% ethanol.
8. If desired, stain the cells en bloc by including 1% uranyl acetate in the first 75% ethanol step. In this case, en bloc stain for 1 hour at 4°C and then take through two 75% ethanol washes (step 7). This does not appear to affect antigenicity adversely (as can fixation) and improves the structural preservation of the cells (Berryman and Rodewald, 1990). Membrane profiles especially benefit from the inclusion of this step, which also enhances the overall contrast of the sections when viewed in the electron microscope (EM).
9. Transfer the chunks of material directly from 75% ethanol into L. R. White resin. Infiltrate with several changes of L. R. White, 1 hour each, followed by overnight at 4°C.
10. Polymerize for 24 hours in gelatin capsules in a 50°C oven. Polyethylene capsules are not satisfactory because they are permeable to oxygen, which inhibits L. R. White polymerization.

Postembedment labeling has several drawbacks, the most significant of which is that only protein epitopes directly exposed on the section surface are available for labeling. For example, a microtubule may be clearly visible within the depth of the section, but cannot be detected unless it is exposed at the surface. Several different methodologies, such as reversible embedments and frozen thin sections, have been developed that should expose more material for labeling. It is often the case that these alternative, labor-intensive strategies result in only about a twofold increase in labeling (Kellenberger *et al.*, 1987). Frozen thin sections, however, do give good membrane preservation (Tokuyasu, 1986) and may be useful for the analysis of flagellar membrane-associated antigens.

C. Flagellar Whole Mounts

In working with flagellar/ciliary axonemes, it is possible to avoid lengthy specimen preparation by simply isolating axonemes, settling them onto electron microscope grids, and immunolabeling them. These whole mounts, in the style of negatively stained preparations, are easily prepared from fresh material. Perhaps most importantly, freed from the masking effects of heavy fixation, dehydration, and embedment, axonemal whole mounts show excellent labeling densities. For example, Fig. 1 shows part of a negatively stained, splayed axoneme immunogold labeled with affinity-purified polyclonal antibodies against radial spoke protein 3 and detected with secondary antibodies conjugated to 10-nm gold particles; note the gold particles associated with each radial spoke pair (arrows). Caution, however, is warranted against overinterpretation of the exact position of the gold particles; as a single IgG molecule is approximately 10 nm long, the primary–secondary links can extend up to 20–30 nm from the antigenic site. The following is a method for the preparation of *Chlamydomonas* axonemal whole mounts modified from Dentler and Rosenbaum (1977).

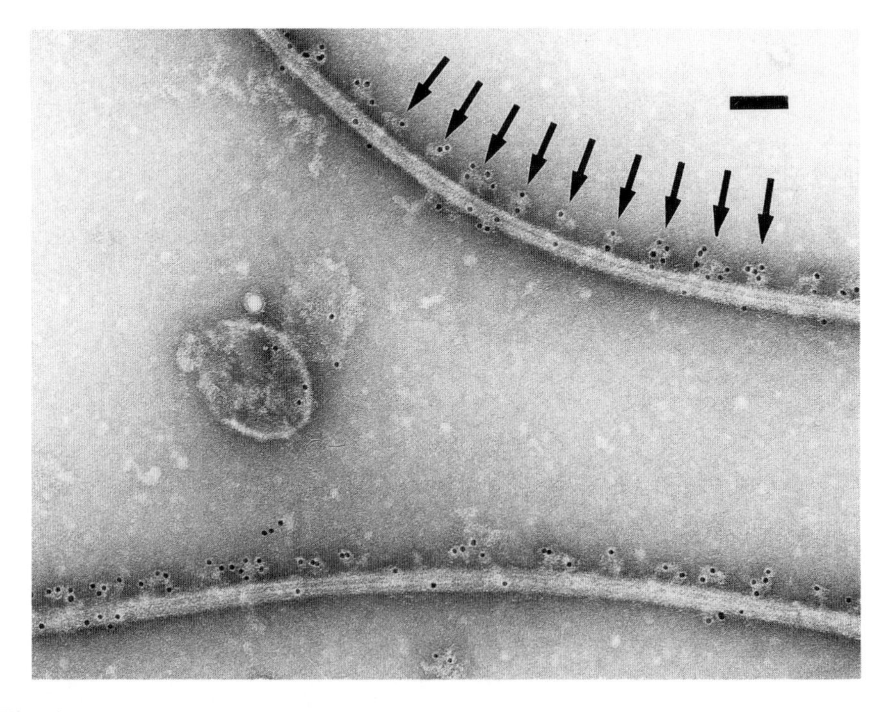

Fig. 1 Partial view of two outer-doublet microtubules of a splayed *Chlamydomonas* axoneme immunolabeled with an affinity-purified polyclonal antiserum against radial spoke protein 3 (Williams *et al.*, 1989), detected with 10-nm gold conjugated to protein A and negatively stained (K. Johnson, unpublished work). Note the gold particles associated with the row of paired radial spoke projections (arrows) that extend from one side of the A tubule of each outer doublet. Bar = 0.1 μm.

1. Materials and Solutions

Carbon-coated, Collodion- or Formvar-covered, nickel EM grids
0.1% w/v poly-L-lysine (Sigma Chemical Co., St. Louis, MO)
Chlamydomonas cells
HMDEK: 30 m*M* 4-(2-hydroxyethyl)-1-piperazineethanesulfonic acid (Hepes), 5 m*M* $MgSO_4$, 1 m*M* DTT, 0.5 m*M* Na_2EDTA, 25 m*M* KCl, pH 7.4 (Witman, 1986)
HMDEK + 1% Nonidet P-40
10 m*M* Hepes, pH 7.4
Anticapillary tweezers

2. Procedure

1. Prepare carbon-coated, Collodion- or Formvar-covered nickel EM grids. Nickel grids are preferred over copper grids as they are less reactive with salt solutions, but copper grids can be used if necessary. Nickel grids are more durable than copper but can be difficult to pick up if one's tweezers become magnetized.

2. Grasping the grid in a pair of anticapillary tweezers, place a droplet of poly-L-lysine solution onto the support surface. Leave the droplet in place for 5 minutes and then rinse with several drops of distilled water. Using a piece of filter paper, wick off most of the excess liquid and allow to air dry.

3. Place a droplet of *Chlamydomonas* cells (in growth medium) onto the support surface and allow the cells to settle onto the grid for 5 minutes.

4. Holding the grid at a slight angle over a beaker, replace the droplet of medium with a droplet of 1% Nonidet-40 in HMDEK. Be careful not to wet the reverse side of the grid. The detergent solution causes flagellar detachment and demembranation in a single step.

5. After a 5-minute incubation, rinse the grid surface with drops of HMDEK until the green of the adherent cell bodies is washed away. At this point a sample grid can be negatively stained to determine if a sufficient number of flagellar axonemes are present on the support surface. Often a small number of the axonemes will be splayed apart, exposing interior axonemal structures such as the radial spokes, inner dynein arms, and central pair apparatus. The number of splayed axonemes can be dramatically increased (to 100%) by adding 1 m*M* ATP to the solution (Dentler and Rosenbaum, 1977) and allowing the grid to dry briefly prior to antibody labeling (Johnson *et al.*, 1994). This can be very important in making structures within the center of the axoneme accessible for labeling. For example, a kinesin-like protein localized along the central pair appatatus could be detected where the axoneme had splayed apart, but not where the axonemal microtubules had remained together in a bundle (Johnson *et al.*, 1994).

6. If desired, fix the axonemes by inverting the grid onto a droplet of 4%

paraformaldehyde, or 1% glutaraldehyde, in 10 m*M* Hepes, pH 7.4, for 15 minutes. If the back of the grid remained dry during steps 4 and 5, the grid will float on the surface of the droplet. Although not essential, fixation preserves the fine structure of the axoneme, especially that of the fragile central pair apparatus.

7. Wash the grids with several drops of 10 m*M* Hepes to dilute unreacted aldehydes and transfer onto blocking solution (for immunolabeling). It may be useful to allow certain grids to air-dry briefly. Drying destroys some of the fine details of the axoneme but can reveal antigenic epitopes that otherwise remain hidden and undetectable (Johnson and Rosenbaum, 1989), perhaps by causing physical collapse of microtubules and other structures.

D. Tips for Incubating EM Grids on Various Solutions

Successful immunolabeling of material on EM grids requires that the same grid be handled many times without damage. One simple way to accomplish this is to use a modified inoculating loop to move grids from solution to solution. The best design is a simple wire loop bent into a "question mark" shape, just larger in diameter than a grid. The loop is brought up under the grid as it is floating inverted on the surface of the solution, and the grid is lifted away with minimal solution carryover. The presence of a small drop of solution below the grid surface is desirable as it prevents surface tension effects and drying during transfer between solutions. Keeping the grid constantly wetted is particularly important in avoiding high background levels, as it is almost impossible to wash away excess antibodies that dry onto the background surface of the grid.

An economical way to set up an immunolabeling experiment is to line a plastic box with a sheet of Parafilm. A few wet paper towels in a box with a snug-fitting lid provide a humidified chamber: grids can be incubated on small drops (20–25 μl) on the Parafilm and moved from drop to drop with an inoculation loop. As long as the back of the grid remains dry, it floats on the center of the drop. An alternative method useful for larger volumes used in multiple wash steps is to fill the wells of a 96-well multititer plate; several grids can be readily transferred well to well in a line through 6 or 12 washes with minimal chance of inadvertently mixing up the grids.

IV. Immunolabeling Technique

A. Choice of Gold Marker

A wide variety of gold conjugates are available from Amersham (Arlington Heights, Il); Biocell (Ted Pella, Reading, CA); E-Y Laboratories (San Mateo, CA); and Zymed Laboratories (South San Francisco, CA). One can use specific secondary antibody conjugates or protein A and protein G conjugates, which are broader in specificity but show different binding affinities based on primary host and immunoglobulin class. More critical is the size of the gold particle: small sizes (5 nm) are very difficult to see at lower working magnifications, and labeling efficiency decreases as larger particles (>15 nm) are used (Ghitescu and Bendayan, 1990). Ten-nanometer gold particles seem to be the best compromise for signal density versus visibility.

Alternatively, 1-nm gold probes have excellent penetration and labeling characteristics and can be enlarged by silver enhancement to 5–15 nm in diameter prior to observation (Burry *et al.*, 1992). Enhancement does require additional grid manipulation, however, and is highly sensitive to developing conditions and contaminants. Signals also can be amplified using protein A–gold conjugates, followed by anti-protein A antibodies and then protein A–gold conjugates again (Bendayan and Duhr, 1986). This can produce three- to fivefold denser labeling but can result in a loss of spatial resolution. Although this method gives a stronger signal than nonamplified detection procedures, it does not significantly enhance the efficiency of detection of nonabundant antigens.

B. Immunogold Labeling of Specimens on EM Grids

1. Materials and Solutions

TBSB: 10 m*M* Tris, 0.9% NaCl, 1% bovine serum albumin, pH 7.4

Blocking buffer: 10% (w/v) nonfat dry milk (Carnation, Los Angeles, CA) in TBSB

Primary antibody, diluted in TBSB

Secondary antibody conjugated to 10-nm gold particles

2. Procedure

1. Block grids bearing material with blocking buffer for 15–30 minutes.
2. Transfer the grids onto drops of primary antibody, diluted in TBSB, and incubate for several hours to overnight. A recommended starting dilution is 10× the antibody concentration that gave a good immunofluorescence signal; it is always best to see a signal and then dilute, rather than to spend several trials at antibody concentrations too low to obtain a reasonable signal.
3. Wash the grids extensively by transferring the grids across 6–12 changes of TBSB, 5 minutes each.
4. Following washing, incubate the grids on a 1:25 to 1:100 dilution of secondary antibody conjugated to 10-nm gold particles for between 30 minutes and 2 hours.
5. Wash again on TBSB and then on distilled water. It is advisable to examine some grids without poststaining, as the gold particles can be difficult to see following contrasting with heavy metals.

6. Following extensive washing on distilled water, poststain grids in conventional ways, i.e., positive staining with uranyl acetate and lead citrate for thin sections or negative staining with uranyl acetate for axonemal whole mounts.

C. Interpretation of Experimental Results

It is extremely useful to examine some of the specimens without poststaining. At first, it can be difficult to interpret the cell/tissue morphology because the only contrast of the specimens is essentially that provided by the inherent biological material itself. Inclusion of the uranyl acetate en bloc staining step in the tissue preparation can help in this regard. Viewing the specimen at lower accelerating voltages (40 kV), using less energetic electrons that are more easily deflected, also improves the image contrast. In the absence of poststaining, the gold particles stand out very well, making it easy to evaluate both levels of specific staining of structures, as well as the levels of background labeling. Again, the initial screening of antibodies by immunofluorescence can be informative in indicating where antigen should be detected.

Careful evaluation of nonspecific background labeling is also facilitated by the examination of nonstained material. For example, levels of gold particles observed on the plastic sections/support surfaces immediately adjacent to material should be low to nonexistent. Control experiments run without primary antibody are also essential to demonstrate the specificity of labeling. An ideal control is provided by performing identical localizations on mutant cells that lack the specific gene product being localized, if such a null is available. The information provided by these observations should be used to modify the blocking time or number of washing steps in subsequent experiments.

Once satisfactory labeling is achieved, immunolabeled material can be stained by conventional techniques. It is important to stress that if the distilled water washing is insufficient, salts picked up during the TBS incubations cause extensive heavy metal precipitation. This can be extremely problematic, though with longer washing times success will be at hand. Postimmunolocalization heavy metal staining can be important in demonstrating the association of label with specific structures; however, as the overall contrast of the material is increased, it often becomes difficult to visualize the gold particles, particularly at lower magnifications. It is often useful, therefore, to present both poststained and non-poststained images for evaluation.

With immunolocalization, patience and perseverence are most important. For example, no one fixation procedure is optimal for all different types of cells or for all different types of antigens (Bendayan *et al.*, 1987). Several variations should be tried to optimize labeling. Perhaps the most valuable advice returns to the evaluation of antisera in immunofluorescence. A positive signal provides a guide, indicating where antigen is localized and suggesting that the antigen is concentrated enough to be detected at the EM level. Often the most successful studies using immunoelectron localization techniques have localized relatively abundant antigens that are concentrated in specific subcellular compartments or structures. A weak or nonexistent immunofluorescence result indicates that either the antigen itself is rare or the antiseurm is not efficiently detecting antigen. In the latter case, it may be possible to raise a better antiserum. When immunoelectron techniques work, they can provide great detail about the localization of particular antigens within the framework of the cell.

Acknowledgments

I thank Bill Dentler, Ursula Goodenough, Mark Mooseker, Joel Rosenbaum, George Witman, and others for numerous, helpful conversations over the years regarding EM technique. I also thank the National Institutes of Health for the support of a postdoctoral fellowship (GM13758) while I was in the laboratory of Joel Rosenbaum (Yale University).

References

Bendayan, M. (1993). Immunoelectron microscopy. *In* "Practical Electron Microscopy" (E. Hunter, ed.), pp. 71–92. Cambridge University Press, Cambridge, United Kingdom.

Bendayan, M., and Duhr, M.-A. (1986). Modification of the protein A-gold immunocytochemical technique for the enhancement of its efficiency. *J. Histochem. Cytochem.* **34,** 569–575.

Bendayan, M., Nanci, A., and Kan, F. (1987). Effect of tissue processing on colloidal gold cytochemistry. *J. Histochem. Cytochem.* **35,** 983–996.

Berryman, M., and Rodewald, R. (1990). An enhanced method for post-embedding immunocytochemical staining which preserves cell membranes. *J. Histochem. Cytochem.* **38,** 159–170.

Burry, R., Vandre, D., and Hayes, D. (1992). Silver enhancement of gold antibody probes in pre-embedding electron microscopic immunocytochemistry. *J. Histochem. Cytochem.* **40,** 1849–1856.

Dentler, W., and Rosenbaum, J. (1977). Flagellar elongation and shortening in *Chlamydomonas* III. Structures attached to the tips of flagellar microtubules and their relationship to the directionality of flagellar microtubule assembly. *J. Cell Biol.* **74,** 747–759.

Ghitescu, L., and Bendayan, M. (1990). Immunolabeling efficiency of protein A-gold complexes. *J. Histochem. Cytochem.* **38,** 1523–1530.

Johnson, K., and Rosenbaum, J. (1989). Accessibility of the acetylated epitope of alpha-tubulin on the *Chlamydomonas* axoneme. *J. Cell Biol.* **111,** 172a.

Johnson, K., and Rosenbaum, J. (1990). The basal bodies of *Chlamydomonas reinhardtii* do not contain immunologically detectable DNA. *Cell* **62,** 615–619.

Johnson, K., and Rosenbaum, J. (1992). Polarity of flagellar assembly in *Chlamydomonas. J. Cell Biol.* **119,** 1605–1611.

Johnson, K., Haas, M., and Rosenbaum, J. (1994). Localization of a kinesin-related protein to the central pair apparatus of the *Chlamydomonas reinhardtii* flagellum. *J. Cell Sci.* **107,** 1551–1556.

Kellenberger, E., Durrenberger, M., Villiger, W., Carlemalm, E., and Wurtz, M. (1987). The efficiency of immunolabel on Lowicryl sections compared to theoretical predictions. *J. Histochem. Cytochem.* **35,** 959–969.

Pringle, J., Preston, R., Adams, A., Sterns, T., Drubin, D., Haarer, B., and Jones, E. (1989). Fluorescence microscopy methods in yeast. *Methods Cell Biol.* **31,** 357–435.

Roth, J. (1989). Postembedding labeling on Lowicryl K4M tissue sections: detection and modification of cellular components. *Methods Cell Biol.* **31,** 513–551.

Sager, R., and Granick, S. (1953). Nutritional studies with *Chlamydomonas reinhardti. Ann. N. Y. Acad. Sci.* **56,** 831–838.

Smith, D., and Fisher, P. (1984). Identification, developmental regulation, and response to heat shock of two antigenically related forms of a major nuclear envelope protein in *Drosophila* embryos: application of an improved method for affinity purification of antibodies using polypeptides immobilized on nitrocellulose blots. *J. Cell Biol.* **99,** 20–28.

Stirling, J. (1990). Immuno- and affinity probes for electron microscopy: a review of labeling and preparation techniques. *J. Histochem. Cytochem.* **38,** 145–157.

Tokuyasu, K. (1986). Application of cryoultramicrotomy to immunocytochemistry. *J. Microsc.* **143,** 139–149.

Williams, B., Velleca, M., Curry, A., and Rosenbaum, J. (1989). Molecular cloning and sequence analysis of the *Chlamydomonas* gene coding for radial spoke protein 3: flagellar mutation pf-14 is an ochre allele. *J. Cell Biol.* **109,** 235–245.

Witman, G. (1986). Isolation of *Chlamydomonas* flagella and flagellar axonemes. *Methods Enzymol.* **134,** 280–290.

CHAPTER 24

Immunofluorescence Microscopy of Cilia and Flagella

M. A. Sanders[*] and J. L. Salisbury[†]

[*]Imaging Center
Genetics and Cell Biology
College of Biological Sciences
St. Paul, Minnesota 55108
[†]Laboratory for Cell Biology
Department of Biochemistry and Molecular Biology
Mayo Clinic Foundation
Rochester, Minnesota 55905

I. Introduction

There are a number of excellent articles on fluorescence microscopy and immunofluorescence methods (Asai, 1993; Beltz and Burd, 1989; Giloh and Sedat, 1982; Matsumoto, 1993; Wang and Taylor, 1989). Our aim in this chapter is to review the application of methods for immunofluorescence staining of motile ciliated and flagellated cells. We present several fixations and sample preparation protocols that we routinely employ when we are using a previously untested antibody or when we are studying a particular cell type for the first time. Also, we present a general strategy for the simultaneous localization of two separate antigens, as well as nuclear DNA staining in the same cells. Although at least one of these methods generally yields successful results, modifications of the procedures are frequently found though empirical trial and error to result in improved localization.

II. General Considerations

The obvious goal of any immunolocalization study is to observe and record the distribution and organization of a particular cellular component in a manner that most closely reflects its distribution *in vivo*. Results should be interpreted cautiously, as preparation of the specimen typically involves harsh manipulations of varying degrees of severity. Chemical fixation, with aldehydes or treatment with cold methanol and/or acetone, is commonly employed to immobilize the antigen of interest. Because antibody probes are large molecules, preparative steps typically involve cell permeabilization to allow access to structures of interest. This step may also extract components that otherwise could contribute to unwanted levels of background labeling, possibly reducing "soluble" pools of the antigen itself. To aid in subsequent manipulations, cells are typically immobilized onto a coverglass. This process may affect cell shape, and can result in a flatter specimen that is more amenable to microscopy; however, changes in cell shape can also alter the organization of the protein of interest. Autofluorescence of the specimen, nonspecific labeling by immune serum and secondary antibodies, and specimen thickness also contribute potential problems. Finally, the selective bias of the microscopist to photograph aesthetic or preconceived results can contribute to interpretation prejudice.

III. Reagents and Solutions

Highest-purity reagents and deionized water should be used throughout. Adjust the pH to the indicated value at room temperature, and filter all solutions (except methanol) through a 0.2-μm filter (Millipore) prior to use.

Microtubule stabilizing buffer (MTSB): Combine 3 m*M* ethylene glycol bis(β-aminoethyl ether)-*N,N'*-tetraacetic acid (EGTA), 1 m*M* $MgSO_4$, 25 m*M* KCl, and 50 m*M* Na-1,4 piperazinediethanesulfonic acid (Pipes), pH 7.2.

MTSB–Triton: Combine MTSB and 0.05% Triton X-100.

Formaldehyde fixative: Formaldehyde (30% stock solution) is made fresh by mixing 3 g paraformaldehyde with ~8 ml of deionized water and warming the solution to 70°C in a water bath located in an exhausted hood. Then add 1 ml of 5 *M* NaOH dropwise until the solution clears. Cool to room temperature and adjust the final volume to 10 ml. Add 3 ml of the formaldehyde stock to 27 ml of MTSB or MTSB–Triton. If localization of cytoplasmic microtubules is desired we use the MTSB–Triton formulation.

Formaldehyde/glutaraldehyde fixative: Add glutaraldehyde (8% stock solution, EM grade, Electron Microscopy Sciences, Fort Washington, PA) to the above fixative to a final concentration of 0.1%.

MTSB–NH_4Cl reducing buffer: Combine MTSB and 50 m*M* NH_4Cl.

Phosphate buffer (PB, 10 m*M*): Add 28 ml KH_2PO_4 and 72 ml K_2HPO_4 to 3900 ml H_2O; check pH 7.2.

Blocking buffer: Combine 10 m*M* phosphate buffer, pH 7.2, 5% normal goat serum (Gibco BRL, Gaithersberg, MD), 5% glycerol, 1% cold water fish gelatin (Sigma), and 0.04% Na-azide. Aliquot and store at −20°C.

Polyethylenimine: Dilute 0.1% polyethylenimine (Sigma) in water.

Secondary antibodies: Commercially available secondary antibodies conjugated to rhodamine or fluorescein can be obtained from a number of sources. We routinely use Cappel products (Organon Tecknika, Durham, NC) goat anti-mouse fluorescein isothiocyanate (FITC)-conjugated IgG (Catalog No. 55493) or rhodamine-conjugated IgG (Catalog No. 55527) as the secondary for mouse monoclonal primary antibodies, and goat anti-rabbit FITC-conjugated IgG (Catalog No. 55494) or rhodamine-conjugated IgG (Catalog No. 55666) as the secondary for rabbit serum primary antibodies. We find consistent results with secondary antibodies diluted 1 : 400 in blocking buffer; however, working dilution should be determined empirically with each new lot or vendor.

IV. Immunofluorescence of Ciliated and Flagellated Cells

A. Immobilization of Cells

We typically immobilize living or fixed cells onto 12-mm No. 1 round coverslips (Fisher Scientific, Catalog No. 12-545-80) or eight-well epoxy-coated microslides (Carlson Scientific, Peotone, IL) for ease of handling in subsequent steps. The coverslips must first be scrupulously cleaned in 1% 7X detergent, thoroughly rinsed with deionized water, treated with 5 m*M* ethylenediaminetetraacetic acid (EDTA) for 5–10 minutes, and finally rinsed in deionized water and air-dried. Multiple cover slips can be processed using a coverslip rack (Thomas Scientific, Catalog No. 8542-E30). Place one drop of 0.1% polyethylenimine onto the center of each cleaned cover slip; after 30 seconds, rinse with deionized water and air-dry. Use the polyethylenimine-treated coverslips within 1 hour. Place a drop of cells (washed in fresh culture medium or, for sperm, in phosphate-buffered Tyrode's salts, Sigma) and allow the cells to settle for 10 minutes. Blot excess suspension and continue immediately to one of the fixation protocols listed below. Care must be taken to note the appropriate side of the coverslip for further processing.

B. Fixation

Two fixation procedures are listed below. The first, −20°C methanol, works well for a majority of the cell types and antibodies that we have tried, and, therefore, it is the general method of choice. Formaldehyde or the combined formaldehyde/glutaraldehyde fixatives work well for particular cell types and epitope/antibody complexes.

1. Cold Methanol Fixation

This method simultaneously fixes and permeabilizes the sample. Precool absolute methanol contained in a Coplin jar to −20°C in a freezer. Fix and permeabilize the specimen by submersing the coverslips in the −20°C methanol for 10 minutes. Remove the coverslip, blot excess methanol, air-dry, and rehydrate in phosphate-buffered saline (PBS), three changes, 5 minutes each. Continue procedure at the antibody labeling step (Section IV,F,1).

2. Aldehyde Fixation

Prewarm the fixative solution to culture temperature. Submerse the coverslip in the appropriate aldehyde fixative and allow the sample to fix for 30 minutes. This fixation step may contain Triton X-100 for simultaneous permeabilization of the cells.

C. Permeabilization

If the sample has not previously been permeabilized, do so now by submersing the coverslip in MTSB–Triton three times for 5 minutes each.

D. Reduction of Free Aldehydes

If the formaldehyde/glutaraldehyde fixative was used, reduce free aldehyde groups by submersing the coverslips in MTSB–NH_4Cl reducing buffer, three changes, 5 minutes each, followed by MTSB, three changes, 5 minutes each. Transfer the coverslips through an increasing series of PB : MTSB (33%, 66%, 100% PB).

E. Blocking

Block specimens by incubating in blocking buffer for 30 minutes at room temperature.

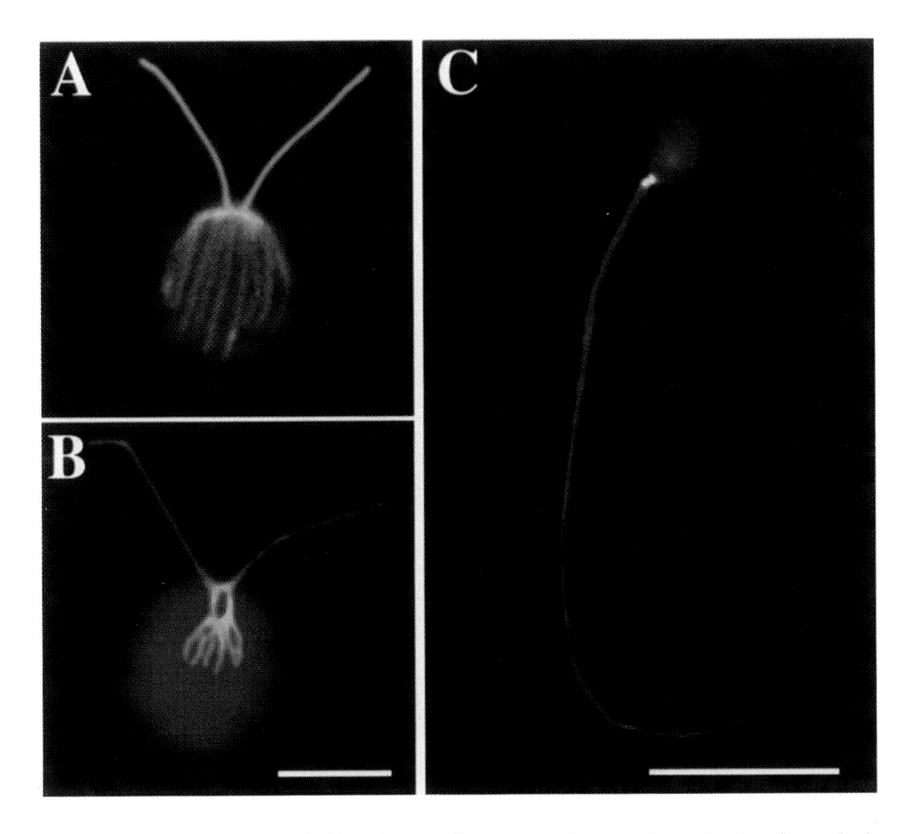

Fig. 1 Three examples of indirect immunofluorescence images obtained using the methods described here. (A) Microtubule-based cytoskeleton of *Chlamydomonas reinhardtii* stained with a 1:400 dilution of an anti-α-tubulin monoclonal antibody (Sigma Chemicals, St. Louis, MO, Cat. No. T9026) and fluorescein-conjugated secondary antibody. The two flagella and the cytoplasmic microtubules stain brightly green. (B) Centrin distribution in *C. reinhardtii* stained with a 1:5000 dilution of an anti-centrin monoclonal antibody (20H5) produced in our laboratory and fluorescein-conjugated secondary antibody. The flagella contain centrin and appear green, and cytoplasmic centrin-based fibers appear yellow on a red background of chlorophyll autofluorescence. (Reprinted with permission of U.W. Goodenough (1992) *Cell* **70**, 533–538.) (C) Human sperm cell triple labeled with a 1:400 dilution of an anti-α-tubulin monoclonal antibody (Sigma Chemicals, Cat. No. T9026) and rhodamine-conjugated secondary antibody, a 1:5000 dilution of an anti-centrin serum (26/14-1) produced in our laboratory, and DAPI. The flagellum stains brightly red (tubulin), the basal body and centriole stain brightly yellow (centrin), and the nucleus stains blue (DNA). The bar in (B) is 5 μm and in (C) is 10 μm.

F. Labeling

1. Antibody Labeling

Dilute the primary antibody of choice in blocking buffer and filter (0.2 μm). Typically, the appropriate primary antibody concentration must be determined empirically. The range for dilution of high-titer sera and monoclonal antibodies is often about 1 : 500 to 1 : 10,000 or greater. Higher serum concentrations may result in unwanted background levels of fluorescence. It is desirable to use preimmune serum produced from the same animal that produced the immune serum, if available, for use as a control treatment to assess the level of nonspecific background labeling. Specific antibodies, affinity purified against the antigen of interest, are also useful for reducing background staining seen in certain immune sera or for particular samples. Incubate the sample in diluted primary antibody for 1 to 4 hours at 37°C, or overnight at 4°C. To avoid dehydration, we typically invert the coverslip over a small drop (~20 μl) of diluted antibody placed on a small piece of Parafilm, and contained in a Petri dish with a moistened filter paper on the inside lid. Following the incubation, carefully remove the coverslip, blot excess antibody solution, and immediately wash in three changes of PBS, 5 minutes each.

2. Secondary Antibody

Incubate, as above, in the appropriate secondary antibody diluted 1 : 400 in blocking buffer, for 1–2 hours at 37°C. Carefully remove the coverslip, blot excess antibody solution, and immediately wash in three changes of PBS, 5 minutes each.

3. Double Labeling

For double labeling, in which two different antigens are to be localized in the same cells, repeat steps 1 and 2 using the second primary and appropriate secondary antibody preparations. It is necessary to use primary antibodies raised in different species (i.e., rabbit and mouse) and the appropriate secondary antibodies conjugated to distinct fluorochromes to achieve the desired discrimination between the two primary targets.

G. Mounting

The mounting medium used must be nonfluorescent, should resist evaporation, and should be formulated to reduce fading of the fluorochrome. A simple antifade mounting medium consisting of 2% *N*-propyl gallate (Sigma), 30% 0.1 *M* Tris buffer, pH 9, 70% glycerin (Giloh and Sedat, 1982), and 0.1 μg/ml 4′,6-diamidino-2-phenylindole hydrochloride (DAPI, Sigma) to stain DNA has proven effective and easy to use and store (aliquot and freeze at −20°C).

Following the final PBS wash in step F, 1, briefly rinse the coverslip in deionized water to remove excess salts. Place the inverted coverslip over a small drop of mounting medium on a clean microscope slide, seal the coverslip with nail polish, and allow the nail polish to dry thoroughly before placing the slide on the microscope stage.

H. Observation

An epifluorescent microscope equipped with the appropriate excitation and barrier filters for each of the fluorochromes used is required. The best result, in terms of image brightness, is generally obtained with the lowest magnification and highest-numerical-aperture oil immersion objective (i.e., 60X, 1.4 NA) necessary to image the structure of interest. For maximum detail and contrast we generally record black and white images on Hypertech film (Microfluor, Stony Brook, NY), exposed at ASA 800 for fluorescein and 3200 for rhodamine and developed in D-19 developer for 6 minutes at 20°C (Fig. 1, see color plate). For color we record images using a CCD video camera system (VI-470, Optronics Engineering, Goleta, CA) and Image I processing hardware and soft ware (Universal Imaging, West Chester, PA).

V. Troubleshooting and Variations

We have found that optimum fixation and success in labeling vary widely for different cell types and antibody probes. The buffer used to wash living cells before fixation and the fixative buffer itself have a dramatic effect on the final quality of labeling. For example, *Chlamydomonas* cells label best after fixation in Hepes buffer, pH 6.8; vertebrate tissue culture cells label best after fixation in Pipes buffer, pH 7.2; and sperm label best after fixation in phosphate buffer, pH 7.2. Also, divalent cations, particularly calcium, have a dramatic effect on the quality of label, in many cases affecting the stability of the structure under investigation. Although, one could try to determine a priori what the optimal divalent cation composition should be, we have just as often been surprised. The reasons for these differences in fixation and quality of label are not obvious. Therefore, the rule of thumb that we generally apply is to try several buffers, adjusting the pH to as near the culture pH as possible, and to fix samples in the presence of either millimolar Ca^{2+} or EGTA. While aldehyde fixation typically results in the best structural preservation, it may also dramatically affect access and the ability of the antibody to bind to the antigen under investigation.

References

Asai, D. (1993). Antibodies in cell biology. *In* "Methods in Cell Biology" p. 452. Academic Press, San Diego.

Beltz, B., and Burd, G. (1989). "Immunocytochemical Techniques, Principles and Practice" p. 181. Blackwell Scientific Publications, Cambridge, MA.

Giloh, H., and Sedat, J. (1982). Fluorescence microscopy: reduced photobleaching of rhodamine and fluorescein protein conjugates by n-propyl gallate. *Science* **217,** 1252–1255.

Matsumoto, B. (1993). Cell biological applications of confocal microscopy. *In* "Methods in Cell Biology" p. 380. Academic Press, San Diego.

Wang, Y.-L., and Taylor, D. (1989). Fluorescence microscopy of living cells in culture. Part A. Fluorescent analogs, labeling cells, and basic microscopy. *In* "Methods in Cell Biology" p. 333. Academic Press, San Diego.

CHAPTER 25

Scanning Transmission Electron Microscopic Analysis of the Isolated Dynein ATPase

Silvio P. Marchese-Ragona[*] and J. S. Wall[†]

[*]TopoMetrix Corporation
Santa Clara, California 95054
[†]Department of Biology
Brookhaven National Laboratory
Upton, New York 11973

I. Introduction

A large number of electron microscopic techniques have been used in the last 30 years to examine the ultrastructure of the isolated dynein arms obtained from a variety of species. The techniques used include metal shadowing (Gibbons and Rowe, 1965; Sale, 1983), rapid-freeze deep-etch techniques (Goodenough and Heuser, 1984; Sale *et al.,* 1985), negative staining (Johnson and Wall, 1983; Marchese-Ragona *et al.,* 1988; Marchese-Ragona and Johnson, 1988; Haiste *et al.,* 1988; King and Witman, 1990; Walczak *et al.,* 1993), and dedicated scanning transmission electron microscope (STEM) analysis (Johnson and Wall, 1983; Witman *et al.,* 1983; Marchese-Ragona *et al.,* 1987, 1988;

Marchese-Ragona and Johnson, 1988; Haiste *et al.*, 1988). Each of these has provided unique, useful, and, occasionally, controversial information concerning the ultrastructure of the dynein ATPase. All of the techniques mentioned are able to provide detailed ultrastructural information and all except STEM analysis require that the specimen be replicated or stained with heavy metal.

The Brookhaven STEM is unique among electron microscopes in that it is able to examine unfixed, uncoated, unstained material. The dynein molecules are rapidly frozen, freeze-dried, and imaged at $-150°C$ with extremely low electron doses to minimize beam damage. Furthermore, the Brookhaven STEM is equipped to determine the mass of individual particles, thus, not only can morphology and molecular weight be determined, but mass analysis allows one to unequivocally identify particles in unfamiliar orientations (Fig. 1).

Johnson and Wall (1983) were the first to use the STEM to analyze isolated outer-arm dynein from *Tetrahymena*. This study represented a landmark in cell motility. For the first time detailed structural information was obtained that could be used as a common basis for the interpretation of images of thin-sectioned and negatively stained axonemes. For a detailed review of STEM analysis of dynein see Marchese-Ragona and Johnson (1988) and Wall and Hainfeld (1988). Here we present the essential elements of specimen preparation that are required for obtaining the morphology and mass analysis of isolated

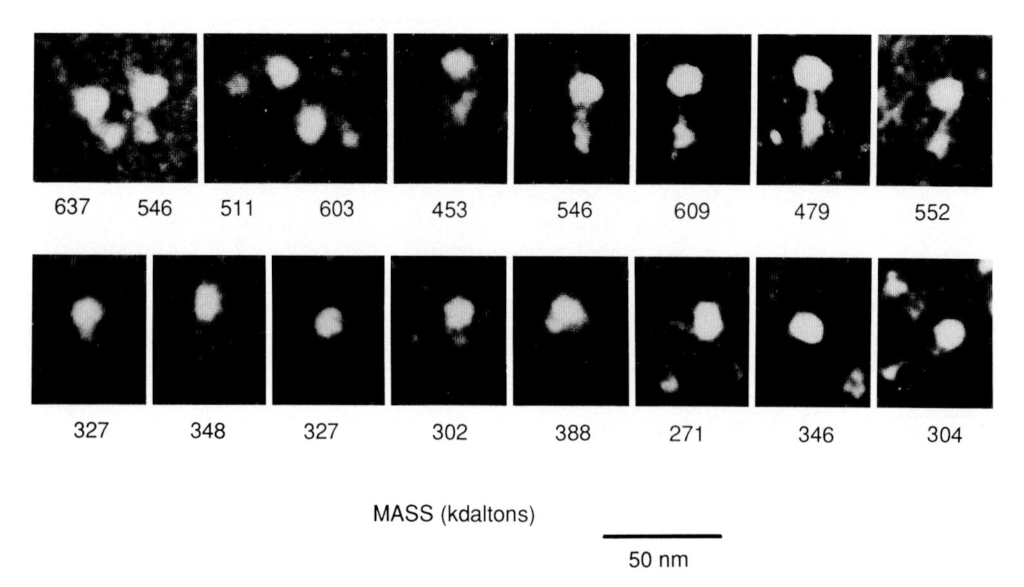

Fig. 1 Scanning transmission electron microscopy (STEM) of unstained, unfixed, 14S dynein revealed the presence of two distinct types of particles that could be distinguished upon the basis of morphology and mass analysis. All of the particles had globular heads and approximately half the particles had tails. The top row shows particles that exhibit tail-like structure, whereas the bottom row shows globular particles without tails. The mass of each particle is given in kdaltons beneath each particle.

dynein particles using the Brookhaven STEM. Similar procedures should be useful for the examination of any suitably sized protein or structure that can be obtained in high purity.

II. Method

A. Dynein Purity

Sucrose gradient-purified dynein should be passed over an A5M (Bio-Rad) column immediately before use to remove sucrose and any proteolytic products. Failure to fully remove the sucrose, proteolyzed particles, and their proteolytic fragments at this stage will result in erroneous mass results. Trace amounts of sucrose around the particles will give artificially high mass readings, whereas the presence of proteolyzed or degraded molecules will give an overall lower mean mass value with a larger standard deviation.

B. Sample Preparation

The dynein samples should be diluted to a concentration of 10–30 μg/ml in 10 m*M* 1,4-piperazinediethanesulfonic acid (Pipes), 0.8 m*M* $MgCl_2$, pH 7.0, and applied to a carbon film by injection of 2.5 μl of sample to 2.5 μl droplet of buffer. The injection method ensures that dynein molecules from solution are preferentially adsorbed to the carbon film instead of the denatured material that may exist in the meniscus of a protein solution (Wall and Hainfeld, 1988). After a 30-second incubation period the grids are washed four times, with tobacco mosaic virus (TMV) particles being in the final wash. The grids are then frozen in nitrogen slush (−210°C) and freeze-dried overnight at −95°C and 10^{-8} Torr. After freeze-drying is complete the samples are transferred under vacuum to the microscope stage and maintained at −150°C.

C. Image Acquisition:

All images are recorded digitally with one sweep of the raster and stored for viewing and subsequent mass analysis. To reduce the effect of beam damage, repeated scanning of areas of interest should be avoided. Images that are to be used for mass analysis should contain several isolated molecules of interest and at least one but preferably two or more TMV particles to be used as mass standards for that image. Images or areas of images that contain aggregated material or particles that are juxtaposed should be avoided.

D. Mass Analysis

The mass of a particle is obtained by the integration of electron scattering intensities over an area containing the specimen. For particles that are less than 50 nm thick (most macromolecular assemblies) the signal arising from elastically scattered electrons is linearly proportional to the mass of the particle. For a more detailed account of the underlying theory of STEM mass analysis the interested reader is referred to Mosesson *et al.* (1981). The mass of a particle is determined in three stages.

1. The mass of the background support is first determined by selecting several areas of clean carbon support. The area size selected should be the same as the area size that will be selected for the particle. For example, a 22 S dynein molecule requires a circle of about 35 nm. Smaller circles would not encompass all of the molecule and would result in a lower mass reading, whereas a circle that is too large will encompass the entire molecule but the signal-to-noise ratio will be lowered, resulting in a larger standard deviation for a population of molecules.
2. Mass calibration of the system is performed using a TMV segment of known length as an internal mass standard.
3. The mass of the particle is then determined by taking the difference in electron scattering intensities between the area containing the particle and the mean value of areas without particles, and multiplying this value by the mass calibration factor.

E. Mass Measurement Errors

Errors in mass measurement arise from four main sources: (1) counting statistics, (2) variations in substrate thickness, (3) mass loss due to beam damage, and (4) sample preparation. The theoretical error expected from the combined effects of counting statistics and variations in substrate thickness is between 4 and 5% for a 22 S dynein molecule. The effects of mass loss due to beam damage tend to be around 1.0% under normal operating conditions (i.e., specimen maintained at $-150°C$ and exposed to 200 electrons/nm^2); however, this small mass loss is largely compensated for by an almost equal mass loss in the TMV particles that are used for the internal mass calibration standard. By far the greatest sources of error are sample heterogeneity due to impurities, proteolysis, and salts or sucrose drying in and around the molecules. To this end all dynein preparations used for STEM analysis need to be rigorously screened by sodium dodecyl sulfate–polyacrylamide gel electrophoresis to check for purity and any resultant proteolysis that may have occurred.

References

Gibbons, I. R., and Rowe, A. J. (1965). Dynein: A protein with adenosine triphosphatase from cilia. *Science* **149,** 424–425.

Goodenough, U. W., and Heuser, J. E. (1984). Structural comparison of purified dynein arms with in situ dynein arms. *J. Mol. Biol.* **180,** 1083–1118.

Hastie, A. T., Marchese-Ragona, S. P., Johnson, K. A., and Wall, J. S. (1988). Structure and mass of the mammalian respiratory ciliary outer arm 19S dynein. *Cell Motil. Cytoskel.* **11,** 157–166.

Johnson, K. A., and Wall, J. S. (1983). Structure and molecular weight of the dynein ATPase. *J. Cell Biol.* **96,** 669–678.

King, S. M., and Witman, G. B. (1990). Localization of an intermediate chain of outer arm dynein by immunoelectron microscopy. *J. Biol. Chem.* **265,** 19807–19811.

Marchese-Ragona, S. P., Gagnon, C., White, D., Belles-Isles, M., and Johnson, K. A. (1987). Structure and mass analysis of 12S and 19S dynein obtained from bull sperm flagella. *Cell Motil. Cytoskel.* **8,** 366–374.

Marchese-Ragona, S. P., and Johnson, K. A. (1988). STEM analysis of the dynein ATPase. *Electron Microsc. Rev.* **1,** 141–153.

Marchese-Ragona, S. P., Wall, J. S., and Johnson, K. A. (1988). Structure and mass analysis of 14S dynein obtained from *Tetrahymena* cilia. *J. Cell Biol.* **106,** 127–132.

Mosesson, M. W., Hainfeld, J., Wall, J., and Haschemeyer, R. H. (1981). Identification and mass analysis of human fibrinogen molecules and their domains by scanning transmission electron microscopy. *J. Mol. Biol.* **153,** 695–718.

Sale, W. S. (1983). Low angle rotary-shadow replication of 21S dynein from sea urchin sperm flagella. *J. Submicrosc. Cytol. Pathol.* **15**(Suppl. 1), 217–221.

Sale, W. S., Goodenough, U. W., and Heuser, J. E. (1985). The substructure of isolated and in situ outer dynein arms of sea urchin sperm flagella. *J. Cell Biol.* **101,** 1400–1412.

Walczak, C. E., Marchese-Ragona, S. P., and Nelson, D. L. (1993). Immunological comparison of 22S, 19S dyneins from *Paramecium* cilia. *Cell Motil. Cytoskel.* **24,** 17–28.

Wall, J. S., and Hainfeld, J. F. (1988). Mass mapping with the scanning transmission electron microscope. *Annu. Rev. Biophys. Biochem.* **15,** 355–376.

Witman, G. B., Johnson, K. A., Pfister, K. K., and Wall, J. S. (1983). Fine structure and molecular weight of the outer dyneins of *Chlamydomonas*. *J. Submicrosc. Cytol. Pathol.* **15,** 193–197.

CHAPTER 26

High-Resolution Negative Staining of the Isolated Dynein ATPase

Silvio P. Marchese-Ragona
TopoMetrix Corporation
Santa Clara, California 95054

I. Introduction

Many electron microscope techniques have been employed successfully in the last three decades to study the ultrastructure of intact axonemes and isolated axonemal proteins. The techniques used include thin sectioning (Satir, 1968), metal shadowing (Gibbons and Rowe, 1965; Sale, 1983), rapid-freeze deep-etch (Goodenough and Heuser, 1984; Sale *et al.,* 1985), dedicated scanning transmission electron microscope (STEM) analysis (Johnson and Wall, 1983; Witman *et al.,* 1983; Marchese-Ragona *et al.,* 1987, 1988), low-dose electron microscopy of frozen hydrated specimens (Murray, 1987), and negative staining (Johnson and Wall, 1983; Toyoshima, 1987; Marchese-Ragona *et al.,* 1988; Marchese-Ragona and Johnson, 1988; Hastie *et al.,* 1988; King and Witman, 1990; Walczak *et al.,* 1993). While each one of these techniques has proved itself to be invaluable, negative staining is perhaps the simplest of all the techniques mentioned above and has the added advantage that the whole technique from initial preparation to final examination takes only minutes and does not require the use of any exotic equipment, other than a transmission electron microscope and a carbon evaporator. As with any technique there are innumerable perturbations and refinements; however the negative staining technique originally described by Valentine (1968) has consistently yielded high-resolution data on isolated dyneins obtained from *Tetrahymena, Chlamydomonas,* mammalian tracheal cilia, and *Paramecium* and on dynein–antibody and kinisin–microtubule complexes (Marchese-Ragona and Harrison, unpublished results). Figure 1 shows a montage of negatively stained dynein arms obtained from *Telsatrymena*.

II. Method

A. Preparation of Mica

Using a vacuum evaporator, mica sheets are coated with a thin layer of carbon. The carbon support film should be as thin as possible (3.0–10.0 nm) and should be prepared on freshly cleaved mica. Once the carbon-coated mica has been prepared it can be stored in a desiccator almost indefinitely. Thicker support films may be used; however, the thicker the support film, the lower the contrast in the final image.

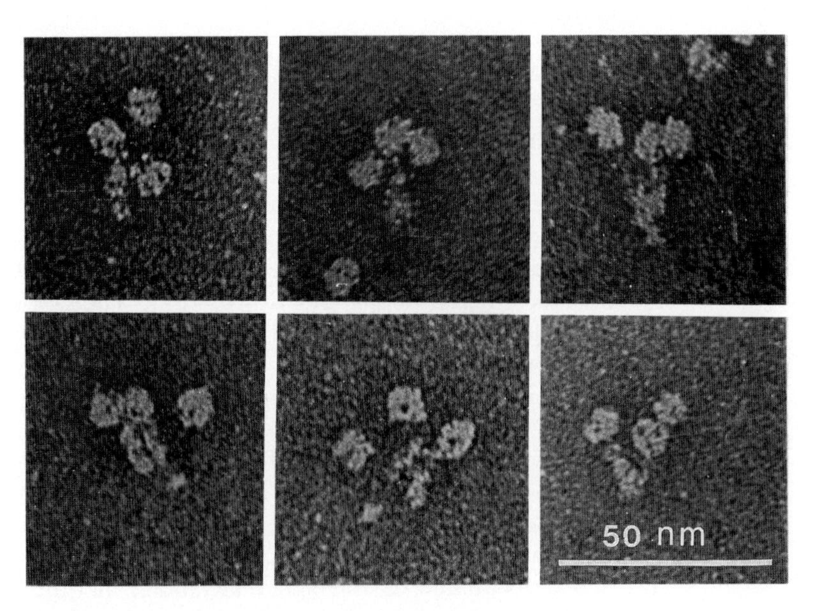

Fig. 1 Selected images of negatively stained 22S dynein. The molecule consists of three globular heads connected by strands to a common base. A stain accumulation near the center of the globular heads suggests the presence of a pit or cleft in the heads.

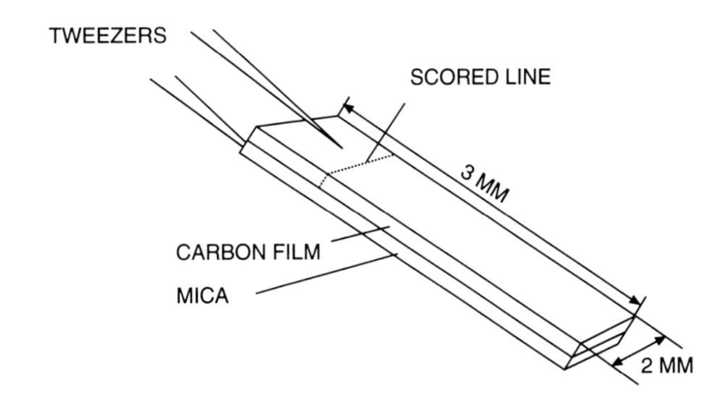

Fig. 2 Shows a 2 × 3 mm piece of carbon coated mica being held by tweezers. A score line is drawn approximately 1 mm from the upper edge, to allow the lower portion of the carbon film to float away.

B. Procedure

1. A 2 × 3-mm piece of carbon-coated mica is cut from larger piece, and a score line is drawn approximately 1 mm from the upper edge, as shown in Fig. 2. The top 1 mm is the region that will be handled by the tweezers and should not come in contact with either the protein solution or the staining solution. The lower 2 × 2 mm will be used as the support substrate for the specimen.

2. The lower portion of the carbon-coated mica is gently lowered at an angle of 30–45° into the dynein solution (Fig. 3A). This action should result in the existing meniscus being pushed away by the carbon film, exposing the underside of the carbon film to molecules from the bulk solution. The mica should be lowered into the solution almost until the score line is reached. Care should be taken to ensure that the tweezers do not enter the solution and that the carbon film does not detach at the score mark and float away freely. If there is any difficulty in seeing the carbon film on the surface, positioning a lamp at an angle of about 30° on the other side of the well may help.

3. The mica and the carbon film are withdrawn and the carbon film is immediately floated onto a 1 ml pool of uranyl acetate by dipping the mica to the score line several times until the carbon film is totally detached from the mica (Fig. 3B).

4. The carbon films should be picked off of the surface of the uranyl acetate solution by gently pressing an uncoated 400-mesh transmission electron microscope grid on to the top side of the floating carbon film and lifting vertically (Fig. 2B). Excess stain solution should be drawn off with the edge of a piece of filter paper, and the grids allowed to air-dry for a few minutes before being examined in the electron microscope.

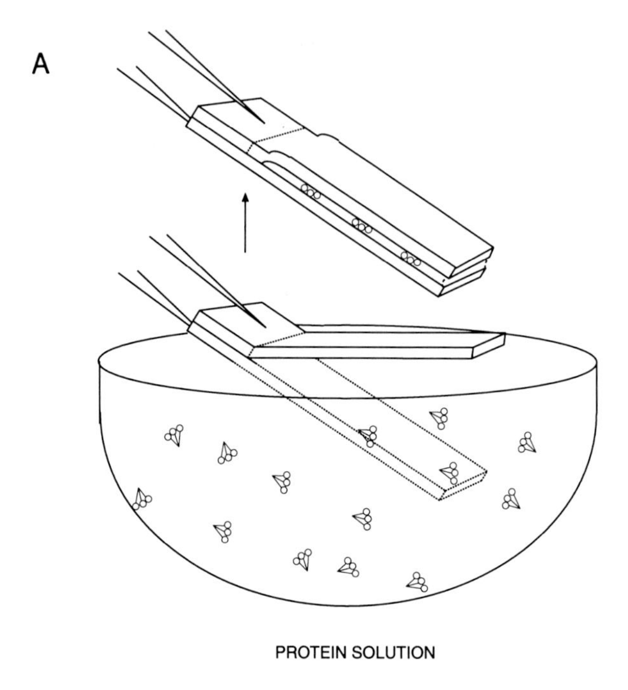

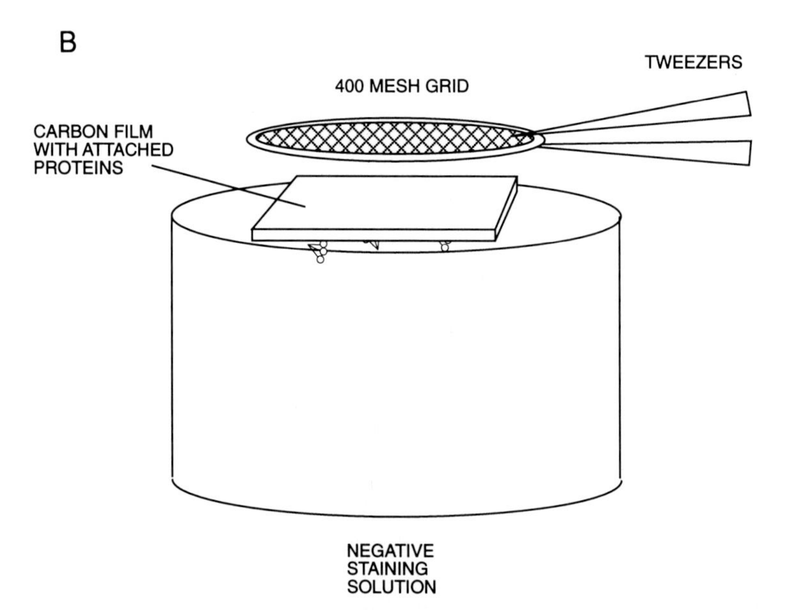

Fig. 3 (A) The mica is lowered into the dynein solution, almost, but not quite to the score line. When the mica is removed, dynein molecules that have adhered to the carbon are withdrawn. The mica is then gently lowered into the negative staining solution, until the score line is reached and the lower portion of carbon floats freely (B). (B) The carbon film is picked off of the surface of the staining solution by gently pressing an uncoated 400 mesh copper grid onto the carbon film and lifting vertically.

C. Precautions

To obtain reproducible and consistently high-resolution data from this technique the following three guidelines should be borne in mind:

1. A 2- to 3-ml aliquot of stain solution should be removed from a 50- to 250-ml stock solution and filtered immediately before use. The 2- to 3-ml aliquot should be carefully taken from below the surface of the stock solution, thus avoiding contaminants from precipitates that may exist at the surface and at the bottom of the stock solution. The precipitates themselves do not, in actual fact, affect the overall quality of the negative staining procedure, but they do result in a local heating of the support film when exposed to the electron beam, which in its most benign form will result in drift and at its worst will result in a ripping and curling of the support film.

2. The carbon support film should be exposed only to proteins that exist in the bulk solution and not to the denatured material that is usually present in the meniscus. This point is particularly important, as the variability seen in other negative staining techniques can be generally attributed to the fact that when a droplet of solution is placed on a support film, the dirt and denatured material present in the meniscus spread across the surface first and are preferentially bound, thus preventing the intact molecules that are in the bulk solution from binding moments later.

3. The greatest potential source of contamination and the one most often ignored in most electron microscopy procedures is the contamination that arises from the tweezers and the support grid. At no time should the tweezers or support grid be dipped into either the protein solution or the negative staining solution.

References

Gibbons, I. R., and Rowe, A. J. (1965). Dynein: A protein with adenosine triphosphatase from cilia. *Science* **149,** 424–425.

Goodenough, U. W., and Heuser, J. E. (1984). Structural comparison of purified dynein arms with *in situ* dynein arms. *J. Mol. Biol.* **180,** 1083–1118.

Hastie, A. T., Marchese-Ragona, S. P., Johnson, K. A., and Wall, J. S. (1988). Structure and mass of the mammalian respiratory ciliary outer arm 19S dynein. *Cell Motil. Cytoskel.* 11:157–166.

Johnson, K. A., and Wall, J. S. (1983). Structure and molecular weight of the dynein ATPase. *J. Cell Biol.* **96,** 669–678.

King, S. M., and Witman, G. B. (1990). Localization of an intermediate chain of outer arm dynein by immunoelectron microscopy. *J. Biol. Chem.* **265,** 19807–19811.

Marchese-Ragona, S. P., Gagnon, C., White, D., Belles-Isles, M., and Johnson, K. A. (1987). Structure and mass analysis of 12S and 19S dynein obtained from bull sperm flagella. *Cell Motil. Cytoskel.* **8,** 366–374.

Marchese-Ragona, S. P., and Johnson, K. A. (1988). STEM analysis of the dynein ATPase. *Electron Microsc. Rev.* **1,** 141–153.

Marchese-Ragona, S. P., Wall, J. S., and Johnson, K. A. (1988). Structure and mass analysis of 14S dynein obtained from *Tetrahymena* cilia. *J. Cell Biol.* **106,** 127–132.

Murray, J. M. (1987). Electron microscopy of frozen hydrated eukaryotic flagella. *J. Ultrastruc. Res.* **95,** 196–209.

Sale, W. S. (1983). Low angle rotary-shadow replication of 21S dynein from sea urchin sperm flagella. *J. Submicrosc. Cytol. Pathol.* **15,** (Suppl. 1), 217–221.

Sale, W. S., Goodenough, U. W., and Heuser, J. E. (1985). The substructure of isolated and in situ outer dynein arms of sea urchin sperm flagella. *J. Cell Biol.* **101,** 1400–1412.

Satir, P. (1968). Studies on cilia. III. Further studies on the cilium tip and a "sliding filament" model of cilliary motility. *J. Cell Biol.* **39,** 77–94.

Toyoshima, Y. Y. (1987). Chymotryptic digestion of *Tetrahymena* 22S dynein. I. Decomposition of three-headed 22S dynein to one- and two-headed particles. *J. Cell Biol.* **105,** 887–895.

Valentine, R., Shapiro, B., and Stadtman, E. (1968). Regulation of Glutamine Synthatase. XII. Electron microscopy of the enzyme from *Escherichia coli. Biochemistry* **7,** 2143–2152.

Walczak, C. E., Marchese-Ragona, S. P., and Nelson, D. L. (1993). Immunological comparison of 22S, 19S and 12S dyneins from *Paramecium* cilia. *Cell Motil. Cytoskel.* **24,** 17–28.

Witman, G. B., Johnson, K. A., Pfister, K. K., and Wall, J. S. (1983). Fine structure and molecular weight of the outer dyneins of *Chlamydomonas. J. Submicrosc. Cytol. Pathol.* **15,** 193–197.

CHAPTER 27

Computer-Assisted Analysis of Flagellar Structure

Eileen O'Toole,[*] David Mastronarde,[*] J. Richard McIntosh,[*] and Mary E. Porter[†]

[*]Laboratory for Three-Dimensional Fine Structure
Department of Molecular, Cellular, and Developmental Biology
University of Colorado
Boulder, Colorado 80309
[†]Department of Cell Biology and Neuroanatomy
University of Minnesota School of Medicine
Minneapolis, Minnesota 55455

I. Introduction

Analysis of flagellar mutants of *Chlamydomonas reinhardtii* by sodium dodecyl sulfate–polyacrylamide gel electrophoresis (SDS–PAGE) and thin-section electron microscopy has led to the identification of several polypeptides as subunits of specific axonemal structures (Luck, 1984). This approach has been particularly useful in sorting out the complexity of the inner dynein arms. The difficulty in interpreting "raw" electron microscope (EM) images has led to the development of image averaging techniques to better visualize axonemal structure and organization. In general, image averaging techniques improve the signal-to-noise ratio over that found in the original electron micrographs, resulting in the detection of structures not seen with the unaided eye. Methods such as cross-correlation analysis and rotational averaging of structures possessing ninefold axial symmetry have been used to average transverse sections of insect sperm flagella (Afzelius *et al.*, 1990; Lanzavecchia *et al.*, 1991). In addition, averages of doublet cross sections have been obtained to make qualitative comparisons between inner and outer dynein arm mutants in *Chlamydomonas* (Kamiya *et al.*, 1991; Sakakibara *et al.*, 1993). Although these techniques result in improved images of axonemal structure, they do not provide a means for quantitative analysis or for making statistical comparisons between axonemes of different strains. We have developed more extensive image averaging techniques and methods for quantitative comparisons of images of mutant and wild-type axonemes (Mastronarde *et al.*, 1992). Described in this chapter are the steps involved in obtaining averages of axonemal components and the methods for making statistical comparisons between them.

II. Specimen Preparation and Electron Microscopy

Axonemes isolated by the method of Porter *et al.* (1992) are prepared for electron microscopy by fixation in 2% glutaraldehyde and 4% tannic acid (Mallinckrodt) in 0.05 *M* Millonig's phosphate buffer (pH 6.9) for 1 hour. The samples are rinsed twice in buffer and fixed in 2% phosphate-buffered glutaraldehyde overnight at 4°C. The samples are then rinsed three times in phosphate buffer and postfixed in 1% OsO_4 for 15 minutes on ice. Following postfixation, the samples are dehydrated through a graded acetone series and embedded in Epon 812–Araldite 502. Sections of 60-nm nominal thickness are cut for cross-sectional analysis; 40-nm sections are used for longitudinal analysis. The sections are placed on bare 200-mesh copper grids, poststained with Reynold's lead citrate and uranyl acetate, and then imaged at 39,000× in a transmission electron microscope operating at 80 kV.

III. Image Analysis

A. Hardware and Software

The image analysis software described in this chapter was written and developed by Dr. David Mastronarde in the Boulder Laboratory for Three-Dimensional Fine Structure, a National Center for Research Resources. The programs involve digitization, display, and quantitative analysis of images and are available in our laboratory to any user. The hardware used for the analysis includes a Dage MTI video camera (Model MTI-81 containing a Saticon tube, Dage-MTI, Michigan City, IN) mounted above a light box. Digitization and quantitative analysis are performed using a Micro VAX III computer containing a Parallax 1280 videographics device (Parallax Graphics, Santa Clara, CA).

B. Analysis of Cross Sections

Images of axonemes containing outer doublets in cross section are found relatively quickly and easily in sections of pellets. As a general rule, axonemes are chosen for analysis if they have a complete set of nine outer-doublet microtubules, an intact central pair, and protofilaments that are visible in at least one outer-doublet microtubule (arrows, Fig. 1a). Axonemes are digitized directly by video within an area of 304 × 300 pixels at a scale of 0.9 nm/pixel. On average, we digitize and store approximately 30 axonemal cross sections per sample. Outer doublets in good cross section are chosen interactively by an operator who places a model point within doublets that show protofilaments in sharp cross section (asterisk, Fig. 1b). Approximately 50–100 outer doublets are selected for a given sample. The image of each selected doublet is then extracted and rotated to an approximate alignment in a 70 × 90-pixel box, using the position of the nine outer doublets as a reference. The individual outer-doublet cross sections are then aligned by applying a general linear transformation to solve for differences in *X* and *Y* translations, rotation, magnification, and stretch along a single axis. An automatic alignment procedure finds these parameters to minimize the difference between corresponding pixels of doublet images. A two-step procedure is performed in which the doublet images are first filtered with a low-pass filter to enhance low spatial frequencies. In this case, the centers of the microtubule subfibers are enhanced and the automatic alignment procedure is applied to align the images to a wild-type template. Then, the images are high-pass filtered to enhance the protofilaments, and the resulting aligned images are averaged (Fig. 1c).

There is significant variability between samples of the same strain due to day-to-day staining differences. Reliable statistical comparisons, therefore, require at least three independent preparations of the same sample; these are averaged together to form a "grand average" of a given sample, with which appropriate statistical comparisons can be made (Mastronarde *et al.*, 1992). To do this, individual sample averages are aligned to one another by marking fiducial points in the centers of 10 protofilaments, and multiple linear regression is used to solve for the transformations that minimize the differences between the fiducial points in successive images. These sample "grand averages" thus contain 150–300 doublets from at least three independent samples (Fig. 1d). Combining images from different samples is, however, complicated by staining differences, which lead to differences in the intensities of different averages, and it is necessary to normalize the intensities to a common scale. This procedure is carried out using a program that scales the background, as determined by the integrated intensity in a model area (Fig. 1e, zn), to zero and the integrated intensity of the outer arm (oa) or the doublet microtubule (dn) to one. We found

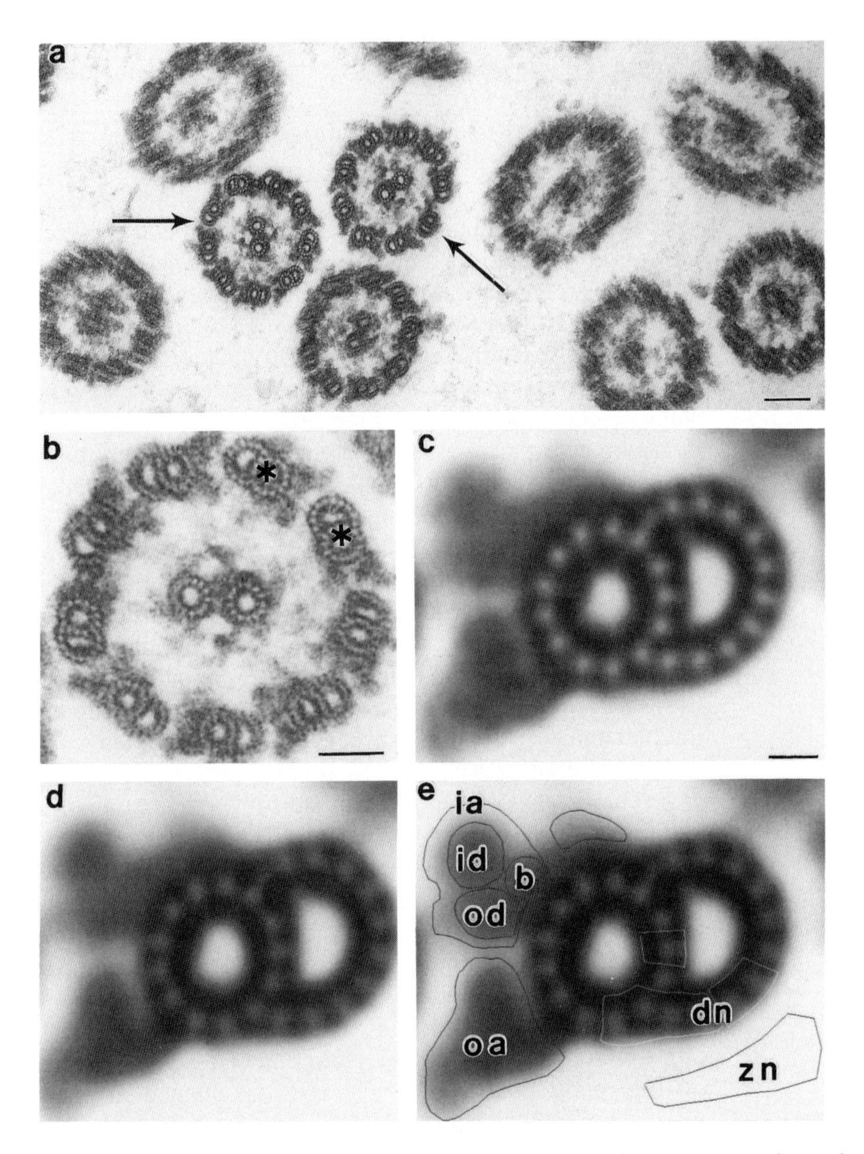

Fig. 1 Analysis of doublet cross sections. (a) Axoneme cross sections are chosen for analysis if the central pair is intact and at least one outer doublet is in sharp cross section (arrows). Bar = 50 nm. (b) Axonemes are digitized and doublets are chosen interactively by placing a model point (asterisk) in doublets with protofilaments in cross section. Bar = 50 nm. (c) Doublets are extracted, aligned, and averaged to obtain a sample average. (d) Three independent sample averages are aligned to form a "grand average." (e) Model areas used for quantitative analysis include the inner arm (ia) with inner domain (id), outer domain (od), and base (b) regions. Outer arm (oa) and doublet (dn) are used for normalization with respect to the zero normalizing area (zn). Bar = 10 nm in (c)–(e).

that outer-arm normalization was a better method for normalizing the intensities between images, and that one needs three to four times the sample size to obtain reliable results with doublet normalization (Mastronarde *et al.*, 1992). For either type of normalization, the program applies the scaling to all doublets in the average and computes a mean and standard deviation in selected areas in the doublet as well as at each pixel.

The above method for obtaining averages of doublet cross sections makes no selection of specific outer doublets nor does it control for position along the length of the axoneme. It is, however, known that several structural asymmetries exist in the axonemes of *Chlamydomonas,* and these asymmetries can be used as markers to identify the outer-doublet positions or the approximate location of the cross section with respect to the proximal or medial/distal position along the length of the axoneme. Outer-doublet asymmetries have been characterized by others (Hoops and Witman, 1983) and include the absence of an outer dynein arm for approximately 90% of the length of doublet 1 (Fig. 2a). Outer-doublet positions are counted clockwise from doublet 1 (Hoops and Witman, 1983). A program has been written that uses the position of doublet 1, as identified and marked by the operator, to extract and group the doublets in the other eight positions. The individual doublets at each position are aligned, averaged, and normalized as described above, and statistical comparisons can then be made to determine if there are doublet-specific differences in dynein structure in a given mutant strain. Outer-doublet asymmetry has been identified as a result of the *bop2-1* mutation (King *et al.*, 1994), resulting in loss of inner-arm structure in a doublet-specific manner. Biochemical and morphological

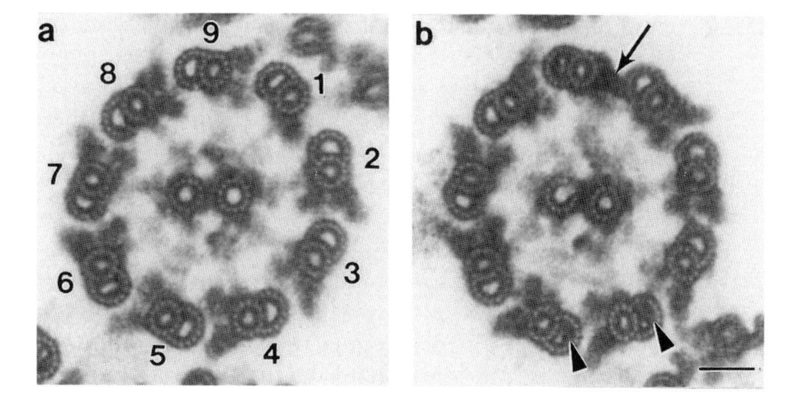

Fig. 2 Structural asymmetries in axoneme cross section allow location of the outer-doublet positions and location along the length of the axoneme. (a) Outer-doublet position is established relative to doublet 1 which is missing the outer arm; doublets 2–9 are numbered clockwise from doublet 1. (b) Proximal cross section showing bridge structure between doublets 1 and 2 (arrow) and B-tubule projections in doublets 1, 5, and 6 (arrowheads). Bar = 50 nm.

differences between the proximal and medial/distal regions of the flagella have also been identified (Piperno and Ramanis, 1991; Dentler and Adams, 1992; Hoops and Witman, 1983). Examples of proximal/distal asymmetry are the location of the B-tubule projections in doublets 1, 5, and 6 (arrowheads, Fig. 2b) in the proximal 40–50% of the flagella and the presence of a two-membered cross-bridge between doublets 1 and 2 (arrow, Fig. 2b) in the proximal one-quarter of the axoneme (Hoops and Witman, 1983). These morphological markers are readily seen in proximal axoneme cross sections (Fig. 2b), and their presence or absence acts as a morphological marker to identify the position of the cross section along the length of the axoneme. Doublets from the proximal or medial/distal regions can be pooled and averages from these regions obtained. Recent data from our laboratory suggest that some mutant strains have structural differences that are localized primarily to the distal regions of the axoneme (Gardner *et al.*, 1994).

C. Analysis of Longitudinal Sections

Analysis of longitudinal sections allows the investigator to obtain detailed information about the organization of the inner- and outer-arm structures along the length of the axoneme. The search for appropriate longitudinal images is more time consuming than the acquisition of cross sections, but suitable views can be obtained using the following criteria. Longitudinal images of the axoneme display two forms of repeating structure: outer dynein arms that show a periodicity of 24 nm, and radial spokes and inner arm structures that repeat every 96 nm. Images of axonemes are chosen for analysis that are fairly straight, have a central pair, and have one outer doublet that shows both radial spokes and outer dynein arms for at least five of the 96-nm repeating units (Fig. 3a). The negatives are digitized at a scale of 0.9 nm/pixel, with the end proximal to the basal body to the left. The areas chosen for averaging (asterisk, Fig. 3b) are marked by the operator at the point midway between the radial spokes, and a program extracts a 168 × 168-pixel box around the model point (Fig. 3b). The extracted image segments are then automatically aligned and averaged to form a sample average, as shown in Fig. 3c. Sample grand averages are obtained using a manual alignment program where sample averages are interactively aligned to a wild-type reference. The aligned images are then averaged together to obtain a sample grand average of a particular strain (Fig. 3d) that contains up to 50–70 individual repeating units from 9 or 10 individual axonemes. The averages in Mastronarde *et al.* (1992) are actually vertical mirror images of the correct orientation of inner-arm-region structures. The images presented here are in correct spatial orientation, with proximal regions of the 96-nm repeat to the left and structures extending out from the microtubule toward the viewer. As shown in Fig. 3, the complexity of inner-arm organization in longitudinal view is revealed as one progresses from raw images, to averages of individual

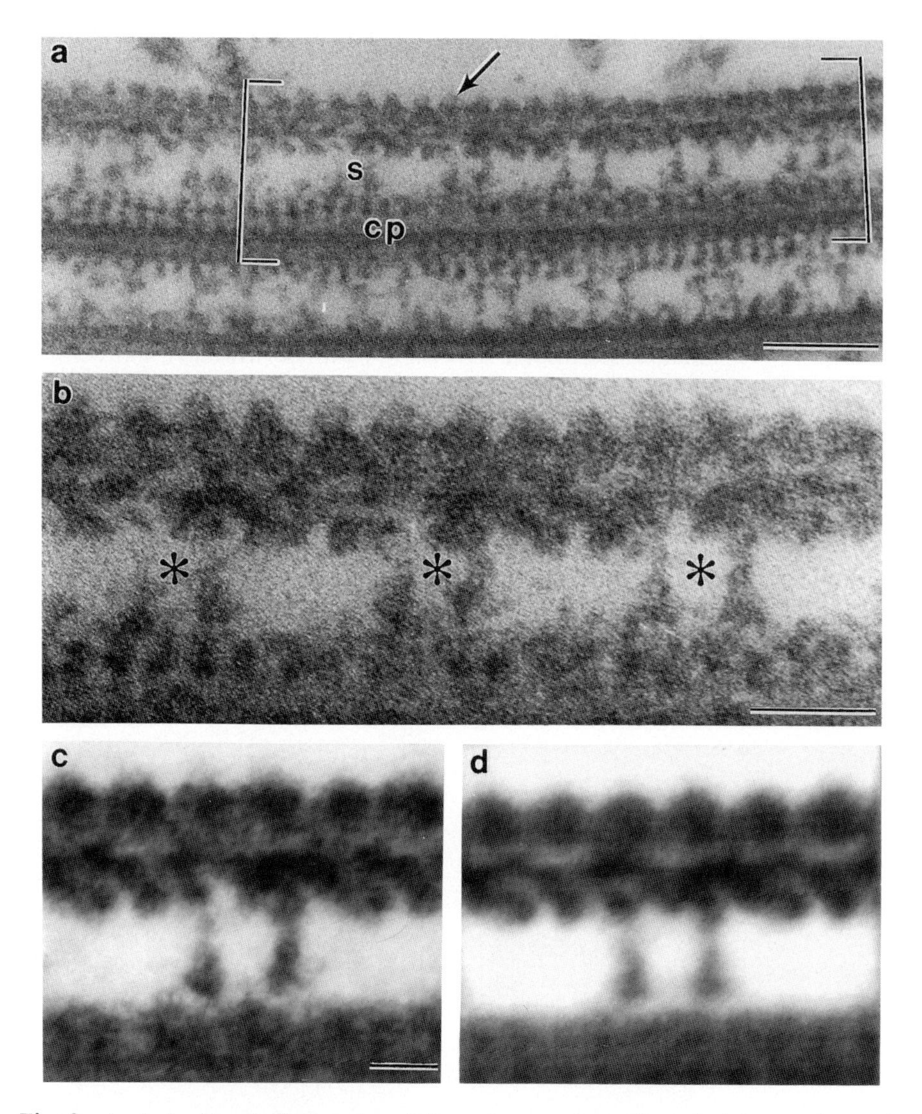

Fig. 3 Analysis of longitudinal sections. This section was chosen for analysis because it contains the central pair (cp) and one outer doublet showing both radial spokes (s) and outer dynein arms (arrow) for at least five of the 96-nm repeating units. The boxed area was digitized. Bar = 100 nm. (b) The digitized area from (a) shows the repeating units selected for averaging (asterisk). A 144-nm square of image surrounding each repeating unit was excised, then aligned for averaging with translations and rotations. Bar = 50 nm. (c, d) Progressively more detail appears as one advances from a sample average containing five repeating units (c) to the grand average containing 60 repeating units from nine axonemes (d). Bar = 25 nm.

axonemes, to a grand average of a particular strain. The principal limitation of imaging longitudinal sections is that information about the specific outer doublet imaged or the position of the image along the length of the axoneme cannot be obtained.

IV. Quantitative Analysis

Statistical comparisons of different samples are performed by a nested analysis of variance, using the sums of squared deviations computed from the means, standard deviations, and sample number for each individual sample. One can then test for the significance of differences in specified areas of the dynein arms (Fig. 1e) or at each pixel in a given pair of images by taking into account the intra- and intersample variability. The difference images shown in Fig. 4 are based on a pixel-by-pixel nested analysis of variance. Figures 4a and b display grand averages for a wild-type and a mutant strain in cross section; Fig. 4c is the difference image, where the differences that are not significant at the 0.05 level are set to zero. Figures 4d, e, and f are the corresponding images for longitudinal sections. We have described image averaging techniques to examine the structure and organization of dynein arms in *Chlamydomonas* axonemes. Combination of biochemical and genetic analysis with such image averaging

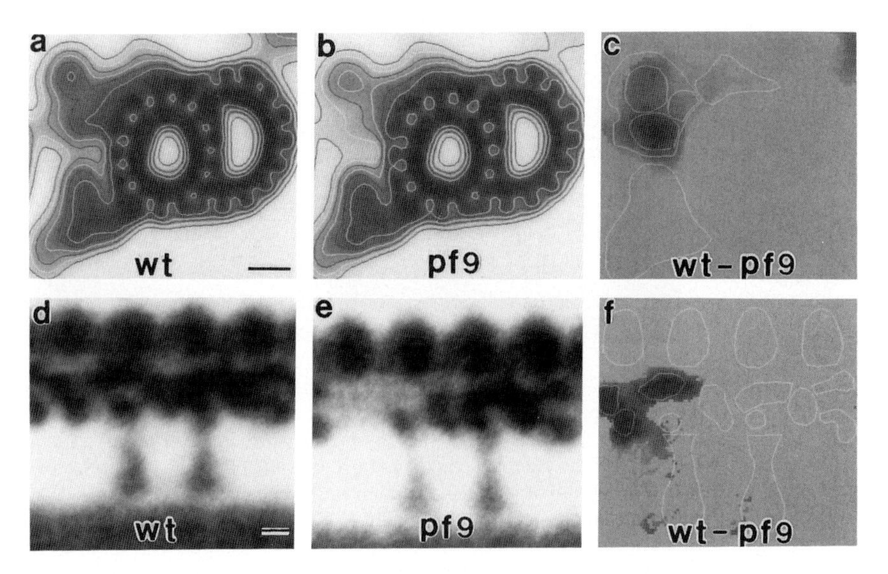

Fig. 4 Quantitative comparisons are made using difference images representing an analysis of variance at each pixel. (a–c) Grand average of wild-type (a) and pf9 (b) cross sections and the difference between them (c). (d–e) Grand average of wild-type (d) and pf9 (e) longitudinal sections and the difference between them (f). Bar = 10 nm.

techniques provides powerful methods for studying structure–function relationships in cell motility.

References

Afzelius, B. A., Bellon, P. L., Dallai, R., and Lanzavecchia, S. (1991). Diversity of microtubular doublets in insect sperm tails: A computer-aided image analysis. *Cell Motil. Cytoskel.* **19,** 282–289.

Dentler, W. L. and Adams, C. (1992). Flagellar microtubule dynamics in *Chlamydomonas:* Cytochalasin D induces period of microtubule shortening and elongation; and colchicine induces disassembly of the distal, but not proximal, half of the flagellum. *J. Cell Biol.* **117,** 1289–1298.

Gardner, L. C., O'Toole, E., Perrone, C. A., Giddings, T., and Porter, M. E. (1994) Components of a "dynein regulatory complex" are located at the junction between the radial spokes and the dynein arms in *Chlamydomonas* flagella. *J. Cell Biol.* **127,** 1311–1325.

Hoops, H. J., and Witman, G. B. (1983). Outer doublet heterogeneity reveals structural polarity related to beat direction in *Chlamydomonas* flagella. *J. Cell Biol.* **97,** 902–908.

Kamiya, R., Kurimoto, E., and Muto, E. (1991). Two types of *Chlamydomonas* flagellar mutants missing different components of inner-arm dynein. *J. Cell Biol.* **112,** 441–447.

King, S. J., Inwood, W. B., O'Toole, E. T., Power, J., and Dutcher, S. K. (1994). The bop-2 mutation reveals radial assymetry in the inner dynein arm region of *Chlamydomonas reinhardtii. J. Cell Biol.* **126,** 1255–1266.

Lanzavecchia, S., Bellon, P. L., and Afzelius, B. A. (1991). A strategy for the reconstruction of structures possessing axial symmetry: Sectioned axonemes in sperm flagella. *J. Microsc.* **164,** 1–11.

Luck, D. J. L. (1984). Genetic and biochemical dissection of the eukaryotic flagellum. *J. Cell Biol.* **98,** 789–794.

Mastronarde, D. N., O'Toole, E. T., McDonald, K. L., McIntosh, J. R., and M. E. Porter (1992). Arrangement of inner dynein arms in wild-type and mutant flagella of *Chlamydomonas. J. Cell Biol.* **118,** 1145–1162.

Piperno, G., and Ramanis, Z. (1991). The proximal portion of *Chlamydomonas* flagella contains a distinct set of inner arms. *J. Cell Biol.* **112,** 701–709.

Porter, M. E., Power, J. M., and Dutcher, S. K. (1992). Extragenic suppressors of paralyzed flagellar mutations in *Chlamydomonas reinhardtii* identify loci that alter the inner dynein arms. *J. Cell Biol.* **118,** 1163–1176.

Sakakibara, H., Takada, S., King, S. M., Witman, G. B., and Kamiya, R. (1993). A *Chlamydomonas* outer dynein arm mutant with a truncated beta heavy chain. *J. Cell Biol.* **122,** 653–661.

CHAPTER 28

Preparation of Cilia and Flagella for Thin-Section Transmission Electron Microscope Analysis

Harold J. Hoops
Department of Biology
State University of New York at Geneseo
Geneseo, New York 14454

I. Introduction

Fixation of cilia, flagella, and axonemes is generally similar to that for other biological material; however, important parts of the flagellum (especially axonemal components) have relatively low contrast. Common modifications to improve contrast include using a mordant such as tannic acid (Tilney *et al.*, 1973), preparation conditions that allow full or partial extraction of the matrix, and embedding in plastics like Epon (or its substitutes) and Epon–Araldite rather than Spurr's.

Because the thickness of a standard "thin section" (50–60 nm) is much greater than the dimensions of many important axonemal components, flagellar images are composed of multiple overlapping structures. The best (clearest) images are the result of nearly exact superimposition of repeating structures and, consequently, are rare. Preparations that give a high density of flagellar sections allow the investigator to find satisfactory images much more quickly.

The first decision the investigator must make is whether to fix isolated axonemes, isolated flagella, or *in situ* flagella. Fixation of isolated axonemes yields a large number of specimens in a given section and allows tannic acid to be used as a mordant. Preparations of isolated flagella also yield a large number of specimens per section and allow observation of the flagellar matrix and membrane, although the matrix may obscure some axonemal structures. Fixation of flagella *in situ* can preserve information about developmental stages of flagellar growth and spatial relationships between the flagella and the underlying cells, and allows portions of the basal body/flagellum proximal to the abscision point to be studied.

The microscopist must also decide on the conditions of fixation. Novel combinations of fixation protocols and specimens are unpredictable, and the beginning investigator should search the literature for a protocol that has worked for a similar specimen. The methods suggested here are based on those that have worked for *Chlamydomonas* and related species (Hoops and Witman, 1983, 1985; Sakakibara *et al.,* 1991), but may be suitable for other specimens as well.

II. Sources of Materials

All chemicals and supplies can be ordered from Electron Microscope Sciences (321 Morris Road, Box 251, Fort Washington, PA 19034). Other sources of chemicals and supplies for electron microscopy include Bio-Rad, Ernest F. Fullam, Ladd Research Industries, Polysciences, and Ted Pella.

III. Procedures

A. Fixation of Isolated Axonemes

1. Isolate axonemes using one of the procedures suggested in this volume.

2. Add suspended axonemes to a 1.5-ml microfuge tube and pellet the axonemes at top speed (16,000*g*) in an Eppendorf (or equivalent) microfuge for 10 minutes at 4°C. An angle-head rotor produces a thinner and flatter pellet on the side of the tube, improving access to the pellet by the fixative and making it easier to tease the pellet off of the side (see below). Samples in microfuge tubes also can be pelleted in a high-speed refrigerated centrifuge using the appropriate adaptors.

3. If the pellet is much thinner than 1 mm, it may fragment; if it is much thicker, penetration of the fixative may be inadequate. If necessary, resuspend the pellet and remove or add axonemes.

4. Prepare the primary fixative. A fixative I have found most suitable for a variety of specimens is 1% glutaraldehyde (EM grade) in 100 m*M* sodium cacodylate, pH 7.2, with or without 1–2% tannic acid. The pH of the solution will have to be readjusted with 2 *N* NaOH if tannic acid is used. The fixative solution should be made immediately before use (Hayat, 1981).

5. Decant the isolation buffers from the pellet and overlay it with the fixative solution. Invert the tube several times to mix.

6. After 15 minutes in fixative, the pellet can be teased off of the side of the microfuge tube with an applicator stick that has been sharpened into a shovel with a razor blade. The free pellet enables the fixatives (and later dehydrating and embedding solutions) to penetrate the sample from all sides. If the sample is much thinner than 1 mm, it should be left in place to reduce the risk of disintegration. Fix for 1–2 hours at room temperature or overnight at 4°C.

7. Rinse axonemes twice, 20 minutes each, in 100 m*M* cacodylate buffer, pH 7.2, at 4°C.

8. Overlay axonemes with 1% OsO_4 with 100 m*M* cacodylate buffer at room temperature. The exact concentration of osmium is not critical, so it is convenient to add 0.5 ml 200 m*M* cacodylate buffer + 0.25 ml distilled H_2O to the tube, then add 0.25 ml (about 7 drops from a Pasteur pipet) of 4% OsO_4. Osmium tetroxide can be purchased in sealed vials and should be pale straw-colored. Use a new vial if the osmium solution is blackish. *Caution:* OsO_4 is both volatile and toxic. Always work with OsO_4 in the hood and dispose of the solutions containing osmium in the proper manner.

9. Rinse the pellet in distilled H_2O twice, 20 minutes each, at 4°C.

10. En bloc stain in freshly made 1% aqueous uranyl acetate overnight at 4°C or at room temperature for 1 hour. Dispose of used uranyl acetate as appropriate for a toxic heavy metal.

11. Rinse pellets in distilled H_2O (quick).

12. Dehydrate in a graded series of acetone/water mixtures (e.g., 25%, 50%, 75%, 90% at 4°C and three changes of 100% acetone at room temperature for 20 minutes each). The final acetone must be anhydrous. Avoid using ice-cold 100% acetone to prevent water condensation in the sample.

13. Prepare the plastic and plastic/acetone mixtures for embedding. For best sectionability make enough plastic to carry out the entire embedding process with the same batch using the recipe given by the supplier. Mix well. Waste plastic or plastic/solvent mixtures should not be dumped down the drain but polymerized after solvents are allowed to evaporate. *Note:* Some of the resins used are toxic and suspected carcinogens. Mix and use the plastics in a functioning fume hood and wear gloves.

14. Infiltrate samples in steps of 25%, 50%, 75%, and 90% plastic in anhydrous acetone for 2 hours each at room temperature, 100% plastic for 2 hours, and a second change of plastic overnight. The exact ratios of plastic to solvent are

not critical and can be made up by adding the plastic and solvent dropwise to shell vials and mixing. Unpolymerized plastic and plastic/solvent mixtures can be stored in the refrigerator until shortly before use, but should be allowed to reach room temperature before uncapping.

15. Remove the pellet from the microfuge tube, place in a flat embedding mold, and surround with fresh plastic. Place the bottom (or top) of the pellet flat against the bottom of the mold to obtain the maximum number of good cross and longitudinal sections.

16. Polymerize at 60°C for 12–36 hours.

17. Section, stain with uranyl acetate and lead citrate, and view as normal.

B. Fixation of Isolated Flagella

For the most part, flagella can be fixed, dehydrated, and embedded using the procedures described above. Flagellar membranes are osmotically active until after postfixation with OsO_4, so the fixation and wash solutions should have the appropriate tonicity. Usually, this has to be determined by trial and error. Fixative solutions that are hypotonic ensure penetration by the fixative, but can also result in swelling of the flagella and sometimes even cause the axoneme to curl up inside of the flagellar membrane (Witman *et al.,* 1972). Hypertonic solutions may result in a dense matrix and obscure details of the axoneme. The fixation and wash solutions described above for axonemes (1% glutaraldehyde in 100 m*M* sodium cacodylate, pH 7.2, 100 m*M* cacodylate washes, and 1% OsO_4 in 100 m*M* cacodylate) generally work well for *Chlamydomonas* flagella without modification. Tannic acid in the glutaraldehyde will not normally be able to penetrate the plasma membrane (Tilney *et al.,* 1973) and, therefore, cannot be used to enhance contrast of axonemal structures in whole flagella.

C. Fixation of Flagella *in Situ*

For the most part, flagella *in situ* can be fixed, dehydrated, and embedded using the procedures described for isolated flagella. For study of axonemal components, partial extraction of the matrix may be helpful. In my experience, fixation of *Chlamydomonas* (or its close relatives) in 1–2% glutaraldehyde in 100 m*M* cacodylate buffer results in a dense matrix. This may resemble the case in living cells, but it tends to obscure axonemal components. Prefixing in 1–2% glutaraldehyde made up with conditioned growth medium for about 15 minutes, followed by a normal cacodylate-buffered glutaraldehyde fixation for the balance of 1–2 hours, results in better axonemal images (Hoops and Witman, 1983, 1985). This procedure may not work for other organisms or different media.

Some cells are rugged enough to be spun into a tight pellet (as was done for axonemes and flagella); however, many specimens are too fragile for this approach. Such cells can be concentrated and suspended in 1% buffered molten low-temperature agarose (e.g., Type XI, Sigma) after fixation. After gelling of the agarose at 4°C, the block should be cut (if necessary) into pieces less than 1 mm thick. These blocks are then postfixed in OsO_4, dehydrated, and embedded as normal.

References

Hayat, M. A. (1981). "Fixation for Electron Microscopy." Academic Press, New York.

Hoops, H. J., and Witman, G. B. (1983). Outer doublet heterogeneity reveals structural polarity related to beat direction in *Chlamydomonas* flagella. *J. Cell Biol.* **97,** 902–908.

Hoops, H. J., and Witman, G. B. (1985). Basal bodies and associated structures are not required for normal flagellar motion or phototaxis in the green alga *Chlorogonium elongatum*. *J. Cell Biol.* **100,** 297–309.

Sakakibara, H., Mitchell, D. R., and Kamiya, R. (1991). A *Chlamydomonas* outer arm dynein mutant missing the α heavy chain. *J. Cell Biol.* **113,** 615–622.

Tilney, L. G., Bryan, J., Bush, D. J., Fujiwara, K., Mooseker, M. S., Murphy, D. B., and Snyder, D. H. (1973). Microtubules: evidence for 13 protofilaments. *J. Cell Biol.* **59,** 267–275.

Witman, G. B., Carlson, K., Berliner, J., and Rosenbaum, J. L. (1972). *Chlamydomonas* flagella I. isolation and electrophoretic analysis of microtubules, matrix, membranes, and mastigonemes. *J. Cell Biol.* **54,** 507–539.

CHAPTER 29

Fixation of Mammalian Spermatozoa for Electron Microscopy

David M. Phillips
The Population Council
New York, New York 10021

I. Introduction

This chapter includes techniques for preparing mammalian spermatozoa *in situ* (e.g., in the epididymis) or in suspension (e.g., in semen) for the transmission electron microscope (TEM). Specialized methods such as freeze-cleaving and immunocytochemistry will be omitted, as detailed discussions can be found in original papers or chapters devoted to those methods.

II. Choice of Fixative

Though one might not intuitively think that the buffer used in the fixative is important, it is. I believe that the best fixation of spermatozoa is obtained when the glutaraldehyde is buffered with collidine. To make up the buffer, add 25 ml of collidine (2,4,6-trimethylpyridine) to 1 liter of distilled water and adjust the pH to 7.4 with 1 *N* HCl. Although I use EM-grade glutaraldehyde, I am not sure it matters. Long ago, when I was a postdoctoral fellow in Don Fawcett's laboratory, his technician guarded a large old bottle of glutaraldehyde, cloudy and brown in color. Rumor had it that because of the impurities, the glutaraldehyde in this bottle provided the best fixation. It was widely believed in the department that this was the only glutaraldehyde that could fix lipid properly. The postdocs used to beg the technician for a few milliliters.

The period of fixation is not critical. I generally leave tissues in fixative overnight, but I do not discern any difference if they are left for a few weeks. I once exmined loris sperm after long-term fixation in glutaraldehyde. My colleague J. Michael Bedford had bagged the loris in Malaysia in 1966, fixed the epididymis, brought the tissue back to New York, and put it in his refrigerator. Dr. Bedford intended to process the loris epididymis, but some other things came up. It was not until 21 years later, while cleaning out his refrigerator in preparation to move to another building, that he found the epididymes. We processed them for the TEM and found the sperm to be reasonably well fixed (Phillips and Bedford, 1987).

III. Fixation of Sperm *in Situ*

Unfortunately both the epididymis and testis are very difficult tissues to fix by immersion. In the case of the testis, seminiferous tubules tend to move apart when the tissue is dissected. Because of this, after the tissue is embedded and cut, thin sections may contain more empty plastic than tissue. The epididymis can be equally problematic because when the tubules are cut up during fixation, spermatozoa may wash out. Not only does this reduce the number of spermatozoa in the tissue, but the displacement of spermatozoa caused by cutting may disturb the orderly arrangement of spermatozoa within the tubules. Tissue damage can be minimized by cutting the tissue very carefully. Alternatively, the intact tissue can be fixed by perfusing the blood vessels with fixative and dissecting the tissue afterward.

A. Fixation by Immersion

Fixation by perfusion is sometimes impractical. One may be working with biopsies or fixing tissue in the field or may lack training in perfusion techniques. If, however, immersion is carried out properly, spermatozoa look as well as or better than tissue that has been perfused. Immersion fixation is also more reliable. Thus, in the case of a very valuable animal, my choice is to fix by immersion.

It is important to begin fixation immediately after the tissue is removed from the animal and to cut the tissue into pieces without compressing it. I do my cutting in a trough made of dental wax, which can be purchased from any supplier of electron microscope accessories. Immerse the tissue in fixative and use two double-edged razor blades to cut in a scissor-like fashion. First cut 3- to 4-mm pieces, making the cuts as clean as possible. With an animal the size of a mouse, the entire epididymis can be placed in fixative. After the pieces have been in the fixative for 10 minutes, cut them into 1- to 2-mm pieces for subsequent dehydration and embedding.

B. Perfusion Fixation

With an animal the size of a mouse, I perfuse the entire animal. I have also perfused an entire rat, although experienced workers often perfuse only the testis and epididymis through the testicular artery. This takes considerable skill and practice. To perfuse an animal the size of a mouse or a rat, I use a small pump, although one can employ gravity or even a syringe. I carry out the following steps in rapid succession with an anesthetized animal:

1. Ligate the neck to prevent perfusing the head.
2. Open the plural cavity and immediately place a needle attached to the perfusion tube into the left ventricle.
3. Cut the right atrium and begin perfusing with saline.
4. After 3 minutes, switch to perfusing with fixative. (Use a Y-connection so you can switch from saline to fixative without removing the needle). Within 5 or 10 minutes, the perfusion is completed and the testes can be removed. The animal should be rigid and the testes yellow and hard.
5. Place the testes in fixative overnight. If properly perfused, the testes should be yellow throughout, hard, and easy to cut. I have found that epididymal spermatozoa may not fix as well with perfusion as with immersion, because the epididymis is not particularly vascular.

IV. Fixation of Sperm in Suspension

When spermatozoa are fixed in suspension, it is important to create a pellet where the cells are tightly packed without compromising fixation or damaging or decapitating the sperm with too high a centrifugal force. It is necessary to pack the sperm tightly, because searching for perfect cross sections of a sperm tail or exact sagittal sections of a sperm head will take less time if the sperm are closer together. It will also be easier to find a field with several sperm cut in the desired orientation. In addition, relatively high centrifugal force tends to pack spermatozoa in register, which may facilitate the location of cross sections of several spermatozoa in the same field. Employ the appropriate-size microfuge tube for the number of spermatozoa. For human spermatozoa, use a 1.8-ml microfuge tube to make a pellet with approximately 5×10^6 spermatozoa or a 400-μl tube with 10^6 spermatozoa. It is possible to make a pellet with as few as 4×10^5 sperm with a Sarstedt 72.707 tube (Sarstedt, Newton, NC), which has a small nipple at the base. The critical number of spermatozoa is different for different species. For example, rat sperm are much larger than human sperm, so fewer spermatozoa result in the same size pellet.

A. Fixation of Spermatozoa in Semen

1. Count the spermatozoa (i.e., in a Mackler chamber or hemocytometer).
2. For human spermatozoa, place semen containing 5×10^6 to 10^7 spermatozoa in a 15-ml plastic test tube. The spermatozoa of rabbits and hoofed animals are a little larger than human sperm so you may need to use fewer spermatozoa.
3. Add an amount of phosphate-buffered saline (PBS) equal to the volume of semen and centrifuge at 400*g* for 4 minutes.
4. Resuspend the pellet in about 1 ml PBS, add a few drops of fixative, and gently mix.
5. Immediately transfer the sperm suspension to a 1.81-ml microfuge tube, and immediately centrifuge at 4000*g* for 2 minutes.
6. Carefully remove the supernatant and slowly add fresh fixative so as not to disturb or dislodge the pellet and refrigerate. The pellet can be processed the next day or several days later.

B. Separation of Motile Sperm from Semen

As the majority of spermatozoa in human semen are either abnormal or dead, you may want to separate these from the motile cells with normal morphology. We employ the following relatively simple protocol to obtain highly motile sperm:

1. Allow semen to liquefy at room temperature for 30 minutes to 1 hour. Do not leave them any longer because seminal plasma is mildly toxic to all cells, including spermatozoa.
2. Gently mix 1 ml of semen with 3 ml of medium in a 5-ml glass test tube. For medium, use either Earle's balanced salt solution or Ham's F-10 containing 3 mg /bovine serum albumin.
3. Centrifuge 10 minutes at 300*g*.
4. Remove the supernate, add 1 ml of medium, and carefully resuspend the pellet.

5. Centrifuge 5 minutes at 300*g*.
6. Remove the supernate and resuspend the pellet by tapping the tube.
7. Place a 5 ml-glass tube containing 1 ml of medium in a 37°C, 5% CO_2 incubator at an angle of about 30°.
8. Using a Pasteur pipet, carefully introduce the sperm suspension to the bottom of a tube.
9. Incubate 2 hours without creating bubbles.
10. Carefully draw off the top 0.7 to 0.8 ml of the medium. This contains the live motile spermatozoa.
11. Place half of the medium in each of two 0.8-ml microfuge tubes, add 0.4 ml fixative (6% collidine-buffered glutaraldehyde), invert the tubes, and immediately spin in a microfuge tube at 4000*g* for 2 minutes. Leave in the refrigerator for a day to a week before dehydration.

V. Conclusion: A Few Hints

The procedures for postosmicating, dehydrating, and embedding spermatozoa are the same as for any other tissue; however, I have listed below several practices I find useful.

1. If you are especially concerned with preserving membrane structure, membranes will look bilaminar if you stain the tissue *in situ* with uranyl acetate. After osmication carry out the following procedure at room temperature. Dehydrate to 30% ethanol. After 15 minutes transfer to 30% methanol for 15 minutes. Next place the tissue in 2% uranyl acetate in 30% methanol for 1 hour. Place in 30% methanol for 10 minutes, then in 50% ethanol, and continue the dehydration. This method does not compromise other aspects of sperm morphology.

2. The subfibers of the flagellar tubes show up best with tannic acid, but I do not generally use this procedure because it may compromise other aspects of sperm structure and make the tissue more difficult to section. In addition, en bloc staining with uranyl acetate cannot be carried out. The procedure is to add 1 to 2% tannic acid to the fixative. Phosphate buffer must be used because tannic acid reacts with collidine.

3. I use very hard plastic when I embed spermatozoa, because the cells look crisper when they are embedded in hard plastic. I make up my embedding medium by mixing 25 ml Epon (Polybed, Polysciences) with 20 ml *N*-methylolacrylamide, 5 ml dodecenylsuccinic anhydride, and 0.8 ml DMP-30 in a 50-ml- Tripour beaker.

4. I embed many pieces of tissue and cut 1-μm-thick sections of 10 or 20 blocks before I cut any thin sections. I stain these sections with toluidine blue, view them in the light microscope, and select the best blocks for thin sectioning.

This ultimately saves time, especially with the epididymis, where the number and orientation of cells vary from block to block.

5. Finally, a sharp diamond knife is exceedingly important. Good luck!

Reference

Phillips, D. M., and Bedford, J. M. (1987). Sperm-sperm associations in the loris epididymis. *Gamete Res*. **18,** 17–23.

PART V

Motors and Motion Analysis

CHAPTER 30

Reactivation of *Chlamydomonas* Cell Models

Cynthia J. Horst[1] **and George B. Witman**

Worcester Foundation for Experimental Biology
Shrewsbury, Massachusetts 01545

I. Introduction

The unicellular biflagellate alga *Chlamydomonas* has been used extensively in the study of flagellar motility. Its utility as an experimental system is readily apparent from the wide variety of approaches to which it is amenable, as is evidenced by an extensive body of literature including numerous chapters in this volume. One such approach is the use of reactivated cell models to assess directly how changes in the surrounding medium can alter axonemal motility (Horst and Witman, 1993; Kamiya and Hasegawa, 1987; Kamiya and Witman, 1984). This chapter describes a procedure for the preparation of these models, with particular reference to their use in examining the effects of free calcium concentration on axonemal motility (Horst and Witman, 1993; Kamiya and Witman, 1984). The basic technique, however, is applicable not only to the study of calcium effects, but to the effect of any soluble substance that acts on a detergent-stable component of the axoneme. Indeed, similar *Chlamydomo-*

[1] Present address: Biology Department, Carroll College, 100 North East Avenue, Waukesha, WI 53186.

nas cell models have been very useful for investigating the functional rebinding of purified dynein to outer-doublet microtubules (Sakakibara and Kamiya, 1989; Takada *et al.,* 1992; Takada and Kamiya, 1994).

In this approach, the cell's plasma membrane is disrupted by the addition of a nonionic detergent (Nonidet P-40), leaving the cell wall, axonemes, and most internal cell components apparently intact. These resulting cell models are then reactivated by dilution into a reactivation buffer containing ATP. Through the addition of ethylene glycol bis(β-aminoethyl ether)-*N,N'*-tetraacetic acid (EGTA) and $CaCl_2$ to the reactivation buffer, the amount of free Ca^{2+} during reactivation can be carefully regulated (Bessen *et al.,* 1980). Dark-field light microscopic observation of reactivated cell models allows the integrity of the axonemes, as well as changes in swimming behavior, to be monitored. In studies comparing cell models reactivated at 10^{-9} and 10^{-7} *M* Ca^{2+}, it has been established that the cell's two axonemes are differentially sensitive to calcium (Kamiya and Witman, 1984). Further studies of a mutant strain defective in phototaxis have established that it is this differential sensitivity of the two axonemes that is responsible for mediating phototaxis (Horst and Witman, 1993).

II. Methods

A. Solutions

Wash buffer: 10 m*M* 4-(2-hydroxyethyl)-1-piperazineethanesulfonic acid (Hepes), pH 7.3, 0.5 m*M* EGTA, 4% sucrose

Demembranation buffer: 30 m*M* Hepes, pH 7.3, 5 m*M* $MgSO_4$, 1 m*M* dithiothreitol (DTT), 1 m*M* EGTA, 50 m*M* potassium acetate, 1% polyethylene glycol (20 kDa), 0.1% Nonidet P-40

Reactivation buffer: 30 m*M* Hepes, pH 7.3, 5 m*M* $MgSO_4$, 1 m*M* DTT, 50 m*M* potassium acetate, 1% polyethylene glycol (20 kDa), 1 m*M* ATP, $CaCl_2$ and EGTA adjusted to give desired Ca^{2+} (Bessen *et al.,* 1980)

	pCa 7	pCa 8	pCa9
$CaCl_2$	2.8 m*M*	0.28 m*M*	0.028 m*M*
EGTA	4.77 m*M*	2.25 m*M*	2.0 m*M*

B. Procedure

1. Small liquid cultures of *Chlamydomonas* are grown synchronously as previously described (Witman, 1986). Because the effects of growth conditions on calcium responsiveness are not known, the cultures are maintained such that they do not reach stationary growth phase. Additionally, because cells maintained in liquid culture undergo subtle changes over time, every 2 months new liquid cultures are started from stocks kept on agar slants.

2. *Chlamydomonas* cells are collected approximately halfway into the light period of the light : dark cycle. Cells are washed twice by centrifugation at 1100*g* and resuspension in wash buffer at room temperature.

3. The final resuspension is in a very small volume such that the density of cells will be appropriate for observation following the 100-fold dilution during demembranation and reactivation. Ten milliliters of cells at 5×10^5 to 2×10^6 cells/ml resuspended in 200–500 μl of wash buffer yields a suitable final cell concentration. Concentrated cells are stored on ice.

4. Demembranation is carried out on ice in a 96-well culture plate. Ten microliters of concentrated cells is added to 100 μl of demembranation buffer and gently stirred with the pipet tip to mix. Use micropipet tips with a reasonably large bore, and stir gently to mix rather than triturating, as the resulting cell models are easily damaged. Cells are left in the demembranation solution for 30–60 seconds.

5. Demembranated cell models can then be reactivated by diluting 10 μl of cell models into 100 μl reactivation buffer on ice, again mixing by very brief, gentle stirring. Although the cell models are immediately reactivated, the Ca^{2+}-dependent axonemal inactivation is time dependent. Initially, most of the cell models swim straight. Over the course of several minutes the number of circling models increases. Therefore, the reactivated cell models are given a 5-minute preincubation before assessing swimming behavior. During this preincubation period, the slide is prepared and the sample allowed to settle.

6. Cell models rapidly stick to untreated glass; therefore, both the slides and coverslips must be silanized before use. This can be accomplished using trimethylchlorosilane under vacuum in a desiccating chamber (Ausubel *et al.*, 1989). (Be sure to vent this in a hood.) The slides and coverslips should be thoroughly washed with detergent and rinsed with water before silanization, and rinsed with water following silanization to remove the resulting cloudy film.

7. Cell model swimming behavior is observed using dark-field light microscopy (see below). The use of dark-field optics greatly simplifies identification of the eyespot and, thereby, the dominant axoneme. Cell models are observed for 10 minutes, after which time a significant percentage of the models are damaged or immotile.

III. Comments

Cell models can be placed into one of five categories based on their behavior: (1) cell models that circle with the *cis* axoneme (that closest to the eyespot) dominant in determining the direction of turning; (2) cell models that circle with the *trans* axoneme (that farthest away from the eyespot) dominant in determining the direction of turning; (3) cell models that circle but in which the eyespot is not evident and, therefore, the dominant axoneme cannot be identified; (4) cell models that continue to swim straight; and (5) cell models that are not reactivated or visibly damaged. By categorizing each observed cell model, changes in the percentage of cell models that were not reactivated, damaged, or continued to swim straight (categories 4 and 5), can serve as internal controls. Using the conditions described here, approximately 70% of the cell models usually are motile, with approximately 30% of the total number of models circling. Despite silanization, some cell models become stuck to the slide or coverslip by either the cell wall or axonemes.

As noted above, cell models are very fragile and easily damaged. Dark-field light microscopy allows the condition of the axonemes to be assessed during the assay. Broken axonemes and axonemes in which the microtubules have splayed apart at the distal end are common forms of damage. Additionally, cell models may be observed in which one axoneme is completely inactivated due to extrusion of the central pair microtubules, which appear coiled at the distal end of the axoneme. These should not be confused with undamaged models circling in response to the Ca^{2+} concentration. Visualization of extruded central pair microtubules probably requires high-quality dark-field optics. We use an Olympus DC dark-field condenser (NA 1.2–1.33) with Zeiss Plan-Neofluar 20× or 40× objectives on a Zeiss Axioskop microscope.

References

Ausubel, F. M., Brent, R., Kingston, R. E., Moore, D. D., Seidman, J. G., Smith, J. A., and Struhl, K. (1986). "Current Protocols in Molecular Biology", John Wiley & Sons, New York, A.3.3–A.3.4.

Bessen, M., Fay, R. B., and Witman, G. B. (1980). Calcium control of waveform in isolated flagellar axonemes of *Chlamydomonas*. *J. Cell Biol.* **86,** 446–455.

Horst, C. J., and Witman, G. B. (1993). ptx1, a nonphototactic mutant of *Chlamydomonas,* lacks control of flagellar dominance. *J. Cell Biol.* **120,** 733–741.

Kamiya, R., and Hasegawa, E. (1987). Intrinsic difference in beat frequency between the two flagella of *Chlamydomonas reinhardtii*. *Exp. Cell Res.* **173,** 299–304.

Kamiya, R., and Witman, G. B. (1984). Submicromolar levels of calcium control the balance of beating between the two flagella in demembranated models of *Chlamydomonas*. *J. Cell Biol.* **98,** 97–107.

Sakakibara, H., and Kamiya, R. (1989). Functional recombination of outer dynein arms with outer arm-missing flagellar axonemes of a *Chlamydomonas* mutant. *J. Cell Sci.* **92,** 77–83.

Takada, S., Sakakibara, H., and Kamiya, R. (1992). Three-headed outer arm dynein from *Chlamydomonas* that can functionally recombine with outer arm-missing axonemes. *J. Biochem.* **111,** 758–762.

Takada, S., and Kamiya, R. (1994). Functional reconstitution of *Chlamydomonas* outer dynein arms from α-β and γ subunits: requirements of a third factor. *J. Cell Biol.* **126,** 737–745.

Witman, G. B. (1986). Isolation of *Chlamydomonas* flagella and flagellar axonemes. *Methods Enzymol.* **134,** 280–290.

CHAPTER 31

Reactivation of Extracted *Paramecium* Models

Yutaka Naitoh
Pacific Biomedical Research Center and Department of Microbiology
University of Hawaii at Manoa
Honolulu, Hawaii 96822

I. Introduction

The swimming behavior of *Paramecium* is dependent mostly on its ciliary motile activity. Ciliary motile activity is regulated by membrane electrogenesis (Naitoh and Eckert, 1969; Eckert, 1972, Naitoh, 1982).

A depolarizing receptor potential evoked by stimulation of the front end of *Paramecium* electrotonically spreads over the entire cell membrane to activate (open) depolarization-sensitive Ca^{2+} channels in the ciliary membrane. Activation of the channels causes regenerative entry of Ca^{2+} ions into the cilia, producing a sudden increase in the intraciliary Ca^{2+} concentration and an action potential. The Ca^{2+} increase activates a mechanism for reversing the direction of the effective power stroke of cilia to cause backward swimming of *Paramecium*. *Paramecium* thereby avoids a noxious stimulus at its front.

A hyperpolarizing receptor potential evoked by stimulation of the posterior end of *Paramecium* also spreads electrotonically over the entire cell membrane. Hyperpolarization of the membrane brings about an increase in the intracellular cAMP concentration, which activates the ciliary beating mechanism to cause an increase in the frequency of ciliary beating in the normal direction (Nakaoka *et al.,* 1984; Majima *et al.,* 1986; Bonini *et al.,* 1986). This brings about rapid forward swimming of the specimen. *Paramecium* thereby escapes from a noxious stimulus at its rear.

To understand the mechanism by which ciliary motile activity is regulated by these chemical messengers, it is important to examine the direct effects of the chemicals on the motile activity of the ciliary apparatus. An external application of these chemicals does not, however, necessarily increase the intracellular concentrations of the chemicals, as the cell membrane is poorly permeable to these chemicals. Moreover, an external application of Ca^{2+} ions will even produce a decrease in the intracellular Ca^{2+} concentration in *Paramecium,* because it deactivates Ca^{2+} channels so that Ca^{2+} influx decreases (Naitoh, 1981). Therefore, permeabilization of the cell membrane is one method essential for examining the direct effects of the chemicals. In a permeabilized cell, functions of the membrane as a signal transducer and a diffusion barrier are disrupted, while the motor function of the ciliary apparatus is kept normal.

The idea of permeabilization of the cell originated from a classic work on the glycerinated muscle fiber by Szent-Györgyi (1947). Hoffman-Berling (1955) applied Szent-Györgyi's technique to protozoan cells to make "ciliary models." Gibbons and Gibbons (1972) first introduced a nonionic detergent, Triton X-100, to obtain permeabilized seaurchin spermatozoa so good that almost 100% of them showed ATP–Mg^{2+}-reactivated flagellar motion.

It should be noted that a detergent destroys not only the membrane but also the ciliary apparatus. The concentration of the detergent, the time of exposure of the cells to the detergent, the ambient temperature, and the ionic conditions all influence the quality of the extracted models.

Tolerance to a given detergent differs from specimen to specimen depending on their culture condition, cell size, species, and strains. Although Triton X-100 is the most popular detergent for permeabilization of *Paramecium,* it is worthwhile examining other detergents as well, such as Nonidet P-40 (Goodenough, 1983) and saponin (Seravin, 1961).

In addition to ATP, Mg^{2+}, and KCl, other chemicals in the reactivation solution are important to keep the reactivated ciliary activity stable. Acetate ions, instead of Cl^- (Gibbons *et al.,* 1982), and polyethylene glycol (Goodenough, 1983) have been used to stabilize the ciliary activity.

It is important to determine your own conditions for permeabilization adequate for your experimental purposes. Therefore, the recipes given in the following sections are subject to your own modification.

II. Methods

A. Triton X-100 Extraction of Paramecium

This standard procedure of Naitoh and Kaneko (1972) for Triton X-100 extraction of *Paramecium caudatum* has been employed, with minor modifications, by many authors for other species of *Paramecium* (*P. tetraurelia:* Kung and Naitoh, 1973; Bonini *et al.*, 1986; Lieberman *et al.*, 1988; *P. multimicronucleatum:* Okamoto and Nakaoka, 1994) and other genera of ciliates, including *Tetrahymena* (Takahashi *et al.*, 1980; Goodenough, 1983) and *Euplotes* (Epstein and Eckert, 1973).

1. Culture of Specimens

Specimens obtained from any kind of culture (bacterized hay, rice straw, wheat straw, lettuce, or milk infusion) can be used. Crude cultures, such as bacterized hay infusion made over a layer of field soil in a ceramic bowl, are preferable to the more sophisticated axenic cultures to obtain good extracted specimens. Such specimens show coordinated and long-lasting ciliary movement when reactivated. Axenic culture, however, is essential for more critical biochemical examinations of the extracted specimens.

2. Stock Solutions

- 0.1 *M* KCl
- 0.1 *M* $CaCl_2$
- 0.1 *M* $MgCl_2$
- 0.1 *M* EDTA (neutralized by KOH)
- 0.1 *M* EGTA (neutralized by KOH)
- 0.1 *M* Tris–maleate pH buffer (pH 7.0)
- 0.1 *M* Tris–HCl pH buffer (pH 7.4)
- 0.1 *M* ATP (neutralized by KOH, to be stored in a freezer)
- 0.1 *M* KOH
- 0.1 *M* HCl
- 1% (by volume) Triton X-100

3. Experimental Solutions

Washing solution I (for living specimens from culture): 2 m*M* $CaCl_2$ + 1 m*M* Tris–HCl (pH 7.2) (in final concentration). Mix 10 ml of 0.1 *M* $CaCl_2$, with 5 ml of 0.1 *M* Tris–HCl and add H_2O to make 500 ml of solution.

Extraction solution: 20 m*M* KCl + 10 m*M* ethylenediaminetetraacetic acid (EDTA) + 10 m*M* Tris–maleate (pH 7.0) + 0.01% (by volume) Triton X-100. Mix 100 ml of 0.1 *M* KCl, 50 ml of 0.1 *M* EDTA, 50 ml of 0.1 *M* Tris–maleate (pH 7.0), and 5 ml of 1% Triton X-100, and add H_2O to make 500 ml of solution.

Washing solution II (for extracted specimens): 50 m*M* KCl + 10 m*M* Tris–maleate (pH 7.0). Mix 250 ml of 0.1 *M* KCl with 50 ml of 0.1 *M* Tris–maleate and add H_2O to make 500 ml of solution.

Test solutions: Basic components of the test solutions are the same as those of the washing solution II. Chemicals to be tested are added to the solution. Two representative test solutions are:

1. Standard reactivation solution: 50 m*M* KCl + 4 m*M* $MgCl_2$ + 4 m*M* ATP + 3 m*M* ethylene glycol bis (B-aminoethyl ether)-*N,N'*-tetraacetic acid (EGTA) + 10 m*M* Tris–maleate (pH 7.0). Mix 5 ml of 0.1 *M* KCl, 0.4 ml of 0.1 *M* $MgCl_2$, 0.4 ml of 0.1 *M* ATP, 0.3 ml of 0.1 *M* EGTA, and 1 ml of 0.1 *M* Tris–maleate and add H_2O to make 10 ml of solution. The extracted specimens swim forward in this solution.

2. Ca series reactivation solutions: Add an amount of 0.1 *M* $CaCl_2$ to be determined by the experiment to the standard reactivation solution. The concentration ratio of Ca^{2+} to EGTA determines the free Ca^{2+} concentration in the solution (Ca^{2+} buffer; Portzehl *et al.*, 1964; Noguchi *et al.*, 1986). The extracted specimens swim backward when the free Ca^{2+} concentration in the solution is above 10^{-6} *M*.

4. Experimental Procedures

a. Collecting and Washing the Specimens

Introduce about 20 ml of culture medium containing specimens of *Paramecium* into a long-necked flask (such as a 50-ml volumetric flask). Fill the flask with washing solution I. Keep the flask still on a table for ca. 4 minutes. Specimens will gather at the neck portion of the flask due to their negative geotaxis. Pipet the specimens into another long-necked flask, then fill the flask with the washing solution. The specimens again gather in the neck portion. Repeat this procedure three times to dilute the culture medium. Pipet the concentrated specimens in the neck portion into a centrifuge tube. Keep the tube in an ice bath for 3 minutes. Centrifuge the tube gently to make a loose pellet of the specimens. Remove the supernatant by suction.

b. Extracting the Specimens

Introduce ca. 5 ml of cold (0–1°C) extraction solution into the tube to suspend the washed specimens in the solution. Keep the tube vertical in an ice bath for ca. 30 minutes. Specimens gradually lose their mobility in the extraction solution and sink toward the bottom of the tube. Remove the extraction solution by suction, leaving the extracted specimens at the bottom.

c. Washing the Extracted Specimens

Introduce ca. 5 ml of cold (0–1°C) washing solution II into the tube to suspend the extracted specimens in the solution. Centrifuge the tube very gently to make a loose pellet of the specimens. Remove the supernatant by suction. Repeat this procedure three times to wash the Triton X-100 and EDTA away from specimens. Keep the extracted specimens cooled in an ice bath, in a minute amount of the washing solution.

d. Reactivating the Extracted Specimens

Pipet about 1 μl of washing solution containing hundreds of extracted specimens into ca. 1 ml of a test solution on a depression slide at room temperature (ca. 24°C). Stir the mixture gently. A few seconds after this mixing, the extracted specimens will begin to swim due to the reactivation of the ciliary movement, providing the test solution contains an adequate amount of ATP and Mg^{2+}.

e. Observing and Recording Reactivated Specimens

Pipet about 100 μl of test solution containing scores of reactivated specimens onto a glass slide to make a droplet. Put several small pieces of crushed coverslip into the droplet so that a thin space is made when a coverslip is put on the droplet. The extracted specimens will swim freely in the thin space between the glass slide and the coverslip. Put white vaseline around the edge of the coverslip to prevent evaporation of the test solution. The specimens are now ready for observation and recording of their motile activity.

Swimming Paths of the Reactivated Specimens. Under dark-field illumination, swimming specimens are easily observed through a conventional binocular dissecting microscope with a final magnification of ×10–×20. Movement of the specimens can be recorded with a conventional video recorder. The swimming path of a specimen is then determined by frame-by-frame analysis of displayed images of the specimen. Various kinds of computerized analyzers are available for this type of analysis (e.g., see Chapter 40 in this volume).

The swimming path of a specimen can be photographed as a curved spiral line on a still film when the exposure time is extended to 1–5 seconds. Swimming direction and swimming velocity can be determined from the swimming path.

Ciliary Motile Activity. A conventional phase-contrast, interference-contrast, or dark-field microscope with a final magnification of ×200–×400 is essential for observing and recording the reactivation of the cilia. High-speed video recording of the cilia is essential to determine the beat frequency, the direction of the effective power stroke, and the beating form of an individual cilium (see Chapter 34 in this volume). The computerized image contrast-enhancement technique is useful for obtaining a good image of a cilium (see Chapter 38 in this volume).

The beat frequency of reactivated cilia of an extracted specimen can be determined from the frequency of change in light intensity of a selected small area of a magnified image of the specimen due to the passage of metachronal waves. This is accomplished by placing a photodiode or a photomultiplier in the path of the image (Naitoh and Kaneko, 1973).

The relative force exerted by the reactivated cilia on a given specimen can be estimated from the swimming path of the specimen (Naitoh and Sugino, 1984). The direction of the force corresponds to the mean direction of the effective power strokes, and the amount of force is proportional to the mean beat frequency of all the cilia.

Ciliary metachronal waves can be observed as darker (or brighter) lines moving on the ciliary surface or as waves moving along the margin of a reactivated specimen. The metachronal wave lines are easily photographed on film. The direction of the effective power stroke can be determined from the metachronal wave line, as the angle between the metachronal wave line and the direction of the effective power stroke is always 90° (Machemer, 1972). The direction in which nonbeating cilia point, which usually corresponds to the direction of the effective power stroke, can be determined from a photograph of an extracted specimen.

B. Modified Triton X-100 Extraction of *Paramecium*

Nakaoka and Toyotama (1979) added Mg^{2+} ions to the Triton X-100-containing extraction solution. Specimens extracted with this solution showed higher sensitivity to Ca^{2+} in modifying beat frequency in Mg^{2+}–ATP-reactivated cilia than those extracted with a solution without Mg^{2+}. The maximum beat frequency under optimum reactivation conditions was higher in the Mg^{2+}-containing extraction solution and close to the maximum beat frequency of living specimens. Formation of metachronal waves on the reactivated ciliary surface was also enhanced (Nakaoka *et al.*, 1984; Nakaoka and Ooi, 1985). The presence of Mg^{2+} in the extraction solution maintains the ciliary motile mechanisms of *Paramecium* in a closer-to-normal condition than when Mg^{2+} is absent during the Triton X-100 treatment.

1. Experimental Solutions

Extraction solution: 20 m*M* KCl + 5 m*M* $MgCl_2$ + 5 m*M* EGTA + 0.008% (by volume) Triton X-100 + 10 m*M* Tris–maleate buffer (pH 7.0) (in final concentration).

Washing solution I: 20 m*M* KCl + 5 m*M* $MgCl_2$ + 3 m*M* EGTA + 10 m*M* Tris–maleate buffer (pH 7.0).

Washing solution II: 20 m*M* KCl + 5 m*M* $MgCl_2$ + m*M* Tris–maleate buffer (pH 7.0).

Reactivation solutions: 10 m*M* KCl + 8 m*M* $MgCl_2$ + 4 m*M* ATP + 1 m*M* EGTA + an amount of $CaCl_2$ needed to provide the appropriate free Ca^{2+} concentration required for the experiment + 10 m*M* Tris–maleate buffer (pH 7.0).

2. Experimental Procedures

a. Collecting and Washing Cultured Specimens

See Section II,A,4,a.

b. Extracting the Specimens

Suspend the washed specimens in the extraction solution in a centrifuge tube for 10–12 minutes at room temperature (25°C). Remove the extraction solution by suction, leaving the extracted specimens at the bottom of the tube.

c. Washing the Extracted Specimens

Introduce cold (0°C) washing solution I into the tube to suspend the extracted specimens in the solution. Keep the tube vertical in an ice bath for 15 minutes. The extracted specimens will sink toward the bottom of the tube. Remove the washing solution by suction, leaving the specimens at the bottom of the tube.

Introduce cold (0°C) washing solution II into the tube to suspend the specimens. Keep the tube in an ice bath. The specimens sink toward the bottom in 15 minutes. Washed specimens thus obtained are ready for experimentation.

d. Reactivating the Extracted Specimens

Pipet a concentrated suspension of the extracted specimens into ca. 1 ml of a reactivation solution at room temperature (20–22°C). Stir the mixture gently for a few seconds. Observations and measurements are made 3–20 minutes after mixing at room temperature.

C. Glycerol Extraction of *Paramecium*

Glycerol-extracted *Paramecium* models do not show ciliary beating in the presence of ATP and Mg^{2+}; however, glycerol extraction supports a change in the direction in which nonbeating cilia point on administration of ATP and Ca^{2+}. That is, cilia swing once to change their pointing direction (ciliary reorientation response). The reorientation response in nonbeating cilia corresponds to a change or reversal of the direction of the effective power stroke (ciliary reversal) in beating cilia (Naitoh, 1966). Glycerol, therefore, specifically destroys the beating mechanism in the ciliary motile apparatus.

1. Experimental Solutions

Washing solution I (for specimens from culture): 2 m*M* $CaCl_2$ + 1 m*M* Tris–HCl (pH. 7.2) (in final concentration).

Extraction solution: 50 m*M* KCl + 10 m*M* EDTA + 10 m*M* Tris–HCl (pH 7.4) + 50% (by volume) glycerol.

Washing solution II (for extracted specimens): 50 m*M* KCl + 10 m*M* Tris–HCl (pH 7.4).

Test solutions: Basic components of a test solution are the same as those of the washing solution II except for pH (pH is 9.0 instead of 7.4 in the washing solution). Chemicals to be tested are added to the basic solution.

2. Experimental Procedures

a. Collecting and Washing the Cultured Specimens

See Section II,A,4,a.

b. Extracting the Specimens

Introduce ca. 5 ml of cold (0–1°C) extraction solution into the centrifuge tube containing a pellet of washed specimens, and stir gently to suspend them in the solution. Keep the tube vertical in a refrigerator for 10–15 days at −15°C. The specimens will sink to the bottom of the tube to form a loose pellet during the extraction.

c. Washing the Extracted Specimens

Remove the glycerol extraction solution from the tube by suction, leaving a pellet of the extracted specimens. Add cold (0–1°C) washing solution II to the tube and stir gently to resuspend the extracted specimens in the solution. Centrifuge the tube gently to make a loose pellet of the extracted specimens. Repeat this procedure three times to remove all glycerol and EDTA from the extracted specimens. The washed specimens are to be kept in the washing solution for at least 15 minutes at 0–1°C prior to experimentation.

d. Reactivating the Extracted Specimens

See Section II,A,4,d.

e. Observing and Recording the Reactivated Specimens

The reactivated specimens are observed and photographed through a phase-contrast objective (×20). The angle between the ciliary axis and the cell surface at the right anterior edge of the specimen (the "right" is at the observer's right hand when the side bearing the oral groove is down and the anterior end points away from the observer) is measured on the print of a photographed specimen (×400) to determine the degree of ciliary orientation response.

D. Triton–Glycerol Extraction of *Paramecium*

As mentioned in the previous section, glycerol destroys the beating mechanism of the ciliary apparatus in *Paramecium*. Noguchi *et al.* (1986) performed successive extractions of *Paramecium,* first by Triton X-100 then by glycerol. Comparison of the reactivated motile activity between Triton-extracted and

Triton–glycerol-extracted specimens is useful for understanding the functional relationship between the reversal and beating mechanisms in the ciliary motile apparatus.

1. Experimental Solutions

Extraction solution I (Triton X-100 extraction solution): This solution is the same as the Triton X-100 extraction solution of Naitoh and Kaneko (1972) (see Section II,A,3).

Extraction solution II (glycerol extraction solution): 50 m*M* KCl + 2 m*M* EDTA + 10 m*M* Tris–maleate (pH 7.0) + 30% (by volume) glycerol.

Washing solution I (for the first Triton-extracted specimens): 50 m*M* KCl + 2 m*M* EDTA + 10 m*M* Tris–maleate (pH 7.0).

Washing solution II (for the second glycerol-extracted specimens): 50 m*M* KCl + 10 m*M* Tris–maleate (pH 7.0) + 30% (by volume) glycerol.

Test solutions: Basic components of the test solutions are the same as those of washing solution II. Chemicals to be tested are added to this solution.

2. Experimental Procedures

a. Collecting and Washing the Specimens

See Section II,A,4,a.

b. Extracting and Washing of Specimens with Triton

The procedures are the same as those for the Triton X-100 extraction described in Sections II,A,4,b and c. The extracted and washed specimens are then kept in cold (0–1°C) washing solution I for 15 minutes.

d. Extracting the Specimens with Glycerol

Centrifuge the washed Triton-extracted specimens gently to make a loose pellet. Resuspend the specimens in ice-cold extraction solution II. Keep them in an ice bath (0°C) for 1 hour. Then centrifuge them gently to make a loose pellet.

e. Washing the Twice-Extracted Specimens

Resuspend the extracted specimens in ice-cold washing solution II. Then centrifuge them gently to make a loose pellet. Repeat this procedure three times to remove EDTA from the extracted specimens. Keep the extracted specimens immersed in the ice-cold washing solution for at least 1 hour prior to experimentation.

f. Reactivating and Observing the Extracted Specimens

See Section II,A,4,e.

E. Modified Triton–Glycerol Extraction of *Paramecium*

To examine the effects of externally applied macromolecules, such as enzymes, antibodies, and polynucleotides, on the motility machinery of the cilia, the membrane of the extracted specimens should be permeable to these molecules. The membrane of Triton X-100-extracted *Paramecium* obtained according to the extraction procedures described in the previous sections is not permeable to these macromolecules. In fact, the external application of immunoglobulin causes osmotic shrinkage of the extracted *Paramecium*.

Noguchi (1987) successfully extracted *Paramecium* with a solution having a higher (1% instead of 0.01%) concentration of Triton X-100. He then examined the effects of externally applied trypsin (trypsin digestion) on the ATP-reactivated motile activity of the extracted specimens.

Highly permeabilized specimens can be used for immunocytological experimentation. In some cases the effects of an antibody to a specific molecule on the ciliary motile activity can be examined by a simple external application of the antibody without the need for intracellular microinjection (Adoutte *et al.*, 1991, Peranen *et al.*, 1993).

1. Experimental Solutions

Washing solution (for both cultured specimens and Triton X-100-extracted specimens): 50 m*M* KCl + 2 m*M* EDTA + 10 m*M* Tris–maleate (pH 7.0) (final concentration).

Extraction solution I (Triton X-100 extraction solution): 20 m*M* KCl + 10 m*M* EDTA + 10 m*M* Tris–maleate (pH 7.0) + 1% (by volume) Triton X-100.

Extraction solution II and equilibration solution (glycerol extraction solution): 50 m*M* KCl + 10 m*M* glycerol + 30% (by volume) glycerol.

Test solutions: Basic components of the test solutions are 50 m*M* KCl, 1 m*M* $MgCl_2$, 1 m*M* ATP, 10 mM Tris-maleate (pH 7.0), and 30% (by volume) glycerol. Chemicals to be tested are to be added to the basic components.

2. Experimental Procedures

a. Collecting and Washing the Specimens

See Section II,A,4,a.

b. Extracting the Specimens with Triton-Containing Solution

Introduce ca. 5 ml of ice-cold extraction solution I into a centrifuge tube containing a loose pellet of washed specimens and resuspend them in the solution for 2 minutes. Centrifuge the specimens gently to make a loose pellet. Remove the extraction solution by suction.

c. Washing the Extracted Specimens
Introduce ca. 5 ml of ice-cold washing solution into the centrifuge tube to resuspend the extracted specimens. Centrifuge the specimens gently to make a loose pellet of the specimens. Remove the washing solution. Repeat this procedure three times to remove Triton from the specimens.

d. Extracting and Equilibrating in Glycerol-Containing Solution
Introduce ca. 5 ml of ice-cold extraction solution II into the tube to resuspend the washed extracted specimens in the solution. Keep the specimens in the solution more than 30 minutes. The centrifuge the specimens gently to make a loose pellet. Remove the solution from the tube. Keep the specimens bathed in a minute amount of the equilibration solution in an ice bath.

e. Reactivating and observating the specimens
See Section II,A,4,e.

F. Isolation of Ciliated Cortical Sheets from Triton-Extracted *Paramecium*

Cilia of *Paramecium* beat in three dimensions (Machemer, 1972; Naitoh and Sugino, 1984). To determine the exact direction of the effective power stroke or the pointing direction of cilia, it is desirable to look down on the ciliated surface. The profile view of cilia gives only approximate values for these directions. A ciliated cortical sheet isolated from Triton X-100-extracted *Paramecium* is particularly suitable for observing and recording the top view of cilia. The quality of microscopic images of cilia is far better in the cortical sheet than when viewed through the whole cell, because the cortical sheet is far thinner so that optical disturbance is far less. Thus Noguchi *et al.* (1991) developed a procedure to isolate ciliated cortical sheets from Triton-extracted *Paramecium*. Glycerol, which was used in their original experimental solutions, can be eliminated if required for your experimental purposes (Noguchi, 1987; Okamoto and Nakaoka, 1994a,b).

1. Experimental Solutions

Washing solution I (both for cultured and extracted specimens): 50 m*M* KCl + 2 m*M* EDTA + 10 m*M* Tris–maleate (pH 7.0) (in final concentration).

2. Extraction solution: 20 m*M* KCl + 10 m*M* EDTA + 10 m*M* Tris–maleate (pH 7.0) + 1% (by volume) Triton X-100.

KCL–glycerol solution: 50 m*M* KCl + 10 m*M* Tris–maleate (pH 7.0) + 30% (by volume) glycerol.

Test solutions: Basic components of the test solutions are 50 m*M* KCl, 1 m*M* $MgCl_2$, 1 m*M* ATP, 10 m*M* Tris–maleate (pH 7.0), and 30% (by volume) glycerol. Free Ca^{2+} concentration is controlled by a Ca–EGTA buffer with 1 mM EGTA.

2. Procedures

a. Collecting and Washing the Cultured Specimens
See Section II,A,4,a.

b. Extracting the Specimens
Suspend washed and concentrated specimens in the ice-cold extraction solution for 2 minutes. Centrifuge them gently to make a loose pellet. Remove the extraction solution by suction.

c. Washing the Extracted Specimens
Resuspend the extracted specimens in ice-cold washing solution. Centrifuge them gently to form a loose pellet. Repeat this procedure three times to remove Triton from the extracted specimens.

d. Fragmenting the Washed Extracted Specimens
Pipet a suspension of the extracted specimens in and out several times to fragment the specimens. The inner diameter of the pipet should be ca. 50 μm. A sharp broken edge of the orifice of the pipet is effective in fragmenting the specimens. Keep fragmented specimens suspended in ice-cold equilibration solution until used.

e. Reactivating and Observing the Cortical Sheets
The following procedures are carried out at a room temperature of 23–25°C. Pipet about 20 μl of the fragment suspension on a glass slide. Place a coverslip with a small amount of vaseline on two opposite edges on the suspension to keep the fragments in a thin space between the coverslip and the glass slide. Adjust the thickness of the space by controlling the amount of vaseline and the force used to press the coverslip onto the glass slide.

Then gently pipet the equilibration solution into the space from its vaseline-free side, and remove excess solution from the opposite side by placing the edge of a piece of filter paper on it. During this perfusion procedure, some of the fragments will attach with their cellular side flat against the glass slide (or coverslip). The ciliated cortical sheets thus obtained are ready for perfusion of a test solution. Images of cilia on the cortical sheet are observed and recorded as described in a previous section (see Section II,A,4,e).

Acknowledgments

I thank Dr. R. D. Allen and Dr. A. K. Fok for their critical reading of the manuscript which was prepared in their laboratory with the support of National Science Foundation grants MCB9017455 and MCB9206097.

References

Adoutte, A., Delgado, P., Fleury, A., Levilliens, N., Laine, M. C., Marty, M. C., Boisvieux-Ulrich, E., and Sandoz, D. (1991). Microtubule diversity in ciliated cells: evidence for its generation by post-translational modification in the axonemes of *Paramecium* and quail oviduct cells. *Biol. Cell* **71,** 227–245.

Bonini, N. M., and Nelson, D. L. (1988). Differential regulation of *Paramecium* ciliary motility by cAMP and cGMP. *J. Cell Biol.* **106,** 1615–1623.

Bonini, N. M., Gustin, M. C., and Nelson, D. L. (1986). Regulation of ciliary motility by membrane potential in *Paramecium:* A role for cyclic AMP. *Cell Motil. Cytoskel.* **6,** 256–272.

Eckert, R. (1972). Bioelectric control of ciliary activity. *Science* **176,** 473–481.

Epstein, M., and Eckert, R. (1973). Membrane control of ciliary activity in the protozoan *Euplotes. J. Exp. Biol.* **58,** 437–462.

Gibbons, B. H., and Gibbons, I. R. (1972). Flagellar movement and adenosinetriphosphate activity in sea urchin sperm extracted with Triton X-100. *J. Cell Biol.* **54,** 75–97.

Gibbons, I. R., Evans, J. A., and Gibbons, B. H. (1982). Acetate anions stabilize the latency of dynein 1 ATPase and increase the velocity of tubule sliding in reactivated sperm flagella. *Cell Motil.* **1**(Suppl.), 181–184.

Goodenough, V. W. (1983). Motile detergent-extracted cells of *Tetrahymena* and *Chlamydomonas. J. Cell Biol.* **96,** 1610–1621.

Hamasaki, T., and Naitoh, Y. (1985). Localization of calcium sensitive reversal mechanism in a cilium of *Paramecium. Proc. Jpn Acad.* **61B,** 140–143.

Hamasaki, T., Barkalow, K., Richmond, J., and Satir, P. (1991). Cyclic AMP-stimulated phosphorylation of an axonemal polypeptide that copurifies with the 22S dynein arm regulates microtubular translocation velocity and swimming speed in *Paramecium. Proc. Natl. Acad. Sci. U.S.A.* **88,** 7918–7922.

Hamasaki, T., Murtaugh, T. J., Satir, B. H., and Satir, P. (1989). In vitro phosphorylation of *Paramecium* axonemes and permeabilized cells. *Cell Motil. Cytoskel.* **12,** 1–11.

Hoffman-Berling, H. (1955). Geiselmodelle und Adenosintriphosphat. *Biochim. Biophys. Acta* **16,** 146–154.

Kung, C., and Naitoh, Y. (1973). Calcium-induced ciliary reversal in the extracted models of "pawn", a behavioral mutant of *Paramecium. Science* **179,** 195–196.

Lieberman, S. J., Hamasaki, T., and Satir, P. (1988). Ultrastructure and motion analysis of permeabilized *Paramecium* capable of motility and regulation of motility. *Cell Motil. Cytoskel.* **9,** 73–84.

Machemer, H. (1972). Ciliary activity and the origin or metachrony in *Paramecium:* Effects of increased viscosity. *J. Exp. Biol.* **65,** 427–448.

Majima, T., Hamasaki, T., and Arai, T. (1986). Increase in cellular cyclic GMP level by potassium stimulation and its relation to ciliary orientation in *Paramecium. Experientia* **42,** 62–64.

Naitoh, Y. (1966). Reversal response elicited in nonbeating cilia of *Paramecium* by membrane depolarization. *Science* **154,** 660–662.

Naitoh, Y. (1969). Control of the orientation of cilia by adenosinetriphosphate, calcium and zinc in glycerol-extracted *Paramecium caudatum. J. Gen. Physiol.* **53,** 517–529.

Naitoh, Y. (1981). Membrane currents in voltage-clamped *Paramecium caudatum. In* "Nerve Membrane Biochemistry and Function of Channel Protein" (G. Matsumoto and M. Kotani, eds.), pp. 113–117. University of Tokyo Press, Tokyo.

Naitoh, Y. (1982). Protozoa. *In* "Electric Conduction and Behaviour in 'Simple' Invertebrates" (G. A. B. Shelton, ed.), pp. 1–48, Clarendon Press, Oxford.

Naitoh, Y., and Eckert, R. (1969). Ionic mechanisms controlling behavioral responses of *Paramecium* to mechanical stimulation. *Science* **164,** 963–965.

Naitoh, Y., and Kaneko, H. (1972). Reactivated Triton-extracted models of *Paramecium:* modification of ciliary movement by calcium ions. *Science* **172,** 523–524.

Naitoh, Y., and Kaneko, H. (1973). Control of ciliary activity by adenosinetriphosphate and divalent cation in Triton-extracted models of *Paramecium caudatum. J. Exp. Biol.* **58,** 657–676.

Naitoh, Y., and Sugino, K. (1984). Ciliary movement and its control in *Paramecium*. *J. Protozool.* **31,** 31–40.

Nakaoka, Y., and Ooi, H. (1985). Regulation of ciliary reversal in Triton-extracted *Paramecium* by calcium and cyclic adenosinemonophosphate. *J. Cell Sci.* **77,** 185–195.

Nakaoka, Y., and Toyotama, H. (1979). Directional changes of ciliary beat effected with Mg^{2+} in *Paramecium*. *J. Cell Sci.* **40,** 207–214.

Nakaoka, Y., Tanaka, H., and Oosawa, F. (1984). Ca^{2+} dependent regulation of beat frequency of cilia in *Paramecium*. *J. Cell Sci.* **65,** 223–231.

Noguchi, M. (1987). Ciliary reorientation induced by trypsin digestion in Triton-glycerol-extracted *Paramecium caudatum*. *Cell Struct. Funct.* **12,** 503–506.

Noguchi, M., Inoue, H., and Kubo, K. (1986). Control of the orientation of cilia by ATP and divalent cations in Triton-glycerol-extracted *Paramecium caudatum*. *J. Exp. Biol.* **120,** 105–117.

Noguchi, M., Nakamura, Y., and Okamoto, K. (1991). Control of ciliary orientation in ciliated sheets from *Paramecium:* differential distribution of sensitivity to cyclic nucleotide. *Cell Motil. Cytoskel.* **20,** 38–46.

Okamoto, K., and Nakaoka, Y. (1994a). Reconstitution of metachronal waves in ciliated cortical sheets of *Paramecium*. I. Wave stabilities. *J. Exp. Biol.* **192,** 61–72.

Okamoto, K., and Nakaoka, Y. (1994b). Reconstitution of metachronal waves in ciliated cortical sheets of *Paramecium*. II. Asymmetry of the ciliary movements. *J. Exp. Biol.* **192,** 73–81.

Peranen, J., Rikkonen, M., and Kaarialnen, L. (1993). A method for exposing hidden antigenic sites in paraformaldehyde-fixed cultured cells applied to initially unreactive antibodies. *J. Histochem. Cytochem.* **41,** 447–454.

Portzehl, H., Caldwell, P. C., and Ruegg, J. C. (1964). The dependence of contraction and relaxation of muscle fibers from the crab *Maya squinado* on the internal concentration of free calcium ions. *Biochim. Biophys. Acta* **79,** 581–591.

Seravin, L. N. (1961). The role of adenosinetriphosphate in the beating of infusorian. *Biokhimiia* **26,** 138–142.

Szent-Györgyi, A. (1947). "Chemistry of muscular contraction". Academic Press, New York.

Takahashi, M., Onimaru, H., and Naitoh, Y. (1980). A mutant of *Tetrahymena* with non-excitable membrane. *Proc. Jpn Acad.* **56B,** 585–590.

Toyotama, H., and Nakaoka, Y. (1979). Effect of temperature on the swimming velocity of Triton-extracted models of *Paramecium caudatum*. *Cell Struct. Funct.* **4,** 34–43.

CHAPTER 32

Isolation and Reactivation of Newt Lung Respiratory Cilia

Robert Hard

Department of Anatomy and Cell Biology
School of Medicine
State University of New York at Buffalo
Buffalo, New York 14214

I. Introduction

Cilia serve as the motile components of the respiratory mucociliary epithelium and thus aid in defense against respiratory tract infections. The constant and coordinated beating of cilia minimizes the adverse effects of inhaled material by exporting out of the respiratory tract any foreign material trapped in the mucous blanket. It is widely known that dysfunction of mucociliary clearance may result either from impaired cilia or from abnormalities in the fluid component. Over the past 10 years, we have developed and used a series of models or simplifications of the respiratory epithelium from newt lungs that possess several advantages for studying mucociliary transport and the regulation of respiratory cilia (Hard and Weaver, 1983; Rieder and Hard, 1990). These models include isolated lungs, isolated epithelial sheets, epithelia in primary tissue culture, isolated ciliated cells, isolated and reactivated ciliary tufts, and populations of demembranated ciliary axonemes. Such models allow one to study the mucociliary transport process at different levels of organization without the regulatory influences of higher levels of organization. The preparation and use of isolated lungs, tissue culture material, and isolated sheets of mucociliary epithelium are outlined elsewhere (Hard and Weaver, 1983; Rieder and Hard, 1990; Weaver and Hard, 1985a,b). In this chapter, the methods used to isolate and reactivate demembranated ciliary axonemes from lungs of the newt, *Taricha granulosa,* are summarized.

II. Methods

A. Solutions

Wash solution: 120 m*M* potassium 1,4-piperazinediethanesulfonic acid (K Pipes), 2.5 m*M* $MgSO_4$, 1 m*M* dithiothreitol (DTT).

Demembranating solution: 120 m*M* K Pipes, 2.5 m*M* $MgSO_4$, 1 m*M* DTT, 1% Triton X-100, 10% sucrose. Add 0.5 m*M* EDTA (adjusted to pH 7.0) to isolated axonemes in demembranating solution.

Reactivating solution: 120 m*M* K Pipes, 2.5 m*M* $MgSO_4$, 1 m*M* DTT, 1.25 m*M* ATP, 0.5 m*M* ethylenediaminetetraacetic acid (EDTA).

High-salt solution: 0.46 *M* KCl, 19.2 m*M* K Pipes, 2.5 m*M* $MgSO_4$, 1 m*M* DTT, 0.5 m*M* EDTA.

B. Isolation and Reactivation of Normal Ciliary Axonemes

With populations of demembranated cilia, one can study ciliary function independent of cellular controls and investigate regulatory mechanisms that are inherent to the axoneme itself. Such demembranated axonemes can be isolated in a simple, one-step process from newt lungs.

1. Newts of the species *Taricha granulosa* are purchased from Carolina Biological Supply and are maintained in the laboratory on a diet of beef liver or meal worms. Animals are sacrificed by cervical dislocation. The lungs of newts are classified as simple lungs, each consisting of a blind, hollow, cylindrical sac. Anteriorly, they communicate into a common pharyngeal region through a slitlike glottis. Lungs are dissected by severing these anterior connections as well as mesenteric connections to organs such as the liver and spleen. Isolated lungs are stored briefly in wash solution. Each lung is placed on a piece of dental wax, and its distal tip removed with a scalpel. A sealed section from a drawn out Pasteur pipet is threaded carefully through the lung (Fig. 1). The sealed tip of the pipet is broken and then inserted into a simple perfusion apparatus (Hard *et al.,* 1988). To isolate populations of ciliary axonemes, the lung is perfused with 1 ml of wash solution to remove mucus and debris. After the wash solution is removed completely, the lung is perfused carefully with 0.5 to 1.0 ml of demembranating solution over a period of 30–60 seconds. The

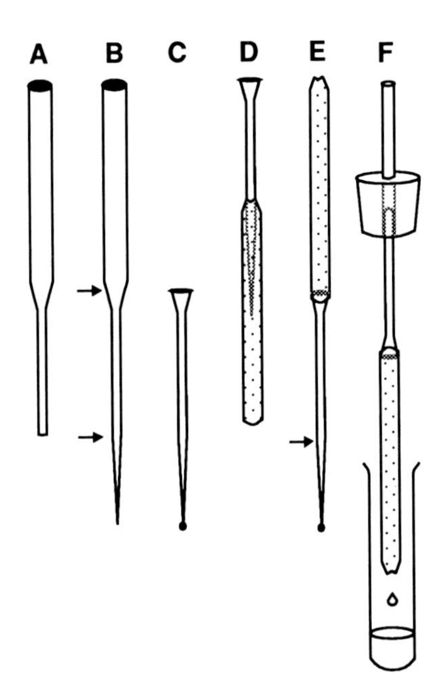

Fig. 1 Preparation of a newt lung for isolating demembranated axonemes. A Pasteur pipet (A) is drawn out in a Bunsen burner (B) and a flanged segment is removed by scoring with a diamond scribe (arrows). The flanged end is fire-polished and the small end is sealed, leaving a small bead on the end (C). The beaded end is threaded through the lung (D,E). Then the tip is broken off (arrow, E) and inserted into the perfusion apparatus (F) (see Hard *et al.,* 1988, for more details on the perfusion apparatus).

first few drops are discarded and the remaining perfusate is collected and placed immediately on ice. EDTA is added from a 100 × stock solution (pH 7.0) to a final concentration of 0.5 m*M*. This suspension of cilia is stored in demembranating solution at 4°C until use.

2. The above procedure removes virtually all the cilia from the intact lung. The stability of this preparation is highly dependent on the DTT, EDTA, and sucrose (Hard *et al.,* 1988). Failure to add the EDTA following isolation causes a significant reduction in reactivated beat frequency. If EDTA is added first to the wash or demembranating solution, the yield of cilia is significantly reduced. Nominal calcium (10^{-6} to 10^{-5} M) must be present for the axonemes to separate from the basal bodies at the basal plate. This conforms to results obtained for a variety of cilia and flagella (Blum, 1971). Other water structuring agents such as glycerol, polyvinylpyrrolidone, Ficoll, and polyethylene glycol can substitute for sucrose in the demembranating solution. In addition, organic buffers other than Pipes (particularly K lactate and K acetate) can be substituted as described elsewhere (Gibbons and Gibbons, 1973). Nonidet P-40 can be substituted for Triton X-100.

Gibbons, B. H., and Gibbons, I. R. (1973). The effect of partial extraction of dynein arms on the movement of reactivated sea urchin sperm. *J. Cell Sci.* **13,** 337–357.

Hard, R., and Weaver, A. (1983). Newt lungs: A versatile system for the study of mucociliary transport. *Tissue Cell* **15,** 217–226.

Hard, R., and Cypher, C. (1992). Reactivation of newt lung cilia: Evidence for a possible temperature- and MgATP-induced activation mechanism. *Cell Motil. Cytoskel.* **21,** 187–198.

Hard, R., Cypher, C., and Schabtach, E. (1988). Isolation and reactivation of highly-coupled newt lung cilia. *Cell Motil. Cytoskel.* **10,** 271–284.

Hard, R., Blaustein, K., and Scarcello, L. (1992). Reactivation of outer-arm-depleted lung axonemes, Evidence for functional differences between inner and outer dynein arms in situ. *Cell Motil. Cytoskel.* **21,** 199–209.

Hennessey, S. J., Wong, L. B., Yeates, D. B., and Miller, I. F. (1986). Automated measurement of ciliary beat frequency. *J. Appl. Physiol.* **60,** 2109–2113.

Kennedy, J. R., and Duckett, K. E. (1981). The study of ciliary frequencies with an optical spectrum analysis system. *Exp. Cell Res.* **135,** 147–156.

Park, J., and Hard, R. (1990). Waveform analysis of reactivated respiratory cilia: Elastic recoil contributes significantly to waveform asymmetry. *J. Cell Biol.* **111,** 171a.

Rieder, C. L., and Hard, R. (1990). Newt lung epithelial cells: Cultivation, use and advantages for biomedical research. *Int. Rev. Cytol.* **122,** 153–220.

Rupp, G., and Hard, R. (1993). Outer arm dynein from newt respiratory cilia. *Mol. Biol. Cell* **4,** 48a.

Sanderson, M. J., and Dirksen, E. R. (1985). A versatile and quantitative computer-assisted photoelectronic technique used for the analysis of ciliary beat cycles. *Cell Motil.* **5,** 267–292.

Verdugo, P., Hinds, T. R., and Vincenzi, F. F. (1979). Laser light-scattering spectroscopy: Preliminary results on bioassay of cystic fibrosis factor(s). *Pediatr. Res.* **13,** 131–135.

Weaver, A., and Hard, R. (1985a). Isolation of newt lung ciliated cell models: Characterization of motility and coordination thresholds. *Cell Motil.* **5,** 355–377.

Weaver, A., and Hard, R. (1985b). New lung ciliated cell models: Effect of MgATP on beat frequency and waveform parameters. *Cell Motil.* **5,** 377–392.

CHAPTER 33

Reactivation of Motility of Demembranated Sea Urchin Sperm Flagella

Charles J. Brokaw
Division of Biology
California Institute of Technology
Pasadena, California 91125

I. Introduction

Demembranated sea urchin spermatozoa provide an excellent *in vitro* system for examining motility of the axoneme—the internal cytoskeleton of flagella. Their length, typically 40 to 45 μm, and planar bending waveforms facilitate detailed analysis of flagellar bending. The earliest work used glycerinated spermatozoa, but high-quality preparations were not achieved until the introduction of Triton X-100 for membrane removal (Gibbons and Gibbons, 1972). Normal motility can be observed with isolated axonemes that have become detached from the sperm head, but there are no reliable procedures for quantitative isolation of detached axonemes that show normal motility. Chapters in earlier volumes in this series provide more information on experimental uses of these methods (Gibbons, 1982; Brokaw, 1986).

II. General Considerations

This chapter describes methods for spermatozoa from two sea urchin species that are readily available in the United States: *Lytechinus pictus* and *Strongylocentrotus purpuratus*. Both are cold water species, and the spermatozoa are damaged by high temperatures. Ideally, reactivation experiments should be carried out in a room maintained at 18°C (or lower). Additional temperature control of the microscope stage at this temperature is desirable. Heat-absorbing glass filters (e.g., No. G4,010, Edmund Scientific, Barrington, NJ) should be used in the illumination path. Of the two species described here, *L. pictus* spermatozoa are less temperature sensitive and probably are not seriously damaged until the temperature exceeds 22°C.

Demembranated spermatozoa can be damaged by impurities in the solutions. The use of high-quality deionized and/or glass-distilled water for solutions and for rinsing glassware is essential. Contamination by iron is a particular concern. It can be minimized by using plastic or Teflon-coated spatulas to transfer chemicals. Iron contamination is a particularly serious problem with light sources that transmit significant power in the UV spectrum. The use of a UV filter (Kodak Wratten 4B) is recommended when individual spermatozoa will be exposed to such illumination for extended observations. Solutions containing labile components such as dithiothreitol (DTT) and ATP should be made fresh for each day's work; the basic buffer solutions can safely be made up and refrigerated for use over 2 to 3 days.

In the following recipes, Tris buffer refers either to trishydroxymethylaminomethane, in base form, with the solution pH lowered to 8.2 with acetic acid, or to an equimolar mixture of Tris base and Tris–HCl that will give the desired pH with minimal adjustment. For quantitative work, the inclusion of chloride anions in the reactivation solutions should be avoided or at least kept at a constant concentration.

The procedures described here are designed to prepare "potentially symmetric" sperm flagella (Gibbons and Gibbons, 1980) that generate reasonably symmetric bending waves at low free Ca^{2+} concentrations. The low-frequency reactivation solutions produce movements with frequencies of 1 to 2 cycles per second, and are particularly useful when observing with continuous illumination. The high-frequency reactivation solutions produce movement with frequencies of 30 to 40 cycles per second, comparable to the frequencies of live spermatozoa. Adequate observation of these high-frequency movements requires stroboscopic illumination. The principal source for stroboscopic illuminators for microscopy is Chadwick–Helmuth (El Monte, CA). In both cases, visualization of the sperm flagellum is greatly facilitated by the use of a dark-field condenser and appropriate objectives.

When a drop of sperm suspension is placed on a microscope slide, the rapidly swimming spermatozoa quickly leave the interior of the drop and accumulate near surfaces, either sticking to the surface or swimming close to the surface with the plane of flagellar bending parallel to the surface. Sticking to the glass surfaces is reduced by inclusion of polyethylene glycol (PEG) in the reactivation solutions, and the concentration of PEG can be safely increased to as much as 2%. Sticking can be further reduced by coating the glass surfaces with Formvar (Gibbons, 1982), siliconizing preparations such as AquaSil (No. 42799, Pierce Chemical), or egg white (expose the surface to raw egg white, pour off the egg white, rinse with the reactivation solution, and use without drying). However, I have never had good success when a thin layer of reactivation solution is compressed between a cover glass and slide. For routine observations, I use an intermediate-range dark-field condenser (Zeiss 0.7–0.85 NA) and a 16×, 0.40 NA objective, without a cover glass. A thin drop of reactivation solution containing demembranated spermatozoa is spread onto the surface of the slide, and spermatozoa swimming at the upper surface of the drop are examined. A plastic shield attached to the microscope objective is used to minimize disturbance by air currents and minimize evaporative cooling of the surface of the drop. This arrangement works best when the microscope stage is cooled, so that water does not evaporate from the drop and condense on the front lens of the objective.

For high-resolution observations, I use a thin hanging drop preparation, suspended in a bath of mineral oil contained in an aluminum well slide with a cover glass forming the lower surface of the well. This is used with a high-NA dark-field condenser and 100× or 40× objectives with iris diaphragms.

III. Methods

A. Reactivation of *Lytechinus pictus* Spermatozoa

1. Materials

After collection of "dry" spermatozoa from the aboral surface of the sea urchin, these spermatozoa should be stored on ice without dilution and used within about 4 hours.

Demembranation solution (A1): 0.25 *M* KCl, 10 m*M* Tris buffer, 2 m*M* $MgSO_4$, 1 m*M* DTT, 1 m*M* ethylene glycol bis(β-aminoethyl ether)-*N*, *N'*-tetraacetic acid (EGTA) (Sigma No. E-4378), and 0.04% Triton X-100.

Activation solution (B1): 0.05 *M* KCl, 10 m*M* Tris buffer, 0.5 m*M* $MgSO_4$ or Mg acetate, 1 m*M* DTT, 0.2 m*M* ATP, and 10 μm cAMP.

Reactivation solutions: 0.25 *M* K acetate, 20 m*M* Tris buffer, 2 m*M* EGTA, 1 m*M* DTT, and 0.5% polyethylene glycol. For low frequencies (10 μM MgATP), add 1.23 m*M* $MgSO_4$ or Mg acetate, 13.5 μM ATP, 2 m*M* Li acetate, and 0.1% methyl cellulose (Fisher No. M-281) (optional). For high frequencies (1 m*M* MgATP), add 2.22 m*M* $MgSO_4$ or Mg acetate and 1.35 m*M* ATP.

2. Procedure

1. Add 1 μl of stock sperm suspension to 50 μl of demembranation solution (A1) in a small glass culture tube. Shake the tube to disperse the spermatozoa uniformly.
2. Thirty seconds later, add 150 μl of activation solution (B1).
3. Incubate for 150 seconds, and then add 2.5 μl of 0.2 *M* $CaCl_2$.
4. After an additional 20 seconds, remove 10 μl and mix into 1 ml of reactivation solution. Transfer a drop of this reactivation mixture to a microscope slide for observation. Examples of the results that should be obtained are illustrated in Figs. 1a and 1b.

B. Reactivation of *Strongylocentrotus purpuratus* Spermatozoa

1. Materials

After collection of "dry" spermatozoa from the aboral surface of the sea urchin, these spermatozoa should be diluted with an equal volume of 0.5 *M* NaCl and stored on ice. These sperm suspensions usually remain healthy for at least 1 day.

Demembranation solution (A2): 0.15 *M* KCl, 10 m*M* Tris buffer, 2 m*M* $MgSO_4$, 1 m*M* DTT, 0.2 m*M* EGTA, 0.2 m*M* ATP, 10 μM cAMP, 0.04% Triton X-100.

Extraction solution (B2): 0.15 *M* KCl, 10 m*M* Tris buffer, 2.2 m*M* $CaCl_2$, 1 m*M* DTT, 0.01% Triton X-100.

Reactivation solution: same as for *L. pictus*.

2. Procedure

1. Add 1 μl of stock sperm suspension to 50 μl of demembranation solution (A2) in a small glass culture tube. Shake the tube to disperse the spermatozoa uniformly.
2. Thirty seconds later, add 450 μl of extraction solution (B2), and incubate for 30 seconds.
3. Remove 10 μl and mix into 1 ml of reactivation solution. Transfer a drop of this reactivation mixture to a microscope slide for observation.

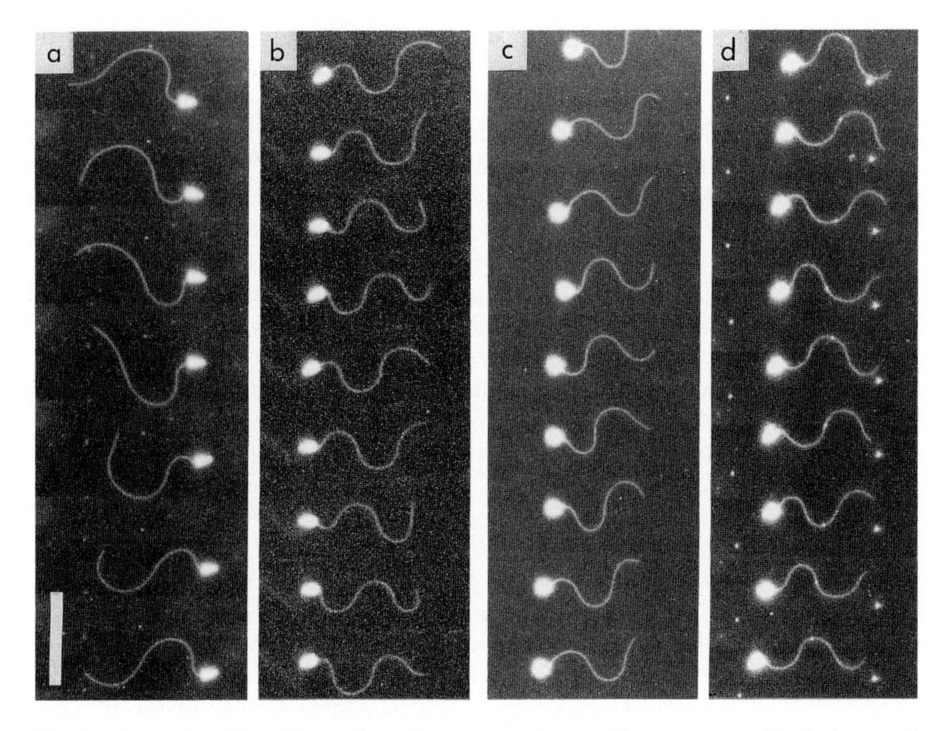

Fig. 1 Examples of MgATP-reactivated movement of sea urchin spermatozoa. Each photograph is a series of images formed on moving film, using stroboscopic flash illumination. These low-resolution photographs were obtained using an open drop of reactivation solution and a 16× objective. (a) *Lytechinus pictus,* low MgATP, 12 flashes per second. (b) *L. pictus,* 1 m*M* MgATP, 150 flashes per second. (c) *Strongylocentrotus purpuratus,* low MgATP, 15 flashes per second. (d) *S. purpuratus,* 1 m*M* MgATP, 150 flashes per second. Bar (in a) = 25 μm.

Examples of the results that should be obtained are illustrated in Figs. 1c and 1d.

IV. Comments

The major factors determining the characteristics of the movement are the concentrations of MgATP and Ca^{2+} in the reactivation solutions. The frequency is largely determined by the MgATP concentration. In general, it is more difficult to obtain high-quality reactivation at higher MgATP concentrations, where the movement is more sensitive to the activation state of the flagella (Brokaw, 1987). If reactivation with 1 m*M* MgATP is poor, solutions with 0.1 or 0.2 m*M* MgATP may give better results. The reactivation solutions given here have excess EGTA and should maintain very low Ca^{2+} concentrations. Increased

Ca^{2+} concentrations will increase the asymmetry of the flagellar beating and cause the spermatozoa to swim in circular paths. At low MgATP concentrations, Li^{+} ions are useful to regulate the amplitude of the bending waves (Brokaw, 1987, 1989), and inclusion of methyl cellulose counteracts the increase in wavelength that occurs as the MgATP concentration is reduced. The concentration of Mg^{2+} has a minor effect on the amplitude and frequency (Okuno and Brokaw, 1979). The reactivation solutions have been calculated to have free Mg^{2+} concentrations of 0.3 m*M*. Recipes for solutions with other concentrations of MgATP, Ca^{2+}, and Mg^{2+} can be calculated using the algorithm presented in Section VI. The availability of dry, highly purified $MgSO_4 \cdot 7H_2O$ argues for the use of $MgSO_4$ to add Mg^{2+} ions to these solutions, even though sulfate anions have a slight inhibitory effect on beat frequency. EGTA from sources other than Sigma has sometimes given poor results. A mixture of 2 m*M* EDTA and 50 μM EGTA solves these problems.

V. Extensions to Other Species

Spermatozoa from another sea urchin, *Arbacia punctulata,* have been successfully reactivated using the procedure given here for *S. purpuratus* spermatozoa. Spermatozoa from the tunicate, *Ciona intestinalis,* can be reactivated well (Brokaw, 1993) using procedures similar to the procedure for *L. pictus* spermatozoa. The major differences between species appear to relate to the cAMP-dependent activation process. In developing procedures for new species, it is a good idea to start with spermatozoa that have been fully activated *in vivo* in an artificial seawater solution with reduced Ca^{2+}, rather than starting with "dry," inactive spermatozoa.

VI. Algorithm for Calculating Recipes for Solutions

This algorithm is in the form of C++ pseudocode. Constants are obtained from references in Brokaw (1986, 1991, 1993). Do not use for pH < 6, pCa < 5. I assume solutions are always contaminated with 20 μM calcium.

```
float B[3], B1, B2, B3, B4[3], B5[3], B7, G8;
// All concentrations are in moles/liter
float pH; cout << "Enter pH"; cin >> pH;
cout << "Enter EGTA concentration"; cin >> B[0];
cout << "Enter EDTA concentration"; cin >> B[1];
cout << "Enter BAPTA concentration"; cin >> B[2];
float Ac; cout << "Enter total Acetate concentration"; cin >> Ac;
float MgATP; cout << "Enter desired MgATP concentration"; cin >> MgATP;
float Ca; cout << "Enter desired free Ca++ concentration"; cin >> Ca;
```

```
float Mg; cout < < "Enter desired free Mg++ concentration"; cin > > Mg;
const float A1[3] = {9.5,10.2,6.36} // H+:EGTA4-, H+:EDTA4-, H+:BAPTA4-
const float A2[3] = {8.9,6.2,5.5};  // H+:HEGTA3-, H+:HEDTA3-, H+:HBAPTA3-
const float A3[3] = {5.2,8.6,1.77}; // Mg:EGTA4-, Mg:EDTA4-, Mg:BAPTA4-
const float A4[3] = {3.4,2.8,1};    // Mg:HEGTA3-, Mg:HEDTA3-, Mg:HBAPTA3-
const float A5[3] = {11,10.7,6.5};  // Ca:EGTA4-, Ca:EDTA4-, Ca:BAPTA4-
const float A6[3] = {5,4,2};        // Ca:HEGTA3-, Ca:HEDTA3-, Ca:HBAPTA3-
const float AP6 = 4;                // Mg:ATP4-
const float AP7 = 3.6;              // Ca:ATP4-
const float AP8 = 6.9;              // H+:ATP4-
const float A9 = 0.51;              // Mg:acetate
for (int k = 0;k < 3;k+ +){
  B1 = 1 + (1 + pow(10,A2[k]-pH))*pow(10,A1[k]-pH);
B2 = Mg*(pow(10,A1[k]-pH)*pow(10,A4[k])+pow(10,A3[k]1));
B3 = Ca*(pow(10,A5[k]));
B4[k] = B[k]*B2/(B1 + B2 + B3);
B5[k] = B[k]*B3/(B1 + B2 + B3);}
B7 = MgATP/(Mg*pow(10,AP6)); // free ATP-4
G8 = Ac*Mg*pow(10,A9)/(1 + Mg*pow(10,A9)); // Mg acetate
Total amount of ATP needed = MgATP + B7*(1 + Ca*pow(10,AP7) + pow(10,AP8-pH));
Total amount of Mg needed  = MgATP + Mg + B4[0] + B4[1] + B4[2] + G8;
Total amount of Ca needed  = Ca + B5[0] + B5[1] + B5[2] + B7*Ca*pow(10,AP7);
```

VII. Health and Safety Considerations

High-intensity xenon arc illuminators produce UV-rich illumination, which can cause formation of ozone in the lamp housing and the illumination path. Appropriate ventilation is required to protect workers in the vicinity of such equipment from exposure to harmful concentrations of ozone.

Acknowledgments

This chapter was prepared with the assistance of Susan Creagh and financial support from the U.S. National Institutes of Health (GM-18711).

References

Brokaw, C. J. (1986). Sperm motility. *Methods Cell Biol.* **27,** 41–56.

Brokaw, C. J. (1987). A lithium-sensitive regulator of sperm flagellar oscillation is activated by cAMP-dependent phosphorylation. *J. Cell Biol.* **105,** 1789–1798.

Brokaw, C. J. (1989). Operation and regulation of the flagellar oscillator. *In* "Cell Movement: The Dynein ATPases" (F. D. Warner, I. R. Gibbons, and P. Satir, eds.), pp. 267–279. A. R. Liss, New York.

Brokaw, C. J. (1991). Calcium sensors in sea urchin sperm flagella. *Cell Motil. Cytoskel.* **18,** 123–130.

Brokaw, C. J. (1993). Microtubule sliding in reduced-amplitude bending waves of *Ciona* sperm flagella: Resolution of metachronous and synchronous sliding components of stable bending waves. *Cell Motil. Cytoskel.* **26,** 144–162.

Gibbons, B. H. (1982). Reactivation of sperm flagella: Properties of microtubule-mediated motility. *Methods Cell Biol.* **25,** 253–271.

Gibbons, B. H., and Gibbons, I. R. (1972). Flagellar movement and adenosine triphosphatase activity in sea urchin sperm extracted with Triton X-100. *J. Cell Biol.* **54,** 75–97.

Gibbons, B. H., and Gibbons, I. R. (1980). Calcium-induced quiescence in reactivated sea urchin sperm. *J. Cell Biol.* **84,** 13–27.

Okuno, M., and Brokaw, C. J. (1979). Inhibition of movement of Triton-demembranated sea-urchin sperm flagella by Mg^{2+}, ATP^{4-}, ADP and P_i. *J. Cell Sci.* **38,** 105–123.

CHAPTER 34

High-Speed Video Microscopy of Flagella and Cilia

Sumio Ishijima
Biological Laboratory
Faculty of Science
Tokyo Institute of Technology
O-okayama, Meguro-ku
Tokyo 152, Japan

I. Introduction

The beat frequency of ciliary and flagellar movements extends up to 90 Hz (Sleigh, 1974; Gibbons *et al.*, 1985). Therefore, precise definition of this movement requires use of a high-speed recording system. High-speed cinemicrography (Hiramoto and Baba, 1978) and similar methods using photographic films (Brokaw, 1986; Chevrier and Dacheux, 1987) have been used for this purpose for a long time. Recent and remarkable developments in video and computer technology have resulted in cinemicrography being largely superseded by videomicrography, which has major advantages over cinemicrography (Inoué, 1986). Images are reviewed on the video monitor as they are recorded, so that adjustments can be made to the focus and the instrument. Images are available for field-by-field analysis immediately after recording. Images can also be recorded at low light levels. Additionally, data recorded on videotape are in ideal form for image analysis by computer-assisted techniques.

One disadvantage of conventional videomicrography is temporal resolution. Here, two aspects of image collection must be considered. First, a sufficiently large number of images must be recorded over a short enough time to provide detailed information on the waveform of a cilium or flagellum throughout a single beat cycle. This is particularly important when studying transient phenomena, such as chemotactic or phototactic steering. Solutions to this problem include the use of video cameras capable of providing multiple images per field or camera and recorder systems capable of capturing greater than 60 fields per second. Second, the image must be captured over a sufficiently short period that a rapidly moving cilium or flagellum is sharply defined. This may be achieved in videomicrography by using either a shuttered camera or stroboscopic illumination.

This chapter briefly describes the advantages and disadvantages of several high-speed video systems that I have used to observe and record ciliary and sperm flagellar movements. As any of these systems may be combined with stroboscopic illumination to obtain sharply defined images, the use of stroboscopic illumination is discussed separately.

II. Video Cameras

A. Video Camera Equipped with a Rapid Shutter

A video camera equipped with a rapid shutter will freeze beating cilia and flagella onto videotape at 60 fields per second using a standard videocassette recorder (VCR). It is important that the VCR have still-field and field-by-field advance capabilities for analysis of images (see VCRs below). Movements of mammalian sperm flagella and most cilia can be recorded using this method (Suarez *et al.*, 1983; Ishijima *et al.*, 1994), although recording of the more rapid flagellar movement of sea urchin spermatozoa and *Chlamydomonas* is less satisfactory (Fig. 1a). A surveillance charge-coupled device (CCD) video camera with a high-speed shutter is inexpensive and easily operated. A video camera with a minimum sensitivity of less than 1 lx is recommended, e.g., Panasonic WT-BD 400 (Panasonic Broadcast & Television Systems, Secaucus, NJ).

As an alternative to a shuttered camera, a standard video camera can be used with stroboscopic illumination to record sharp images of cilia and flagella at 60 Hz. Although this has certain advantages (see Section III, A), expensive modifications must be made to the camera to couple the strobe to it, and the strobe system itself may be quite expensive.

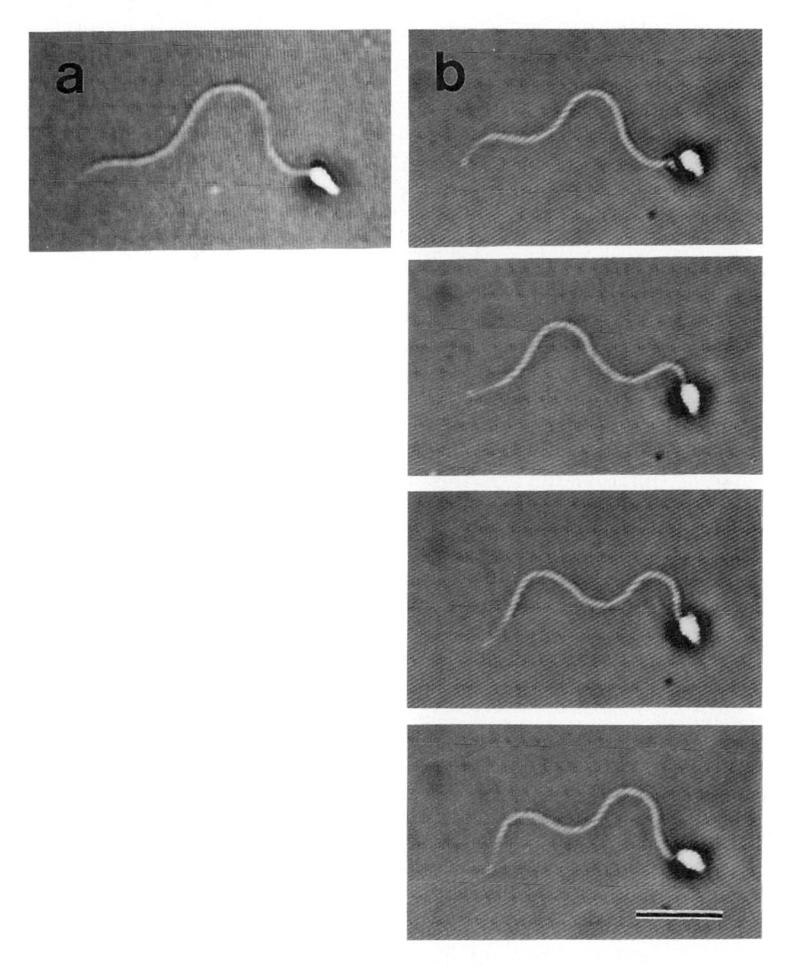

Fig. 1 Phase-contrast videomicrographs of the flagellar movement of sea urchin spermatozoa. (a) The sperm flagellum was illuminated with a tungsten lamp and recorded using a CCD video camera with a rapid shutter. The shutter speed was 1/1000 s. Only one image of the sperm flagellum (beating at 45.9 Hz at 23°C) could be recorded per beat cycle. (b) The sperm flagellum was recorded with the Nac MHS-200 using Nac's integral stroboscopic illumination system. The time interval between successive images was 1/200 s. Four different images of the sperm flagellum (beating at 47.2 Hz at 23°C) could be recorded during a single beat cycle. Bar = 10 μm.

B. Video Camera with Four Vertical Scanning Frequencies

A video camera with four vertical scanning frequencies of 60, 120, 180, and 240 Hz produces one, two, three, and four separate images per field, yielding up to 240 successive images per second, even though a standard VCR enabling still-field display is used (Ishijima and Witman, 1987). Such a camera is useful for recording rapidly beating flagella and flagella that are undergoing transient changes in beat pattern. There is, however, a loss of vertical resolution because the number of TV lines scanned per image is reduced. Several companies (Xybion Electronic Systems, Miralani Drive, San Diego, CA; Hamamatsu Photonics Systems, Foothill Road, Bridgewater, NJ) will modify video cameras in this way. These cameras may be either shuttered or coupled to a stroboscope to obtain sharp images.

C. High-Speed Videotape Recorder with Camera

There are several camera/recorder systems in which the recording itself is sped up. Most of them, except the Nac MHS-200, use a solid-state pickup device with a sensitivity too low to capture beating cilia and sperm flagella. The Nac MHS-200 uses a Plumbicon tube as a pickup device, so that sharp images of beating cilia and flagella can be obtained using a phase-contrast microscope and the strobe supplied with the system (Fig. 1b). The resolution in the playback picture is 280 lines, similar to what one gets with a standard VCR recording at 60 Hz.

III. Other Equipment

A. Stroboscope

Stroboscopic illumination is admirably suited to applications where the object being recorded is moving rapidly and must be frozen in time to obtain a sharp image (Brokaw, 1986). Excellent systems are manufactured by Chadwick–Helmuth (El Monte, CA). The strobe systems available from this company produce repetitive flashes (up to 256 Hz) with a flash duration of 20 to 40 microseconds, which is short enough to yield very sharp images of rapidly beating cilia and flagella. Another advantage of stroboscopic illumination is that the very short exposure times are less apt to cause damage to the specimen (Ishijima and Witman, 1987). The Nac MHS-200 comes equipped with its own strobe.

B. Videocassette Recorder

For recording and analyzing movement of cilia and flagella with a standard VCR, a VCR having still-field and field-by-field advance capabilities is recommended because it yields 60 successive images per second. Not all VCRs permit still-field playback; some display only every other field in single-frame advance mode, thus limiting data analysis to 30 Hz. This is presently the case even for optical disk recorders, which otherwise would have some advantages over VCRs. Therefore, this parameter should be carefully checked prior to purchasing a VCR. Three-quarter-inch U-format VCRs have been used widely in video microscopy, but the new Super-VHS (S-VHS) recorders and tapes are much less expensive and are capable of high resolution (400–430 lines) and image quality comparable to that obtained with $\frac{3}{4}$-in. VCRs.

C. Microscope

For high-speed video microscopy of cilia and sperm flagella, a phase-contrast or differential-interference-contrast microscope is suitable

IV. Procedure for Recording

1. Mount a video camera without a lens on a microscope with a C-mount adaptor and connect it with a shielded coaxial cable to the VCR. It is recommended that a time–date generator be connected between the camera and the VCR to mark every field.
2. Place the specimen on the microscope stage and focus the specimen.
3. While viewing the image on the monitor, adjust the gain and black level controls of the video camera to enhance image contrast.

Acknowledgments

The author is grateful to Dr. Y. Hamaguchi for many helpful discussions and advice. This work was supported by the Association of Livestock Technology, Japan (Bioscience Research for Livestock Technology 91-3).

References

Brokaw, C. J. (1986). Sperm motility. *In* "Methods of Cell Biology" (T. E. Schroeder, ed.), Vol. 27, pp. 41–56. Academic Press, Orlando.

Chevrier, C., and Dacheux, J. L. (1987). Analysis of the flagellar bending waves of ejaculated ram sperm. *Cell Motil. Cytoskel.* **8,** 261–273.

Gibbons, B. H., Baccetti, B., and Gibbons, I. R. (1985). Live and reactivated motility in the 9+0 flagellum of *Anguilla* sperm. *Cell Motil.* **5,** 333–350.

Hiramoto, Y., and Baba, S. A. (1978). A quantitative analysis of flagellar movement in echinoderm spermatozoa. *J. Exp. Biol.* **76,** 85–104.

Inoué, S. (1986). "Video Microscopy." Plenum, New York.

Ishijima, S., and Witman, G. B. (1987). Flagellar movement of intact and demembranated, reactivated ram spermatozoa. *Cell Motil. Cytoskel.* **8,** 375–391.

Ishijima, S., Ishijima, S. A., and Afzelius, B. A. (1994). Movement of *Myzostomum* spermatozoa: Calcium ion regulation of swimming direction. *Cell Motil. Cytoskel.* **28,** 135–142.

Sleigh, M. A. (1974). Patterns of movement of cilia and flagella. *In* "Cilia and Flagella" (M. A. Sleigh, ed.), pp. 79–92. Academic Press, London.

Suarez, S. S., Katz, D. F., and Overstreet, J. W. (1983). Movement characteristics and acrosomal status of rabbit spermatozoa recovered at the site and time of fertilization. *Biol. Reprod.* **29,** 1277–1287.

CHAPTER 35

Micromanipulation of Sperm and Other Ciliated or Flagellated Single Cells

Sumio Ishijima
Biological Laboratory
Faculty of Science
Tokyo Institute of Technology
O-okayama, Meguro-ku
Tokyo 152, Japan

I. Introduction

Ciliary and flagellar movements are generally three-dimensional (Ishijima *et al.*, 1992). It is therefore important to determine the three-dimensional shape of beating cilia and flagella to understand not only the mechanisms of cilairy and flagellar movements but also their behavior. Information about the three-dimensional waveforms of cilia and flagella is not easily accumulated, however, because conventional light microscopy is unsuitable for such observations. To overcome this difficulty, a micromanipulative technique has been developed. A cell is captured and held with a sucking micropipet, and its ciliary or flagellar movement is observed from various directions under the microscope (Ishijima and Hiramoto, 1982; Ishijima *et al.*, 1986).

In this chapter, the technique of holding a spermatozoon is described in detail. This method has been used in various experiments (Ishijima and Hiramoto, 1982; Gibbons *et al.*, 1987) and may be applied to other motile or nonmotile cells (Uehara and Yanagimachi, 1976; Rüffer and Nultsch, 1987).

II. Methods

A. Fabrication of Micropipets

1. Instruments

A 50-μl Microcaps capillary glass tubing with a 1-mm outside diameter (Drummond Scientific); a micropipet puller (PN-3, Narishige Scientific Instrument Laboratory); and a microforge (MF-79, Narishige Scientific Instrument Laboratory) with a platinum–iridium heating filament having a 30 μm diameter are used.

2. Procedure

1. A micropipet is drawn from a Microcaps tube using the micropipet puller so that the entire tapered portion of the micropipet is about 1.5 cm long.
2. The micropipet is mounted in the micropipet clamp of the microforge. The tip of the micropipet is broken by pushing it against the heating filament of the microforge and then fire-polished by bringing the tip close to the heating filament until the inside diameter of the tip is about 70% of the diameter of the sperm head or cell body.

3. Discussion

The outer diameter of the Microcaps tube should be 1.0 mm to fit the microinstrument holder.

The most important point for successful manipulation is to make suitable micropipets. Most sperm heads are firmly held with a micropipet having a circular cross section; however, for spermatozoa with flat heads, such as ram spermatozoa, a micropipet having a rectangular aperture should be devised to prevent rotation of the sperm heads within the micropipet (Ishijima and Witman, 1987). Rodent spermatozoa have hook-shaped heads; thus a micropipet with a lateral orifice is used to immobilize these spermatozoa (Ishijima and Mohri, 1985).

To hold a spermatozoon or a cell for a long time, a braking micropipet with a constriction near the tip may be helpful (Hiramoto, 1974).

Micropipets formed by a micropipet puller in the horizontal orientation exhibit some sagging of the tip portion, which produces trouble when the micropipet is rotated around its axis (see below). A vertical puller such as Model PA-81 (Narishige Scientific Instrument Laboratory) or Model 750 (David Kopf Instruments, Tujunga, CA) is used to eliminate this problem (Brown and Flaming, 1986).

B. Holding Spermatozoa and Other Cells

1. Instruments

A phase-contrast or differential-interference-contrast microscope equipped with a long-working-distance condenser; a hydraulic micromanipulator (MO-102, Narishige Scientific Instruments Laboratory); a custom-built microinstrument holder that permits rotation of the micropipet; a microinstrument sleeve (e.g., Leitz No. 520 145); a micropipet (see above); a connecting tube; a micrometer syringe (Microinjector IM-4, Narishige Scientific Instrument Laboratory); and an observation chamber 1.5 mm deep, 10 mm wide, and 17 mm long, filled with sperm suspension, are used.

2. Procedures

1. The focal plane of the microscope is adjusted on the sperm (or cells) swimming under the coverslip in the observation chamber.

2. Without changing the focus, the observation chamber is removed from the field of view using the mechanical stage drive.

3. The rear end of a microinstrument sleeve is connected to a micrometer syringe with connecting tube, and the space from the tip of the microinstrument sleeve to the micrometer syringe is filled with distilled water by applying negative pressure to the micrometer syringe. Air bubbles are allowed only in the shaft of the micropipet. The micropipet, mounted in the microinstrument sleeve, which itself is mounted in a micropipet holder held in a hydraulic micromanipulator, is angled slightly higher than horizontal so that the tip of the micropipet reaches the lower surface of the coverslip (Fig. 1).

4. With the hydraulic micromanipulator, the tip of the micropipet is brought into focus in the center of the microscope field and then moved slightly downward. The observation chamber is moved back into place using the mechanical stage drive, and the micropipet tip is then brought back into focus using the micromanipulator controls. Thus, the micropipet tip and the sperm (or cells) are both in focus in the same microscope field.

5. A spermatozoon (or a cell) is captured by using the mechanical stage drive to bring it to the tip of the micropipet and then gently sucking it into the micropipet by applying negative pressure with the micrometer syringe.

6. The orientation of the spermatozoon is adjusted with the hydraulic micromanipulator to bring the entire length of the flagellum into the focal plane of the microscope. To observe the flagellar movement of the spermatozoon from different angles relative to the beating plane, the microinstrument sleeve is rotated around its axis (Fig. 1).

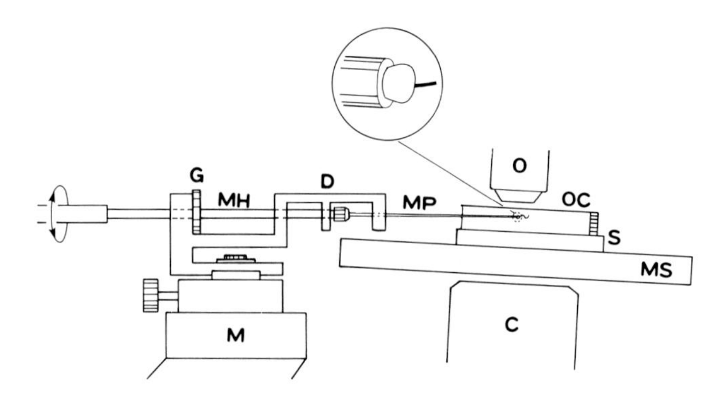

Fig. 1 Schematic diagram of apparatus for observing flagellar movement of a spermatozoon held by its head with a micropipet. C, Long-working-distance condenser; D, custom-built holder that permits rotation of the microinstrument sleeve while keeping the tip of the micropipet in focus and within the microscopic field; G, graduated ring for measuring rotation angle of micropipet; M, hydraulic micromanipulator; MH, microinstrument sleeve; MP, micropipet; MS, microscope stage; O, objective; OC, observation chamber; S, spacer. Modified, with the publisher's permission, from Ishijima (1992).

3. Discussion

In my experiments, I have found it very useful to mount the microinstrument sleeve in a custom-built holder (Ishijima and Mohri, 1985), which allows the micropipet to be rotated around its longitudinal axis (Fig. 1). The holder is designed so that the microinstrument sleeve is rotated precisely around its longitudinal axis, thus ensuring that the specimen at the tip of the micropipet remains within the microscope field during the rotation.

A long-working-distance condenser or long-working-distance objective is convenient for micromanipulation. The microscope stage should be small enough so as not to obstruct the micropipet movement. If this is impossible, a spacer should be inserted between the microscope stage and the observation chamber (Fig. 1).

Any micromanipulator with similar features is suitable; however, when working with a microscope having a stage that moves up and down during focusing, it is preferable to use a small micromanipulator that can be attached directly to the movable stage (Kiehart, 1982).

For coarse positioning, the coarse controls of the micromanipulator (Narishige MN-1) are used to maneuver the instrument holder until the micropipet tip intercepts the light from the condenser.

Acknowledgments

The author is grateful to Professor Y. Hiramoto for introducing him to micromanipulation. This work was supported by the Association of Livestock Technology, Japan (Bioscience Research for Livestock Technology 91-3).

References

Brown, K. T., and Flaming, D. G. (1986). Instruments for pulling micropipettes. *In* "Methods in the Neurosciences", Vol. 9, pp. 10–22. John Wiley & Sons, Chichester.

Gibbons, I. R., Shingyoji, C., Murakami, A., and Takahashi, K. (1987). Spontaneous recovery after experimental manipulation of the plane of beat in sperm flagella. *Nature (London)* **325,** 351–352.

Hiramoto, Y. (1974). A method of microinjection. *Exp. Cell Res.* **87,** 403–406.

Ishijima, S. (1992). A method for measuring the sperm motility. *In* "Spermatology" (H. Mohri, M. Morisawa, and M. Hoshi, eds.), pp. 212–224. University of Tokyo Press, Tokyo (in Japanese).

Ishijima, S., and Hiramoto, Y. (1982). Mechanical properties of sperm flagella. *Cell Motil* **1,** *Suppl.,* 149–152.

Ishijima, S., and Mohri, H. (1985). A quantitative description of flagellar movement in golden hamster spermatozoa. *J. Exp. Biol.* **114,** 463–475.

Ishijima, S., and Witman, G. B. (1987). Flagellar movement of intact and demembranated, reactivated ram spermatozoa. *Cell Motil. Cytoskel.* **8,** 375–391.

Ishijima, S., Oshio, S., and Mohri, H. (1986). Flagellar movement of human spermatozoa. *Gamete Res.* **13,** 185–197.

Ishijima, S., Hamaguchi, M. S., Naruse, M., Ishijima, S. A., and Hamaguchi, Y. (1992). Rotational movement of a spermatozoon around its long axis. *J. Exp. Biol.* **163,** 15–31.

Kiehart, D. P. (1982). Microinjection of echinoderm eggs: apparatus and procedures. *In* "Methods in Cell Biology" (L. Wilson, ed.), Vol. 25, pp. 13–31. Academic Press, New York.

Rüffer, U., and Nultsch, W. (1987). Comparison of the beating of cis- and trans-flagella of *Chlamydomonas* cells held on micropipettes. *Cell Motil. Cytoskel.* **7,** 87–93.

Uehara, T., and Yanagimachi, R. (1976). Microsurgical injection of spermatozoa into hamster eggs with subsequent transformation of sperm nuclei into male pronuclei. *Biol. Reprod.* **15,** 467–470.

CHAPTER 36

Preparation and Reactivation of Demembranated, Cytosol-Free Ram Spermatozoa

Jovenal T. San Agustin and George B. Witman

Male Fertility Program
Worcester Foundation for Experimental Biology
Shrewsbury, Massachusetts 01545

I. Introduction

Reactivation refers to the ATP-mediated restoration of movement of cilia or flagella, or flagellated or ciliated cells, after they have undergone a process of demembranation. Demembranation is usually accomplished by treatment with a detergent such as Triton X-100 or Nonidet P-40. After exposure to detergent, movement is reinitiated either by adding ATP directly to the demembranating solution or, as is more often done, by transferring an aliquot of the demembranated sample to a medium containing ATP (see San Agustin and Witman, 1993, for selected references on sperm). In both cases detergent and cell matrix are present during reactivation. Rarely, demembranated cells have been washed before exposure to the reactivation medium (Torres *et al.*, 1977; Bonini *et al.*, 1986; Lieberman *et al.*, 1988). Washing of demembranated sperm is especially desirable to prevent prolonged exposure of the models to proteolytic enzymes released from the ruptured acrosome. Also, because washing removes cytosolic components, the question of whether these factors are necessary for reactivation can be more properly addressed. Unfortunately, demembranated models of the sperm of some species, such as ram, are sensitive to mechanical damage during washing and cannot be reactivated after centrifugation into a pellet (Ishijima and Witman, unpublished results).

To overcome this problem, we have developed a rapid and gentle procedure for preparing demembranated, cytosol-free ram sperm models that can be reactivated (San Agustin and Witman, 1993). The procedure uses centrifugation steps to introduce the sperm into the demembranating solution and to remove the sperm from the same solution. The sperm are collected on a Percoll cushion. In sperm models thus prepared, 86% of tail cross sections did not have plasma membranes; assay of a marker enzyme indicated at least 98% of the cytosolic protein was removed. Percentage reactivation of these models is comparable to percentage motility of the intact sperm from which the models are prepared. These models have been very useful for investigating the effect of cytosolic proteins and cAMP on reactivation (San Agustin and Witman, 1993, 1994).

II. Methods

A. Solutions and Materials

Wash buffer: 137 m*M* NaCl, 2.7 m*M* KCl, 8.1 m*M* Na_2HPO_4/1.5 m*M* KH_2PO_4 (pH 7.5), 0.5 m*M* $MgCl_2$, 5 m*M* D-glucose (Ishijima and Witman, 1987)

10× solution A: 2 *M* sucrose, 250 m*M* monopotassium glutamate, 100 m*M* 4-(2-hydroxyethyl)-1-piperazineethanesulfonic acid (Hepes), pH 7.9, stored frozen as 3-ml aliquots; thaw out on day of use and add 4.6 mg dithiothreitol (DTT) (10 m*M*)

500 μ*M* cAMP

10 m*M* DTT

Solution A: 1 ml 10× solution A + 9 ml water

90% Percoll: 9 ml Percoll + 1 ml 10× solution A

60% Percoll: 6.7 ml 90% Percoll + 3.3 ml solution A

Demembranation medium (DM): 0.2% Triton X-100, 200 m*M* sucrose, 25 m*M* monopotassium glutamate, 1 m*M* ethylenediaminetetraacetic acid (EDTA), 2.5 m*M* 4-aminobenzamidine, 0.2 m*M* phenylmethylsulfonyl fluoride (PMSF), 15% w/v Ficoll type 70 (Sigma), 10 m*M* Hepes, pH 7.9

Solution B: 1 ml 10× solution A diluted into a final volume of 10 ml, and containing 1 m*M* EDTA, 0.2 m*M* PMSF, and 2.5 m*M* 4-aminobenzamidine

2× reactivation medium (RM): 0.38 *M* sucrose, 47.5 m*M* monopotassium glutamate, 2.2 m*M* $MgSO_4$, 2.2 m*M* ATP, 79 m*M* Hepes, pH 7.9, stored frozen as 1-ml aliquots

Acid-washed glass beads: glass beads (Sargent-Welch, 0.25–0.30 mm) stirred in 10 vol of 0.1 *N* HCl for 1 hour, rinsed with 10 vol of distilled water until the pH is about 7, then dried in an oven at low heat

Nalgene filter funnel, 15-ml capacity, 1-cm stem

B. Washing of Ram Sperm

This procedure is modified from that of Ishijima and Witman (1991).

1. Take 1 ml ram semen and mix it with 9 ml of wash buffer in a 12-ml polypropylene centrifuge tube (Sarstedt No. 57,513).
2. Invert the tube a few times and then centrifuge at 1500 rpm for 12–15 minutes (Model CL Clinical Centrifuge, IEC, Boston, MA).
3. Resuspend the loose sperm pellet in 9 ml of wash buffer.
4. Prepare a glass bead filter as follows: Wet a small piece of cotton with the wash buffer and place it over the opening of the filter funnel stem where it meets the cone. Put 4 g acid-washed glass beads into the funnel and then wet the beads with 8 ml wash buffer. Check that the buffer flows through the glass beads and out of the funnel.
5. Pass the sperm suspension through the glass bead filter. Dead and slow-moving sperm are retained in the glass beads. If the proportion of dead to live sperm is quite high, more glass beads may be needed.
6. Centrifuge as in step 2 and resuspend in wash buffer to a density of 1.5×10^9 sperm/ml (about 2–2.5 ml final volume of washed sperm from starting volume of 1 ml semen).

C. Demembranation of Ram Sperm

1. Pipet 800 μl of 60% Percoll into a 1.5-ml microfuge tube.
2. Prepare two 1-ml syringes fitted with 18-gauge needles.
3. Aspirate 90% Percoll into the first syringe.
4. With the needle tip touching the bottom of the tube containing 60% Percoll, gently push 200 μl of 90% Percoll out of the syringe into the tube. The interface between the two Percoll solutions should be visible when the tube is viewed against the light.
5. Using the second syringe, layer 200 μl of DM on top of the 60% Percoll and then place the tube in the microfuge (Fisher Microcentrifuge Model 235 A).
6. Measure out 50 μl of washed sperm (1.5×10^9/ml) and carefully pipet it on top of DM.
7. Push the power knob of the microfuge on and off quickly. This will disperse the flagella in DM. The rotor stops after about 15 seconds. Keep the sperm flagella in DM for another 20 seconds (total demembranation time of 35 seconds).

8. Turn the microfuge on and spin for 35 seconds. The demembranated sperm collect on top of the 90% Percoll layer. Solubilized plasma membrane and cytosolic components remain in the demembranating solution.

9. Take out 400 μl of the top solution and discard or store as desired. This contains DM and the solubilized components of the sperm, mostly plasma membrane.

10. Replace it gently with 400 μl of solution B to rinse the sides of the tube, and then discard.

11. Remove the rest of the 60% Percoll.

12. Collect the demembranated sperm using a disposable transfer pipet (Bio-Rad) (about 75–100 μl).

D. Reactivation of Demembranated Sperm

The above demembranation procedure usually takes 3–3.5 minutes. Reactivation of the demembranated ram sperm must then be initiated as quickly as possible. A lower rate of reactivation results if the demembranated sperm are left to stand before use. We usually start reactivation at $t = 4$ minutes, where $t = 0$ is the start of the 35-second demembranation.

1. Make up 200 μl of reactivation medium in a tissue culture plate (Falcon Model 3047, 24 wells) as follows: 2× RM, 90 μl; 10 m*M* DTT, 20 μl; 500 μM cAMP, 10 μl; water (including other additions, i.e., inhibitors) to make a total volume of 190 μl. The final composition of the reactivation medium is the same as that of Ishijima and Witman (1987).

2. Demembranate washed sperm as described above.

3. At $t = 4$ minutes, add 10 μl of demembranated sperm to the reactivation medium. Swirl gently and then transfer 20 μl to a 3 × 1-in. plastic slide (Rinzl, Carolina Biological Supply) and top with a plastic coverslip (22 × 22 mm, Carolina Biological Supply).

The reactivated models are then viewed through the microscope and may be videotaped for quantitative assessment of reactivation and beat parameters at a later time (see Chapter 34 in this volume, and Ishijima and Witman, 1987). If video recording is done, it is advisable to adhere to a fixed time frame because the percentage of motile models decreases with time (San Agustin and Witman, 1994). We usually record the movement of reactivated ram sperm 1–7 minutes into the reactivation process.

References

Bonini, N. M., Gustin, M. C., and Nelson, D. L. (1986). Regulation of ciliary motility by membrane potential in *Paramecium:* a role for cyclic AMP. *Cell Motil. Cytoskel.* **6,** 256–272.

Ishijima, S., and Witman, G. B. (1987). Flagellar movement of intact and demembranated, reactivated ram spermatozoa. *Cell Motil. Cytoskel.* **8,** 375–391.

Ishijima, S., and Witman, G. B. (1991). Demembranation and reactivation of mammalian spermatozoa from golden hamster and ram. *Methods Enzymol.* **196,** 417–428.

Lieberman, S. J., Hamasaki, T., and Satir, P. (1988). Ultrastructure and motion analysis of permeabilized *Paramecium* capable of motility and regulation of motility. *Cell Motil. Cytoskel.* **9,** 73–84.

San Agustin, J. T., and Witman, G. B. (1993). Reactivation of demembranated, cytosol-free ram spermatozoa. *Cell Motil. Cytoskel.* **24,** 264–273.

San Agustin, J. T., and Witman, G. B. (1994). Role of cAMP in the reactivation of demembranated ram spermatozoa. *Cell Motil. Cytoskel.* **27,** 206–218.

Torres, L. D., Renaud, F. L., and Portocarrero, C. (1977). Studies on reactivated cilia. II. Reactivation of ciliated cortices from the oviduct of *Anolis crisstatellus*. *Exp. Cell Res.* **108,** 311–320.

CHAPTER 37

Microscopic Assays of Flagellar Dynein Activity

Winfield S. Sale and David R. Howard

Department of Anatomy and Cell Biology
Emory University School of Medicine
Atlanta, Georgia 30322

I. Introduction

One crucial advance in the study of molecular motors has been the development of *in vitro* microscopic assays. Such assays were originally applied to the study of myosin function and subsequently adapted for study of microtubule-based motors (see Scholey, 1993). The wonder of these assays is that large, multicomponent complexes referred to as molecular motors can be adsorbed to inert surfaces yet retain the ability to generate force and translocate or move along microtubules. The development of videomicroscopy led, in part, to the discovery of kinesin and cytoplasmic dynein (reviewed in Bloom, 1992). Furthermore, videomicroscopy has resulted in several novel observations about the mechanism of flagellar dynein. For example, microscopic assays have permitted the observation of the rotation of microtubules during translocation for certain flagellar dyneins (Vale and Toyoshima, 1988; Kagami *et al.*, 1990, 1992), segregation of functions to specific domains within dynein (Sale and Fox, 1988;

METHODS IN CELL BIOLOGY, VOL. 47

Vale and Toyoshima 1989; Moss *et al.*, 1992a,b), previously unrecognized fragile or weak dynein/microtubule binding states (Vale *et al.*, 1989), and regulation of flagellar dynein activity (Hamasaki *et al.*, 1991). *In vitro* microscopic assays offer many advantages including use of small amounts of purified proteins, rapid and precise assessment of motility parameters, and convenient changes in the experimental variables.

The goal of this chapter is to provide a description of the methods used for *in vitro* assay of flagellar dyneins. In general, the assay involves videomicroscopic observation of microtubule translocation across a slide surface to which purified dynein or dynein subunits have been adsorbed. These assays have proven useful in the study of dynein derived from *Tetrahymena* cilia (Vale and Toyoshima, 1988, 1989), sea urchin sperm flagella (Paschal *et al.*, 1987; Sale and Fox, 1988; Moss *et al.*, 1992a,b; Vale *et al.*, 1989), *Paramecium* cilia (Hamasaki *et al.*, 1991), and *Chlamydomonas* flagella (Smith and Sale, 1991; Kagami *et al.*, 1990, 1992).

II. Methods

A. Microtubules

Most assays described to date make use of taxol-stabilized microtubules assembled from purified tubulin from either bovine or porcine brain. Purification involves both cycles of assembly and disassembly and purification by chromatography (see "Methods in Cell Biology," Vol. 24). Typically we store purified tubulin in a 1,4-piperazinediethanesulfonic acid (Pipes) polymerization buffer [0.1 *M* Pipes, 2 m*M* ethylene glycol bis(β-aminoethyl ether)-*N,N'*-tetraacetic acid (EGTA), 1 m*M* $MgSO_4$, 0.2 m*M* GTP, pH 6.85] at about 5–8 mg/ml protein, frozen by dropping from a Pasteur pipet onto liquid nitrogen and then stored in liquid nitrogen. The frozen pellets are conveniently thawed, diluted to 2–3 mg/ml protein in polymerization buffer, and incubated at 37°C for 10–20 minutes. An equal volume of polymerization buffer, containing 20 μM taxol, is then added, and stabilized microtubules are stored at room temperature until needed. For most assays, microtubules are sedimented in a Beckman Airfuge (80,000*g*) for 5 minutes and gently resuspended in motility buffer (20 μM taxol, 10 m*M* Tris, pH 7.3, 2.0 m*M* $MgSO_4$, 1.0 m*M* EGTA, 0.1 *M* potassium acetate, 1.0 m*M* dithiothreitol, 1.0 m*M* ATP). This process generally results in microtubules of appropriate length; however, excessively long microtubules can result when the sample sits for a prolonged period at room temperature. In this case, microtubules of a shorter and more useful length can be produced by one pass of the microtubule stock through a 26-gauge, $\frac{1}{2}$-in.-long needle attached to a tuberculin syringe.

Microtubules can also be assembled with a "tag" on one end to determine the polarity of movement or detect microtubule rotation. The details of these methods can be found in Vale and Toyoshima (1988) and Howard and Hyman (1993).

B. Videomicroscopy

The key technical advance which made these assays possible is the development of light microscopic methods by which single microtubules can be detected. The approaches include (1) video- and computer-enhanced, differential-interference-contrast microscopy; (2) fluorescent microscopy; and (3) dark-field light microscopy, first used by Summers and Gibbons (1971) to observe microtubule sliding in isolated flagellar axonemes. Our method has been to use dark-field microscopy coupled to a silicon-intensified target camera to record microtubule movement on videotape. We use both upright and inverted microscopes with equal facility. The microscope is equipped with a 100-W mercury arc lamp and a high-numerical-aperture (1.2/1.4) dark-field condenser (Zeiss, Ultracondensor). Heat and interference filters are placed in the light path between the light source and the condenser.

The specimen is contained in a glass perfusion chamber constructed with a cover glass and slide. We have not found the source of the glass to make a difference as long as the glass is clean and high quality. Dark-field microscopy is very demanding in that any interference from defects in the glass or debris will totally obscure visualization of microtubules. Some investigators have found it useful to clean the slides (e.g., Moss *et al.*, 1992a; Kuo and Sheetz, 1993). To construct the chamber for the upright microscope, an 18-mm-square coverslip is attached to the slide using two narrow strips of double-stick tape, forming a channel 3–4 mm wide. (Some investigators use vacuum grease rather than double-stick tape.) The chamber contains a volume of about 8–10 μl. For the inverted microscope, the chamber is constructed with a 20 $\times$ 40-mm coverslip mounted perpendicular to the slide so that the ends of the cover slip project beyond the edges of the slide and act as a surface on which perfusion solutions can be exchanged.

For typical observations we use either a 40$\times$ Plan APO lens on the inverted microscope or a 40$\times$ Plan-Neofluar objective on the upright microscope. In both microscopes the image is projected through a 10$\times$ or 20$\times$ eye piece onto the target of a silicon-intensified target camera, and the output is passed through a time–date generator and recorded by a VHS recorder equipped with jog-shuttle device for field-by-field analysis of movement as described below.

C. Experimental Protocol

In a typical experiment, a dynein fraction at about 0.06 mg/ml protein is added to the perfusion chamber. This concentration has been found to be a lower threshold, probably indicating that the surface needs to be nearly covered with dynein for the microtubule translocation to proceed (Vale and Toyoshima,

1988; Sale and Fox, 1988). We have found that for more dilute samples (e.g., 0.01 mg/ml), the sample can be perfused into the chamber multiple successive times, apparently concentrating the dynein fraction on the surface (Sale and Fox, 1988). In contrast to studies of kinesin-driven microtubule motility in which other proteins can be used to coat the surface and reduce the amount of kinesin needed (e.g., Howard *et al.*, 1989; Block *et al.*, 1990), similar attempts with bovine serum albumin or casein have not resulted in a reduced threshold of dynein needed for the assays. Following adsorption for about 1 minute, the chamber is rinsed with about 50 μl of motility buffer (see above), 10 μl of microtubules (diluted to about 0.1–0.2 mg/ml in motility buffer containing 1 m*M* ATP) is added, and motility buffer is added to rinse out unbound microtubules. Microtubule movement is then recorded for a 1- to 2-minute period or in some cases much longer. It is important to note that at lower MgATP concentrations, microtubules will not stick and translocate across dynein derived from sea urchin flagella (see Moss *et al.*, 1992a). In some cases an ATP-regenerating system is required and may be introduced by supplementation of the motility buffer (e.g., 20 U/ml creatine kinase, 1.0 m*M* phosphocreatine; Vale and Toyoshima, 1988; Moss *et al.*, 1992a; Sale *et al.*, 1993). Additionally, for flagellar dynein from sea urchin sperm tails, we have found microtubule movement generally occurs on the slide surface. In contrast, very few microtubules move across the cover glass surface. We have not investigated the basis for this difference of movement on the slide versus the coverslip.

D. Data Analysis

The primary measurement is the velocity of microtubule translocation. Velocity varies as a function of [MgATP], temperature, source of dynein, and functional state of dynein. Velocity, however, is generally independent of microtubule length (e.g., Vale and Toyoshima, 1988; Sale *et al.*, 1993). Translocation is measured manually from the video screen, and time is measured using the time–date generator or by counting fields (60 fields/s). Velocity is expressed as the change in distance as a function of time. Alternatively, automated methods of velocity measurement may be used (e.g., Cohn *et al.*, 1993; Sellers *et al.*, 1993).

E. Future

Major questions that may now be addressed using these methods include (1) further assessment of weak-versus-strong binding states (see Vale *et al.*, 1989; Moss *et al.*, 1992a); (b) determination of whether torque plays a role in flagellar movement (see Vale and Toyashima, 1988; Kagami and Kagami, 1992); description of the properties of distinct dynein subtypes (Smith and Sale, 1991; Kagami and Kagami, 1992); and regulation of dynein activity (e.g., Hamasaki *et al.*, 1991). These methods also offer the opportunity to explore the force–velocity relationship (see Kuo and Sheetz, 1993; Hall *et al.,* 1993; Svoboda et al., 1993); however, before such measurements can be made, methods must first be developed to generate movement with single dynein molecules. The assay is also ideally suited for analysis of individual dynein heavy chains and of motor domains within dynein heavy chains (cf. Sale and Fox, 1988; Vale *et al.,* 1989; Moss *et al.,* 1992a,b). Therefore, it may be anticipated that the assay will be valuable for the characterization of modified dyneins derived from expression of cloned dynein heavy chain genes. Thus, combined with molecular genetic and biophysical approaches, these assays have a promising future for dissection of molecular mechanisms and regulation of flagellar dyneins.

References

Block, S., Goldstein, L., and Schnapp, B. (1990). Bead movement by single kinesin molecules studied with optical tweezers. *Nature* **348:** 348–352.

Bloom, G. (1992). Motor proteins for cytoplasmic microtubules. *Curr. Opin. Cell Biol.* **4,** 66–73.

Cohn, S., Saxton, W., Lye, J., and Scholey, J. (1993). Analyzing microtubule motors in real time. *Methods Cell Biol.* **39,** 76–88.

Hall, K., Cole, D., Yeh, Y., Scholey, J., and Baskin, R. (1993). Force-velocity relationships in kinesin-driven motility. *Nature* **364,** 457–459.

Hamasaki, T., Barkalow, K., Richmond, J., and Satir, P. (1991). cAMP-stimulated phosphorylation of an axonemal polypeptide that copurifies with the 22S dynein arm regulates microtubule translocation velocity and swimming speed in *Paramecium. Proc. Natl. Acad. Sci. U.S.A.* **88,** 7918–7922.

Howard, J., Hudspeth, A. J., and Vale, R. (1989). Movement of microtubules by single kinesin molecules. *Nature* **342,** 154–158.

Howard, J., and Hyman, A. (1993). Preparation of marked microtubules for the assay of polarity of microtubule based motors by fluorescent microscopy. *Methods Cell Biol.* **39,** 105–113.

Kagami, O., Takada, S., and Kamiya, R. (1990). Microtubule translocation caused by three subspecies of inner-arm dynein from *Chlamydomonas* flagella. *FEBS Lett.* **264,** 179–182.

Kagami, O., and Kamiya, R. (1992). Translocation and rotation of microtubules caused by multiple species of *Chlamydomonas* inner arm dynein. *J. Cell Sci.* **103,** 653–664.

Kuo, S., and Sheetz, M. P. (1993). Force of single kinesin molecules measured with optical tweezers. *Science* **260,** 232–234.

Moss, A. G., Gatti, J., and Witman, G. B. (1992a). The motile B/IC1 subunit of sea urchin outer arm dynein does not form rigor bonds. *J. Cell Biol.* **118,** 1177–1188.

Moss, A. G., Sale, W. S., Fox, L. A., and Witman, G. B. (1992b). The α subunit of sea urchin sperm outer arm dynein mediates structural and rigor binding to microtubules. *J. Cell Biol.* **118,** 1189–1200.

Paschal, B. M., King, S. M., Moss, A. G., Collins, C. A., Vallee, R. B., and Witman, G. B. (1987). Isolated flagellar outer arm dynein translocates brain microtubles *in vitro. Nature* **330,** 672–674.

Sale, W. S. and Fox, L. A. (1988). Isolated β-heavy chain subunit of dynein translocates microtubules *in vitro. J. Cell Biol.* **107,** 1793–1798.

Sale, W., Fox, L. A., and Smith, E. F. (1993). Assays of axonemal dynein driven motility. *Methods Cell Biol.* **39,** 89–103.

Scholey, J. (1993). Motility assays for motor proteins. *In* "Methods in Cell Biology" Vol. 39, Academic Press, San Diego.

Sellers, J., Cuda, G., Wang, F., and Homsher, E. (1993). Myosin specific adaptations of the motility assay. *Methods Cell Biol.* **39,** 24–48.

Smith, E. F., and Sale, W. S. (1991). Microtubule binding and translocation by inner dynein arm subtype I1. *Cell Motil. Cytoskel.* **18,** 258–268.

Summers, K. E., and Gibbons, I. R. (1971). Adenosine triphosphate-induced sliding of tubules in trypsin-treated flagella of sea-urchin sperm. *Proc. Natl. Acad. Sci. U.S.A.* **68,** 3092–3096.

Svoboda, K., Schmidt, C., Schnapp, B., and Block, S. (1993). Direct observation of kinesin stepping by optical trapping interferometry. *Nature* **365,** 721–727.

Vale, R. D., and Toyoshima, Y. Y. (1988). Rotation and translocation of microtubules *in vitro* induced by dyneins from *Tetrahymena* cilia. *Cell* **52,** 459–469.

Vale, R. D., and Toyoshima, Y. Y. (1989). Microtubule translocation properties of intact and proteolytically cleaved dyneins from *Tetrahymena* cilia. *J. Cell Biol.* **108,** 2327–2334.

Vale, R. D., Soll, D. R., and Gibbons, I. R. (1989). One dimensional diffusion of microtubules bound to flagellar dynein. *Cell* **59,** 915–925.

Witman, G. B. (1992). Axonemal dyneins. *Curr. Opin. Cell Biol.* **4,** 74–79.

CHAPTER 38

High-Resolution Imaging of Flagella

Keith G. Kozminski
Department of Biology
Yale University
New Haven, Connecticut 06520

I. Introduction

Light microscopy has been used for almost three centuries to study cilia and flagella. In a discussion of flagellated or ciliated cells, motility is equated, even today, with dynein-based ciliary or flagellar beating. This perspective is understandable given that beating is the most obvious motility of cilia and flagella. Yet in addition to beating, the eukaryotic flagellum displays four motilities that are independent of flagellar dynein. These dynein- and beat-independent motilities (reviewed in Bloodgood, 1990, 1992; Goodenough, 1991) continue to be studied in the biflagellate, unicellular, green alga, *Chlamydomonas*. The motilities studied by conventional light microscopy (i.e., phase, fluorescence, differential interference contrast) include the gliding of whole cells across surfaces by means of their flagella (Lewin, 1952; Bloodgood, 1981), the saltatory bidirectional movement of polystyrene beads attached to the flagellar membrane (Bloodgood, 1977; Bloodgood *et al.,* 1979), and the movement (tipping) of glycoproteins from positions along the length of the flagellum to the flagellar tip during mating (Goodenough and Jurivich, 1978). Recently, the bidirectional movement of granule-like particles was observed along the length of the flagellum beneath the flagellar membrane (Kozminski *et al.,* 1993). Visualization of this motility, termed *intraflagellar transport* (IFT), was possible only with the high-resolution imaging afforded by video-enhanced differential-interference-contrast (DIC) microscopy.

This chapter focuses on video-enhanced DIC microscopy techniques used to study the beat-independent flagellar motilities of *Chlamydomonas*. Many of the points outlined, such as the choice of cell lines and sample preparation, are useful for the study of these motilities by conventional microscopy as well. In fact, only the visualization of intraflagellar transport requires the high-resolution imaging provided by video-enhanced DIC microscopy; both phase and standard DIC microscopy are perfectly adequate for visualizing the other motilities. For a more thorough treatment of video-enhanced microscopy, see Allen (1981a,b; 1985) and Inoué (1981, 1986).

II. Methods

A. Choice of Specimen

The most important criterion in choosing a specimen for high-resolution imaging is obtaining immotile cilia/flagella. Excessive movement of cilia/flagella causes them to change focal planes, giving no opportunity to adjust the optics and video camera for maximum resolution and optimal contrast. Immotility can be ensured by three general methods that may be used alone or in combination: biological, chemical, and mechanical.

The method of choice for obtaining an immotile specimen is biological. That is, one should take advantage of naturally immotile cilia/flagella (e.g., apical tuft cilia of sea urchin blastulas, *Limulus* sperm tails) or use organisms in which motility mutants exist. Be cautioned that all motility mutants are not optimal for high-resolution imaging. Among the paralyzed-flagella strains of *Chlamydomonas reinhardtii,* flagellar rigidity varies (Huang, 1986). Strains lacking the central pair microtubules, such as *pf18,* have very rigid flagella that rarely become flush to the coverslip along their length, resulting in loosely attached flagella that may be in several focal planes. When use of a central pairless mutant is not obligatory, one should use a strain with supple flagella, such as *pf1* (no radial spoke heads), that bend to lay flush against the coverslip along their entire length.

The second method for obtaining immotile specimens is by chemical modification of the medium. With *Chlamydomonas,* for instance, the addition of 1 m*M* ethylene glycol bis(β-aminoethyl ether)-*N,N'*-tetraacetic acid (EGTA) to the medium or the omission Ca^{2+} inhibits gliding, leaving the cells attached to the coverslip (Bloodgood *et al.,* 1979; Kozminski *et al.,* 1993). If the focus of the experiment is gliding motility, this technique has an obvious disadvantage, though it is an excellent method for completely immobilizing cells for the study of intraflagellar transport. One should, however, be warned that modification of the medium may adversely affect the physiology of the cilium/flagellum. For example, omitting Ca^{2+} from the medium while raising the monovalent cation concentration will result in flagellar resorption (Lefebvre *et al.,* 1978).

Specimens may be also immobilized mechanically. Cells can be squeezed between two coverslips, held apart with Vaseline, by gently tapping down the top coverslip. This approach provides a very small chamber volume, making perfusion impractical, and requires a fine touch to prevent compression damage to the specimen. Even if the specimen is immobilized, the cilia and flagella, which are thinner than the whole cell, may continue to move, unless a paralyzed cell type is used. If maximum image resolution is not required, the specimen can be immobilized in an agarose block (see Section II,B).

As an alternative, a poly-amino acid can be used to impart a uniform positive or negative charge across the surface of the coverslip; specimens will adhere to this surface electrostatically. In this method, a solution of poly-amino acid (e.g., polylysine) is pipetted onto a clean coverslip and allowed to stand for 3–5 minutes. The poly-amino acid solution is then removed, and the coverslip rinsed thoroughly with double-distilled water before allowing it to air-dry. For *Chlamydomonas,* Bloodgood (1981) reports that the application of 0.1 mg/ml polylysine will favor cell adhesion and allow gliding to occur; higher concentrations (up to 10 mg/ml) will result in tight adhesion and the cessation of gliding motility. At the higher concentrations of polylysine, the flagella of *Chlamydomonas* often excise or ball up with the axoneme curling within the flagellar membrane. Use of polyaspartate has proven ineffectual with *Chlamydomonas,* although the application of a negative charge to the coverslip may be useful with other organisms.

Two other related considerations in the choice of a specimen for high-resolution imaging are species density and ciliary/flagellar density. To obtain the most information from a specimen, one must maximize image resolution and contrast by appropriate adjustment of the optical and video systems. Regardless of the adjustments made, information cannot be extracted from a poorly chosen specimen. For example, cell bodies are very bright under the optical conditions employed for the high-resolution imaging of flagella. Therefore, observations will be difficult in the proximal region of the cilium/flagellum or with those cilia/flagella that are short or held close to the cell body. This difficulty can be alleviated in *Chlamydomonas* by changing from the commonly studied species *C. reinhardtii* with 10-μm flagella to a related species, *C. moewusii,* in which the flagellar are 50–100% longer. Optical interference from light scattering can also be reduced by choosing specimens that have a low ciliary/flagellar density.

B. Sample Preparation

1. Coverslip Chambers for High-Resolution Microscopy

Specimens are prepared in chambers made with two 22-mm^2 No. 1 coverslips (Corning, Corning, NY). Coverslips are placed in mini-Coplin dishes (Wheaton Scientific Products, Millville, NJ) and washed overnight in 6 *N* HCl, followed by a thorough rinse with double-distilled water and a final rinse in 95% ethanol. Although extensive treatment to clean the coverslips is not essential for obtaining an image initially, this preparatory step is recommended when optimal image quality is required.

A thin chamber (200 μl) that allows perfusion but not rapid solute concentration by evaporation can be made from two coverslips (Fig. 1). Chambers should be assembled on a surface that is free of dirt and hand oils (e.g., a piece of lens paper). To facilitate perfusion, 3–4 mm is cut from the side of one coverslip with a diamond pen knife. This coverslip forms the top of the chamber. A thin track of Vaseline should be applied along the shorter sides of this coverslip, using a syringe fitted with a blunted 18-gauge needle. Though commonly used as a substitute for Vaseline, silicon grease is toxic to *Chlamydomonas*. Plastic shims, 3 × 18 mm^2, are cut from 0.5-mm-thick, 6-in. plastic rulers (Macalaster Bicknell, New Haven, CT) and placed on the Vaseline tracks. Another line of Vaseline is applied on top of the shims. A dilute (ca. 10^5 cells/ml) sample of specimen (150–200 μl) can be applied to the middle of a second uncut coverslip, allowing time for the cells to settle. Excessive sample density will cause light scattering and degrade image quality. The chamber is completed by inverting the smaller coverslip, complete with adhering plastic shims, with tweezers onto the second coverslip containing the sample. The top coverslip should be gently tapped down, forcing the specimen to completely fill the chamber. The chamber can then be affixed to a coverslip holder with Scotch tape or silicon grease. In simplest form, the coverslip holder is a custom machined metal plate with the same dimensions as a standard glass microscope slide, but with a circular or rectangular hole smaller than the dimensions of the coverslip.

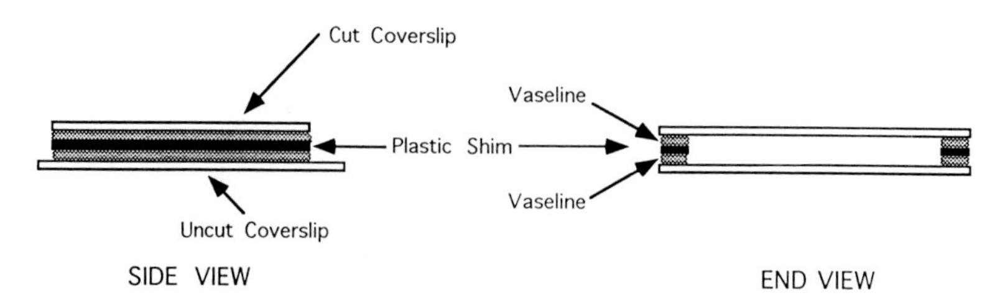

Fig. 1 Coverslip chamber used for high-resolution microscopy.

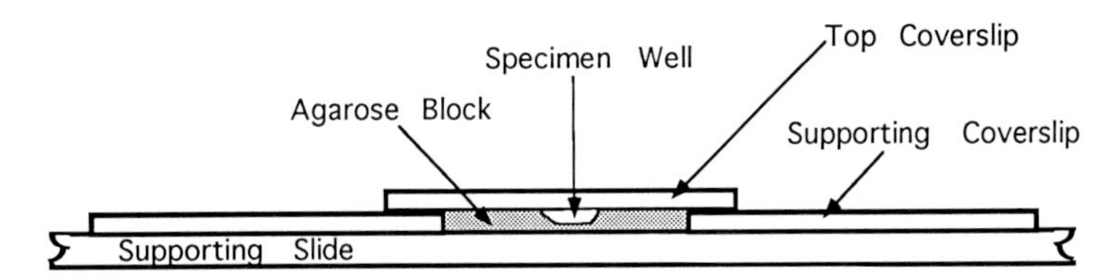

Fig. 2 Side view of an agarose block chamber used to immobilize cells.

2. Agarose Block Immobilization

This simple and efficient method of mechanically immobilizing cells was devised by Dr. Jeremy Pickett-Heaps specifically for time-lapse microcinematography. Similar methods have been described by Reize and Melkonian (1989). Although intraflagellar transport can be observed in motile strains of *Chlamydomonas* immobilized by this technique, the agarose does create a noticeable decrease in resolution. As shown in Fig. 2, a glass slide or coverslip of similar dimensions is used to support two 22-mm^2 square coverslips placed approximately 15 mm apart from each other. A spot of VALAP (1 : 1 : 1 Vaseline/lanolin/paraffin) on the slide beneath the coverslips will ensure proper adhesion of the coverslips to the slide. Warm 0.3% agarose (type V, Sigma), made in the culture medium, is pipetted into the space between the coverslips to form the agarose block; seepage of the agarose beneath the coverslips also helps to seal the coverslips to the supporting slide. As the agarose solidifies, gentle suction through a pipet is used to form a well in the agarose block. Specimen is then added to the well. A chamber is formed by placing a third 22-mm^2 coverslip on top of the agarose block and supporting coverslips. The chamber can be sealed by placing VALAP along the edges of the third or top coverslip. Immobilized cells will ring the specimen well where the agarose meets the top coverslip.

C. Equipment

1. Optics

High-resolution imaging of flagellar motility in *Chlamydomonas* (Kozminski *et al.,* 1993) was successfully performed with an Axiovert inverted microscope (Carl Zeiss, Thornwood, NJ) fitted with DIC optics. Specifically, the microscope was equipped with a Zeiss 63 × /1.4 NA Planapo objective, a 1.4 NA condenser, a 75-mm-focal-length convex auxiliary lens (Oriel, Stratford, CT), and a green glass filter to minimize photodamage (Oriel). High numerical aperture (NA) optics are essential for high-resolution imaging.

To obtain full and even illumination of the condenser back focal plane, which is required for high-resolution imaging, light from a 100-W Hg arc lamp should be transmitted through a fiberoptic scrambler. Fiberoptic scramblers can be purchased commercially (E. Leitz, Rockleigh, NJ) or assembled from individual parts. The scambler apparatus used for imaging the beat-independent flagellar motilities of *Chlamydomonas* consisted, in series, of the Hg arc lamp, a 0.5-cm KG1 glass for UV and IR light absorption (Schott Glass, Duryea, PA), a 60-nm-bandpass interference filter centered at 540 nm (Carl Zeiss), an iris aperture (Newport, Irvine, CA), and a Zeiss 16×/0.35 NA/phase 2 Planapo objective to focus the collimated light onto a 1-mm-diameter single fiberoptic cable (General Fiber Optics, Cedar Grove, NJ). The light emitted from the cable was collected by a No. 3 bright-field objective (E. Leitz) to recollimate the beam before entry into the optical train of the microscope.

2. Video and Recording

A 10× or 16× lens (Carl Zeiss) should be used to project the image of the specimen into a Nuvicon camera (Hammatsu Photonic Systems, Bridgewater, NJ). The analog image is viewed on a video monitor (Panasonic, Secaucus, NJ; Sony, Paramus, NJ), with the video signal monitored on a waveform monitor (Tektronix, Beaverton, OR). The image can also be digitized with an image processor. Video signals of the beat-independent flagellar motilities were often digitized, using a Series 151 image processor (Imaging Technology Systems, Woburn, MA) controlled by an IBM-compatible Everex 386/25 host computer (Everex, Fremont, CA). Other commercially available systems should prove satisfactory. A super-VHS recording of either the analog or digital image, offering higher resolution than standard VHS tape, can be made with a video recorder (Panasonic).

D. Obtaining an Image

These basic steps for obtaining high-resolution video images of the beat-independent flagellar motilities of *Chlamydomonas* should serve as a guide and will vary with each microscope. The key to good images is good optics. Although the enhanced contrast provided by video reveals the fine details of a specimen, true resolution can only be increased by optimization of the optics (i.e., high-NA optics, condenser and objective oiled to the coverslip chamber).

1. A practice sample should be made, as described above, for the initial adjustment of the microscope. Before turning on any electronic devices, it first may be necessary to turn on the mercury arc lamp and wait for steady illumination. Electromagnetic pulses generated by lamp ignition can damage other active electronic devices not properly shielded with surge suppressors.

2. A good starting point for high-resolution imaging of cilia and flagella is Koehler illumination.

3. To obtain video images that use the full NA of the condenser lens, the condenser aperture should be completely open. The dimensions of the video image on the monitor can be determined with a stage micrometer.

4. Initially, the video controls should be set at minimum sensitivity with the dynamic range (gain) and background level (offset) at one-third to one-half of maximum; video boost and shading correction should be off. These adjustments will vary with each video system. Essentially, the video camera should be near the point of light saturation. Oversaturation, often indicated by a light on the video control panel, will damage the camera photodetector. The amount of light entering the system, therefore, should be reduced by closing the scrambler iris diaphragm. Conversely, if the video image remains dark even after increasing the video gain and background level, the camera is light-starved and the amount of light entering the system should be increased by opening the scrambler iris diaphragm. If one is using *Chlamydomonas,* the amount of light can be judged by the brightness of the cell bodies. With the proper amount of light the cell bodies should look bright, making detail difficult, but not impossible to see. If the cell bodies appear as orbs of light that have flagella, too much light is present and the level should be reduced.

5. The flagellum/cilium is the structure of interest, so time should not be spent obtaining a good image of the cell body. Once a suitable light level has been found, the video background level (offset) should be brought to 40%, as read on the waveform monitor, and the dynamic range (gain) of the video signal corresponding to the flagella stretched from 0 to 90%. If the contrast is properly adjusted, the image should be set against an evenly lit, soft gray background without noise introduced by excessive gain. When oriented parallel to the shear axis of the first (condenser-side) Wollaston prism, the flagellum will have low contrast compared with background. A 90° rotation of the flagellum, placing it perpendicular to the shear axis, will produce a shadowcast image; this orientation is most common for obtaining DIC images of good contrast. It is at the first position, however, where the flagella are oriented *parallel* to the shear axis and *not* shadowcast, that the granule-like particles constituting IFT remain shadowcast and are best observed. The orientation of the flagella is not critical for viewing the other beat-independent motilities.

A common problem is excessive contrast of the flagella that cannot be adjusted by a manipulation of the background level and gain, regardless of the orientation of the flagellum with respect to the shear axis of the first Wollaston prism. Contrast can be reduced by adjusting the second Wollaston prism, mounted with the objective lens, toward the direction of increasing light intensity. By adjusting this prism, the amount of light transmitted will also vary; therefore, readjustment of the scrambler iris diaphragm may be necessary to avoid oversaturation of the video camera. These adjustments also require a readjustment of the video controls as described above.

6. Depending on the video system, shading of the analog image can be corrected by adjusting the shading controls on the video camera. Before making this adjustment, however, one should check for the proper height and centering of the condenser for the position of the auxiliary lens, and for the presence of bubbles in the immersion oil, all of which can introduce shading anomalies.

7. Once an analog image of acceptable contrast and shading has been obtained, the video signal can be digitized and background anomalies subtracted to enhance the details within the image. To background subtract, a digital image of an out-of-focus specimen is recorded and then subtracted from a digital image of the in-focus specimen.

8. The final digitized image can be recorded on super-VHS tape or optical disk. Though of lower recording and playback quality than optical disk, VHS tape is more economical for recording dynamic events, which require longer recording times.

III. Conclusion

Each major advance in light microscopy has allowed us to obtain new information about cilia and flagella. The visualization of intraflagellar transport, for example, was made possible only with the high-resolution imaging provided by video-enhanced DIC microscopy; whereas the other beat-independent motilities (e.g., gliding, bead movement) can be observed with conventional optics. In the case of gliding motility, additional information was also obtained with interference-reflectance microscopy (Bloodgood, 1990). Clearly, cilia and flagella, observed for decades as beating featureless rods, have additional motilities with motors and functions yet to be understood.

Acknowledgments

I thank Dennis Diener, Joel Rosenbaum, and Corey Thompson for making critical comments on this manuscript and, most especially, Paul Forscher for teaching me many of the microscopy techniques described here. This work was supported by National Institutes of Health Grant GM-14642 and National Science Foundation Grant 45147 to Joel Rosenbaum.

References

Allen, R. D. (1985). New observations on cell architecture and dynamics by video-enhanced contrast optical microscopy. *Annu. Rev. Biophys. Biophys. Chem.* **14,** 265–290.

Allen, R. D., Allen, N. S., and Travis, J. L. (1981a). Video-enhanced contrast, differential interference contrast (AVEC-DIC) microscopy, a new method capable of analyzing microtubule-related motility in the reticulopodial network of *Allogromia laticollaris*. *Cell Motil.* **1,** 291–302.

Allen, R. D., Travis, J. L., Allen, N. S., and Yilmaz, H. (1981b). Video-enhanced contrast polarization (AVEC-POL) microscopy: a new method applied to the detection of birefringence in the motile reticulopodial network of Allogromia laticollaris. *Cell Motil.* **1,** 275–289.

Bloodgood, R. A. (1977). Motility occurring in association with the surface of the *Chlamydomonas* flagellum. *J. Cell Biol.* **75,** 983–989.
Bloodgood, R. A. (1981). Flagella-dependent gliding motility in *Chlamydomonas. Protoplasma* **106,** 183–192.
Bloodgood, R. A. (1990). Gliding motility and flagellar glycoprotein dynamics in *Chlamydomonas. In* "Ciliary and Flagellar Membranes" (R. A. Bloodgood, ed.) pp. 91–128. Plenum, New York.
Bloodgood, R. A. (1992). Directed movements of ciliary and flagellar membrane components: a review. *Biol. Cell* **76,** 291–301.
Bloodgood, R. A., Leffler, E. M., and Bojczuk, A. T. (1979). Reversible inhibition of *Chlamydomonas* flagellar surface motility. *J. Cell Biol.* **82,** 664–674.
Goodenough, U. W. (1991). Chlamydomonas mating interactions. *In* "Microbial Cell-Cell Interactions" (M. Dworkin, ed.) pp. 71–112. American Society for Microbiology, Washington, D.C.
Goodenough, U. W., and Jurivich, D. (1978). Tipping and mating-structure activation induced in *Chlamydomonas* gametes by flagellar membrane antisera. *J. Cell Biol.* **79,** 680–693.
Harris, E. H. (1989). "The *Chlamydomonas* Sourcebook", pp. 780. Academic Press, New York.
Huang, B. P.-H. (1986). *Chlamydomonas reinhardtii:* A model system for the genetic analysis of flagellar structure and motility. *Int. Rev. Cytol.* **99,** 181–215.
Inoué, S. (1981). Video image processing greatly enhances contrast, quality, and speed in polarization-based microscopy. *J. Cell Biol.* **89,** 346–356.
Inoué, S. (1986). "Video microscopy", pp. 584. Plenum, New York.
Kozminski, K. G., Johnson, K. A., Forscher, P., and Rosenbaum, J. L. (1993). A motility in the eukaryotic flagellum unrelated to flagellar beating. *Proc. Natl. Acad. Sci. U.S.A.* **90,** 5519–5523.
Lefebvre, P. A., Nordstrom, S. A., Moulder, J. E., and Rosenbaum, J. L. (1978). Flagellar elongation and shortening in *Chlamydomonas* IV. Effects of flagellar detachment, regeneration, and resorption on the induction of flagellar protein synthesis. *J. Cell Biol.* **78,** 8–27.
Lewin, R. A. (1952). Studies on the flagella of algae. I. General observations of *Chlamydomonas moewusii* gerloff. *Biol. Bull.* (*Woods Hole, MA*) **103,** 74–79.
Reize, I. B., and Melkonian, M. (1989). A new way to investigate living flagellated ciliated cells in the light microscope: Immobilization of cells in agarose. *Botanica Acta* **102,** 145–151.

II. Description of Gliding Motility

Gliding motility is exhibited by all flagellated stages of *Chlamydomonas:* vegetative cells, gametic cells, and quadriflagellate heterokaryons that result from fertilization. Most observations of gliding motility have been made using *C. reinhardtii* and *C. moewusii.* Lewin (1952) described gliding motility in *C. moewusii,* which he referred to as ''creeping.'' Bloodgood (1981) provided the first detailed description of gliding motility in *C. reinhardtii.* A more complete description of gliding motility is provided in a recent review (Bloodgood, 1990). Gliding cells exhibit a characteristic morphology: the two flagella are oriented 180° from each other and the direction of gliding correlates with the axes of the flagella. During the course of gliding motility, the flagella can appear to be absolutely straight and immotile, exhibiting no obvious bends or undulations. On occasion, however, the distal 1 μm of the leading flagellum can be seen to wiggle; this observation, coupled with the fact that cells can glide in curved trajectories (see Fig. 4 in Bloodgood, 1981, and Fig. 3 in Bloodgood, 1987), suggests the possibility that *Chlamydomonas* can regulate the direction of gliding motility, perhaps even in response to external stimuli, such as light.

Cells can stop and then resume gliding in the same or the opposite direction. A number of observations suggest that the lead flagellum is providing the motive force for cell movement (although both flagella are capable of performing the role of the lead flagellum). While a wild type cell is gliding, it is occasionally observed to lift one of its flagella off the substrate, beat the flagellum once, and then reattach it to the substrate. If the leading flagellum is involved in this activity, the cell stops moving while the flagellum is not in contact with the substrate; if the trailing flagellum is involved, the cell continues to move. Uniflagellar cells move continuously in one direction, always with the flagellum leading the cell body. Gliding motility occurs at 1–2 μm/s. A wide variety of paralyzed flagellar strains (exhibiting defects in outer dynein arms, inner dynein arms, radial spokes, and central pair structures) exhibit normal gliding behavior (Lewin, 1954; Bloodgood, 1981; Kozminski *et al.,* 1993), suggesting that gliding motility is not coupled to axonemal motility and is not dependent on the motor responsible for axonemal motility and flagellar beating (inner and outer dynein arms). Gliding behavior suggests the involvement of a minus end-directed microtubule-associated motor.

III. Procedures

A. Observation of Gliding Motility

In general, gliding motility is best observed in the phase or differential-interference-contrast (DIC) microscope on well-cleaned glass coverslips or glass slides using paralyzed flagellar mutants (strain pf18 has been used most often in the case of *C. reinhardtii,* and strain M-475 has been used most often in the case of *C. moewusii*). Clean slides are soaked in 0.5 m*M* ethylenediaminetetraacetic acid (EDTA) and then washed thoroughly in deionized water; these slides are stored dried and rinsed in distilled water just before use. If soap is used to clean slides, 7× detergent is recommended (Linbro Scientific). Coverslips are cleaned with distilled water just before use. After putting a drop of cell culture on a glass slide and *carefully* applying the coverslip, it is best to wait a few minutes to give the cells a chance to settle and to interact with the glass surfaces. Some variability is observed from culture to culture in the percentage of cells that adhere to the substrate and exhibit gliding. In general, paralyzed-flagella mutants exhibit a much higher percentage of gliding cells and it is not unusual to see cultures of paralyzed-flagella strains in which virtually all of the cells either are exhibiting gliding motility or are in the characteristic gliding configuration but not exhibiting gliding motility. It is assumed that these latter cases represent situations where both flagella are mechanically interacting with the substrate and the balanced forces exerted by the two flagella result in no net movement. As careful observations of cell attachment to a glass surface suggest that gliding activity of the flagellar tips is necessary for establishing the characteristic gliding configuration (Bloodgood, 1981), one can probably score gliding cells in a population merely by scoring the percentage of cells in the gliding configuration, thus speeding up the assay considerably. The composition of the medium can affect flagellar adhesion to the substrate and gliding motility.

Kozminski and Rosenbaum (personal communication) observed better adhesion and gliding in minimal medium (medium I of Sager and Granick, 1953) than in minimal medium supplemented with acetate. Pretreating glass slides or coverslips with a 0.1 mg/ml solution of poly-L-lysine promotes interaction of the flagellar surfaces with the substrate and increases the percentage of the cells in the preparation that express gliding behavior, but treating glass surfaces with 10 mg/ml poly-L-lysine promotes such tight adhesion of cells (perhaps by the cell wall as well as the flagella) to the substrate that no gliding motility is observed (Bloodgood, 1981).

Gliding motility is reversibly inhibited at 4°C by lowering the free calcium concentration in the medium with ethylene glycol bis(β-aminoethyl ether)-*N,N*-tetraacetic acid (EGTA) and by adding 100 m*M* NaCl (Kozminski *et al.,* 1993); however, it is sometimes difficult to obtain reproducible, quantitative data on the effects of treatments on gliding motility. We prefer the polystyrene microsphere movement assay (see below) in cases where quantitation is desirable.

B. Polystyrene Microsphere Movements

If a population of *Chlamydomonas* (preferably a paralyzed-flagella mutant strain) is mixed with polystyrene microspheres (with a diameter in the range 0.25–0.40 μm, Polysciences) and then examined in the phase or DIC microscope, it is readily apparent that the microspheres adhere to and are transported in a bidirectional manner along the surfaces of the flagella at velocities similar to that of gliding motility (1–2 μm/s). Several polystyrene microspheres can be observed in motion on the same flagellum; in such a case, the movements of the microspheres are independent of one another, arguing for the ability of the flagellar machinery to transduce force locally at the flagellar surface. Microsphere movements have never been observed on flagella that have detached from cells. Most of the published work on microsphere movement has been performed using polystyrene microspheres obtained from Polysciences. Although a variety of latex microspheres (including ones derivatized with concanavalin A, bovine serum albumin, amino groups, carboxyl groups, and hydroxyl groups) and other small inert objects (including bacteria) are capable of being transported along the surface of the flagella (Bloodgood, 1977), the underivatized polystyrene microspheres (of about 0.35 μm in diameter) appear to be the most useful for routine observations. The amount of beads added to medium containing cells is rather empirical; one wants sufficient microspheres bound to the flagella without having too high a density of free microspheres in solution which can interfere with observations in the microscope. Using polystyrene microspheres from Polysciences (which usually come as a suspension of 2.5% solids), we usually mix the beads with medium at a ratio of 1/10 to 1/50 (v/v). Bead binding is concentration dependent (Reinhart and Bloodgood, 1988) and conditions can be found where multiple microspheres can be seen to be attached to and moving along individual flagella. Lowering the ionic strength of the medium tends to promote microsphere adhesion.

Microsphere movement is more easily quantitated than gliding motility and provides a better system for assaying the effects of various treatments (Bloodgood *et al.*, 1979; Hoffman and Goodenough, 1980; Snell *et al.*, 1982; Detmers and Condeelis, 1986). Snell *et al.* (1982) demonstrated that 1 m*M* lidocaine (xylocaine) almost completely inhibits both polystyrene microsphere movements and fertilization in *C. reinhardtii* gametic cells. Detmers and Condeelis (1986) demonstrated that the calmodulin antagonists trifluoperazine (TFP) and W-7 inhibit both polystyrene microsphere movements and fertilization in *C. reinhardtii* in a similar dose-dependent manner. All three of these reagents (lidocaine, TFP, W-7) have modes of action related to calcium. Flagellar surface motility is quantitated by determining the percentage of polystyrene microspheres attached to a flagellum that are in motion at the time of observation. Almost all attached microspheres, if observed long enough, will exhibit periods of motion, so it is important to assay for microsphere movement immediately on observing a particular microsphere. It is fairly easy to recognize microspheres that are mechanically coupled to the flagellar surface because they no longer exhibit the vigorous Brownian movement shown by microspheres free in the medium. The control figure for microsphere movements on the flagella of *C. reinhardtii* pf18 cells in medium I of Sager and Granick (1953) at room temperature is approximately 60% (Bloodgood *et al.*, 1979); Snell *et al.* (1982) found a similar figure for *C. reinhardtii* pf17 cells.

An important question is whether microsphere movements along the flagellar membrane reflect the same force transduction system responsible for whole-cell gliding motility. There are four arguments for this point: (1) Velocity is similar for both phenomena. (2) Both are inhibited by removal of calcium from the external medium (Bloodgood *et al.,* 1979; Kozminski *et al.,* 1993). (3) A number of environmental treatments (low calcium, low temperature, high salt) that inhibit microsphere movement also inhibit gliding motility (Bloodgood, 1990; Kozminski *et al.*, 1993). (4) Certain mutant cell strains exhibit defects in both gliding motility and microsphere movement (see below). As is the case with gliding motility, a wide range of mutant cell lines with defects in outer dynein arms, inner dynein arms, radial spokes, and central pair structures exhibit normal microsphere movement (Bloodgood, 1977; Hoffman and Goodenough, 1980; Bloodgood, 1990; Kozminski *et al.,* 1993).

IV. Nongliding Mutants

Lewin (1982) isolated the first nongliding mutants of *Chlamydomonas;* he used UV mutagenesis of a paralyzed-flagella mutant (M-475) of *C. moewusii.* Mutagenized cells were spread onto very soft agar and small colonies were selected and screened for gliding motility. Six nongliding (or *fg*−) cell lines (designated fg−1, fg−2, fg−3, fg−5, fg−6, and fg−7) were obtained although it was not demonstrated that these were independent isolates. The nongliding mutants were unable to achieve the characteristic gliding configuration described above. Both Lewin (1982) and Reinhart and Bloodgood (1988) demonstrated that these nongliding cell strains were unable to exhibit microsphere movement along the flagellar surface, suggesting that microsphere movements are a manifestation of the same force transduction system responsible for whole-cell gliding motility. Reinhart and Bloodgood (1988) demonstrated that Lewin's fg− mutants could be divided into subclasses based on the level of adhesion of polystyrene microspheres (0.35 μm diameter, Polysciences) to the flagellar surface and whether the level of bound microspheres was dependent on microsphere concentration. Strain fg−3 cells exhibited a much lower level of microsphere adhesion than the cells of the M-475 parent strain, while strain fg−2 cells exhibited a much higher level of microsphere adhesion than cells of the parent strain. These observations suggest that at least some of the Lewin nongliding mutants may have alterations in their flagellar surface adhesive properties. The Lewin (1982) nongliding mutants are available from the *Chlamydomonas* Genetics Center. The culture collection numbers are CC-1838 (*C. moewusii* fg−1), CC-1839 (*C. moewusii* fg−2), CC-1840 (*C. moewusii* fg−3), CC-1841 (*C. moewusii* fg−5), CC-1842 (*C. moewusii* fg−6), and CC-1843 (*C. moewusii* fg−7). The paralyzed-flagella parent cell line for these nongliding mutants is available as CC-957 (*C. moewusii* M-475). The culture collection also lists a gf−1 mutant (used by Kozminski *et al.,* 1983), which is probably the same mutant strain as the fg−1 (CC-1838) mutant listed above.

Keith G. Kozminski (Kozminski and Rosenbaum, 1994; personal communication), working in Dr. Joel Rosenbaum's laboratory at Yale University, isolated a large number of nongliding mutants in *C. reinhardtii* using the procedure of insertional mutagenesis (Tam and Lefebvre, 1993). A paralyzed-flagella strain (pf − 1), which was also deficient for nitrate reductase (nit − 1) and thus unable to use nitrate, was transformed with the nitrate reductase gene. In *Chlamydomonas,* exogenously added DNA inserts randomly into the genome, tagging endogenous genes while at the same time disrupting them. Cells expressing nitrate reductase after transformation were selected on nitrate-containing soft agar and simultaneously screened for abnormal colony morphology following Lewin's (1982) original screen for nongliding mutants. A considerable number of nongliding mutants were isolated and characterized in terms of adhesion to glass, binding of polystyrene microspheres, and movement of polystyrene microspheres. Some nongliding strains were affected in their adhesion to glass surfaces and/or polystyrene microspheres, whereas others retained normal flagellar surface adhesive properties, these being candidates for defects in the motor machinery or the signal transduction pathway.

V. Perspectives

Gliding motility is a unique form of substrate-dependent whole-cell locomotion that is dependent on the glycoprotein dynamics of the flagellar membrane (Bloodgood and Salomonsky, 1989; Bloodgood, 1992). *Chlamydomonas* is convenient to grow and to manipulate and gliding motility is easy to observe in the light microscope using phase or DIC optics. Attachment to and movement along the flagellar surface of polystyrene microspheres provides an easy, quantitative assay for the force transduction system responsible for gliding motility. Availability of nongliding mutants, especially ones that have been insertionally tagged, provides the opportunity to identify genes whose products are involved in this form of whole-cell locomotion. Nongliding cell lines could be defective in flagellar membrane components necessary for substrate adhesion, in components of the motor responsible for force transduction, or in components of the signaling pathway that couples the sensory and motor properties of the flagellar surface.

References

Bloodgood, R. A. (1977). Rapid motility occurring in association with the *Chlamydomonas* flagellar membrane. *J. Cell Biol.* **75,** 983–989.

Bloodgood, R. A. (1981). Flagella-dependent gliding motility in *Chlamydomonas. Protoplasma* **106,** 183–192.

Bloodgood, R. A. (1987). Glycoprotein dynamics in the *Chlamydomonas* flagellar membrane. *In* "Advances in Cell Biology" (K. R. Miller, ed.), Vol. 1, pp. 97–130. JAI Press, Greenwich, CT.

Bloodgood, R. A. (1990). Gliding motility and flagellar dynamics in *Chlamydomonas. In* "Ciliary and Flagellar Membranes" (R. A. Bloodgood, ed.), pp. 91–128. Plenum Press, New York.

Bloodgood, R. A. (1992). Directed movements of ciliary and flagellar membrane components: A review. *Biol. Cell* **76,** 291–302.

Bloodgood, R. A., and Salomonsky, N. L. (1989). Use of a novel *Chlamydomonas* mutant to demonstrate that flagellar glycoprotein movements are necessary for the expression of gliding motility. *Cell Motil. Cytoskel.* **13,** 1–8.

Bloodgood, R. A., Leffler, E. M., and Bojczuk, A. T. (1979). Reversible inhibition of Chlamydomonas flagellar surface motility. *J. Cell Biol.* **82,** 664–674.

Detmers, P. A., and Condeelis, J. S. (1986). Trifluoperazine and W-7 inhibit mating in *Chlamydomonas* at an early stage of gametic interaction. *Exp. Cell Res.* **163,** 317–326.

Hoffman, J. L., and Goodenough, U. W. (1980). Experimental dissection of flagellar surface motility in *Chlamydomonas. J. Cell Biol.* **86,** 656–665.

Kozminski, K. G. (1994). Gliding mutants in *Chylamydomonas reinhardtii. Mol. Biol. Cell* **5,** 170a.

Kozminski, K. G., Johnson, K. A., Forscher, P., and Rosenbaum, J. L. (1993). A motility in the eukaryotic flagellum unrelated to flagellar beating. *Proc. Natl. Acad. Sci. U.S.A.* **90,** 5519–5523.

Lewin, R. A. (1952). Studies of the flagella of algae. I. General observations of *Chlamydomonas moewusii* Gerloff. *Biol. Bull.* (*Woods Hole, MA*) **103,** 74–79.

Lewin, R. A. (1954). Mutants of *Chlamydomonas moewusii* with impaired motility. *J. Gen Microbiol.* **11,** 358–363.

Lewin, R. A. (1982). A new kind of motility (non-gliding) in *Chlamydomonas. Experientia* **38,** 348–349.

Reinhart, F. D., and Bloodgood, R. A. (1988). Gliding defective mutant cell lines of *Chlamydomonas moewusii* exhibit alterations in a 240 kDa surface-exposed flagellar glycoprotein. *Protoplasma* **144,** 110–118.

Sager, R., and Granick, S. (1953). Nutritional studies with *Chlamydomonas reinhardtii. Ann. N.Y. Acad. Sci.* **56,** 831–838.

Snell, W. J., Buchanan, M., and Clausell, A. (1982). Lidocaine reversibly inhibits fertilization in *Chlamydomonas:* a possible role in sexual signaling. *J. Cell Biol.* **94,** 607–612.

Tam, L.-W., and Lefebvre, P. A. (1993). Cloning of flagellar genes in *Chlamydomonas reinhardtii* by DNA insertional mutagenesis. *Genetics* **135,** 375–384.

CHAPTER 40

Assay of *Chlamydomonas* Phototaxis

Anthony G. Moss,* Gregory J. Pazour,† and George B. Witman†

†Worcester Foundation for Experimental Biology
Shrewsbury, Massachusetts 01545
*Department of Zoology and Wildlife Sciences
Auburn University
Auburn, Alabama 36849

I. Introduction

Chlamydomonas responds phototactically to blue illumination with a peak response at 503 nm (Nultsch *et al.*, 1971; Foster and Smyth, 1980). The response can be either positive (cells swim toward light source) or negative (cells swim away from light source) depending on actinic beam intensity, history of cells' exposure to light, genetic background of cells, and other factors. Phototaxis depends on slight light-dependent changes in the waveform of the *cis* and *trans* flagella, as revealed by step-up and step-down illumination of immobilized cells (Rüffer and Nultsch, 1990, 1991). For reviews of *Chlamydomonas* phototaxis, the reader is directed to Feinleib (1985) and Witman (1993). Investigation of the basis for phototaxis in *Chlamydomonas* promises to yield much new information on the molecular mechanisms involved in the regulation of flagellar movement. In addition, the assay of phototaxis offers the most stringent test of *Chlamydomonas* flagellar function currently available, as it measures not only motility but also the ability of the flagella to respond to intracellular signals. This chapter describes a method for analysis of the swimming direction of a population of individual cells in response to light. The method, based on a phototaxis assay originally developed by Morel-Laurens and Feinleib (1983), has been particularly useful in our laboratory for the characterization of mutants with defects in their ability to phototax (Horst and Witman, 1993; Pazour *et al.*, manuscript in preparation).

II. Culture and Handling of *Chlamydomonas* Cells

Cells are grown in 250-ml Erlenmeyer flasks containing 125 ml of medium (modified from Sager and Granick, 1953) aerated with air and 5% CO_2 as previously described (Witman, 1986), or in 13 × 100-mm culture tubes containing 5 ml of the same medium supplemented with 0.2% sodium acetate. In either case, cells are maintained on a 14-hour light: 10-hour dark cycle. Cells typically are used at 0.5–1 × 10^6 ml^{-1} in culture medium; higher concentrations may result in self-shading, which affects the assay. *Chlamydomonas* phototaxis is strongest during logarithmic growth (Stavis and Hirschberg, 1973) and shows diurnal periodicity (Bruce, 1970). Because phototactic responsiveness varies with physiological state and degree of dark adaptation, cultures to be compared are grown and handled identically.

III. Phototaxis Room Setup

A room used for phototaxis measurements must meet several criteria. Dark-adapted *Chlamydomonas* are quite light-sensitive. Therefore, the room must be lighttight when the door is closed to prevent biasing of light-dependent swimming. Our room is diffusely illuminated with a red-filtered 15-W gooseneck lamp aimed at the white ceiling. All extraneous light sources should be masked with red acetate barrier filters (Edmund Scientific, Barrington, NJ, Catalog No. J35, 137). Test the room to be sure that cells do not display any directed swimming in the absence of a stimulus beam. The room should be well-vented, as the laser generates considerable heat.

The microscope and beam positioning equipment (Fig. 1 and see below) must be located on the same table or desktop for accurate beam adjustment. Optimally, they would be mounted on a precision optical bench or weigh table. Equipment can be held in place by bolting or hot glue (Parker Industries, Worcester, MA). The laser must be mounted on a separate table to prevent degradation of the image by vibration from the laser cooling fan.

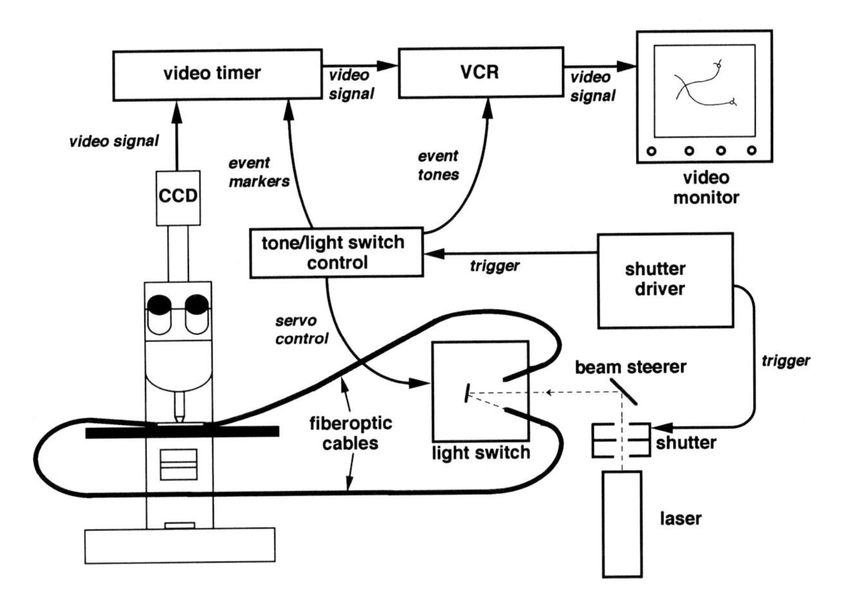

Fig. 1 Setup for measuring phototaxis.

IV. Equipment for Actinic Illumination of Cells during Assay

Blue actinic light (488 nm) is provided by a 10-mW argon laser (Model 162, Spectra-Physics, Piscataway, NJ), operated in the light-controlled mode, using the laser tube current (maintained at 4.3 A) as a guide to illumination intensity. Laser light intensity is adjusted by a dual shearing attenuator (Model 925B, Newport, Fountain Valley, CA). Intensity is determined with a calibrated platinum-black thermopile (The Eppley Laboratory, Newport, RI). The beam typically provides 4.5 mW/mm^2 at the entry into the light switch (below). The light stimulus (0.6-millisecond rise time) is controlled by an electronic shutter and driver (Models 26L2ADX5 and SD1000, respectively, Vincent Associates, Rochester, NY). The SD1000 is sensitive to electrical transients and requires careful attention to cable routing and circuit isolation. The shutter and beam attenuator are located immediately in front of the laser beam exit.

The laser beam is directed by a first-surface dual-mirror beam steering instrument (Newport Corp., model 670) into a custom-made dual fiberoptic light switch box (Fig. 2A). Within this box, a rotatable front-surface glass mirror, controlled by a solenoid, reflects the beam into one or the other of two fiberoptic cables leading to opposing sides of the phototaxis chamber (see below). This allows the investigator to alternately apply the actinic light from opposite directions during the course of an experiment, thus preventing excessive clustering

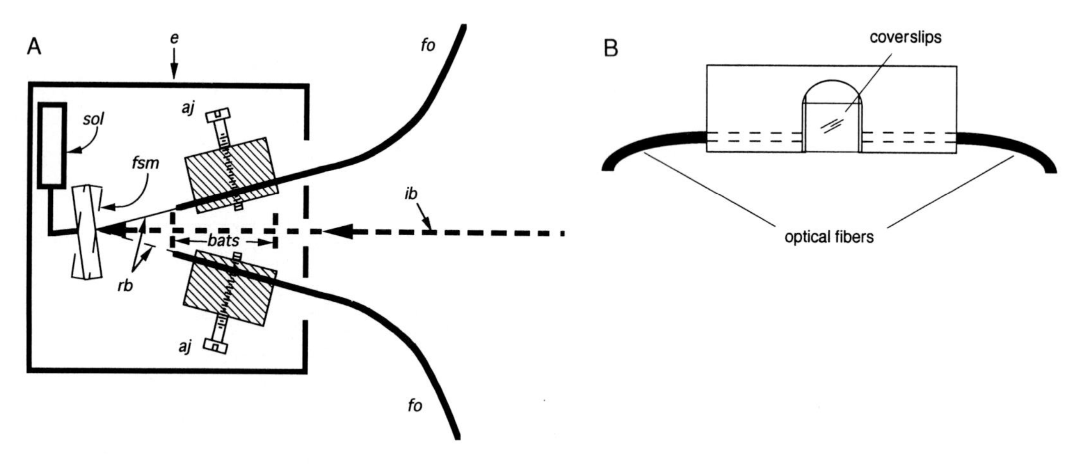

Fig. 2 (A) Dual fiberoptic light switch for reflecting laser light into the fiberoptic guides used to bring actinic light to the phototaxis chamber. e, lighttight enclosure; sol, solenoid; fo, fiberoptic cable; aj, adjustment screw for aligning fiberoptic cable to beam; ib, incident beam; bats, beam alignment targets; fsm, front surface mirror; rb, reflected beam. The two positions of the mirror are shown: one solid with a solid reflected beam, and one dashed with a dashed reflected beam. (B) Phototaxis chamber.

of cells on one side of the chamber. Beam intensity is attenuated nearly fivefold by the fiberoptic cable.

V. Phototaxis Chamber

Phototaxis chambers are constructed of black Plexiglas (Fig. 2B). The two plastic fiberoptic cables (Edmund Scientific, Catolog No. F2536), which are exactly the same length, are mounted in milled grooves and stabilized with TackyWax (CSC Softseal, Central Scientific, Chicago, IL, Catalog No. 11444-000), which allows fiberoptic replacement. The fiber itself coincides precisely with the chamber gap, so that the actinic light spans the chamber. Coverslips (Corning, No. $1\frac{1}{2}$; 18 × 18 mm) are mounted with VALAP (Vaseline:paraffin:beeswax 1 : 1 : 1) or silicon high-vacuum grease (Dow Corning, Midland, MI) on the top and bottom of 0.8-mm-thick ledges. During the phototaxis assay, it is critical that the microscope objective be aligned with an imaginary line running between the outlets of the two fiberoptic cables so that the actinic light enters the field of view strictly from the left or right.

VI. Microscopy

Chlamydomonas cells are placed in the phototaxis chamber and imaged in nonactinic (red) light, using a Corning 2418 glass bandpass filter (625-nm cutoff).

Dark-field microscopy is preferred because the cells are then brightly illuminated against a dark background. We use a Zeiss Universal microscope that has a turret condenser equipped with a 1.25 NA cap and phase 3 annulus, and 6.3 × (0.16 NA), 10× (0.22 NA), and 16× (0.40 NA) objectives. A 0.9 NA dry dark-field condenser also should work well. We use an 8× projection ocular to project the image to the camera.

Bright-field illumination also is suitable; one need only have highly contrasted, clearly outlined cells on a uniform background. Differential-interference-contrast, Hoffmann modulation contrast, and phase-contrast microscopy is not recommended because the complex images may lead to inaccurate paths during video processing.

VII. Video Equipment

Images of cells are captured with a CCD camera (Nippon Electronics Corporation America, Melville, NY, Model TI-22A), recorded using a high-resolution monochrome ¾-in. videocassette recorder (Sony Corporation of America, San Diego, CA, Model VO-5800H), and displayed on a 12-in. monochrome monitor (Fig. 1). The records of the cells then can be played back to the video processor (see below) for computer-assisted tracking and analysis of cell movements. Virtually any moderately sensitive video camera is acceptable except for cameras (e.g., SIT or intensified CCD or CID cameras) with unusually noisy signals that might lead to excessive thresholding difficulties in the video processor; any monochrome monitor is also acceptable. A CCD or lead oxide tube camera is preferable because such cameras do not superimpose one image onto the next as a result of decay lag. Many inexpensive CCD cameras are electronically gated and so provide sharp images of moving objects.

A homemade control box (circuit diagram available on request) is triggered by the electronic shutter driver and places on the videotape both visual and audio signals that precisely mark onset and cessation of stimuli. The visual marker uses digital clock characters from a video timer (FOR.A, West Newton, MA, Model VTG-55), which also provides a record of the time course of the experiment. The audio tone is generated by a switchable tone generator.

VIII. Analysis and Presentation

Data analysis begins with the operation of the ExpertVision software/hardware package from Motion Analysis (Santa Rosa, CA). This package consists of a dedicated high-speed video processor (Motion Analysis VP100) and a software package which, in our case, runs on a PC/AT style 386-33 MHz computer. Computer tape backup is important as many data files are generated and must be archived. Phototaxis assays could be run directly from the live image; however, as a matter of practicality we analyze sequences captured on videotape. The video processor allows easy adjustment of thresholding, with minor image processing. Data capture is set typically at 10–15 interlaced frames per second.

The general procedure for data acquisition is as follows: (1) The videotape is played into the VP100. (2) ExpertVision is used to acquire the data to the hard disk as video threshold images via the vide command. (3) Centroids of each cell are calculated for each video frame via the cent command. (4) Centroid positions are linked to make intact paths via the path command. (5) The operator edits the data sets as needed to determine the most accurate representation of the cell's activities. An interpolation function allows the investigator to fill in short gaps in the data. Within a few minutes of playback it is possible to print a screen image of the paths (Figs. 3A,B). As all functions can be optimized and then strung together as part of a built-in user batch program, analysis can be very fast.

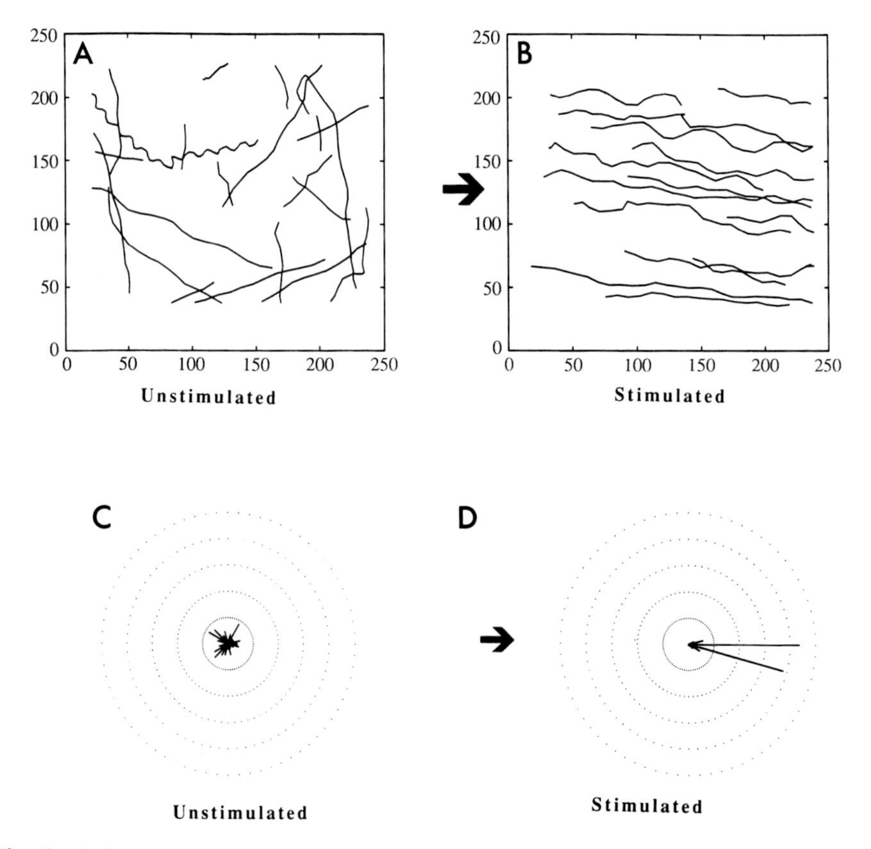

Fig. 3 Cell paths (A,B) plotted by ExpertVision and accompanying polar histograms (C,D) of swimming direction of wild-type *Chlamydomonas*. (A,C) Nonstimulated cells; (B,D) stimulated cells. Each annulus equals 10% of the total number of cells.

We use the list/re command of ExpertVision to generate a time series of paths as (x,y) positions for each cell and to print it to the disk as a simple ASCII file. We then use our program MEANANGL (available on request) to capture the time series of X, Y coordinates for each of the paths. The program uses the circular statistical methods of Batschelet (1981) to determine the mean angle of the path taken by the cell as it swims through the field. The program then sorts the mean angle of movement for each cell into one of 24 bins, each of which encompasses a 15° sector. MEANANGL outputs these data to the disk as a second ASCII file. The output is then read by our program FOTOPLOT (also available on request), which generates polar coordinate graphs on a plotter (Hewlett–Packard, Boise, ID, Model 7440) (Figs. 3C,D). The plot displays the data as percentage of cells in each bin.

Acknowledgments

We are very grateful to Dr. Mary Ella Feinleib and Dr. Nicole Morel-Laurens for the kind gift of the laser, beam-positioning equipment, attenuator, and thermopile. Dr. Morel-Laurens provided us with much practical advice regarding the construction of the system.

References

Batschelet, E. (1981). *In* "Circular Statistics in Biology" pp. 371. Academic Press, New York.

Bruce, V. (1970). The biological clock in *Chlamydomonas reinhardtii. J. Protozool.* **17,** 328–334.

Feinleib, M. E. H. (1985). Behavioral studies of free-swimming photoresponsive organisms. *In* "Sensory Perception and Transduction in Aneural Organisms" (G. Colombetti and P. S. Song, eds.), pp. 119–146. Plenum, New York.

Foster, K., and Smyth, R. D. (1980). Light antennas in phototactic algae. *Microbiol. Rev.* **44,** 571–630.

Horst, C. J., and Witman, G. B. (1993). ptxl, a nonphototactic mutant of *Chlamydomonas,* lacks control of flagellar dominance. *J. Cell Biol.* **120,** 733–741.

Morel-Laurens, N. M. L., and Feinleib, M. E. (1983). Photomovement in an "eyeless" mutant of *Chlamydomonas. Photochem. Photobiol.* **37,** 189–194.

Nultsch, W., Throm, G., and vonRimscha, I. (1971). Phototaktische Untersuchungen on *Chlamydomonas reinhardtii* Dangeard in homokontinurerlichen Kultur. *Arch. Mikrobiol.* **80,** 351–369.

Rüffer, U., and Nultsch, W. (1990). Flagellar photoresponses of *Chlamydomonas* cells held on micropipettes: I. Change in flagellar beat frequency. *Cell Motil. Cytoskel.* **15,** 162–167.

Rüffer, U., and Nultsch, W. (1991). Flagellar photoresponses of *Chlamydomonas* cells held on micropipettes: II. Change in flagellar beat pattern. *Cell Motil. Cytoskel.* **18,** 269–278.

Sager, R., and Granick, S. (1953). Nutritional studies with *Chlamydomonas reinhardtii. Ann. N.Y. Acad. Sci.* **56,** 831–838.

Stavis, R. L., and Hirschberg, R. (1973). Phototaxis in *Chlamydomonas reinhardtii. J. Cell Biol.* **59,** 367–377.

Witman, G. B. (1986). Isolation of *Chlamydomonas* flagella and flagellar axonemes. *Methods Enzymol.* **134,** 280–290.

Witman, G. B. (1993). *Chlamydomonas* phototaxis. *Trends Cell Biol.* **3,** 403–408.

CHAPTER 41

Quantification of Ciliary Beat Frequency and Metachrony by High-Speed Digital Video

Michael J. Sanderson* and Ellen R. Dirksen†

*Department of Physiology
University of Massachusetts Medical Center
Worcester, Massachusetts 01655
†Department of Anatomy and Cell Biology
UCLA School of Medicine
University of California, Los Angeles
Los Angeles, California 90024

I. Introduction

The principle on which the quantification of ciliary activity rests is the detection and analysis of the rhythmic variation in light intensity of a microscope image induced by ciliary movement. This approach requires the simplification and optimization of signal contrast and waveform and its success depends on the photodetection method, the microscope optics, and the type of cell preparation used. For beat frequency measurements, a single detector system consisting of a photomultiplier, phototransistor, or fiberoptic device is sufficient. For the analysis of ciliary metachrony, spatial information is essential and a system requires a multipoint detector such as a digital video camera. This chapter highlights the advantages and limitations of earlier techniques for measuring ciliary activity (reviewed by Sanderson and Dirksen, 1985; Braga, 1988a,b; and Romet *et al.*, 1991) and describes the principles and methodology of an advanced high-speed digital video technique developed in our laboratory.

II. Optimization of the Light Signal Waveform Produced by Beating Cilia

A large light signal amplitude is important and can be obtained from coordinated groups of active cilia of isolated or cultured cells with phase-contrast optics (Sanderson and Dirksen, 1985). Unfortunately, because of "halo" effects, these optics are less useful for imaging thick tissue or resolving individual cilia. It is also equally important that the signal waveform be simple and easily interpretable; however, cilia of isolated cells or epithelial sheets generally beat in a metachronal manner and this can result in a complicated signal waveform due to the orientation of the ciliary beat cycle with respect to the detector and the number of cilia simultaneously contributing to the signal. The extent of this problem will vary with the tissue preparation and ciliary alignment. For example, the study of human cilia is commonly achieved by observing profiles of ciliary activity of cells isolated by nasal brushing (e.g., Braga, 1988a). As a result, a full range of ciliary beat forms from multiple cilia contribute simultaneously to the signal. This, together with motion artifacts associated with isolated cells, results in a complex waveform. These complex signals can be analyzed with fast Fourier transform (FFT) or autocorrelation techniques, but only an average beat frequency can be determined (e.g., Eshel *et al.*, 1985). If isolated cell preparations must be used, an improvement in signal waveform can be achieved by monitoring only the effective stroke by recording close to the ciliary tips.

A better solution for simplifying the signal waveform is the measurement of ciliary activity of cells in a monolayer culture (Sanderson and Dirksen, 1985; Braga, 1988a; Lamiot *et al.*, 1990; Romet *et al.*, 1991). Signal complexity is reduced by monitoring light changes within the plane of ciliary synchrony (i.e., from above or below) rather than along the plane of metachrony. This is achieved by matching the photodetector size to the width of the zone of synchronously beating cilia to prevent out-of-phase cilia from contributing to the light signal. With the correct alignment and size limitation of the detector and a rapid sampling rate (250 Hz), the timings of the effective, recovery, and rest phases of the ciliary beat cycle can be determined from a highly repetitive photoelectronic signal (Sanderson and Dirksen, 1985). Additional advantages of cultured cells are that they exhibit increased cell viability, cell immobilization, convenient availability, and compatibility with experimental manipulation.

III. Quantification of Ciliary Metachronism

The isolation or culture of ciliated cells may not change the beat frequency or cycle of individual cilia, but this procedure disrupts the normal arrangement of fields of contiguous cilia and, consequently, the associated metachronism. This is a major disadvantage because metachronal activity is an integral part of the physiological function of cilia and its measurement is important for understanding ciliary activity. The study of metachronism requires intact ciliated tissues that are compatible with light microscopy and in which the interciliary spacing remains unchanged. Good examples of these are found in the frog palate or esophagus (Eshel and Priel, 1987; Braga, 1988b) or in the tracheal epithelium (Sanderson and Sleigh, 1981; Braga, 1988b). The increased tissue thickness and dense ciliary packing require that differential-interference-contrast (DIC) or equivalent optics be used to observe the cilia, but the low contrast of these optics is a disadvantage for photodetection. Because metachrony is a two-dimensional phenomenon, its analysis requires spatial as well as temporal information. The traditional choice of high-speed cinematography is expensive and data collection cannot be verified during an experiment. Another approach for metachronal wave analysis is the use of multiple photodetectors (Eshel and Priel, 1987). A photodetector, consisting of a fiberoptic connected to an electronic amplifier, is placed at the image planes of the microscope eyepieces or photoport. Beat frequency information can be obtained simultaneously from two or three locations with an orientation of 90 degrees to each other (Gheber and Priel, 1994). The phase relationship between cilia parallel and perpendicular to the direction of the effective stroke can be determined and geometrical analysis of these data provides the metachronal wavelength, travel direction, and velocity. A disadvantage of this approach is that dynamic changes in metachrony cannot be easily documented because multiple measurements, with different sensor spacings, must be made to characterize the metachronal waveform. Variations in beat frequency at different locations also complicate the interpretation of phase differences in terms of metachronal wave parameters (Eshel *et al.,* 1985). These problems are particularly relevant to mucus-transporting cilia, where the existence of multiple metachronal sequences of ciliary activity are not common due to the presence of a rest phase, in which the cilia are immotile or moving slowly (Sanderson and Sleigh, 1981).

An alternative to *in vitro* studies is the analysis of ciliary beat frequency and metachronism *in vivo*. Laser light is introduced and detected in the airways by a rigid-optic or fiberoptic system. The ciliary motion is monitored by the analysis of variations in reflected light. The signal obtained is complex because it originates from multiple surfaces and numerous cilia but it can be analyzed with an FFT or autocorrelation to determine beat frequency (Braga, 1988b; Verdugo and Goldborne, 1988; Wong *et al.,* 1988). To examine metachrony, dual-point recordings can be performed *in vivo* by rapidly scanning the illuminator/detector between two points (Wong *et al.,* 1993). In view of the complexities of *in vivo* methods, they should be considered only when the relevance of the *in vivo* condition to the ciliary activity is important.

IV. Video Techniques for Measuring Ciliary Activity

Video technology provides quantification methods with excellent sensitivity and spatial resolution at a relatively inexpensive cost and an ease of use (Hennessy *et al.,* 1986; Teichtahl *et al.,* 1986; Braga, 1988a; Ben-Shimol *et al.,* 1991; Lamiot *et al.,* 1990; Romet *et al.,* 1991; Teunis *et al.,* 1992; Nguyen *et al.,* 1994). Videotape recordings of ciliary activity can be analyzed offline by manually counting beat cycles or by following variations in image intensity induced by ciliary motion from the video monitor with a photosensitive device. (Intensity variations associated with the video raster scan are separated from variations induced by the cilia by FFT analysis.) A multiple-point analysis is simply achieved by repetitive replays; however, the major disadvantage of current video techniques is the slow sample rate (30 frames per second, NTSC; 25 frames per second, PAL). Ciliary beat frequencies above 15 Hz, as well as the faster components of each beat cycle, cannot be accurately measured because of artifacts induced by signal aliasing. Temporal resolution can be doubled, at the expense of spatial resolution, if video fields can be analyzed (Ben-Shimol *et al.,* 1991; Romet *et al.,* 1991), but even at 60 frames per second, a cilium beating at 20 Hz would only have three images/data points to describe each beat cycle. Furthermore, extended sampling periods are required to obtain sufficient data points to perform an FFT analysis (Ben-Shimol *et al.,* 1991).

V. A High-Speed Digital Video Photodetection Technique

We have developed a digital video and analysis technique with a high temporal and spatial resolution that can be used for the measurement of ciliary beat frequency and metachrony. The combination of this technique with a fluorescence imaging system that can simultaneously measure intracellular ions or messengers provides a unique opportunity for investigation of the physiological control of ciliary activity (Roos and Parker, 1990; Sanderson *et al.,* 1993). Our video system is custom-fabricated and a general protocol cannot be provided, but by understanding the technique and design principles, researchers familiar with microscopy, video, and digitization methods will be able to construct their own system tailored to their specific needs.

A. Preparation of Cells or Tissue

We recommend the use of cultured cells for simultaneous recording of beat frequency and intracellular ions or messengers. For the analysis of metachrony, intact epithelial sheets are required and can be obtained from the tracheal mucosa of rabbits (Sanderson and Sleigh, 1981). Methods for culturing airway epithelial cells are described in detail in Chapters 10 and 11 of this volume.

B. Basic Microscope System

An inverted microscope forms the basis of our system because it is well suited for fluorescence microscopy and micromanipulation of cells (Sanderson *et al.*, 1993). If these facilities are not available, an upright microscope serves equally well. The microscope should be capable of phase-contrast and/or DIC optics. On an inverted microscope, the use of an oil immersion lens necessitates that the microscope stage is fitted with a recessed adapter to mount the cell-bearing coverslips.

C. High-Speed Video Camera

Efficient quantification of ciliary beat frequency or metachrony requires rapid sampling at multiple sites. This can be achieved with a high-speed video camera that can record up to 240 images per second (Roos and Parker, 1990; Roos and Taylor, 1993). This camera is relatively inexpensive and consists of a charge-coupled device (CCD) modified so that a reduced section of the photosensitive array is scanned multiple times during the time it would take to normally scan the complete photosensitive area. During the period of a single video field (16 milliseconds), one, two, or four scans of all, one-half, or one-fourth of the photosensitive area, respectively, can be obtained. This is equivalent to 60, 120, or 240 images per second. The vertical dimension of the area of interest is reduced proportionally at higher recording speeds, but a reduction in magnification can compensate to some extent for the reduced viewing area. This modified camera (Sanyo VDC 3800) is available from Academic Products Group (Las Vegas, NV).

D. Recording of Video Data

The video signal generated by the modified CCD camera is fully compatible with conventional video recording equipment (Roos and Taylor, 1993; Sanderson *et al.*, 1993). An optical memory disk recorder (OMDR) is currently recommended for the recording. This device has all the characteristics of a tape recorder and the additional advantages of archival data storage and computer compatibility that enables automation of experiments and their analysis. More importantly, an OMDR has the ability to record or play back random, single, or multiple frames, an operating mode essential for compatibility with digital video frame grabbers (Sanderson *et al.,* 1993).

E. Digital Analysis of Data

The format of high-speed recordings generally requires further digital processing. Video images or frames consist of two interlaced fields. In the high-speed mode the even field is constructed of early sequential images (i.e., 1–4, recording at 240 fps), while the odd field is composed of subsequent sequential images (i.e., 5–8). The video frame recorded by the OMDR is an interlaced representation of these two fields, but these may be separated and arranged consecutively for analysis with the aid of a digital frame grabber (DT2861, Data Translation, Marlboro, MA). This technique is software, but not hardware dependent. After digitization of the appropriate frame, the pixel points of the even-numbered lines are removed and odd lines are closed up to form a continuous image. The even lines are restored in a contiguous format below the even lines to reproduce the real temporal sequence of the images. The images have a resolution of approximately 512 × 120 pixels at 240 fps.

Beat frequency data are obtained from the images by a gray-scale analysis at selected points or regions with respect to frame number (Lamiot *et al.,* 1990; Ben-Shimol *et al.,* 1991; Romet *et al.,* 1991). Analysis of recordings of ciliary activity made at 5- to 10-second intervals provides temporal measurement of beat frequency. If beat frequency alone is required, the digital reconstruction of the original images is not essential. All that is required are the coordinates of the pixel points at the sites of interest. By adjusting the pixel coordinates for the offsets imposed by the odd or even field placement and correlating the gray intensity with the correct temporal timing, a simple pixel point analysis of the frame sequence can be performed. This approach can be extended to advanced hardware systems that are capable of extracting pixel point data directly from camera CCD arrays in real time (Nguyen *et al.,* 1994). Digital imaging boards with a large memory capacity can be used to store a series of images from which the required data can be extracted before capturing a further series of images.

F. Analysis of Metachrony

The same principle of high-speed video recording and digital analysis also applies to the quantification of metachrony; however, these observations must be made on intact epithelia and may require electronically enhanced DIC optics. Metachronal analysis requires video-field separation and restoration, and the selection of "lines of interest" rather than "points of interest." The geometric analysis of the waveform represented by gray intensities along lines of interest, perpendicular and parallel to the effective stroke direction, provide the metachronal wavelength and direction of wave travel. In addition, gray-scale intensities along a line parallel to the direction of the metachronal wave will reveal the profile of the metachronal wave and the timings of the phases (effective, recovery, and rest) of the ciliary beat cycle. Changes in wave profile with respect to time provide wave velocity and the extent of wave travel, as well as dynamic responses of metachrony to various experimental conditions.

G. Simultaneous Imaging Applications

A major advantage of video techniques for the measurement of ciliary activity is that they can be combined with fluorescence imaging systems used to quantify intracellular signals or messengers. Photomultipliers, instead of video cameras, may also be used for this purpose but this has the major disadvantage that spatial information is not available (Korngreen and Priel, 1994). The digital fluorescence microscope used in our laboratory is described by Sanderson *et al.* (1993) and the integration of fluorescence and conventional light microscopy is simply achieved by incorporating the high-speed video camera into the existing image path with the aid of a beam splitter. A beam splitter consisting of multiple reflectors and dual camera mounts is available from Nikon. A simpler configuration consists of a single dichroic mirror, placed at 45° in the image path, to reflect an image into a second camera. Dual imaging of a fluorescence probe and a phase-contrast image is obtained by manipulating the wavelengths of the excitation and illumination light in combination with the bandpass characteristics of the dichroic mirrors and filters. This principle of dual imaging is easily illustrated by the example of the measurement of $[Ca^{2+}]_i$ with fura-2. The phase-contrast images are generated with light wavelengths greater than 600 nm (by filtration or from a photodiode), whereas the wavelength of light emitted by fura-2 is centered around 510 nm. By using a dichroic mirror with a reflection limit of 550 nm, the fura-2 fluorescence image is reflected, via a 510-nm-bandpass filter, to a low-light video camera, while the phase-contrast image is transmitted through the dichroic mirror to the high-speed camera. The images viewed by each camera are aligned by adjusting the orientation of the dichroic mirror. A magnification adjustment may also be required, to match CCD chip arrays or image tubes of different sizes. The use of a single reflecting surface also results in one image being transposed left to right, but this can be electronically corrected by reversing the line scan of either one of the video cameras or of the display monitor. If the monitor is modified, recorded signals will remain transposed. Image magnification and orientation may also be digitally corrected.

VI. Summary

In the past, the complexity and speed of ciliary activity have severely limited the success of a variety of quantification techniques; however, high-speed digital video techniques have addressed the shortcomings of earlier methods and the further development of this approach promises to simplify the dynamic investigations of ciliary activity and its regulation.

Acknowledgments

Michael J. Sanderson was supported by National Institutes of Health Grant HL-49288. Ellen R. Dirksen was supported by the Smokeless Tobacco Research Council, Inc., and the Tobacco Related Disease Research Program of the University of California.

References

Ben-Shimol, Y., Dinstein, I., Meisels, A., and Priel, Z. (1991). Ciliary motion features from digitized video photography. *J. Comput. Assist. Microsc.* **3,** 103–116.

Braga, P. C. (1988a). *In vitro* observation and counting methods for ciliary motion. *In* "Methods in Bronchial Mucology" (P. C. Braga and L. Allegra, eds.), pp. 257–268. Raven Press, New York.

Braga, P. C. (1988b). *In vivo* observation and counting methods for ciliary motion. *In* "Methods in Bronchial Mucology" (P. C. Braga and L. Allegra, eds.), pp. 269–276. Raven Press, New York.

Eshel, D., Grossman, Y., and Priel, Z. (1985). Spectral characterization of ciliary beating: variations of frequency with time. *Am. J. Physiol.* **249**(Cell Physiol. 18) C160–C165.

Eshel, D., and Priel, Z. (1987). Characterization of metachronal wave of beating cilia on frog's palate epithelium in tissue culture. *J. Physiol.* (*Lond.*) **388,** 1–8.

Gheber, L., and Priel, Z. (1994). Metachronal activity of cultured mucociliary epithelium under normal and stimulated conditions. *Cell Motil. Cytoskel.* **28,** 333–345.

Hennessy, S. J., Wong, L. B., Yeates, D. B., and Miller, I. R. (1986). Automated measurement of ciliary beat frequency. *J. Appl. Physiol.* **60,** 2109–2113.

Korngreen, A., and Priel, Z. (1994). Simultaneous measurement of ciliary beating and intracellular calcium. *Biophys. J.* **67,** 377–380.

Lamiot, E., Zahm, J. M., Pierrot, D., and Puchelle, E. (1990). Analysis of ciliary beating frequency by video-image processing. *J. Comput. Assist. Microsc.* **2,** 91–95.

Nguyen, T., Escobar, A., Vergara, J., and Verdugo, P. (1994). Ca^{2+} sub-compartmentalization in ciliated cells. *Biophys. J.* **66,** A275.

Romet, S., Schoevaert, D., and Marano, F. (1991). Dynamic image analysis applied to the study of ciliary beat on cultured ciliated epithelial cells from rabbit trachea. *Biol. Cell* **71,** 183–190.

Roos, K. P., and Parker, J. M. (1990). A low cost two-dimensional digital image acquisition subsystem for high speed microscopic motion detection. *Proc. Soc. Photo-Opt. Instrum. Eng.* **1205,** 134–141.

Roos, K. P., and Taylor, S. R. (1993). High-speed video imaging and digital analysis of microscopic features in contracting striated muscle cells. *Opt. Eng.* **32,** 306–313.

Sanderson, M. J., and Sleigh, M. A. (1981). Ciliary activity of cultured rabbit tracheal epithelium: beat pattern and metachrony. *J. Cell Sci.* **47,** 331–347.

Sanderson, M. J., and Dirksen, E. R. (1985). A versatile and quantitative computer-assisted photoelectronic technique used for the analysis of ciliary beat cycles. *Cell Motil.* **5,** 267–292.

Sanderson, M. J., Charles, A. C., and Dirksen, E. R. (1993). Measurement of the temporospatial dynamics of intercellular calcium signaling with digital fluorescence microscopy. *Am. Lab.* **25,** 29–36.

Teichtahl, H., Wright, P. L., and Kirsner, R. L. G. (1986). Measurement of *in vitro* ciliary beat frequency: a television-video modification of the transmitted light technique. *Med. Biol. Eng. Comput.* **24,** 193–196.

Teunis, P. F. M., Bretschneider, F., and Machemer, H. (1992). Real-time three-dimensional tracking of fast-moving microscopic objects. *J. Microsc.* **168,** 275–288.

Verdugo, P., and Goldborne, C. E. (1988). Remote detection of ciliary movement by fiber-optic laser-doppler spectroscopy. *IEEE Trans. Biomed. Eng.* **35,** 303–307.

Wong, L. D., Miller, I. F., and Yeates, D. B. (1988). Stimulation of ciliary beat frequency by autonomic agonists: *in vivo*. *J. Appl. Physiol.* **65,** 971–981.

Wong, L. D., Miller, I. F., and Yeates, D. B. (1993). Nature of the mammalian ciliary metachronal wave. *J. Appl. Physiol.* **75,** 458–467.

PART VI

Cytoskeletal Preparations

CHAPTER 42

Isolation and Fractionation of the *Tetrahymena* Cytoskeleton and Oral Apparatus

Norman E. Williams* **and Jerry E. Honts**†

* Department of Biological Sciences
University of Iowa
Iowa City, Iowa 52242
† Department of Molecular and Cellular Biology
University of Arizona
Tucson, Arizona, 85721

I. Introduction

Tetrahymena has upward of 600 cilia, depending on cell cycle stage. About 20% of these are in the oral apparatus, and they show regional differentiation. All ciliary basal bodies are firmly anchored in a cortically disposed cytoskeleton, which can be isolated intact as outlined in Method 1 (Section II,A). Physical integration is provided by a globally distributed membrane skeleton (sometimes called ''epiplasm''). The major basal body-associated structures in this preparation are (1) cortical microtubules, (2) tetrin filaments, and (3) kinetodesmal fibers (striated rootlets). The structures present in this preparation have been characterized and some of the major proteins have been identified as to structure

(Williams *et al.*, 1979, 1986, 1987). The membrane skeleton contains three major proteins (125, 135, and 235 kDa), and the tetrin filaments are composed of four subunit proteins (79–89 kDa). α- and β-tubulin are present. It has recently been found that an antibody to the striated rootlets recognizes a 26-kDa protein, and anticentrin recognizes a 20-kDa protein in this preparation (unpublished).

Method 2 (Section II,B) can be used when microtubule protein is not required. Microtubules, including basal bodies, are absent from this preparation, and centrin is not present. On the other hand, the membrane skeleton proteins, the striated rootlet protein, and the tetrin filament proteins are enriched. Special proteins (K-antigens) that anchor basal bodies in the membrane skeleton have been identified in this preparation (Williams *et al.*, 1990).

Method 3 (Section II,C) is used for the direct isolation of assembly-purified tetrin from *Tetrahymena* (Honts and Williams, 1990; Dress *et al.*, 1992). Tetrin has also been found in *Paramecium* (Keryer *et al.*, 1990) and *Trypanosoma* (Hemphill and Williams, unpublished), where it is also basal body associated. There may be tetrin-related proteins in higher cells (Honts and Williams, 1990).

II. Methods

A. Method 1: Purification of the Cortical Cytoskeleton

1. Solutions

Solution 1: 0.25 *M* sucrose

Solution 2: 1 *M* sucrose, 1 m*M* ethylenediaminetetraacetic acid (EDTA), 0.1% β-mercaptoethanol, 10 m*M* Tris–HCl, pH 9.0

Solution 3: 10% Triton X-100

Solution 4: 10 m*M* sodium phosphate buffer, pH 6.9, 0.1% β-mercaptoethanol

2. Procedure

1. Chill 5×10^6 to 1×10^7 *Tetrahymena* cells in an ice bath.
2. Collect the cells in one 50-ml conical polycarbonate centrifuge tube by low-speed centrifugation (300–400*g* for 3–5 minutes) at 5°C.
3. Aspirate the supernatant and add 5 ml of cold 0.25 *M* sucrose. Mix gently with a spatula.
4. Transfer to a precooled 100-ml beaker with a stirring bar.
5. While mixing at moderate speed, add 15 ml cold solution 2 and continue mixing for 30 seconds.
6. Add 2.5 ml of 10% Triton X-100 and mix 30 seconds longer. Observe microscopically at this stage.
7. Transfer to a cold 50-ml round-bottomed polycarbonate centrifuge tube and centrifuge 4000*g* for 20 minutes at 5°C.

8. Remove supernatant, leaving the pellet in 1–2 ml of solution. Mix and add 10 ml of the phosphate buffer solution (No. 4). Transfer to a cold 15-ml tube and centrifuge 15,000*g* for 20 minutes at 5°C.
9. Aspirate the supernatant, and drain and wipe the inside walls of the tube. For electrophoresis, solubilize the pellet in 500 μl Laemmli sample buffer in the standard manner.

3. Comments

Mechanical shear using something like a Logeman homogenizer after step 6 will free the oral apparatuses, and these can be harvested separately (Rannestad and Williams, 1971; Williams, 1986). To protect against proteolysis, a stock solution of 10 mg/ml leupeptin can be made in water (store frozen) and used at 10 μ*M* in the lysate and subsequent washes.

B. Method 2: Isolation of Membrane Skeletons with Striated Rootlets and Tetrin Filaments

1. Solutions

Solution 1: 1.5 *M* KCl, 1% Triton X-100, 1 m*M* EDTA

Solution 2: 1.5 *M* KCl, 1% Triton X-100, 1 m*M* ethylene glycol bis(β-aminoethyl ether)-*N*,*N*′-tetraacetic acid (EGTA)

Solution 3: 137 m*M* NaCl, 2.7 m*M* KCl, 1.5 m*M* KH_2PO_4, 8 m*M* Na_2HPO_4, 1 m*M* EDTA, 1 m*M* EGTA

Solution 4: 1% leupeptin in H_2O (store frozen)

2. Procedure

1. Prepare THS (Triton X-100/high salt)–EDTA working solution (will not store) by adding 88 μl β-mercaptoethanol (25 μ*M* final) and 25μl leupeptin stock (10 μ*M* final) to 50 ml of the THS–EDTA stock solution (No. 1). Do this for each sample to be processed.

2. Prepare THS–EGTA working solution (will not store) by adding 88 μl β-mercaptoethanol and 25 μl leupeptin stock as above, this time to 50 ml of the THS–EGTA stock solution (No. 2). Do this for each two samples to be processed.

3. Add 88 μl β-mercaptoethanol to 50 ml of the phosphate-buffered saline plus chelators described as stock solution 3.

4. Pellet 2 × 10^7 *Tetrahymena* in their own medium using a 50-ml conical polycarbonate centrifuge tube. The cells may be washed first in 10 m*M* Tris–HCl, pH 7.4, at room temperature.

5. Add 40 ml cold THS–EDTA working solution to the cell pellet. Transfer to a beaker and mix 3 minutes with a stirring bar. Observe microscopically at this point. Can be dried down for staining (see Comments).

6. Transfer to a 50-ml round-bottomed polycarbonate centrifuge tube and centrifuge at 10,000*g* for 20 minutes.

7. Remove as much supernatant as possible (some pellets are slippery), add 20 ml THS–EGTA working solution, mix well for several minutes, and then centrifuge at 10,000*g* for 15 minutes.

8. Remove supernatant, take up pellet in 3–6 ml of phosphate-buffered saline solution, transfer to a 15-ml tube, and centrifuge at 10,000*g* for 10 minutes.

9. For electrophoresis, remove the supernatant and boil the pellet 2 minutes in 600 μl Laemmli sample buffer.

3. Comments

Other protease inhibitors may be added to the THS solutions. Pepstatin is stored at 1 μg/ml in methanol and used at 0.7 μg/ml. Phenylmethylsulfonyl fluoride (PMSF) is stored in 2-propanol at 100 m*M* and used at a final concentration of 0.2 m*M*. The pellicles can be used for immunofluorescence studies, but only prior to centrifugation. Take 2.5 ml of the initial lysate in THS–EDTA, dilute to 15 ml with water, spot on clean slides, and air-dry. Follow with standard procedures for immunofluorescence.

C. Method 3: Isolation of Tetrin Filaments

1. Solutions

Solution 1: 1.4 *M* sucrose, 4 m*M* EGTA, 4 m*M* EDTA, 2% Triton X-100, 20 m*M* Tris–HCl, pH 9.0

Solution 2: 1 m*M* EDTA, 1 m*M* EGTA, 0.1% Nonidet P-40, 10 m*M* sodium phosphate, pH 7.0

Solution 3: 0.6 *M* KI, I m*M* dithiothreitol (DTT); 1 m*M* EDTA, 1 m*M* EGTA, 10 m*M* Tris–HCl, pH 9.0

Solution 4: 1.0 *M* KI, 1 m*M* EDTA, 1 m*M* EGTA, 1 m*M* DTT, 10 m*M* Tris–HCl, pH 9.0

Solution 5: 0.1 m*M* EDTA, 0.1 m*M* DTT, 2 m*M* Tris–HCl, pH 8.0

Solution 6: 1% leupeptin in H_2O, store at −70°C

Solution 7: 200 m*M* phenanthroline in ethanol, store at −20°C

2. Procedure

1. Add protease inhibitors to all working solutions just prior to use. Add leupeptin and phenanthroline at 20 μg/ml and 0.4 m*M*, respectively, to the 2×

lysis buffer (solution 1). Add leupeptin at 10 μg/ml and phenanthroline at 0.2 m*M* to solutions 2, 3, and 4, and make solution 5 0.5 μg/ml in leupeptin.

2. Chill from 2×10^7 to 1×10^8 *Tetrahymena* cells for 5 minutes in an ice bath. Harvest by centrifugation at 300*g* for 5 mintues. This and all subsequent steps are to be performed at 0–4°C.

3. Resuspend the pelleted cells in cold 1 m*M* Tris–HCl, pH 7.4 (final volume 40–80 ml).

4. Mix cell suspension with an equal volume of 2× lysis buffer (solution 1) and rapidly mix for 1 minute in a beaker with a stirring bar.

5. Pellet the cytoskeletal residues by centrifugation at 16,000*g* for 10 minutes.

6. Carefully discard supernatant and gently but thoroughly resuspend pellet in 40–80 ml of phosphate buffer wash (solution 2).

7. Pellet the cytoskeletal residues by centrifugation at 16,000*g* for 5 minutes.

8. Discard the supernatant and resuspend the pellet in 40–80 ml of 0.6 *M* KI extraction buffer (solution 3). Gently mix with a spatula for 5 minutes. The cytoskeletal residues should completely disintegrate.

9. Recover the insoluble residue, which contains tetrin filaments, by centrifuging at 16,000*g* for 10 minutes.

10. Discard the supernatant and resuspend the pellet in 1 *M* KI extraction buffer (solution 4). Use 1 ml per 2×10^7 cells. Mix gently with a spatula for 5 minutes to solubilize the tetrin proteins.

11. Centrifuge the extract at 30,000*g* for 5 minutes. Recover the supernatant, transfer to a new tube, and spin again at 30,000*g* for 30 minutes.

12. Recover the supernatant and transfer it to a dialysis bag.

13. Dialyze against 1.0 liter of low-salt dialysis buffer (solution 5) with rapid stirring for 1 hour.

14. Dialyze again against a second liter of low-salt dialysis buffer for 1 hour. A fine precipitate should appear.

15. Recover the contents of the dialysis bag and centrifuge at 30,000*g* for 15 minutes. The resulting pellet is highly enriched in tetrin filament proteins.

References

Dress, V., Yi, H., Musal, M. R., and Williams, N. E. (1992). Tetrin polypeptides are colocalized in the cortex of *Tetrahymena*. *J. Struct. Biol.* **108,** 187–194.

Honts, J. E., and Williams, N. E. (1990). Tetrins: Polypeptides that form bundled filaments in *Tetrahymena*. *J. Cell Sci.* **96,** 293--302.

Keryer, G., Iftode, F., and Bornens, M. (1990). Identification of proteins associated with MTOCs and filaments in the oral apparatus of the ciliate, *Paramecium tetraurelia*. *J. Cell Sci.* **97,** 553–563.

Rannestad, J., and Williams, N. E. (1971). The synthesis of microtubule and other proteins of the oral apparatus in *Tetrahymena*. *J. Cell Biol.* **50,** 709–720.

Williams, N. E. (1986). The nature and organization of filaments in the oral apparatus of *Tetrahymena*. *J. Protozool.* **33,** 352–358.

Williams, N. E., Vaudaux, P. E., and Skriver, L. (1979). Cytoskeletal proteins of the cell surface in *Tetrahymena*. I. Identification and localization of major proteins. *Exp. Cell Res.* **123,** 311–320.
Williams, N. E., Honts, J. E., and Graeff, R. W., (1986). Oral filament proteins and their regulation in *Tetrahymena pyriformis*. *Exp. Cell Res.* **164,** 295–310.
Williams, N. E., Honts, J. E., and Jaeckel-Williams, R. F. (1987). Regional differentiation of the membrane skeleton in *Tetrahymena*. *J. Cell Sci.* **87,** 457–463.
Williams, N. E., Honts, J. E., and Kaczanowska, J. (1990). The formation of basal body domains in the membrane skeleton of *Tetrahymena*. *Development* **109,** 935–942.

CHAPTER 43

Release of the Cytoskeleton and Flagellar Apparatus from *Chlamydomonas*

Bruce E. Taillon[*] and Jonathan W. Jarvik[†]

[*]Department of Biology
Washington University
St. Louis, Missouri 63130
[†]Department of Biological Sciences
Carnegie Mellon University
Pittsburgh, Pennsylvania 15213

I. Introduction

Chlamydomonas reinhardtii has been a productive model system for studies of the flagellar apparatus for more than three decades. About a decade ago (Wright *et al.*, 1985), we showed that the flagellar apparatus can be released from cells in a complex that includes the flagella, the basal bodies, the rootlet microtubules, the structures linking the basal bodies to the nucleus (nucleus–basal body connectors), and a chromatin-containing nuclear remnant. We called this complex the nucleoflagellar apparatus, or NFAp.

Studies of isolated *Chlamydomonas* NFAps, in both wild-type and mutant organisms, have led to a better understanding of the structures that make up this basic unit of the basal apparatus and have allowed for detailed analysis of its components at the cytological and biochemical levels (Wright *et al.*, 1985, 1989; Jarvik and Suhan, 1991; Taillon *et al.*, 1992; Sanders and Salisbury, 1994). For example, by working with isolated NFAps, one eliminates background autofluorescence from chloroplast pigments that can seriously interfere with immunofluorescence analysis. In another example, isolation of NFAps has made it possible to observe by electron microscopy the nucleus–basal body connector, a structure that is not easily seen in whole cells (Wright *et al.*, 1985; Jarvik and Suhan, 1991; Taillon *et al.*, 1992). Furthermore, the juxtaposition of mitochondria, nucleus, and basal body apparatus makes it problematic to determine, in a whole cell, whether a protein is localizing to the basal body apparatus, whereas the distinction can readily be made using isolated NFAps. For biochemical studies, the fact that isolated NFAps have a greatly simplified protein composition as compared with whole cells has obvious analytical and preparative advantages (Snell *et al.*, 1974). Finally, and more generally, the recognition that the cytoskeleton, the flagellar apparatus, and the nucleus exist as a physically integrated complex gives the investigator a new vantage point, and a new perspective, from which to view these fascinating and complex organelles.

II. Strains and Reagents

Two strains have been used for most of our basal body apparatus isolations: the cell wall-less strain *cw*92 (CC503) and the wild-type strain 137c (CC125). *cw*92 cells are easily lysed with nonionic detergents, facilitating separation of the NFAps from the other cellular components, while detergent extraction of walled cells allows for the isolation of cell "ghosts" with the NFAps trapped inside. Wild-type cells can also be treated with autolysin to remove their cell walls (see Harris, 1989). *cw*15 (CC400) is another frequently used cell wall-less mutant that, in our hands, does not yield preparations as clean as those from *cw*92. On standard minimal medium, in agar or in liquid, cell wall-less strains do not grow as well as wild type. Addition of acetate, or acetate plus yeast extract, promotes better growth, probably because carbonic anhydrase, which facilitates growth on dissolved carbon dioxide, is a periplasmic enzyme and may not be retained at the surface of cell wall-less cells (see Harris, 1989, for general *Chlamydomonas* growth requirements).

We typically maintain our strains in confluent patches on agar medium. Twelve to twenty-four hours prior to preparing NFAps, a small loopful of cells is removed from the patch and suspended in 2 ml of sterile liquid medium in a 13 × 100-mm culture tube. The cells are often not flagellated in the patch, and they may exist as "palmelliod" clusters of cells, but they separate from one another and become flagellated in the liquid medium. If the cultures are old they may contain considerable quantities of debris that can interfere with the generation of clean NFAp preparations. Centrifugation of the culture through Percoll (Pharmacia) is advised. Overlaying the culture onto a sterile 25% Percoll cushion and centrifugation for 10 minutes in a clinical centrifuge have worked well for us. The Percoll should be washed away from the cleaned up samples. Overall, it has been our experience that healthy, flagellated cells give the best NFAps.

III. Solutions

10× Nucleus buffer minus Ca^{2+} (NB): 25 m*M* KCl, 100 m*M* $MgCl_2$, 37 m*M* ethylene glycol bis(β-aminoethyl ether)-*N,N'*-tetraacetic acid (EGTA), 67 m*M* Tris–Hcl, pH 7.3. Dilute 10-fold for working concentrations. Sterilize and store at 4°C.

10% Nonidet P40 (NP-40, Sigma) in deionized H_2O. Dilute 10-fold for working concentrations. Sterilize and store at 4°C.

Freshly prepared 30% paraformaldehyde (Electron Microscopy Services). To 10 ml deionized H_2O preheated to near boil, add 3 g paraformaldehyde and stir. Add one drop (~50 μl) of 10 *N* NaOH, allow solution to cool, and use at a working concentration of 2%. This reagent is a variation of Dingle's (Larson and Dingle, 1981) fixative.

Cold (−20°C) acetone stored over desiccant.

10% Polyethylenimine (Sigma), 10 g/100 ml deionized H_2O. Dilute 100-fold for working concentrations. Store at −20°C.

IV. Procedures

A. Preparation of NFAps on Glass Slides for Light Microscopy

1. At least 12 hours prior to NFAp preparation, place a small loopful of cells into 2 ml of liquid medium. If you are using a cell wall-less mutant, disperse the cells by tapping the culture tube with your finger; strong shear forces, as in a Vortex mixer, should be avoided, as these cells can be quite fragile.

2. Place the tube in the light. If illuminated from above, you can get a reasonable impression of the culture's readiness by visual inspection of the tube; if the medium is distinctly green throughout (and at the meniscus), motile cells are present, whereas if the cells remain at the bottom of the tube they are probably not flagellated (assuming that the strain does not have a motility defect).

3. We typically use Teflon-coated 10-well slides purchased from Carlson Scientific (Peotone, IL). Slides are washed in a mild detergent such as Sparkleen (Fisher), rinsed thoroughly in deionized water, and air-dried. Place a drop of

0.1% polyethylenimine in each well and remove with a Pasteur pipet. Gently rinse the slide by running deionized water onto the uncoated portion of the slide and then letting it flow over the rest of the slide. Place slides on a Kimwipe, coated side up, to dry.

4. Concentrate cells from the 2-ml cultures by centrifugation, either in a clinical centrifuge for 3–5 minutes at 3000 rpm or in a microfuge with a 30-second pulse. To avoid dead cells and debris, use the upper 1 ml of the culture, which contains largely motile cells.

5. Wash cells once in fresh medium and resuspend in 1× NB. This should be done just prior to application of the cells to the slide. The volume in which the cell pellet is resuspended depends on how many slides or wells are needed, with each well requiring 10 μl of cell suspension. Our rule of thumb is that the drops applied to the wells appear medium to dark green, and we usually resuspend each culture in 110 μl, sufficient for 10 wells.

6. Place polyethylenimine-coated slides in a wet box, and apply 10μl of cell suspension to each well. Cover the wet box and incubate for 10 minutes. It is important that the slide be incubated in the dark so that the cells do not accumulate at the top of the drop by phototaxis. Moisture in the chamber is important so that the drops do not evaporate. Wet boxes are commercially available but can also be made from standard items such as an empty micropipet tip box or a Petri dish. The lids can be wrapped in foil, or the entire assembly can be incubated in a closed drawer. A dampened Kimwipe inside the box will keep the air saturated with moisture and prevent drying.

7. Following incubation, the cells that have not adhered to the wells are removed by placing the entire slide in a Coplin jar containing 1× NB. In and out of the jar is enough.

8. Transfer the slides to a second Coplin jar containing 1× NB + 1% NP-40. Incubate for 2 minutes with vigorous swirling of the jar. We have found that keeping the lysis solution in motion is critical for a clean preparation, presumably because it prevents cell debris from sticking to the wells as the cells lyse.

9. Rinse again in 1× NB in a Coplin jar.

10. Transfer the slide to a jar containing 1× NB + 2% freshly prepared paraformaldehyde for 10 minutes.

11. Rinse again in 1× NB in a Coplin jar.

12. Transfer the slide to a jar containing −20°C acetone, stored over a desiccant to keep it water free. Incubate for 2 minutes; then air-dry the slide face up.

13. The slide can now be viewed by phase-contrast or Nomarski optics to determine how successful the preparation was. The dry slide should be covered with a coverslip for observation. Even though the NFAps are dry, they can be visualized as blobs (the nuclei) with projecting arms (the axonemes). The slides

can be stored dry for several days prior to processing for immunofluorescence at room temperature.

B. Immunofluorescence Microscopy

We present here an outline for processing slides for immunofluorescence. Elsewhere in this volume is a chapter on immunofluorescence techniques (see Chapter 24).

1. Process the slides for immunofluorescence by incubating in blocker (2.5% bovine serum albumin, 10% fetal calf serum) for 15 minutes.
2. Add a drop of primary antibody, at the appropriate working dilution, directly to the blocker solution. Usually 10-μl drops are used, as a total of 20 μl will not leave the confines of the Teflon-encircled well.
3. Incubate with primary antibody for 1 hour at room temperature.
4. Rinse the slide twice in 0.05% Tween 20 in phosphate-buffered saline (T-PBS) and once in PBS. Carefully dry the slides around the wells, keeping the samples themselves wet, and then reblock the wells.
5. Application of secondary antibody follows the same protocol as for the primary.
6. Following the final rinse, add a small drop of antifade mounting medium to each well, apply a coverslip, and seal around the edges with clear nail polish. Colored nail polish should not be used, as some dyes in the polish fluoresce; likewise, only pencil, and not colored markers, should be used to write on the slide.

C. Preparation of NFAps for Electron Microscopy

Isolated NFAps have been extremely informative subjects for electron microscopy. For thin-selection transmission electron microscopy, samples can be attached to glass coverslips or prepared from pellets of NFAps and fixed as described previously (Jarvik and Suhan, 1991; Wright *et al.*, 1983; also see Chapter 28 in this volume).

To attach the NFAps to coverslips, simply follow the procedure above using a coverslip in place of the multiwell slide up until the fixation step. We avoid awkward Coplin jars with coverslips and float coverslips on drops of the reagents set out on sheets of Parafilm. One can use jars or beakers if the coverslip is firmly grasped with a hemostat or tweezers. After fixation and dehydration, the material on coverslips is infiltrated with Epon–Araldite following standard procedures. Coverslips are drained for 15 minutes by placing them at 45° angle and are placed, specimen side up, on an empty BEEM capsule (Electron Microscopy Services). Approximately 250 μl of Epon–Araldite is placed on the raised coverslip and polymerized in a 60°C oven for 25 hours. The coverslip is removed from the oven, placed under a dissecting microscope, specimen side down, and held with forceps. Hot air is blown across the coverslip with a hair dryer set to high heat and, as the resin softens, the tip of a second pair of forceps is used to gently pry the coverslip and sample apart. A $1 \times 1\text{-mm}^2$ of resin is removed and glued, specimen side up, to a blank BEEM capsule using two-part epoxy cement. Thin sections are cut directly from this sample and processed for transmission electron microscopy using standard methods.

To produce a pellet of NFAps for electron microscopy, take cells from Section IV,A, step 4, and add an equal volume of 1× NB + 2% NP-40. Incubate for 10 minutes or longer, as the cells lyse. The progress of the lysis can be monitored by removing small samples at intervals and observing them by light microscopy. When lysis is complete, collect the NFAps by microcentrifugation for 5 minutes and wash with 1× NB. Resuspend the pellet in 1× NB, fix, and then process for electron microscopy.

To prepare whole mounts of NFAps for observation by negative or positive stain electron microscopy or other techniques, wall-less cells can be lysed directly on Formvar- and carbon-coated electron microscope grids (Wright *et al.*, 1985).

D. Isolation of NFAps

For biochemical analysis of NFAps, wall-less cells are lysed as described above for electron microscopy, washed in 1× NB, and resuspended in the buffer of choice. The procedure can be scaled up to meet most needs, although our yields from scaled up preparations have generally been smaller than expected. A more complex procedure for NFAp isolation is described by Dutcher in Chapter 45.

E. Preparation of Cell Ghosts Containing NFAps for Light or Electron Microscopy

To prepare cell ghosts, begin with walled cells and proceed in an identical manner as for wall-less cells, with one significant exception: the lysis step (which in this case is actually an extraction step as the cells retain their walls and, strictly speaking, do not lyse) must be prolonged, typically to 30 minutes or longer.

V. Additional Treatments

Flagella can be removed prior to sample preparation, using methods described elsewhere in this volume. This may reduce the number of apparatuses that will be recovered attached to the slide or coverslip, but the procedure will work.

Removal of Mg^{2+} from the NB will result in the preparation of flagellar apparatuses without attached nuclei possibly as a result of loss of the nucleus–basal body connectors (Sanders and Salisbury, 1994).

We have found that specific detergent treatments can differentially solubilize NFAp components. Triton X-100 and NP-40 work essentially identically; however 1% deoxycholate solubilizes all but the basal bodies and the proximal portion of the rootlet microtubules. Addition of 0.05% to 0.1% sodium dodecyl sulfate will remove some associated structures. In all these cases the investigator needs to determine empirically the conditions required for the experiment at hand. To help stabilize the cytoplasmic microtubules, one can add 0.02 m*M* taxol to the fixative.

References

Harris, E. H. (1989). "The Chlamydomonas Sourcebook". Academic Press, San Diego.

Jarvik, J. W., and Suhan, J. P. (1991). The role of the flagellar transition region: Inferences from the analysis of a *Chlamydomonas* mutant with defective transition region structures. *J. Cell Sci.* **99,** 731–740.

Larson, D. E., and Dingle, A. D. (1981). Isolation, ultrastructure, and protein composition of the flagellar rootlet of *Naegleria gruberi. J. Cell Biol.* **89,** 424–432.

Sanders, M. A., and Salisbury, J. L. (1994). Centrin plays an essential role in microtubule severing during flagellar excision in *Chlamydomonas reinhardtii. J. Cell Biol.* **124,** 795–805.

Snell, W., Dentler, W., Haimo, L., Binder, L., and Rosenbaum, J. L. (1974). Assembly of chick brain tubulin onto isolated basal bodies of *Chlamydomonas reinhardti. Science* **185,** 357–360.

Taillon, B. E., Adler, S. A., Suhan, J. P., and Jarvik, J. W. (1992). Mutational analysis of centrin: An EF-hand protein associated with three distinct contractile fibers in the basal body apparatus of *Chlamydomonas. J. Cell Biol.* **119,** 1613–1624.

Wright, R. L., Adler, S. A., Spanier, J. G., and Jarvik, J. W. (1989). Nucleus-basal body connector in *Chlamydomonas:* Evidence for a role in basal body segragation and against essential roles in mitosis or in determining cell polarity. *Cell Motil. Cytoskel.* **14,** 516–526.

Wright, R. L., Chojnacki, B., and Jarvik, J. W. (1983). Abnormal basal-body number, location, and orientation in a striated fiber-defective mutant of *Chlamydomonas reinhardtii. J. Cell Biol.* **96,** 1697–1707.

Wright, R. L., Salisbury, J., and Jarvik, J. W. (1985). A nucleus-basal body connector in *Chlamydomonas reinhardtii* that may function in basal body localization or segregation. *J. Cell Biol.* **101,** 1903–1912.

CHAPTER 44

Purification of SF-Assemblin

Angela Bremerich, Karl-Ferdinand Lechtreck, and Michael Melkonian

Botanisches Institut der Universität zu Köln
D-50931 Köln
Federal Republic of Germany

I. Introduction

The basal apparatuses of many flagellate organisms are associated with non-contractile cross-striated fibers. In the green algae these fibers have been termed *system I fibers* (Melkonian, 1980; Lechtreck and Melkonian, 1991a), or striated microtubule-associated fibers (Lechtreck and Melkonian, 1991b). The system I fibers run parallel to flagellar root microtubules.

System I fibers have been isolated from the biflagellate green alga *Spermatozopsis similis* (Lechtreck and Melkonian, 1991b). Detergent treatment of cells yields cytoskeletons (see Chapter 46 in this volume). Axonemes and basal body-associated microtubules are largely removed after homogenization, and basal apparatuses can be enriched by differential centrifugation. The isolated basal apparatuses contain two system I fibers which are unequal in length and are connected at their proximal ends in such a way as to appear as a single continuous fiber (Patel *et al.*, 1992).

High-salt extraction and mechanical disintegration of isolated basal apparatuses yield single system I fibers. The fibers exhibit an overall polarity. They have a hooked, proximal end (near the basal bodies) and a pointed, distal end. The fibers consist of several layers of 2-nm filaments arranged in cross-striated bundles with a periodicity of 28 nm (Lechtreck and Melkonian, 1991b; Patel *et al.*, 1992). They can be dissolved in 2 *M* urea, and dialysis of these extracts against 150 m*M* KCl yields paracrystals which closely resemble the native fibers in filament arrangement and cross-striation pattern (Lechtreck and Melkonian, 1991b).

The major structural protein of the system I fibers has an apparent molecular weight of 34 kDa and has been named SF-assemblin (SF for striated fiber) (Lechtreck and Melkonian, 1991b; Weber *et al.*, 1993).

A polyclonal antiserum raised against SF-assemblin from *S. similis* labels the system I fibers in *S. similis* and several other flagellate green algae (*Chlamydomonas reinhardtii, Polytomella parva,* and *Dunaliella bioculata*) by indirect immunofluorescence. Proteins of the same apparent size as SF-assemblin are detected by immunoblotting in *C. reinhardtii, P. parva,* and *Nephroselms olivacea,* whereas in *D. bioculata* a 31-kDa protein is immunoreactive.

Analysis of the amino acid sequence and the predicted secondary structure of SF-assemblin shows that the protein has two structural domains: a 31-residue-long NH_2-terminal, nonhelical domain rich in proline, and an α-helical rod domain of 253 residues which forms a segmented coiled coil with a 29-residue repeat pattern based on four heptads followed by a skip residue (Weber *et al.*, 1993). A similar structure has been predicted for β-giardin, a protein from the microtubular apparatus of the parasitic archezoan flagellate *Giardia lamblia*. SF-assemblin and β-giardin may thus belong to a widespread protein family forming microtubule-associated fibers of 2-nm filaments in eukaryotic cells.

SF-assemblin can be enriched after several cycles of disassembly and reassembly: five cycles yield almost pure (>90%) polymers of the 34-kDa protein (Lechtreck and Melkonian, 1991b). A purification method should take as little time as possible; we therefore developed an alternative purification procedure using anion-exchange chromatography.

II. Methods

A. Materials

Spermatozopsis similis (strain SAG 1.85; Schlösser, 1986) may be obtained from the Sammlung von Algenkulturen, Pflanzenphysiologisches Institut, Nikolausberger Weg 18, D-37073 Göttingen, Federal Republic of Germany.

1. Equipment

Centrifuges
- Sorvall Dupont RC 28S; rotors GS3 and SS34
- Sorvall Dupont OTC-Combi; rotor TST 60.4
- Low-speed bench-top centrifuge

Tangential-flow filtration system (0.45 μm, Millipore)
FPLC (Mono Q column, HR 5/5, Pharmacia-LKB)
25-ml tissue homogenizer (Kontes Glass, Vineland, NJ)
Transmission electron microscope (TEM)
Dialysis membrane

2. Chemicals

Dithiothreitol (DTT)
Ethylene glycol bis (β-aminoethyl ether)-*N,N'*-teraacetic acid (EGTA)
Ethylenediaminetetraacetic acid (EDTA)
4-(2-Hydroxyethyl)-1-piperazineethanesulfonic acid (Hepes)
KCl
4-Morpolineethanesulfonic acid (MES)
NaCl
Tris–HCl
Triton X-100
Urea

3. Buffers

MT buffer: 30 m*M* Hepes, 5 m*M* EGTA, 15 m*M* KCl, pH 7
Assembly buffer: 150 m*M* KCl, 10 m*M* Mes, 2 m*M* EDTA, 1 m*M* DTT, pH 6.25
Disassembly buffer: 2 *M* urea in MT buffer, pH 7
Urea buffer: 8 *M* urea, 20 m*M* Tris–HCl, 1 m*M* DTT, pH 7.8

B. Preparation of Basal Apparatuses

1. Concentrate 20–100 liters of culture of *S. similis* (10^7 cells/ml) with a tangential-flow filtration system[1] and centrifuge the remaining 2 liters at 600*g* for 15 minutes at 4°C.

[1] Concentration by centrifugation (600*g*, 15 minutes, 4°C) also may be possible.

2. Wash cells in cold MT buffer and centrifuge as above.

3. Resuspend cells in cold MT buffer and lyse cells by addition of an equal volume of MT buffer containing 3% (v/v) Triton X-100.[2] Centrifuge the cytoskeletons at 1500*g* for 15 minutes at 15°C, resuspend in MT buffer, and wash several times with lysis buffer with decreasing concentrations of Triton X-100 (1%, 0.5%, 0.25%, 0% v/v) in MT buffer.

4. Prepare basal apparatuses by homogenization of the cytoskeletons (40 strokes) and centrifuge at 1500*g* for 30 minutes. Basal apparatuses, when analyzed by one-dimensional sodium dodecyl sulfate–polyacrylamide gel electrophoresis (SDS–PAGE) reveal more than 30 polypeptides with tubulin as the major protein (Fig. 1, lane 1). Basal apparatuses can be stored for up to 2 months at −20°C.

[2] Frozen cell material (−20°C) also may be used.

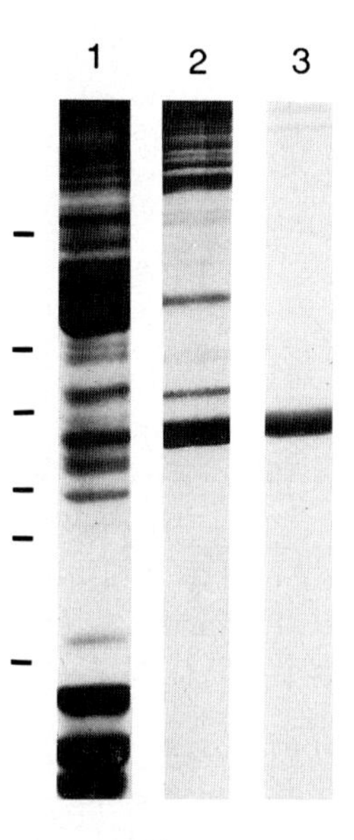

Fig. 1 Sodium dodecyl sulfate–polyacrylamide gel electrophoresis analysis (12.5% polyacrylamide gel) of basal apparatuses, paracrystals, and purified SF-assemblin from *Spermatozopsis similis*. Lane 1: isolated basal apparatuses; lane 2: paracrystals after one cycle of disassembly; lane 3: paracrystals after anion-exchange chromatography. Molecular weight markers indicated at the left are, from top to bottom, 66, 45, 36, 29, 24, and 20.1 kDa.

C. Isolation of System I Fibers

1. Resuspend the basal apparatuses (freshly prepared or as frozen pellets) in 25 ml MT buffer plus 2 *M* NaCl by homogenization (20–60 strokes), and incubate the suspension for 5–10 hours at 4°C with agitation.

2. Centrifuge the suspension at 48,500*g* for 30 minutes at 4°C. The pellet consists of pairs of system I fibers with remnants of distal connecting fibers at their proximal ends, as seen by negative staining of whole mounts in the TEM.

D. *In Vitro* Reassembly of System I Fibers

1. Resuspend isolated system I fibers in disassembly buffer and incubate the suspension for 3–5 hours at 4°C.[3]

2. Centrifuge at 200,000*g* for 60 minutes at 4°C.

3. Dialyze the 200,000*g* supernatant against reassembly buffer for 8–16 hours at 4°C. After a few hours the solution turns turbid because paracrystals begin to form. Paracrystals can be observed directly by whole-mount electron microscopy.

4. Collect paracrystals by centrifugation (48,500*g* for 30 minutes at 4°C). SDS–PAGE of preparations of paracrystals (first reassembly) shows an enrichment of SF-assemblin (Fig. 1, lane 2).

E. Anion-Exchange Chromatography

1. Dissolve the reconstituted paracrystals in 2 ml 8 *M* urea buffer and incubate for 1–2 hours at 15°C.

2. Equilibrate a Mono Q column (HR 5 Pharmacia-LKB, Uppsala, Sweden) with 2 column-volumes of urea buffer.

3. Apply the solution of the dissolved paracrystals to the column with a flow rate of 0.5 ml/min. Elute SF-assemblin from the column with a 42-ml linear salt gradient (0–300 m*M* KCl in urea buffer). Collect 1-ml fractions.

4. Pool the fractions containing SF-assemblin (SF-assemblin elutes from the column at a salt concentration of 120–150 m*M* KCl) and dialyze against reassembly buffer as above. Whole-mount electron microscopy documents the extensive formation of paracrystals with a striation pattern identical to that of system I fibers (Fig. 2).

5. Centrifuge the paracrystals at 48,500*g* for 30 minutes at 4°C. Preparations of paracrystals after anion-exchange chromatography consists almost exclusively of SF-assemblin when analyzed by one-dimensional SDS–PAGE (Fig. 1, lane 3).

[3] The system I fibers also can be solubilized in other dissociating agents (e.g., 1 *M* KI) and in low-salt buffer ($<$10 m*M*, pH 8).

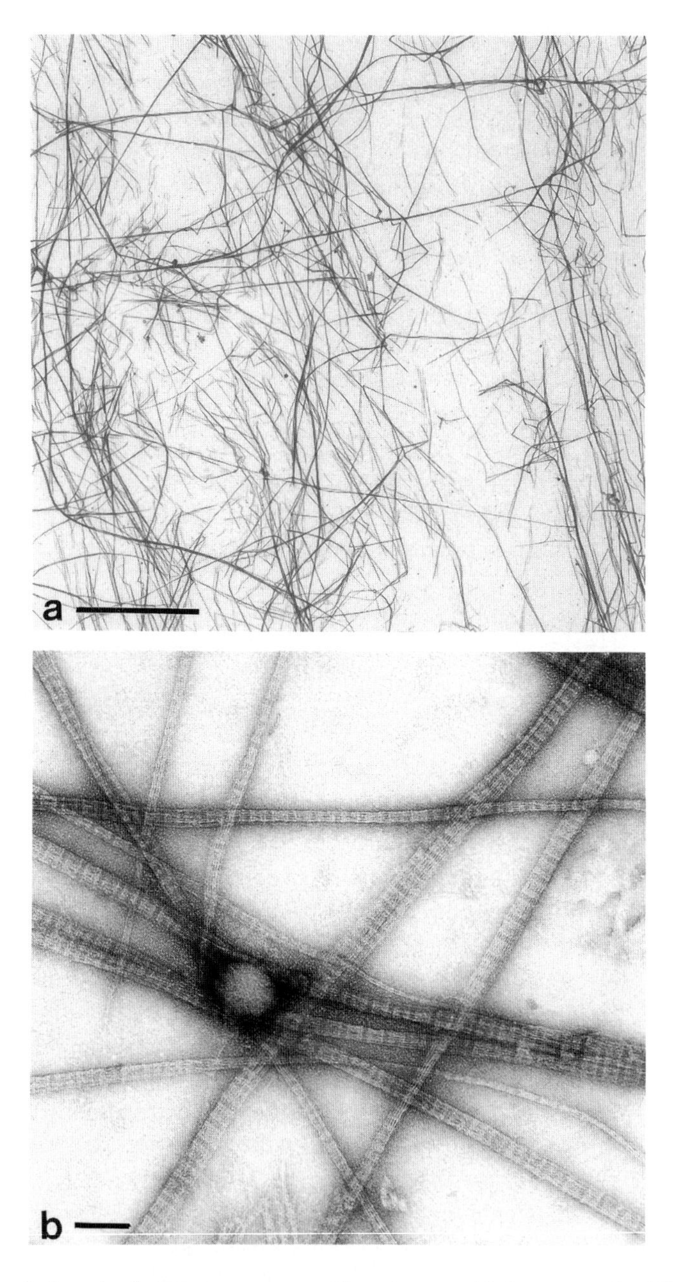

Fig. 2 Negatively stained whole-mount preparation of reassembled paracrystals of SF-assemblin after anion-exchange chromatography. (a) Extensive formation of paracrystals; bar = $2 \mu m$. (b) Paracrystals at higher magnification revealing a regular cross-striation pattern; bar = $0.4 \mu m$.

III. Comments

Using this purification protocol, a 100-liter culture of *S. similis* yields 15–20 mg of basal apparatus protein. After *in vitro* reassembly of paracrystals and anion-exchange chromatography, we routinely obtain 80–90 μg of purified SF-assemblin.

References

Lechtreck, K.-F., McFadden, G. I., and Melkonian, M. (1989). The cytoskeleton of the naked green flagellate *Spermatozopsis similis:* Isolation, whole mount electron microscopy, and preliminary biochemical and immunological characterization. *Cell Motil. Cytoskel.* **14,** 552–561.

Lechtreck, K.-F., and Melkonian, M. (1991a). An update on fibrous flagellar roots in green algae. *Protoplasma* **164,** 38–44.

Lechtreck, K.-F., and Melkonian, M. (1991b). Striated microtubule-associated fibers: Identification of assemblin, a novel 34 kDa protein that forms paracrystals of 2-nm filaments *in vitro*. *J. Cell Biol.* **115,** 705–716.

Melkonian, M. (1980). Ultrastructural aspects of basal body associated fibrous structures in green algae: a critical review. *BioSystems* **12,** 85–104.

Patel, H., Lechtreck, K.-F., Melkonian, M., and Mandelkow, E. (1992). Structure of striated microtubule-associated fibers of flagellar roots. Comparison of native and reconstituted states. *J. Mol. Biol.* **227,** 698–710.

Schlösser, U. G. (1986). Sammlung von Algenkulturen Göttingen: additions to the collection since 1984. *Ber. Dtsch. Bot. Ges.* **99,** 161–168.

Weber, K., Geisler, N., Plessman, U., Bremerich, A., Lechtreck, K.-F., and Melkonian, M. (1993). SF-assemblin, the structural protein of the 2-nm filaments from striated microtubule associated fibers of algal flagellar roots, forms a segmented coiled coil. *J. Cell Biol.* **121,** 837–845.

CHAPTER 45

Purification of Basal Bodies and Basal Body Complexes from *Chlamydomonas reinhardtii*

Susan K. Dutcher
Molecular, Cellular, and Developmental Biology
University of Colorado
Boulder, Colorado 80309

I. Introduction

The *Chlamydomonas reinhardtii* basal body complex organizes the microtubules of both the interphase and the mitotic cell; it is the microtubule organizing center (MTOC) of the cell. The MTOC in *Chlamydomonas* is defined as the pair of basal bodies, the rootlet microtubules, the distal and proximal striated fibers, and the nucleobasal body connectors (NBBCs) (Fig. 1). The morphology

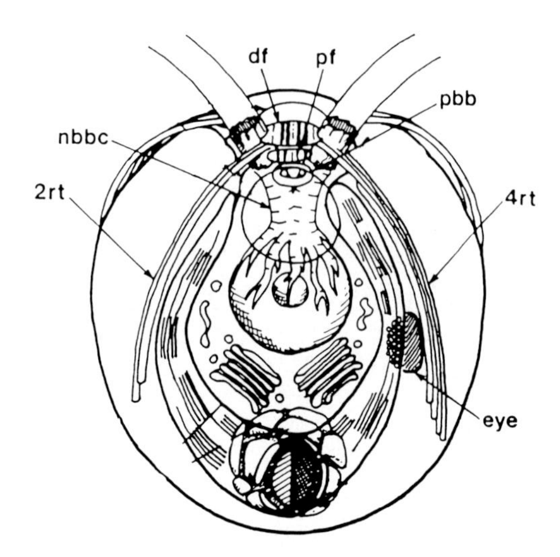

Fig. 1 Schematic diagram of the cytoskeletal elements in the basal body complex of *Chlamydomonas*. These include the distal striated fiber (df), the proximal striated fiber (pf), the nucleobasal body connectors (nbbc), the two-membered rootlets (2rt), the four-membered rootlets (4rt), and the basal bodies (pbb).

of the MTOC has been studied extensively, but the functions and composition of the MTOC are not known. In *Chlamydomonas,* three different protocols can be used to obtain isolated basal body complexes. All of the isolated complexes include basal bodies and proximal fibers, but they differ in whether they contain other structures. These preparations may be used for studying flagellar function, calcium-mediated contraction, polymerization of microtubules, and composition. In addition to the preparations described here, the preparation of *Chlamydomonas* cytoskeleton and flagellar apparatuses for light and electron microscopic analysis is described in Chapter 43.

II. Methods

A. Preparation of Autolysin

Removal of the cell wall is a required step in preparing basal bodies. Autolysin is an enzyme produced by vegetative and gametic cells for removing the cell wall (Jaenicke *et al.*, 1987). Vegetative autolysin is used in the shedding of the mother cell wall following cell division. Gametic autolysin is used in the shedding of the cell wall prior to cytoplasmic fusion of mating cells. Although both autolysins can be purified, the preparation of the gametic autolysin is easier as highly synchronous mating can be obtained using two strains that mate at high efficiency.

1. Strain cc124 (mating type minus) and strain cc125 (mating type plus) are used for the preparation of crude autolysin (Harris, 1989). To prepare cells for mating, 5×10^6 cells of each mating type in exponential growth are used to inoculate each of ten 100-mm Petri plates containing gametic medium (Chapter 76 in this volume).

2. Cells are grown at 25°C with 17.5 μEinstein/m^2/s continuous light for 3 days; the cells are then moved to a lower-light-intensity environment (6 μEinstein/m^2/s) for 2 days.

3. Mating type plus and minus cells are collected separately from the plates either by scraping with a bent plastic pipet tip into 100 ml of M-N/5 medium (see Chapter 76 in this volume) or by flooding the plates with 10 ml M-N/5 medium and scraping the cells with a bent glass rod. The cells are collected in two separate, sterile, 1-liter flasks. The cells are incubated overnight at 21°C at 17.5 μEinstein/m^2/s continuous light.

4. The two cultures are mixed together for 15 minutes. The mating mixture is examined microscopically (640×) for signs of aggregation; large clumps of cells should be observed after 5 minutes. The cultures are centrifuged at 4°C for 10 minutes at 800*g*. The supernatant is collected and filter-sterilized with 0.8-μm-pore filter. The filtrate can be kept at 4°C for up to 2 hours before using.

5. Autolysin that is active will have a pinkish tint.

B. Isolation of Basal Body Complexes

The most important step in the isolation of basal body complexes is removal of the cell wall. Failure to remove the cell wall dramatically decreases the yield of basal bodies in this protocol adapted from Snell *et al.* (1974). The yield from 2 liters of cells is 300–400 μg of protein. Gametic cells are generally used. Therefore, the basal body apparatus comes primarily from cells in the G_0 phase of the cell cycle.

1. Solutions

Protease inhibitors: 2.5 μg/ml aprotinin, 2.5 μg/ml phenylmethylsulfonyl fluoride (PMSF), 2.5 μg/ml leupeptin, 2.5 μg/ml pepstatin

HSD buffer 1000 ml: 10 m*M* 4-(2-hydroxyethyl)-1-piperazineethanesulfonic acid (Hepes), pH 7.4, 4 m*M* $SrCl_2$, 1 m*M* dithiothreitol (DTT)

Nonidet P-40 nonionic detergent (NP-40): 10% (v/v) in HSD buffer

Sucrose solutions (w/v, all filtered through 0.8-μm filter and made with HSD buffer): 25%, 250 ml; 40%, 100 ml; 50%, 100 ml; 55%, 50 ml; 60%, 50 ml

Percoll solution: (autoclaved prior to dilution with HSD buffer; 60% (v/v))

Resuspension buffer I: 30 m*M* Hepes, pH 7.4, 5 m*M* $MgSO_4$, 1 m*M* ethylene glycol bis(β-aminoethyl ether)-*N*,*N*′-tetraacetic acid (EGTA), 0.1 m*M* ethylenediaminetetraacetic acid (EDTA), 1 m*M* DTT

2. Procedure

1. Eighty 100-mm Petri plates with gametic medium (2 liters) are each inoculated with 1×10^6 cells and grown for 3 days. The cells are harvested as described above for the crude autolysin preparation and left in 1000 ml M-N/5 medium in a 2-liter flask overnight. Approximately 1×10^{10} cells are collected.

2. The cells are washed two times by centrifugation for 10 minutes at 800*g* at 18°C in M-N/5 medium to remove cell wall debris. The cells are gently resuspended in 200 ml M-N/5 medium. Do not vortex or use streams of liquid to resuspend the cells to avoid cell lysis. After the second centrifugation the cells are resuspended in M-N/5 minus the calcium. The cells are deflagellated using the pH shock method and monitored with a pH meter (Witman, 1972). The pH of the medium is lowered to 4.5 using 0.5 *N* acetic acid and mixed constantly on a magnetic stirplate. After 45 seconds, the pH is raised to 7.0 using 1 *M* potassium bicarbonate. Deflagellation is assessed microscopically and the pH shock treatment is repeated if complete deflagellation is not achieved.

3. The mixture is centrifuged at 4°C at 800*g* for 15 minutes to remove the detached flagella. The cells are resuspended in 200 ml M-N/5 medium (4°C) and the centrifugation step is repeated.

4. The cell pellet is resuspended in the autolysin. The cells are stirred continuously for 30 minutes at room temperature. Ten percent NP-40 detergent is added to a final concentration of 1%. Add protease inhibitor mixture. The mixture is homogenized in a Dounce homogenizer with 15 strokes. The homogenized mixture is returned to continuous stirring for 15 minutes at 4°C. All further steps are performed at 4°C. When examined microscopically, more than 99.9% of the cells should have lysed.

5. The lysed cells are gently layered onto a 25% sucrose cushion (25 ml) in eight 50-ml disposable conical tubes. The tubes are centrifuged for 15 minutes at 900*g*.

6. The supernatant above the interphase is collected with a wide-bore needle (10 gauge) attached to a 50-ml syringe and it is layered onto four 20-ml 40%–60% step gradients in 50-ml tubes. They are centrifuged in a swinging bucket rotor for 75 minutes at 18,200*g*.

7. The 40% step is collected as above and layered onto four 9-ml 50%–55%–60% step gradients in 30-ml Corex tubes. The samples are centrifuged for 60 minutes as in step 6.

8. The 50% fraction is collected and layered onto two 10-ml 60% Percoll gradients in 15-ml Corex tubes and centrifuged for 90 minutes at 18,200*g* in a swinging bucket rotor. A single band is formed and collected. This fraction is centrifuged in a fixed-angle rotor at 34,500*g* for 60 minutes. The pellet is placed in resuspension buffer (50–100 μl).

C. Isolation of Flagellum–Basal Body Complexes

Gould (1975) was the first to develop a method for isolating complexes that contained the flagella, the basal bodies, the striated fibers, and perhaps the NBBC. The isolated complexes can be treated with pressure and then centrifuged to separate basal bodies and proximal fibers from the flagella and other fiber systems. The presence of EDTA in this protocol is important to retaining the flagella and basal bodies together in the early steps. We have found the yields from this method (<200 μg protein) are lower than from the first protocol. Although we have tried to use autolysin in this protocol, we have not succeeded. Thus, this protocol has the disadvantage that the defective cell wall strain *cw15* is required and this phenotype does not behave as a single gene mutation through meiotic crosses. Very gentle handing is needed to avoid detaching the flagella from the cells.

1. Solutions

Polyethylene glycol (PEG) 400, 100 ml

TE buffer, 2000 ml: 15 m*M* Tris–HCl, pH 7.8, 10 m*M* EDTA

TES buffer, 250 ml: TE buffer, 500 m*M* sucrose

SET buffer: 25 ml TE buffer, 1.5 *M* sucrose, 5% PEG 400

PET buffer, 1000 ml: TE buffer, 10% PEG 400

Sucrose solutions (all filtered through 0.8-μm filters and made with PET buffer): 25%, 250 ml; 40%, 100 ml; 50%, 100 ml; 55%, 50 ml; 60%, 50 ml

HET buffer: TE buffer, 10% hexylene glycol

2. Procedure

1. Cells are grown in three 4-liter flasks containing 2 liters of liquid R medium (Harris, 1989) with gentle stirring at 21°C with 17.5 μEinstein/m^2/s continuous light. At a density of 2 × 10^6/ml, the cells are collected by centrifugation at 800*g* for 10 minutes. The pellet is washed in 5% unbuffered sucrose, centrifuged as above, and resuspended in 15 ml unbuffered 5% sucrose. All subsequent manipulations are performed at 4°C.

2. The cells are slowly added to 100 ml PEG 400. The mixture is swirled for 30 seconds, avoiding turbulence.

3. The mixture is diluted into 250 ml of cold TES buffer.

4. The solution is layered onto a 2-ml cushion of SET buffer in eight 50-ml round-bottom polycarbonate tubes. It is then spun at 600*g* in a swinging bucket rotor for 10 minutes.

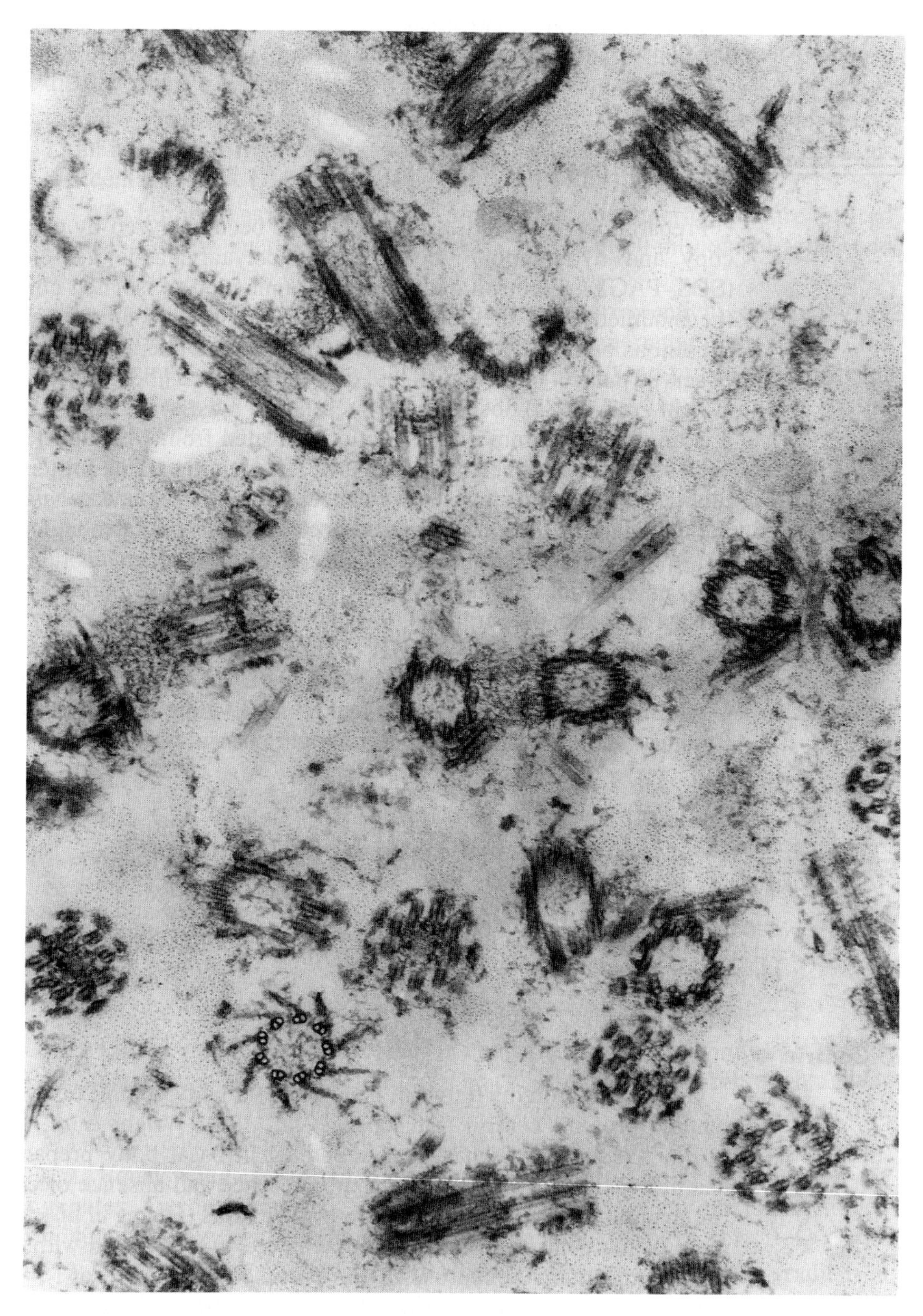

Fig. 2 Thin-section electron micrograph through the pellet obtained in method I (Section II,B) for preparing basal bodies.

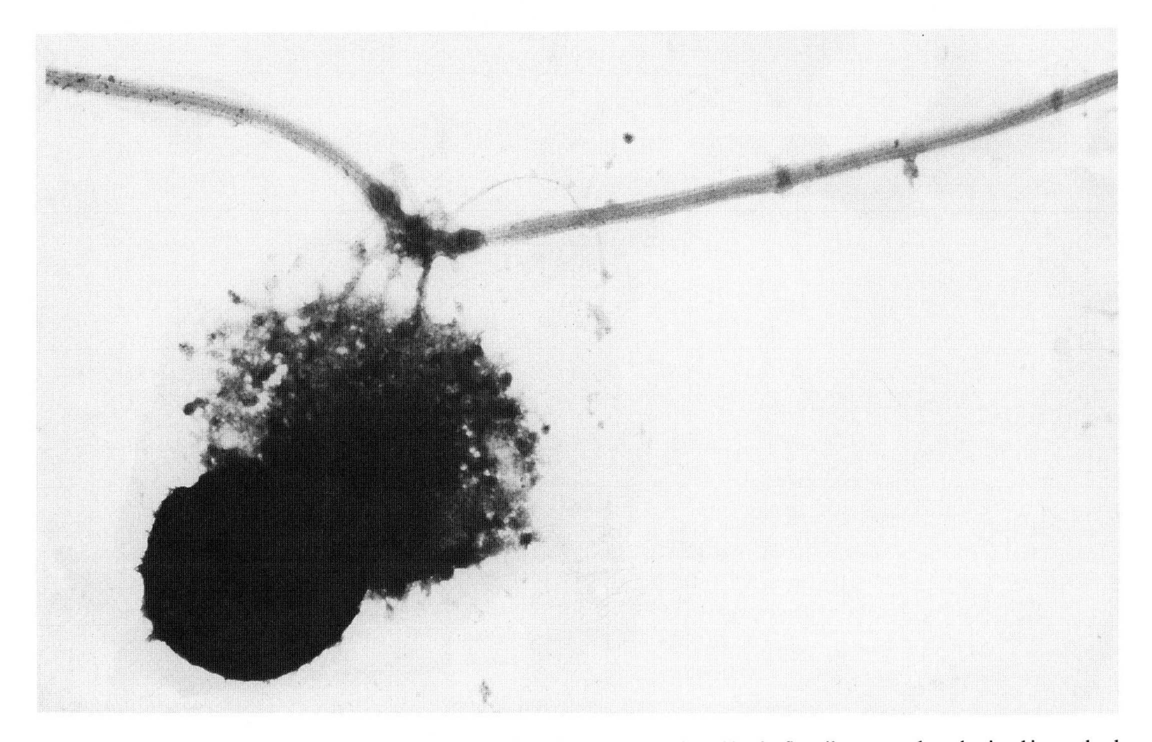

Fig. 3 Electron micrograph of an isolated nucleobasal body-flagellum complex obtained in method III (Section II,D). Picture courtesy of Dr. Robin Wright.

B. Assemblin

A second class of fibers associated with the basal body complex are the system I or 2-nm fibers (Chapter 44; Lechtreck and Melkonian, 1991). They have been isolated from a related alga, *Spermatozopsis similis*. The protein has been sequenced and is likely to form a coiled coil molecule (Weber *et al.*, 1993). There is an immunologically related polypeptide in *Chlamydomonas* (Lechtreck and Melkonian, 1991).

C. Tektins

Tektins were first characterized in sperm flagella. They are a class of microtubule-associated proteins found in three or four of the protofilaments of the A tubule of flagella (Linck and Laevgin, 1985). Immunologically related polypeptides are found in basal bodies (Chang and Piperno, 1987). Extraction of isolated basal bodies with 8 *M* urea leaves a subset of polypeptides with apparent molecular weights of 40,000 to 60,000 that have apparent molecular weights and isoelectric points similar to those of a group of polypeptides in flagella that are not solubilized by urea. These are likely to be related to tektins (Dutcher, 1986).

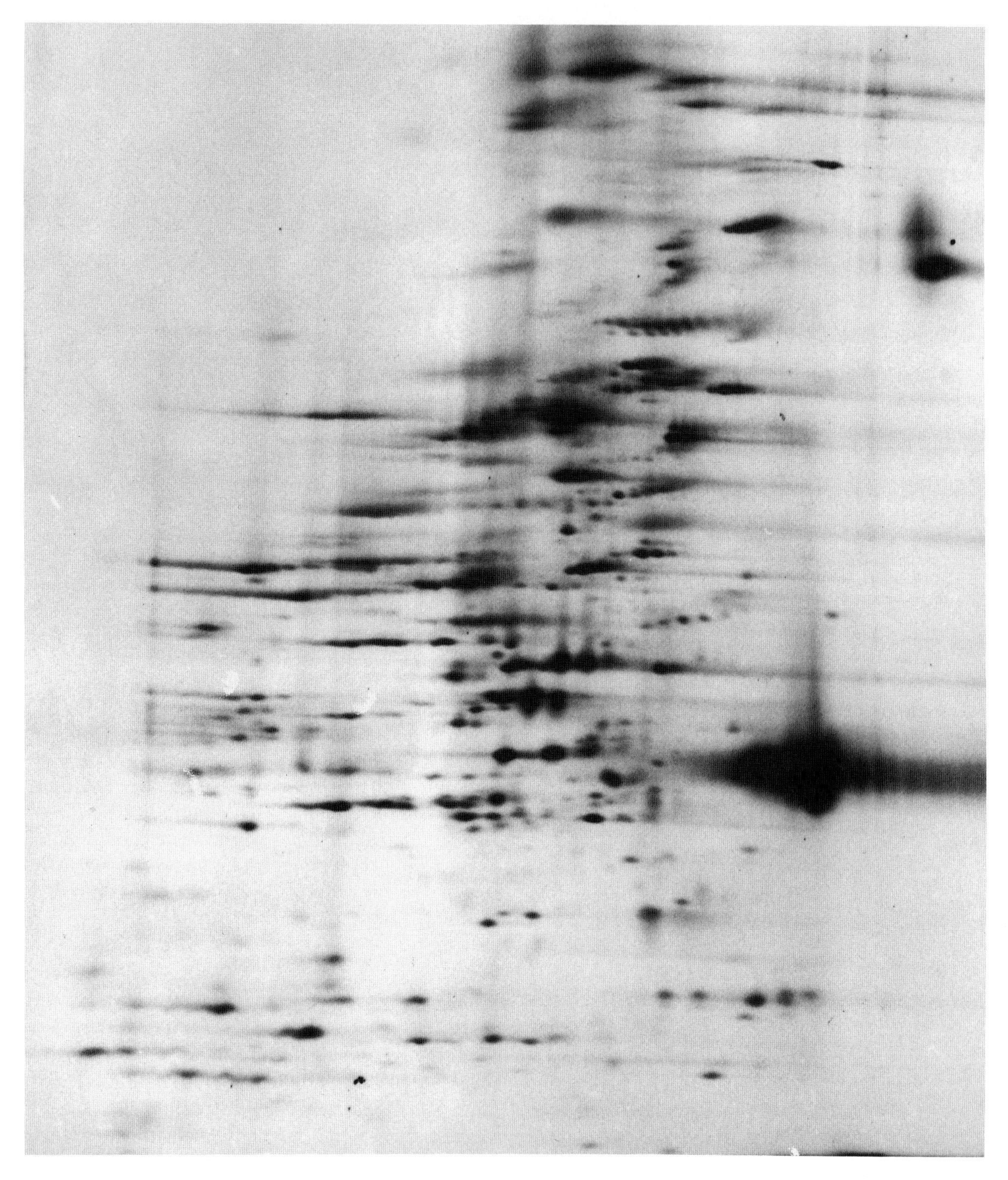

Fig. 4 Autoradiogram obtained after two-dimensional SDS–PAGE of a preparation of basal bodies isolated by method I (Section II,B) from [^{35}S]sulfuric acid-labeled cells. The gel resolves polypeptides with apparent molecular weights of 200,000 to 12,000. The pH is from acidic (~4.5) on the right to basic (~8.7) on the left. The major components are the α- and β-tubulins.

D. *PF10* Locus

The *PF10* locus is defined by a single mutant allele that exhibits altered swimming (Dutcher *et al.,* 1988). We examined two-dimensional SDS–PAGE of isolated basal body preparations from the *pf10* mutant strain and compared them with basal bodies isolated from wild-type cells. We find that two polypeptides with apparent molecular weights of 65,000 are missing in multiple preparations from *pf10-1* cells that are present in wild-type preparations (S. K. Dutcher and J. Power, work in progress).

V. Learning about Function

Two primary ways may be available in *Chlamydomonas* to begin to learn the function of various MTOC components. The first is the application of molecular and classic genetics. The second is the use of *in vitro* assays.

The isolation of mutations that affect the MTOC may be accomplished by classic means. Potential mutant phenotypes may include flagellar assembly defects and conditional lethality. The advent of insertional mutagenesis (Tam and Lefebvre, 1993) makes it possible to screen for mutations that are tagged with transforming DNA, which should facilitate the cloning of the locus of interest. Sodeinde and Kindle (1993) reported that homologous integration of transforming DNA also occurs, but at a reduced frequency. This should make it possible to disrupt genes for MTOC proteins that have been identified in other organisms or for proteins that have been identified from isolated basal body complexes. Centrosome preparations from mammalian cells are possible (Bornens *et al.,* 1987) and could be compared with the *Chlamydomonas* preparations. Experiments to reconstitute a functional MTOC from sperm centrioles and cytoplasmic extracts from *Xenopus* (Felix *et al.,* 1994; Stearns and Kirschner, 1994) should be possible with *Chlamydomonas* basal bodies and cell extracts in the future to analyze function.

References

Bornens, M., Paintrand, M., Berges, J., Marty, M.-C., and Karsenti, E. (1987). Structural and chemical characterization of isolated centrosomes. *Cell Motil. Cytoskel.* **8,** 238–249.

Chang, X., and Piperno, G. (1987). Cross-reactivity of antibodies specific for flagellar tektins and intermediate filament subunits. *J. Cell Biol.* **104,** 1563–1568.

Dutcher, S. K. (1986). Genetic properties of linkage group XIX in *Chlamydomonas reinhardtii. In* "Extrachromosomal Elements in Lower Eukaryotes" (R. B. Wickner, A. Hinnebusch, A. M. Lambowitz, I. C. Gunsalus, and A. Hollaender, eds.), pp. 303–325. Plenum Press, New York.

Dutcher, S. K., Gibbons, W., and Inwood, W. B. (1988). A genetic analysis of suppressors of the *PF10* mutation in *Chlamydomonas reinhardtii. Genetics* **120,** 965–976.

Felix, M.-A., Antony, C., Wright, M., and Maro, B. (1994). Centrosome assembly in vitro: Role of γ-tubulin recruitment in *Xenopus* sperm aster formation. *J. Cell Biol.* **124,** 19–31.

Goodenough, U. W., and St. Clair, H. S. (1975). Bald-2: A mutation affecting the formation of doublet and triplet sets of microtubules in *Chlamydomonas reinhardtii. J. Cell Biol.* **66,** 480–491.

Gould, R. R. (1975). The basal bodies of *Chlamydomonas reinhardtii*. *J. Cell Biol.* **65,** 65–74.

Harris, E. H. (1989). "The Chlamydomonas Sourcebook. A Comprehensive Guide to Biology and Laboratory Use". Academic Press, New York.

Holmes, J. A., and Dutcher, S. K. (1988). Cellular asymmetry in *Chlamydomonas reinhardtii*. *J. Cell Sci.* **94,** 273–285.

Huang, B., Watterson, D. M., Lee, V. D., and Schibler, M. J. (1988). Purification and characterization of a basal-body associated Ca^{2+}-binding protein. *J. Cell Biol.* **107,** 121–131.

Jaenicke, L., Kuhne, W., Spessert, R., Wahle, U., and Waffenschmitt, S. (1987). Cell-wall lytic enzymes (autolysins) of *Chlamydomonas reinhardtii* are (hydroxy) proline-specific proteases. *Eur. J. Biochem.* **170,** 485–491.

Linck, R. W., and Langevin, G. L. (1982). Structure and chemical composition of insoluble filamentous components of sperm flagellar microtubules. *J. Cell Biol.* **58,** 1–22.

Lechtreck, K. F., and Melkonian, M. (1991) Striated microtubule-associate fibers: identification of assemblin, a novel 34-kD protein that forms paracrystals of 2-nm filaments *in vitro*. *J. Cell Biol.* **115,** 705–716.

Salisbury, J. L., Baron, A., Surek, B., and Melkonian, M. (1984). Striated flagellar roots: isolation and partial characterization of a calcium-modulated contractile organelle. *J. Cell Biol.* **99,** 962–970.

Snell, W. J., Dentler, W. L., Haimo, L. T., Binder, L. I., and Rosenbaum, J. L. (1974). Assembly of chick brain tubulin onto isolated basal bodies of *Chlamydomonas reinhardtii*. *Science* **185,** 357–359.

Sodeinde, O. A., and Kindle, K. A. (1993). Homologous recombination in the nuclear genome of *Chlamydomonas reinhardtii*. *Proc. Natl. Acad. Sci. U.S.A.* **90,** 9199–9203.

Stearns, T., and Kirschner, M. (1994). *In vitro* reconstitution of centrosome assembly and function: The central role of γ-tubulin. *Cell* **76,** 623–637.

Tam, L.-W., and Lefebvre, P. A. (1993). Cloning of flagellar genes in *Chlamydomonas* by DNA insertional mutagenesis. *Genetics* **135,** 375–384.

Taillon, B. E., Adler, S. A., Suhan, J. P., and Jarvik, J. W. (1992). Mutational analysis of centrin: An EF-hand protein associated with three distinct contractile fibers in the basal body apparatus of *Chlamydomonas*. *J. Cell Biol.* **119,** 1613–1624.

Witman, G. B., Carlson, K., Berliner, J., and Rosenbaum, J. L. (1972). *Chlamydomonas* flagella. I. Isolation and electrophoretic analysis of microtubules, matrix, membranes, and mastigonemes. *J. Cell Biol.* **54,** 507–539.

Wright, R. L., Salisbury, J., and Jarvik, J. W. (1985). Nucleus-basal body connector in *Chlamydomonas reinhardtii* that may function in basal body localization or segregation. *J. Cell Biol.* **101,** 1903–1912.

Wright, R. L., Adler, S. A., Spanier, J. G., and Jarvik, J. W. (1989). Nucleus-basal body connector in *Chlamydomonas:* Evidence for a role in basal body segregation and against essential roles in mitosis or in determining cell polarity. *Cell Motil. Cytoskel.* **14,** 516–526.

CHAPTER 46

Preparation and Reactivation of *Spermatozopsis* Cytoskeletons

Karl-Ferdinand Lechtreck and Michael Melkonian

Botanisches Insitut
Universität zu Köln
D-50931 Köln
Federal Republic of Germany

I. Introduction

The flagellate green alga *Spermatozopsis similis* Preisig et Melkonian (Chlorophyceae) was originally isolated from a small pond near Madingley, Cambridge, England, in April 1980. The crescent and spirally twisted cells are naked and 4–6 μm long. They bear two flagella of subequal length and contain a single chloroplast with an eyespot but lacking a pyrenoid (Preisig and Melkonian, 1984). Cells reproduce asexually in the flagellate state by longitudinal division into two progeny cells; sexual reproduction is unknown. *S. similis* can be grown in mass cultures in a modified Waris medium to cell density of 1–3 $\times$ 10^7 cells/ml (McFadden and Melkonian, 1986).

The basal apparatus resembles that of *Chlamydomonas reinhardtii* in comprising two basal bodies interconnected by distal and proximal connecting fibers and four microtubular flagellar roots. In addition, two types of fibrous flagellar roots occur in *S. similis:* a single small nucleus–basal body connector (NBBC), which contains the calcium-modulated EF-hand protein centrin (also known as caltractin), and two striated microtubule-associated fibers (SMAFs or system I fibers) which accompany the two-membered microtubular flagellar roots over most of their length (Melkonian and Preisig, 1984; McFadden *et al.,* 1987; Patel *et al.,* 1992).

The complete absence of a cell wall or glycocalyx allows the isolation of structurally intact cytoskeletons from *S. similis* by extraction with nonionic detergents (Nonidet P-40, Triton X-100) using minimal shear forces. The cytoskeletons retain the unique shape of intact cells and consist of the two axonemes, the interconnected basal bodies, the fibrous roots, the four microtubular flagellar roots, and 6–12 secondary cytoskeletal microtubules (SCMTs). The root microtubules and most of the SCMT are bound together posteriorly by a fibrous cap-like structure, the rhizosyndesmos (Fig. 1) (Lechtreck *et al.,* 1989). The addition of 5 m*M* $MgSO_4$ to the isolation buffer yields nucleoflagellar apparatus complexes in which a stable nuclear remnant (the karyoskeleton) remains attached to the basal bodies by the NBBC (Figs. 2,3). Isolated cytoskeletons can be used as a source for biochemical analysis of cytoskeletal proteins such as tubulin and centrin (Höhfeld *et al.,* 1994) or to isolate subcomponents of the flagellar apparatus such as basal apparatuses and SMAFs (Lechtreck *et al.,* 1989). The coiled-coil forming protein SF-assemblin, which is the major component of SMAF in various green flagellates, was first isolated from the cytoskeletons of *S. similis* (Lechtreck and Melkonian, 1991; Weber *et al.,* 1993; see also Chapter 44 in this volume).

Some flagellar apparatus functions of *S. similis* have been reconstituted *in vitro,* namely, axonemal shedding, axonemal motility, and reorientation of the basal bodies (McFadden *et al.,* 1987). Cytoskeletons isolated and reactivated at $<10^{-8}$ *M* Ca^{2+} swim forward with an asymmetrical beat pattern (breast-stroke mode) and with basal bodies oriented antiparallel [angle between basal bodies 180° (Fig. 2)]. Calcium concentrations $\geq 5 \times 10^{-8}$ *M* induce centrin-mediated contraction of the distal connecting fiber, resulting in a reorientation of the basal bodies to the parallel orientation (angle between basal bodies 0°; Figs. 3,4). At 10^{-7} *M* Ca^{2+} in reactivation buffer all cytoskeletons have reoriented their basal bodies to the parallel configuration; about 30% of the axonemes have changed their beating pattern to the symmetrical undulatory-type pattern (cytoskeletons swim backward; Fig. 3), while the remainder still beat in the breast-stroke mode. At $\geq 10^{-6}$ *M* Ca^{2+}, all axonemes adopt the symmetrical beat form (Fig. 4). *In vivo* basal body reorientation in *S. similis* can be elicited as a phobic shock response by very high photon flux rates or mechanical stimulation (McFadden *et al.,* 1987; Kreimer and Witman, 1994). The cell biology of phototaxis in *S. similis* have been extensively studied *in situ* and in isolated eyespot apparatuses (Kreimer, 1994; Kreimer *et al.,* 1991a,b).

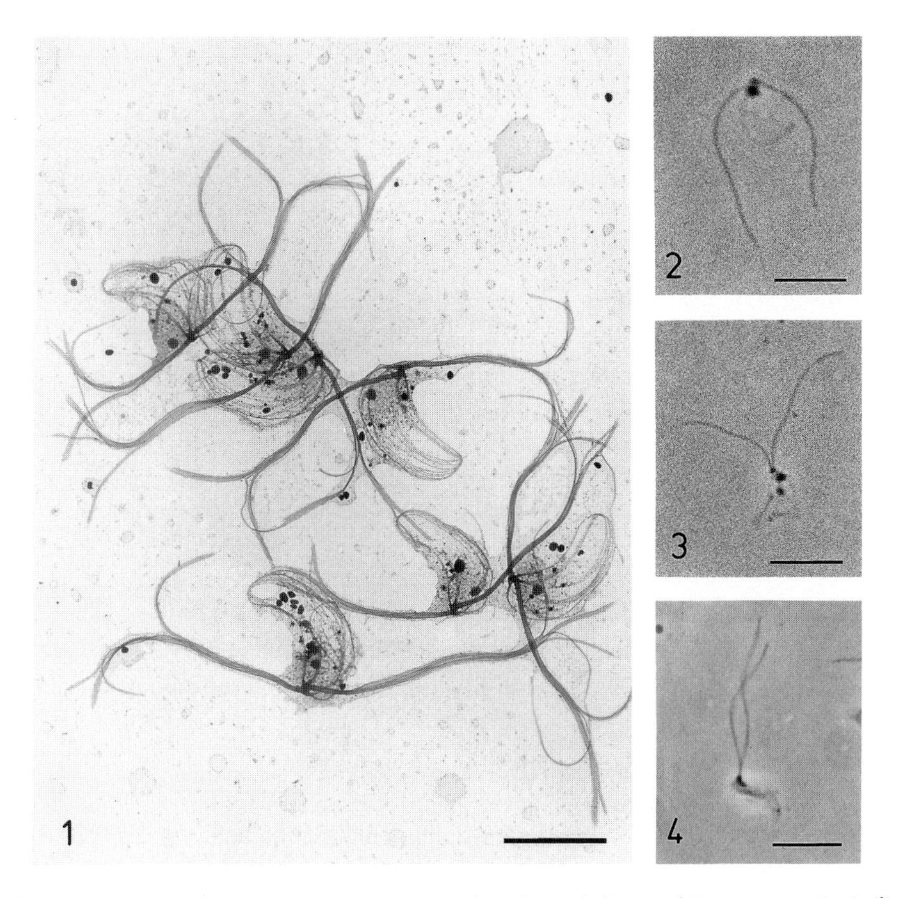

Fig. 1 Negatively stained whole-mount preparation of cytoskeletons of *Spermatozopsis similis* isolated in MT buffer, illustrating the antiparallel basal body orientation. Cytoskeletons were allowed to adhere to pioloform-coated copper grids for 5 minutes and then stained with 1% (w/v) uranyl acetate for 2 minutes. Bar = 5 μm. Adapted with permission from Lechtreck and Melkonian (1991, Fig. 1a, p. 707).

Fig. 2 Phase-contrast light micrograph of isolated cytoskeleton in reactivation buffer (without ATP) at 10^{-9} *M* Ca^{2+}. Bar = 5 μm.

Fig. 3 At 10^{-7} *M* Ca^{2+}, the basal bodies and axonemes are reoriented to the parallel orientation. Phase-contrast light micrograph of isolated cytoskeleton. Bar = 5 μm.

Fig. 4 Phase-contrast light micrograph of isolated cytoskeleton resuspended in 10^{-5} *M* Ca^{2+}. Bar = 5 μm. Figures 2 – 4 are adapted with permission from McFadden *et al.* (1987, Figs. 3b – d, p. 906).

II. Methods

A. Materials

Spermatozopsis similis (strain SAG 1.85; Schlösser, 1986) may be obtained from Sammlung von Algenkulturen, Pflanzenphysiologisches Institut, Nikolausberger Weg 18, D-37073 Göttingen, Federal Republic of Germany.

1. Equipment

Low-speed bench-top centrifuge
Light microscope with phase-contrast optics
Perfusion chamber (Hyams and Borisy, 1978)

2. Chemicals

ATP
$CaCl_2$
Dithiothreitol (DTT)
Ethylene glycol bis(β-aminoethyl ether)-*N,N′*-tetraacetic acid (EGTA)
4-(2-Hydroxyethyl)-1-piperazineethanesulfonic acid (Hepes)
KCl
KOH
$MgSo_4$
Disodium ethylenediaminetetraacetic acid (Na_2EDTA)
Nonidet P-40 (NP-40) or Triton X-100
Paraformaldehyde
Polyethylene glycol (PEG, M_r 20,000)
Sigmacote

3. Buffers

MT buffer: 30 m*M* Hepes, 5 m*M* $MgSO_4$, 25 m*M* KCl, 5 m*M* EGTA, adjusted with KOH to pH 7

Lysis buffer: MT buffer plus 0.2–2% (v/v) NP-40

Reactivation buffer: 30 m*M* Hepes, 5 m*M* $MgSO_4$, 25 m*M* KCl, 0.5% (w/v) PEG, 1 m*M* DTT, 0.5 m*M* Na_2EDTA, 1 m*M* ATP, solution adjusted to pH 7.3 at 22°C [A detailed description of Ca^{2+}-buffered reactivation solutions is given in Bessen *et al.* (1980). Buffers containing defined Ca^{2+} concentrations can also be calculated using various programs (e.g., the program EQCAL, L. Backman, Biosoft, Cambridge, U.K.).]

B. Isolation of *Spermatozopsis* Cytoskeletons

1. Concentrate cells by centrifugation at 500*g* in a bench-top centrifuge for 5 minutes.
2. Wash cells once in MT buffer and centrifuge as above.
3. Resuspend cells in ice-cold MT buffer and after 5 minutes add an equal volume of lysis buffer [0.2–2% NP-40 (v/v) in MT buffer]. Use MT buffer lacking $MgSO_4$ to isolate cytoskeletons without a karyoskeleton (Fig. 1).
4. For ultrastructural or immunological studies of isolated cytoskeletons, fix cytoskeletons with 3% (v/v) formaldehyde in MT buffer for at least 20 minutes at room temperature.
5. For mass isolation of cytoskeletons, lyse cells with 1% NP-40 or Triton X-100 (v/v, final concentration) in MT buffer and pellet the cytoskeletons at 3000*g* for 20 minutes. Wash the cytoskeletons several times with decreasing concentrations of detergent (0.5, 0.25, 0.1%, v/v) and twice without detergent.

C. Reactivation of Isolated Cytoskeletons

1. Precool all buffers and equipment to 4°C.
2. Concentrate, wash, and lyse cells as described in Section II,B, steps 1–3, but use MT buffer plus 1% (w/v) PEG and 1 m*M* DTT.
3. Centrifuge lysate at 1000*g* in microtubes for 5 minutes.
4. Discard the green supernatant and resuspend the upper milky portion of the biphasic pellet in MT buffer plus PEG and DDT.
5. Reactivate cytoskeletons within 6 hours of isolation by suspending a 5-μl aliquot of cytoskeletons in 2 ml reactivation buffer.
6. Observe reactivated cytoskeletons in phase contrast in a perfusion chamber such as that described by Hyams and Borisy (1978) using siliconized glass slides.
7. To induce forward swimming (Fig. 2), use calcium concentrations at or below 10^{-8} *M* in the standard reactivation buffer.
8. To induce basal body reorientation and backward swimming of cytoskeletons, adjust the calcium concentration to $\geq 10^{-6}$ *M* (Fig. 4). Calcium concentrations $\geq 10^{-4}$ *M* lead to shedding of axonemes from the cytoskeletons.

References

Bessen, M., Fay, R. B., and Witman, G. B. (1980). Calcium control of waveform in isolated axonemes of *Chlamydomonas*. *J. Cell Biol.* **86,** 446–455.

Höhfeld, I., Beech, P. L., and Melkonian, M. (1994). Immunolocalization of centrin in *Oxyrrhis marina* Dujardin (Dinophyceae). *J. Phycol.* **30,** 474–489.

Hyams, J. S., and Borisy, G. G. (1978). Isolated flagellar apparatus of *Chlamydomonas:* characterization of forward swimming and alteration of waveform and reversal by calcium *in vitro*. *J. Cell Sci.* **33,** 235–253.

Kreimer, G. (1994). Cell biology of phototaxis in flagellate algae. *Int. Rev. Cytol.* **148,** 229–310.

Kreimer, G., Brohsonn, U., and Melkonian, M. (1991a). Isolation and partial characterization of the photoreceptive organelle for phototaxis of a flagellate green alga. *Eur. J. Cell Biol.* **55,** 318–327.

Kreimer, G., Marner, F.-J., Brohsonn, U., and Melkonian, M. (1991b). Identification of 11-cis and all-trans-retinal in the photoreceptive organelle of a flagellate green alga. *FEBS Lett.* **293,** 49–52.

Kreimer, G., and Witman, G. B. (1994). A novel touch-induced, Ca^{2+}-dependent phobic response in a flagellate green alga. *Cell Motil. Cytoskel.* **29,** 97–109.

Lechtreck, K.-F., McFadden, G. I., and Melkonian, M. (1989). The cytoskeleton of the naked green flagellate *Spermatozopsis similis:* Isolation, whole mount electron microscopy, and preliminary biochemical and immunological characterization. *Cell Motil. Cytoskel.* **14,** 552–561.

Lechtreck, K.-F., and Melkonian, M. (1991). Striated microtubule-associated fibers: Identification of assemblin, a novel 34-kD protein that forms paracrystals of 2-nm filaments *in vitro*. *J. Cell Biol.* **115,** 705–716.

McFadden, G. I., and Melkonian, M. (1986). Use of Hepes buffer for microalgal culture media and fixation for electron microscopy. *Phycologia* **25,** 551–557.

McFadden, G. I., Schulze, D., Surek, B., Salisbury, J. L., and Melkonian, M. (1987). Basal body reorientation mediated by a Ca^{2+}-modulated contractile protein. *J. Cell Biol.* **105,** 903–912.

Melkonian, M., and Preisig, H. R. (1984). Ultrastructure of the flagellar apparatus in the green flagellate *Spermatozopsis similis*. *Plant Syst. Evol.* **146,** 145–162.

Patel, H., Lechtreck, K.-F., Melkonian, M., and Mandelkow, E. (1992). Structure of striated microtubule-associated fibers of flagellar roots. Comparison of native and reconstituted states. *J. Mol. Biol.* **227,** 698–710.

Preisig, H. R., and Melkonian, M. (1984). A light and electron microscopical study of the green flagellate *Spermatozopsis similis* sp. nov. *Plant Syst. Evol.* **146,** 57–74.

Schlösser, U. G. (1986). Sammlung von Algenkulturen Gottingen: additions to the collection since 1984. *Ber. Dtsch. Bot. Ges.* **99,** 161–168.

Weber, K., Geisler, N., Plessman, U., Bremerich, A., Lechtreck, K.-F., and Melkonian, M. (1993). SF-assemblin, the structural protein of the 2-nm filaments from striated microtubule associated fibers of algal flagellar roots, forms a segmented coiled coil. *J. Cell Biol.* **121,** 837–845.

CHAPTER 47

Centrin-Based Contractile Fibers: Chromatographic Purification of Centrin

Andre T. Baron, Ramesh Errabolu, Jacquelyn Dinusson, and Jeffrey L. Salisbury

Laboratory for Cell Biology
Department of Biochemistry and Molecular Biology
Mayo Clinic Foundation
Rochester, Minnesota 55905

I. Introduction

Centrin is an acidic, low-molecular-weight (~21,000 M_r), calcium-binding phosphoprotein that belongs to the EF-hand superfamily of calcium-binding proteins (reviewed by Bazinet *et al.,* 1990; Lee and Huang, 1990; Salisbury, 1992). Centrin is a component of an assortment of calcium-modulated cytoskeletal fiber systems found associated with flagellar basal apparatus, centrosomes, and mitotic spindle poles of eukaryotic cells. Centrin-based systems include striated flagellar roots (Salisbury *et al.,* 1984; Wright *et al.,* 1985), distal fibers (McFadden *et al.,* 1987), transition zone fibers (Sanders and Salisbury, 1989), fibrous bundles (Koutoulis *et al.,* 1988), and pericentriolar lattice (Baron and

Salisbury, 1988; Baron *et al.,* 1992). Immunofluorescence microscopic studies demonstrate that centrin is also a component of the flagellar axoneme in *Chlamydomonas reinhardtii* (Huang *et al.,* 1988a; Goodenough, 1992) and paraflagellar rod of transverse flagella in *Peridinium inconspicuum* (Hohfeld *et al.,* 1988). Centrin-based fibers play a role in anchoring the flagellar basal apparatus to the nucleus (Wright *et al.,* 1985; Salisbury *et al.,* 1987), in positioning of basal bodies and centrioles during the cell cycle (Salisbury *et al.,* 1988), in severing microtubules at the time of flagellar excision (Sanders and Salisbury, 1989, 1994), and in coiling the paraflagellar rod of transverse flagella (Hohfeld *et al.,* 1988).

Here, we present detailed methods for generating milligram quantities of pure centrin. These methods involve expression of a cDNA clone encoding *Chlamydomonas* centrin in *Escherichia coli* (Maniatis *et al.,* 1982) and chromatographic purification of the expressed protein (Harris and Angal, 1990). The chromatographic method described here is similar to schemes already described for purifying centrin and other calcium-binding proteins from cultured cells and tissues (Marshak *et al.,* 1984; Davis *et al.,* 1986; Fulton *et al.,* 1986; Salisbury *et al.,* 1986; McDonald *et al.,* 1987; Coling and Salisbury, 1992; Weber *et al.,* 1994), and works well for the purification of centrin from isolated *Chlamydomonas* flagella.

II. Methods

A. Construction of *E. coli* BL21 pT7-5Cen

A cDNA clone of 1049 bp encoding *C. reinhardtii* centrin, designated pCen, was obtained by screening a λgt11 cDNA library. Sequence analysis of pCen revealed an open reading frame of 507 bases, beginning at an initiation codon (ATG) 44 bp from the 5′ end of the cDNA insert. The open reading frame predicts an encoded protein of 169 amino acids having a calculated molecular mass of 19,459 kDa. With the exception of a single codon polymorphism, resulting in the substitution of a serine for a threonine at amino acid 39, and seven single-base-pair differences in untranslated regions, the *Chlamydomonas* centrin cDNA clone isolated in our laboratory is identical to the sequence reported by Huang and co-workers (1988b). *Chlamydomonas* centrin shows a high degree of sequence identity with the CDC31 gene product of *Saccharomyces cerevisiae* (Baum *et al.,* 1986), as well as human (Lee and Huang, 1993; Errabolu *et al.,* 1994), *Atriplex* (Zhu *et al.,* 1992), *Xenopus,* and *Naegleria* (see Errabolu *et al.,* 1994) centrin. Each of these centrin sequences shares lower, yet significant, similarities with other members of the EF-hand family of calcium-binding proteins.

To produce milligram quantities of purified centrin, we have subcloned the *Chlamydomonas* centrin open reading frame into plasmid pT7-5, which contains the inducible T7 promoter (Tabor and Richardson, 1985). We subsequently transfected the resultant bacterial expression plasmid pT7-5Cen into *E. coli* BL21. This bacterial expression system is available on request (J. L. Salisbury, Mayo Foundation, Rochester, MN 55905).

B. Induction and Lysis of *E. coli* BL21 pT7-5Cen

1. Materials

LB plate containing 50 μg/ml ampicillin (Maniatis *et al.*, 1982)

LB broth containing 50 μg/ml ampicillin

Isopropylthio-β-glactoside (IPTG)

Cell lysis buffer: 50 m*M* Tris, pH 7.4, 0.5 m*M* ethylenediaminetetraacetic acid (EDTA), 0.5 *M* NaCl, 0.1% Nonidet P-40 (NP-40), 0.04% NaN_3

Lysozyme: 10 mg/ml stock in 10 m*M* Tris, pH 8.0

Protease inhibitor cocktail: 2 mg/ml aprotinin, 0.5 mg/ml leupeptin, 1.0 mg/ml pepstatin A in dimethyl sulfoxide

1 *M* $CaCl_2$

1 *M* $MgCl_2$

2. Procedure

1. Grow *E. coli* BL21 containing plasmid pT7-5Cen on an LB plate containing 50 μg/ml ampicillin overnight at 37°C. Check ampicillin activity by streaking out another LB–ampicillin plate with the parent bacterium *E. coli* BL21. *E. coli* BL21 are not ampicillin resistant and will die if they are not transfected with pT7-5Cen, which contains an ampicillin resistance gene.

2. Select a single colony of *E. coli* BL21 pT7-5Cen and inoculate 50 ml of LB broth that contains 50 μg/ml ampicillin. Grow the culture on a shaker overnight at 37°C, then add the "overnight" to 950 ml of LB–ampicillin broth. Grow this culture until the cells are in logarithmic growth ($\sim 5 \times 10^8$ cells/ml), then induce the expression of centrin by adding 2 g/liter (8.4 m*M*) isopropylthio-β-galactoside (IPTG). Monitor cell density by measuring culture OD at an absorbance of 600 nm; $0.1\text{OD}_{600} = \sim 1 \times 10^8$ cells/ml. Continue to grow the culture until it attains stationary phase ($1\text{–}2 \times 10^9$ cells/ml).

3. Prepare sodium dodecyl sulfate–polyacrylamide gel electrophoresis (SDS–PAGE) samples of the bacterial culture before and after IPTG induction (see Troubleshooting). Adjust protocol appropriately to grow more than 1 liter of bacteria. Harvest cells by centrifugation at ~10,000*g* (Beckman JA-10 rotor, 7500 rpm, or JA-17 rotor, 8500 rpm) for 15 minutes. Store final cell pellet at −70°C or continue with protocol.

4. Resuspend the cell pellet, frozen or otherwise, with four times the amount (w/v) of cold (4°C) cell lysis buffer containing lysozyme and protease inhibitor

cocktail. Add 1 μl lysozyme solution and 1 μl of protease inhibitor cocktail per milliliter of cell lysis buffer to cell lysis buffer just prior to use. Keep the preparation cold until beginning the phenyl-sepharose chromatography.

5. Transfer cell suspension to a 50-ml conical centrifuge tube(s) and incubate at 4°C for 30 minutes on a tumbler. The peptidoglycan cell wall will be digested by the lysozyme during this period and the cells will also be partially extracted, thus making the solution viscous with DNA.

6. Transfer this whole-cell lysate to a sonication vessel in an ice bath and sonicate (2–5 $\times$ 30-second pulses at full power with 1-minute cooling periods; use either a marcoprobe or microprobe) to further break the cells and DNA. Evaluate the whole-cell lysate after sonication; if necessary, continue to sonicate until the lysate is no longer viscous with DNA. Prepare a sample of whole-cell lysate for SDS–PAGE (see Troubleshooting).

7. Centrifuge whole-cell lysate at 30,000*g* (Beckman JA-17 rotor, 15,000 rpm) for 1 hour to obtain supernatant S1 and pellet P1. Discard P1 after taking samples of both S1 and P1 for SDS–PAGE (see Troubleshooting).

8. Add $CaCl_2$ and $MgCl_2$ (1.0 *M* stock solutions) to S1 to give final concentrations of 2.0 m*M* Ca^{2+} and 4.0 m*M* Mg^{2+}, respectively. Centrifuge S1 at 100,000*g* (Beckman Ti-60 rotor, 31,000 rpm) for 1 hour to obtain S2 and P2 fractions. Discard P2 after taking samples of both S2 and P2 for SDS–PAGE (see Troubleshooting).

9. Begin phenyl-Sepharose CL-4B chromatography immediately to fractionate centrin from proteases or store S2 at −20°C if necessary.

C. Phenyl-Sepharose CL-4B Chromatography

1. Materials

Chromatography column packed with phenyl-Sepharose CL-4B hydrophobic interaction resin (Pharmacia Biotech, Catalog No. 17-0810-01) according to the manufacturer's instructions

CL-4B Eluant A: 50 m*M* Tris, pH 7.4, 0.5 m*M* EDTA, 0.5 *M* NaCl, 4.0 m*M* $MgCl_2$, 2.0 m*M* $CaCl_2$, 0.04% NaN_3

CL-4B eluant B: 50 m*M* Tris, pH 7.4, 0.5 m*M* EDTA, 0.5 *M* NaCl, 4.0 m*M* $MgCl_2$, 5 m*M* ethylene glycol bis(β-aminoethyl ether)-*N,N′*-teraacetic acid (EGTA), 0.04% NaN_3

CL-4B eluant C: 50 m*M* Tris, pH 7.4, 0.5 m*M* EDTA, 5.0 m*M* EGTA, 0.04% NaN_3

Use chromatography-grade salts of high purity for all solutions including the cell lysis buffer (see above) to achieve stable low-noise baselines. Filter all chromatography solutions with a 0.22-μm filter (Nalgene), degas, and store at room temperature.

2. Procedure

1. Attach the phenyl-Sepharose column to your chromatography system. We use G10 × 250 column (Amicon) and automated low-pressure EconoSystem chromatography workstation (Bio-Rad) with the chromatography parameters for a 1-liter culture of bacteria set as follows: flow rate of 1.5 ml/min; absorbance of 280 nm; absorbance units full scale (AUFS) of 0.2; chart speed of 1.0 cm/h; and fraction size of 8 ml.

2. Equilibrate phenyl-Sepharose CL-4B column with CL-4B eluant A until a stable baseline at A_{280} is achieved.

3. Filter S2 through a 0.22-μm filter (Nalgene) to remove any debris and precipitated protein carried over from P2 just before applying to phenyl-Sepharose CL-4B column.

4. Apply filtrate to column with CL-4B eluant A and collect the flow-through (void) until baseline returns to zero.

5. Elute and collect centrin and other calcium-binding proteins with CL-4B eluant B until baseline returns to zero.

6. Elute and collect other hydrophobically bound proteins with CL-4B eluant C until baseline returns to zero.

7. Elute and collect tightly bound proteins with 3 bed volumes of 30% aqueous isopropanol; this step also regenerates the column for next use.

8. Follow isopropanol elution with 3 bed volumes of deionized water. Either equilibrate the column with 3 bed volumes CL-4B eluant A for another chromatographic run or 20% ethanol for long-term storage. If column performance begins to deteriorate (poor peak resolution, high back-pressure due to precipitated proteins, tightly bound proteins, lipids, and lipoproteins), clean column according to the procedures recommended by the manufacturer.

9. Prepare electrophoresis gel samples of the column fractions and assay for centrin by SDS–PAGE (see Troubleshooting). Figure 1 shows the separation of centrin into different fractions from whole-cell lysate through phenyl-Sepharose chromatography.

D. Superose-12 Gel Filtration Chromatography

1. Materials

0.22-μm Millex-GV4 filter unit (Millipore)

Bicinchoninic acid assay reagent (Pierce Chemical, Catalog No. 23225)

FPLC Superose-12 HR 10/30 gel filtration column (Pharmacia Biotech) equilibrated with Mono Q eluant A

Mono Q eluant A: 20 m*M* Tris, pH 7.4, 1.0 m*M* $CaCl_2$, 1.0 m*M* dithiothreitol, 0.04% NaN_3

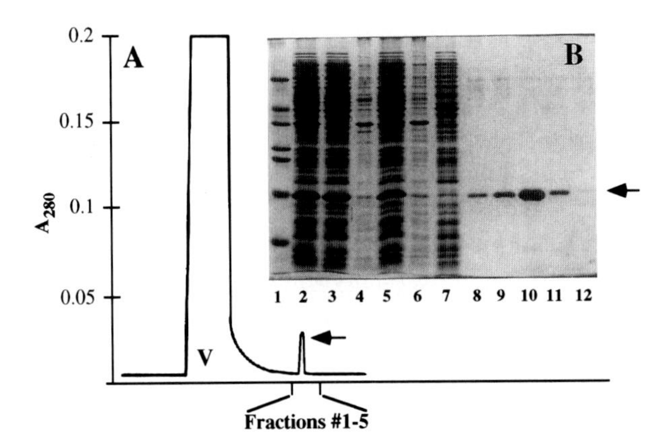

Fig. 1 (A) Chromatogram of S2 run over a phenyl-Sepharose CL-4B hydrophobic interaction column, and (B) SDS–polyacrylamide gel electrophoretogram of *E. coli* BL21 pT7-5Cen cellular fractions and corresponding phenyl-Sepharose CL-4B column fractions. The chromatogram shows the void (V) and centrin elution peak (arrow). The position of centrin in the gel is indicated by an arrow. Gel samples: lane 1, low-molecular-weight protein standards; lane 2, whole-cell lysate; lane 3, S1; lane 4, P1; lane 5, S2; lane 6, P2; lane 7, CL-4B column void; lanes 8–12, CL-4B column fractions 1–5. Protein standards from top to bottom are bovine albumin, 66 kDa; egg albumin, 45 kDa; glyceraldehyde-3-phosphate dehydrogenase, 36 kDa; carbonic anhydrase, 29 kDa; trypsinogen, 24 kDa; trypsin inhibitor, 20.1 kDa; and α-lactalbumin, 14.2 kDa (Sigma).

2. Procedure

1. Pool purest centrin-containing fractions from phenyl-Sepharose chromatography. Concentrate centrin-containing fractions to 200 μl by ultrafiltration with Centriprep and Centricon devices having a molecular weight cutoff of 10 kDa (Amicon).

2. Filter retentate with the Millex-GV4 filter unit and estimate protein concentration with bicinchoninic acid assay reagent using bovine serum albumin as a standard. Knowing the protein concentration is necessary not to overload the gel filtration column.

3. Fractionate and exchange the buffer of 200-μl aliquots of the centrin retentate (maximum load of 5–10 mg total protein) on the Superose-12 column with Mono Q eluant A. Typical chromatography parameters are flow rate of 1.0 ml/min, absorbance at 280 nm, AUFS of 0.2, chart speed of 1.0 cm/min, and fraction size of 2 ml.

4. Evaluate peak fractions by SDS–PAGE (Fig. 2).

E. Mono Q Anion-Exchange Chromatography

1. Materials

FPLC Mono Q HR 5/5 anion-exchange column (Pharmacia)

Mono Q eluant A: 20 m*M* Tris, pH 7.4, 1.0 m*M* $CaCl_2$, 1.0 m*M* dithiothreitol, 0.04% NaN_3

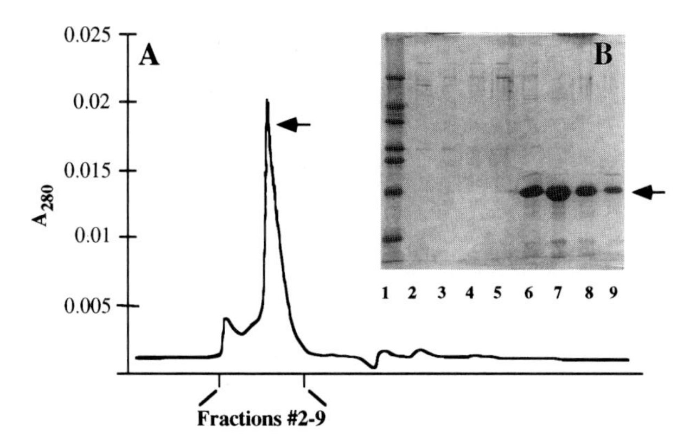

Fig. 2 (A) Chromatogram of centrin-containing CL-4B column fractions run over a Superose-12 gel filtration column, and (B) SDS–polyacrylamide gel electrophoretogram of the corresponding S-12 column fractions. The location of centrin on the chromatogram and in the gel is indicated by an arrow. Gel samples: lane 1, low-molecular-weight protein standards; lanes 2–9, S-12 column fractions 2–9. Protein standards from top to bottom are bovine albumin, 66 kDa; egg albumin, 45 kDa; glyceraldehyde-3-phosphate dehydrogenase, 36 kDa; carbonic anhydrase, 29 kDa; trypsinogen, 24 kDa; trypsin inhibitor, 20.1 kDa; and α-lactalbumin, 14.2 kDa (Sigma).

Mono Q eluant B: 20 m*M* Tris, pH 7.4, 1.0 m*M* $CaCl_2$, 0.4 *M* NaCl, 1.0 m*M* dithiothreitol, 0.04% NaN_3

1.0 *M* NaCl

2. Procedure

1. To ensure that no contaminating proteins are adsorbed to the FPLC Mono Q HR 5/5 anion-exchange column (Pharmacia Biotech) from previous use, equilibrate with 3 bed volumes Mono Q eluant A, elute with 3 bed volumes of Mono Q eluant B, and reequilibrate with Mono Q eluant A until a stable baseline is achieved. This procedure also reveals whether the eluants are well matched; well-matched eluants show little or no difference in absorbance. Poorly matched buffers have different absorbances and cause a drifting baseline during gradient elution, an artifact that can obscure peaks.

2. Pool purest centrin-containing fractions from Superose-12 chromatography and load on Mono Q column with Mono Q eluant A. The chromatography parameters we use are flow rate of 2.0 ml/min, absorbance of 280 nm, AUFS of 0.1, chart speed of 0.25 cm/min, and fraction size of 2 ml. Collect the void and continue to run Mono Q eluant A over the column until the baseline is reestablished.

3. Elute centrin from column with a 0–100% gradient of Mono Q eluant B over 10 minutes. Centrin will elute at approximately 70% Mono Q eluant B in the gradient. Continue to run Mono Q eluant B over column for 5 minutes.

4. Wash column of other adsorbed proteins with 1.0 ml of 1.0 *M* NaCl and

reequilibrate with Mono Q eluant A for another run. Store the column in 20% ethanol when not in use, and clean periodically to maintain column performance as recommended by the manufacturer. Evaluate peak fractions by SDS–PAGE (Fig 3.).

5. Pool final Mono Q centrin-containing fraction(s), exchange into a buffer appropriate for experimentation, and evaluate the protein concentration by the bicinchoninic acid assay reagent. We exchange centrin into an appropriate experimental buffer by concentrating the centrin-containing fraction(s) to 500 μl by ultrafiltration with Centriprep and Centricon devices having a molecular weight cutoff of 10 kDa, and running the retentate over an FPLC Fast Desalting HR 10/10 column (Pharmacia Biotech). The chromatography parameters we use are flow rate of 6.0 ml/min, absorbance of 280 nm, AUFS of 0.1, chart speed of 1.0 cm/min, and fraction size of 2 ml. For long-term storage, we exchange the final centrin-containing fraction(s) into 20 mM ammonium bicarbonate buffer, lyophilize, and store the powder at −70°C.

F. Troubleshooting

As with all scientific endeavors, protein purification is subject to Murphy's law: anything that can go wrong will go wrong. Potential problems we have encountered include (1) inactive ampicillin, which may result in the selection and growth of bacteria that have lost the pT7-5Cen plasmid; (2) failure of

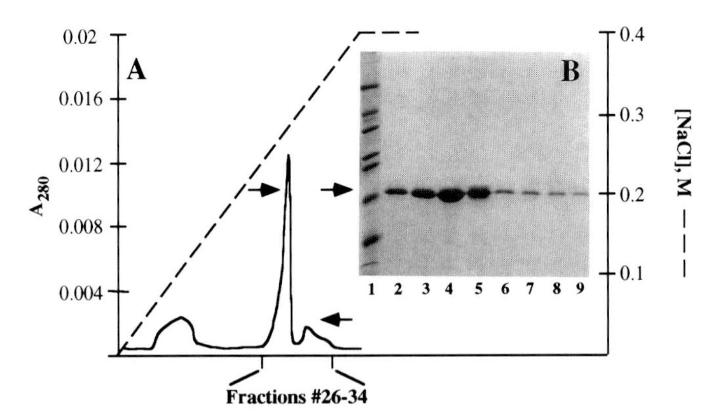

Fig. 3 (A) Chromatogram of centrin-containing S-12 column fractions run over a Mono Q anion-exchange column, and (B) SDS–polyacrylamide gel electrophoretogram of the corresponding Mono Q column fractions. The location of centrin on the chromatogram and in the gel is indicated by an arrow. Gel samples: lane 1, low-molecular-weight protein standards; lanes 2–5, Mono Q column fractions 26–29; lanes 6–9, Mono Q column fractions 31–34. Note that centrin has eluted from the Mono Q column as a major and minor peak of approximately 70% and 80% Mono Q eluant B, respectively. Protein standards from top to bottom are bovine albumin, 66 kDa; egg albumin, 45 kDa; glyceraldehyde-3-phosphate dehydrogenase, 36 kDa; carbonic anhydrase, 29 kDa; trypsinogen, 24 kDa; trypsin inhibitor, 20.1 kDa; and α-lactalbumin, 14.2 kDa (Sigma).

transcription induction due to inactive IPTG or T7 promoter; (3) induction of protein products that are larger or smaller than centrin due to improper bacterial transcription and/or translation; (4) inefficient digestion of the bacterial cell wall with lysozyme; and (5) inefficient binding of centrin to a fouled phenyl-Sepharose column.

To troubleshoot and assess the efficiency of the purification procedure, it is advisable to prepare SDS–PAGE samples of *E. coli* BL21 pT7-5Cen before and after IPTG induction, whole-cell lysate following sonication, P1, S1, P2, S2, column voids, and column elution fractions. SDS–PAGE samples of *E. coli* BL21 pT7-5Cen before and after IPTG induction are easily prepared by transferring 1 ml of bacterial culture to a microfuge tube, making a cell pellet by centrifugation for 10 seconds, resuspending the pellet with 1 ml of 1× Laemmli sample buffer (see below), boiling for 2 min, and sonicating to shear released DNA. Gel samples of whole-cell lysate, S1, P1 (resuspend P1 with cell lysis buffer to equal the volume of S1; sonicate to disrupt pellet), S2, P2 (resuspend P2 with cell lysis buffer to equal the volume of S2; sonicate to dissociate pellet), column voids, and column elution peaks are prepared by adding 90 μl of each fraction to 30 μl of 4× Laemmli sample buffer: 250 m*M* Tris, pH 6.8, 8% w/v SDS, 40% v/v glycerol, 20% v/v β-mercaptoethanol, 4 m*M* EDTA, 0.08% w/v bromphenol blue (Bio-Rad Catalog No. 161-0404). All samples are boiled for 2 minutes prior to SDS–PAGE (Laemmli, 1970).

III. Perspectives

The method described here yields a quantity of purified centrin sufficient to perform numerous types of experiments. Experiments designed to characterize the biochemical and biophysical properties of centrin, to address centrin filament assembly, to identify and purify centrin-binding proteins by preparing centrin affinity columns, and to study the dynamic *in vivo* behavior of centrin with fluorophore-conjugated analogs are now possible. In addition, the method described can be used to purify centrin from isolated *Chlamydomonas* flagella. In conclusion, the availability of milligram quantities of purified centrin will allow the development of several novel experimental approaches to further elucidate centrin's fundamental role at the flagellar basal body apparatus, centrosome, and mitotic spindle poles.

References

Baron, A. T., Greenwood, T. M., Bazinet, C. W., and Salisbury, J. L. (1992). Centrin is a component of the pericentriolar lattice. *Biol. Cell* **76,** 383–388.

Baron, A. T., and Salisbury, J. L. (1988). Identification and localization of a novel, cytoskeletal, centrosome-associated protein in PtK_2 cells. *J. Cell Biol.* **107,** 2669–2678.

Baum, P., Furlong, C., and Byers, B. (1986). Yeast gene required for spindle pole body duplication: homology of its product with Ca^{2+}-binding proteins. *Proc. Natl. Acad. Sci. U.S.A.* **83,** 5512–5516.

Bazinet, C. W., Baron, A. T., and Salisbury, J. L. (1990). Centrin: a calcium-binding protein associated with centrosomes. *In* "Stimulus Response Coupling: The Role of Intracellular Calcium-binding Proteins" (V. L. Smith and J. R. Dedman, eds.), pp. 39–56. CRC Press, Boca Raton, FL.

Coling, D. E., and Salisbury, J. L. (1992). Characterization of the calcium-binding contractile protein centrin from *Tetraselmis striata* (Pleurastrophyceae). *J. Protozool.* **39,** 385–391.

Davis, T. N., Urdea, M. S., Masiarz, F. R., and Thorner, J. (1986). Isolation of the yeast calmodulin gene: calmodulin is an essential protein. *Cell* **47,** 423–431.

Errabolu, R., Sanders, M. A., and Salisbury, J. L. (1994). Cloning of a cDNA encoding human centrin, an EF-hand protein of centrosomes and mitotic spindle poles. *J. Cell Sci.* **47,** 9–16.

Fulton, C., Cheng, K.-L., and Lai, E. Y. (1986). Two calmodulins in *Naegleria* flagellates: characterization, intracellular segregation, and programmed regulation of mRNA abundance during differentiation. *J. Cell Biol.* **102,** 1671–1678.

Goodenough, U. W. (1992). Green yeast. *Cell* **70,** 533–538.

Harris, E. L. V., and Angal, S. (1990). Protein purification methods: a practical approach. *In* "The Practical Approach Series" (D. Rickwood and B. D. Hames, eds.), Oxford University Press, New York.

Hohfeld, I., Otten, J., and Melkonian, M. (1988). Contractile eukaryotic flagella: centrin is involved. *Protoplasma* **147,** 16–24.

Huang, B., Watterson, D. M., Lee, V. D., and Schibler, M. J. (1988a). Purification and characterization of a basal body-associated Ca^{2+}-binding protein. *J. Cell Biol.* **107,** 121–131.

Huang, B., Mengersen, A., and Lee, V. D. (1988b). Molecular cloning of cDNA for caltractin, a basal body-associated Ca^{2+}-binding protein: homology in its protein sequence with calmodulin and the yeast CDC31 gene product. *J. Cell Biol.* **107,** 133–140.

Koutoulis, A., McFadden, G. I., and Wetherbee, R. (1988). Spine-scale reorientation in *Apedinella radians* (*Pedinellales, Chrysophyceae*): the microarchitecture and immunocytochemistry of the associated cytoskeleton. *Protoplasma* **147,** 25–41.

Laemmli, U. K. (1970). Cleavage of structural proteins during the assembly of the head of bacteriophage T4. *Nature* (*London*) **227,** 680–685.

Lee, V. D., and Huang, B. (1990). Caltractin: a basal body-associated calcium-binding protein in *Chlamydomonas*. *In* "Calcium as an Intracellular Messenger in Eucaryotic Microbes" (D. H. O'Day, ed.), pp. 245–257. American Society for Microbiology, Washington, DC.

Lee, V. D., and Huang, B. (1993). Molecular cloning and centrosomal localization of human caltractin. *Proc. Natl. Acad. Sci. U.S.A.* **90,** 11039–11043.

Maniatis, T., Fritsch, E. F., and Sambrook, J. (1982). "Molecular cloning: a laboratory manual". Cold Spring Harbor, Cold Spring Harbor Laboratory Press, New York.

Marshak, D. R., Clarke, M., Roberts, D. M., and Watterson, D. M. (1984). Structural and functional properties of calmodulin from the eukaryotic microorganism *Dictyostelium discoideum*. *Biochemistry* **23,** 2891–2899.

McDonald, J. R., Walsh, M. P., McCubbin, W. D., and Kay, C. M. (1987). Isolation and characterization of a novel 21-kDa Ca^{2+}-binding protein from bovine brain. *Methods Enzymol.* **139,** 88–105.

McFadden, G. I., Schulze, D., Surek, B., Salisbury, J. L., and Melkonian, M. (1987). Basal body reorientation mediated by a Ca^{2+}-modulated contractile protein. *J. Cell Biol.* **105,** 903–912.

Salisbury, J. L. (1992). Centrin-based calcium-sensitive contractile fibers of the green algae. *In* "The Cytoskeleton of the Algae" (D. Menzel, ed.), pp. 393–410. CRC Press, Boca Raton, FL.

Salisbury, J. L., Aebig, K. W., and Coling, D. E. (1986). Isolation of the calcium-modulated contractile protein of striated flagellar roots. *Methods Enzymol.* **134,** 408–414.

Salisbury, J. L., Baron, A. T., and Sanders, M. A. (1988). The centrin-based cytoskeleton of *Chlamydomonas reinhardtii:* distribution in interphase and mitotic cells. *J. Cell Biol.* **107,** 635–641.

Salisbury, J. L., Baron, A. T., Surek, B., and Melkonian, M. (1984). Striated flagellar roots: isolation and partial characterization of a calcium-modulated contractile organelle. *J. Cell Biol.* **99,** 962–970.

Salisbury, J. L., Sanders, M. A., and Harpst, L. (1987). Flagellar root contraction and nuclear movement during flagellar regeneration in *Chlamydomonas reinhardtii*. *J. Cell Biol.* **105,** 1799–1805.

Sanders, M. A., and Salisbury, J. L. (1989). Centrin-mediated microtubule severing during flagellar excision in *Chlamydomonas reinhardtii*. *J. Cell Biol.* **108,** 1751–1760.

Sanders, M. A., and Salisbury, J. L. (1994). Centrin plays an essential role in microtubule severing during flagellar excision in *Chlamydomonas reinhardtii*. *J. Cell Biol.* **124,** 795–805.

Tabor, S., and Richardson, C. C. (1985). A bacteriophage T7 RNA polymerase/promoter system for controlled exclusive expression of specific genes. *Proc. Natl. Acad. Sci. U.S.A.* **82,** 1074–1078.

Weber, C., Lee, V. D., Chazin, W. J. and Huang, B. (1994). High level expression in Escherichia coli and characterization of the EF-hand calcium-binding protein caltractin. *J. Biol. Chem.* **269,** 15795–15802.

Wright, R. L., Salisbury, J. L., and Jarvik, J. W. (1985). A nucleus-basal body connector in *Chlamydomonas reinhardtii* that may function in basal body localization or segregation. *J. Cell Biol.* **101,** 1903–1912.

Zhu, J.-K., Bressan, R. A., and Hasegawa, P. M. (1992). An *Atriplex nummularia* cDNA with sequence relatedness to the algal caltractin gene. *Plant Physiol.* **99,** 1734–1735.

PART VII

Flagellar Fractionation and Membrane Characterization

CHAPTER 48

Isolation of Flagellar Paraxonemal Rod Proteins

Huân M. Ngô and G. Benjamin Bouck

Department of Biological Sciences
University of Illinois at Chicago
Chicago, Illinois 60607

I. Introduction

A major component of some lower eukaryotic flagella is the paraxonemal rod (PR), a highly ordered lattice of cytoskeletal filaments with properties distinct from those of microtubules, microfilaments, and intermediate filaments (Cachon *et al.,* 1988; Saborio *et al.,* 1989; Schlaeppi *et al.,* 1989). Because it is directly coupled to both the tip-assembled axonemal doublets (Johnson and Rosenbaum, 1993) and to the base-assembled surface mastigonemes (Rogalski and Bouck, 1982), the paraxonemal rod may provide the pivotal link between what were previously assumed to be two relatively independent events in flagellar biogenesis.

The paraxonemal rods of *Euglena* and *Trypanosoma* appear to be analogous as suggested by immunological cross-reactivity (Gallo and Schrével, 1985) and by the similarity in partial PR amino acid sequences (Ngô *et al.,* 1993), but there are significant differences between the two rods. The more ordered paracrystalline PR of *Trypanosoma* is composed of two major highly helical proteins, which are present in approximately equal molar ratios and are resistant to extraction by nonionic detergent and low salt (Beard *et al.,* 1992; Saborio *et al.,* 1989; Schlaeppi *et al.,* 1989). In contrast, the major proteins of the coiled-sheet PR found in the *Euglena* emergent flagellum exhibit a greater degree of heterogeneity; i.e., 6–12 isoelectric isoforms migrating in the range 64–75 kDa are recognized by an anti-PR monoclonal antibody (unpublished), and biochemical complexity is further suggested from the heterogeneous population of cDNA clones (Ngô and Bouck, unpublished). In addition, *Euglena* PR proteins are readily solubilized with detergents, salts, or chaotropes and are extremely sensitive to trypsin and Pronase digestion (Hyams, 1982; Rogalski, 1981). Limited protease digestion of *Trypanosoma* PR proteins, on the other hand, consistently produces a 65-kDa protease-resistant core (Schlaeppi *et al.,* 1989).

Although attempts to isolate the paraxonemal rod itself have seen limited success in both systems, the more stable PR proteins from *Trypanosoma cruzi* have been successfully urea-extracted and purified by column chromatography (Saborio *et al.,* 1989). For *Euglena,* we have developed thermal fractionation protocols that can produce PR-enriched flagellar fractions for *in vitro* flagellar reassembly studies and PR protein binding assays. Procedures for the thermal enrichment of *Euglena* PR proteins (unpublished) and purification of *Trypanosoma* PR proteins as described in Beard *et al.* (1992) are summarized here.

II. Methods

A. Thermal Fractionation of *Euglena* Paraxonemal Rod Proteins

The major proteins of the complex *Euglena* flagellum include the high-molecular-weight mastigoneme glycopolypeptides (>200 kDa), α- and β-tubulins (56 and 52 kDa, respectively), and the paraxonemal rod polypeptides (75-kDa PR1 and 64-kDa PR2) as visualized by Coommassie blue-stained one-dimensional sodium-dodecyl sulfate–polyacrylamide gel electrophoresis (SDS–PAGE). These flagellar proteins can be selectively solubilized at various temperature with nonionic detergents, urea, or salt (see Table I).

1. Enrichment of Paraxonemal Rod Proteins

Enrichment of paraxonemal rod proteins with minimal contamination of other flagellar protein can be accomplished by treatment with 1 *M* urea at 42°C, or 1 *M* NaCl at 42°C, or 1% octylglucoside at 30°C, all in the presence of 100 m*M* β-mercaptoethanol.

a. Materials

PET: 0.1 *M* piperazine-*N,N'*-bis(2-ethanesulfonic acid), 1 m*M* ethylene glycol bis(β-aminoethyl ether)-*N,N'*-tetraacetic acid (EGTA), 10 m*M* *p*-tosyl-L-arginine methyl ester–HCl (TAME), pH 7.0, 100 m*M* β-mercaptoethanol, 10 m*M* phenylmethylsulfonyl fluoride

Table I
Thermal Fractionation of *Euglena* Flagella[a]

	1% Octylglucoside					1 *M* Urea					1 *M* NaCl				
Flagellar proteins	4°C	30°C	42°C	50°C	65°C	4°C	30°C	42°C	50°C	65°C	4°C	30°C	42°C	50°C	65°C
Mastigonemes				+	+			+	+	+	+	+		+	+
PR1 (75K)		+				+	+				+	+	+	+	
PR2 (64K)	+	+				+	+				+	+	+	+	
P60	+					+					+	+			
P58	+	+				+									
α-Tubulin	+	+	+	+	+	+	+	+	+	+	+	+	+	+	+
β-Tubulin	+		+	+	+	+	+	+	+	+	+	+	+	+	+
P50	+	+	+								+	+		+	+
P40	+	+	+			+	+	+	+		+	+	+	+	

[a] Proteins solubilized in the treatment are indicated by a +. 100 m*M* β-mercatoethanol is included in all treatments. PR proteins, mastigonemes, and tubulins are verified by immunoblotting with corresponding antibodies. P60, P58, P50, and P40 are uncharacterized flagellar proteins.

PET + 1 *M* urea
PET + 1 *M* NaCl
PET + 1% octylglucoside
PET + 100 m*M* β-mercaptoethanol
10 m*M* dithiothreitol
100 m*M* iodoacetamine

b. Procedure

1. Harvest and deflagellate 15 liters of *Euglena* by cold shock as described in Chapter 5 of this volume.

2. Resuspend flagellar pellet with 500 μl of 1 *M* urea, 1 *M* NaCl, or 1% octylglucoside, dissolved in PET preequilibrated at appropriate temperature. Extract flagella for 30 minutes, centrifuge at 12,000 rpm in a microcentrifuge for 15 minutes, and collect the soluble fraction.

3. Extracted proteins are routinely reduced and alkylated prior to SDS–PAGE by adding 400 μl of 10 m*M* dithiothreitol to the soluble fraction, and the sample is boiled for 3 minutes. Add 100 μl of 100 m*M* iodoacetamine and incubate at 50°C for 15 minutes. If untreated, the PR proteins will migrate as a broad protein smear in the range 63–75 kDa, whereas reduced and alkylated proteins can be resolved into two bands at ~64 and ~75 kDa.

2. Fractionation of PR1 and PR2

PR1 usually copurifies with PR2 but can be differentially extracted from PR2 using thermal fractionation.

a. Extraction of PR2

Collect flagella as outlined in Section II,A,1,b, but extract flagella at 4°C with 1% octylglucoside and 100 m*M* β-mercaptoethanol in PET. Sequential extraction with the same treatment will enhance the PR2 yield.

b. Extraction of PR1

Using the same extraction solution as in a, extract first at 30°C for 30 minutes followed by a second treatment at 42°C for an additional 30 minutes. The first soluble fraction (30°C) will contain both proteins, whereas the second extraction (42°C) will selectively solubilize the 75-kDa PR1.

3. Comment

The central pair microtubules are readily solubilized from the *Euglena* flagellum (unpublished observation) and most likely represent the small amount of tubulin found in all extracts.

B. Purification of *Trypanosoma* Paraxonemal Rod Proteins

In contrast to *Euglena, Trypanosoma* PR proteins exhibit less heterogeneity; i.e., the *T. brucei* paraxonemal rod is reported to contain only one major PR protein that migrates at 69 kDa under reducing conditions (Schlaeppi *et al.*, 1989), and the apparently more complex *T. cruzi* PR is made up of two major proteins: a 68-kDa PAR1 with a p*I* of 6.5 and a 70-kDa PAR2 with a slightly more acidic p*I* (Saborio *et al.*, 1989). The more homogeneous and less soluble *T. cruzi* PR proteins can be purified using sequential urea extractions and anion-exchange column chromatography as developed by Beard and co-workers (1992). This isolation procedure is summarized below, whereas methods for *T. cruzi* culturing and flagellar fractionation can be found in Beard *et al.* (1992) and Saborio *et al.* (1989).

1. Materials

2 *M* urea in 10 m*M* tricine, pH 8.5
6 m*M* urea in 10 m*M* tricine, pH 8.5
Mono Q column equilibrated with 6 *M* urea, 10 m*M* tricine, pH 8.5
500 m*M* NaCl in 10 m*M* tricine, pH 8.5

2. Enrichment of PR Protein

1. Collect flagella from 10^{10} epimastigote cells (Saborio *et al.*, 1989).
2. Extract flagella with 2 m*M* urea in 10 m*M* tricine, pH 8.5, on ice for 30 minutes, and centrifuge at 12,000 rpm in a microcentrifuge for 20 minutes.

Collect the supernatant, which contain 80% tubulins and 20% PR proteins.

3. Sequentially extract flagellar pellet with 6 m*M* urea also in 10 m*M* tricine, pH 8.5, on ice for 30 minutes. Spin at 12,000 rpm in a microcentrifuge, and collect the supernatant, which contains 50% tubulins and 50% PR proteins.

3. Ion-Exchange Chromatography

1. Equilibrate Mono Q column with 6 *M* urea in the same buffer and apply proteins extracted with 6 *M* urea.

2. Elute bound proteins with a 0–500 m*M* NaCl gradient in the same buffer; this will separate PR protein from tubulins.

4. Comments

1. PAR1 and PAR2 copurify in all steps of the isolation procedure although different fractions eluted from the Mono Q column are reported to be enriched for either PAR1 (fractions 25 and 26) or PAR2 (fraction 21) (Beard *et al.*, 1992).

2. It is critical that the tricine buffer should be at pH 8.5 to inhibit SH-dependent protease digestion of PR proteins (Saborio *et al.*, 1989).

Acknowledgments

This work was supported in part by National Science Foundation Grant MCD9105226 to G.B.B. and by a fellowship to H.M.N. from the Laboratory for Molecular Biology of the University of Illinois at Chicago.

References

Beard, C. A., Saborio, J. L., Tewari, D., Krieglstein, K. G., Henschen, A. H., and Manning, J. E. (1992). Evidence for two distinct major protein components, PAR1 and PAR2, in the paraflagellar rod of *Trypanosoma cruzi*. *J. Biol. Chem.* **267,** 21656–21662.

Cachon, J., Cachon, M., Cosson, M.-P., and Cosson, J. (1988). The paraflagellar rod: a structure in search of a function. *Biol. Cell* **63,** 169–181.

Gallo, J.-M., and Schrével, J. (1985). Homologies between paraflagellar rod proteins from trypanosomes and euglenoids revealed by a monoclonal antibody. *Eur. J. Cell Biol.* **36,** 163–168.

Johnson, K. A., and Rosenbaum, J. L. (1993). Flagellar regeneration in *Chlamydomonas:* a model system for studying organelle assembly. *Trends Cell Biol.* **3,** 156–161.

Hyams, J. S. (1982). The *Euglena* paraflagellar rod: structure, relationship to other flagellar components and preliminary biochemical characterization. *J. Cell Sci.* **55,** 199–210.

Ngô, H. M., Levasseur, P. J., and Bouck, G. B. (1993). Identification in *Euglena* of a putative third member of the flagellar paraxial rod gene family. *Mol. Biol. Cell* **4,** 273a.

Rogalski, A. A. (1981). Flagellar surface complex of *Euglena gracilis*. Ph.D Thesis. University of Illinois at Chicago.

Rogalski, A. A., and Bouck, G. B. (1982). Flagellar surface antigens in *Euglena:* Immunological evidence for an external glycoprotein pool and its transfer to the regenerating flagellum. *J. Cell Biol.* **93,** 758–766.

Saborio, J. L., Hernandez, J. M., Narayanswami, S., Wrightsman, R., Palmer, E., and Manning, J. (1989). Isolation and characterization of paraflagellar proteins from *Trypanosoma cruzi. J. Biol. Chem.* **264,** 4071–4075.

Schlaeppi, K., Deflorin, J., and Seebeck, T. (1989). The major component of the paraflagellar rod of *Trypanosoma brucei* is a helical protein that is encoded by two identical, tandemly linked genes. *J. Cell Biol.* **109,** 1695–1709.

CHAPTER 49

Preparation of Ciliary and Flagellar Remnants

Raymond E. Stephens
Department of Physiology
Boston University School of Medicine
Boston, Massachusetts 02118

I. Introduction

It has been known for some time that gentle heating will selectively depolymerize or "melt" the microtubules of axonemes *in vivo* (Behnke and Forer, 1967) and *in vitro* (Stephens, 1970). In both cases, considerable mass remains after thermal treatment. Furthermore, when the proper conditions are chosen, the tubulin derived from thermal fractionation is polymerization competent (Linck and Langevin, 1981), suggesting that selective tubulin removal is not simply a denaturation artifact.

Only recently, however, has much attention been paid to the material remaining after solubilization of most of the tubulin of the 9 + 2 structure. The ninefold symmetry of the organelle is retained, as is its full length, from basal body to tapering tip (Stephens *et al.*, 1989). This "ciliary remnant" retains most of the structural or architectural (nontubulin, nondynein) proteins of the axoneme, perhaps most significant among them being the tektins, the integral microtubule proteins that form the A–B junctional protofilaments of the outer doublets (Linck and Langevin, 1982) (see Chapter 51 in this volume). One of these tektins is synthesized coincident with and in proportion to ciliary elongation in sea urchin embryos (Stephens, 1989), suggesting that it is a length-limiting element, coordinately coassembling with tubulin in each outer doublet.

In the case of flagella, the ninefold cylindrical structure is generally not retained, although under low-shear conditions, sheets of nine doublets will fractionate to sheets of nine singlets (Stephens, 1970) and further fractionation will yield insoluble long, fibrous material that will retain the same kinds of proteins as in cilia but having no ordered cylindrical structure. Molluscan gill cilia versus sperm flagella Linck (1973) and sea urchin embryonic cilia versus sperm flagella (Stephens, 1978, 1986) show similar organelle-specific differences with dialysis fractionation: cilia fractionate to cylinders of nine A-singlets, but flagella fractionate to sheets of nine AB-doublets. The implication of this is that the outer doublets must be tied together differently in the two otherwise morphologically identical organelles and/or there are differences in the B-tubule and A–B junction.

It may be significant that centrioles and basal bodies also have a similar sort of scaffolding or skeletal array. Fais *et al.* (1986) described a "centriolar rim" structure that retained the basic ninefold configuration after removal of the triplet microtubules from isolated centrioles by either high ionic strength or extremes of pH. Furthermore, Gavin (1984) demonstrated the initial formation of ninefold symmetric structures preceding the formation of centrioles *in vitro*.

Considering that both the ninefold symmetry and the final length may be a function of the proteins that form these remnants or skeletons, study of remnants may provide some insight into the mechanism of centriole or basal body formation, ciliogenesis, and length regulation. Identification of remnant-specific proteins has already proven useful for exploring pathways for their differential incorporation into growing cilia (Stephens, 1994).

II. Remnant Fractionation

Temperature and time course details for particular molluscan and sea urchin species may be found in the primary work on which this procedure is based (Stephens *et al.,* 1989). Trial-and-error must be used with any new species, but general guidelines are given below.

A. Solutions

Tris–EDTA–mercaptoethanol (10TESH): 10 m*M* Tris–HCl, pH 8, 1 m*M* ethylenediaminetetraacetic acid (EDTA), 0.1% 2-mercaptoethanol [0.25 m*M* phenylmethylsulfonyl fluoride (PMSF) is optional, but may be necessary in some cases]

100 m*M* PMSF stock: 17 mg/ml PMSF in isopropanol (sensitive to moisture)

B. Procedure

1. Beginning with demembranated and washed ciliary or flagellar axonemes, prepared as outlined in Chapter 7, suspend the pellet uniformly in 10 vol of 10TESH, to a concentration of 1–2 mg/ml. Transfer to a thin-walled glass tube to improve heat conduction.

2. Place the tube in a water bath warmed to the appropriate temperature, generally 40–45°C for cold-water molluscan or sea urchin material and 45–50°C for tropical marine or vertebrate species.

3. Using a plastic pipet tip, initially draw the suspension in and out rapidly, a process that quickly brings the material to bath temperature. Pipet periodically to keep the suspension uniformly mixed.

4. Within a few minutes, the cloudy suspension will become translucent. Note the approximate time. Continue pipetting no more than about twice this length of time for cilia, or four times this time for flagella, taking samples from zero time onward to judge the time course of tubulin depolymerization by sodium dodecyl sulfate–polyacrylamide gel electrophoresis (SDS–PAGE) analysis or protein determination of the supernatant obtained in step 6.

5. Place the tube on ice and pipet the material as above to stop further melting, or place each timed aliquot immediately into a chilled centrifuge tube.

6. Transfer the main preparation to a centrifuge tube, if appropriate, and spin at 40,000*g* for 15 minutes. High-speed centrifugation is necessary to sediment any small remnant fragments that may arise from shear, especially in the case of flagella.

7. To remove residual, interstitial soluble proteins, resuspend the pellet thoroughly in fresh 10TESH and repeat step 6, discarding the wash supernatant.

The initial melt supernatant(s) will contain most of the tubulin, dynein, and some microtubule-associated proteins. The pellet consists of the ninefold ciliary remnants (or flagellar fibrous skeletons). The pellets should be resuspended back to the original volume with 10TESH for stoichiometric SDS–PAGE analysis. When SDS–PAGE sample buffer is added directly to remnant pellets they are difficult to resuspend.

Once the optimum time and/or temperature are established this procedure may be scaled up with little change except to compensate for the increased time needed to attain the temperature maximum and cool-down. Regardless of the size of the preparation, two complications may arise. The first is that surface denaturation may take place if the pipetting is too vigorous. Constant reexposure of the solution to the glass surface of the tube and the introduction of air bubbles are the causes; the solution is obvious. The second problem is proteolysis. Cilia from the species that we have studied are free of contaminating proteolytic enzymes, generally washed away at the demembranation step; however, other cilia or flagella, such as those from protozoan or mammalian sources, may require the presence of PMSF or more elaborate inhibitor cocktails.

The characteristic change in turbidity corresponds mainly to B-tubule depolymerization in both organelles. In cilia, the A-tubule quickly follows but the A-tubule depolymerization is generally slower in flagella. This fact is what permitted the clean separation of flagellar B-subfiber and A-singlet tubulin in earlier thermal fractionation work (Stephens, 1970). Similarly, the flagellar B-subfiber is far more stable than the ciliary B-subfiber to low-ionic-strength dialysis (Linck, 1973; Stephens, 1978). Consequently, well-defined ciliary remnants are far easier to prepare than flagellar remnants as one can depolymerize ciliary microtubules more easily.

References

Behnke, O., and Forer, A. (1976). Evidence for four classes of microtubules in individual cells. *J. Cell Sci.* **2,** 169–192.

Fais, D. A., Nadezhdina, E. S., and Chentsov, Y. S. (1986). The centriolar rim. The structure that maintains the configuration of centrioles and basal bodies in the absence of their microtubules. *Exp. Cell Res.* **164,** 27–64.

Gavin, R. H. (1984). In vitro reassembly of basal body components. *J. Cell Sci.* **66,** 147–154.

Linck, R. W. (1973). Chemical and structural differences between cilia and flagella from the lamellibranch mollusc, *Aequipecten irradians*. *J. Cell Sci.* **12,** 951–981.

Linck, R. W., and Langevin, G. L. (1981). Reassembly of flagellar B($\alpha\beta$) tubulin into singlet microtubules: consequences for microtubule structure and assembly. *J. Cell Biol.* **89,** 323–337.

Linck, R. W., and Langevin, G. L. (1982). Structure and chemical composition of insoluble filamentous components of sperm flagellar microtubules. *J. Cell Sci.* **58,** 1–22.

Stephens, R. E. (1970). Thermal fractionation of outer doublet microtubules into A- and B-subfiber components: A- and B-tubulin. *J. Mol. Biol.* **47,** 353–363.

Stephens, R. E. (1978). Primary structural differences among tubulin subunits from flagella, cilia, and the cytoplasm. *Biochemistry* **17,** 2882–2891.

Stephens, R. E. (1986). Isolation of embryonic cilia and sperm flagella. *Methods Cell Biol.* **27,** 217–227.

Stephens, R. E. (1989). Quantal tektin synthesis and ciliary length in sea urchin embryos. *J. Cell Sci.* **92,** 403–413.

Stephens, R. E., Oleszko-Szuts, S., and Linck, R. W. (1989). Retention of ciliary ninefold structure after removal of microtubules. *J. Cell Sci.* **92,** 391–402.

Stephens, R. E. (1994). Tubulin and tektin in sea urchin embryonic cilia: pathways of protein incorporation during turnover and regeneration. *J. Cell Sci.* **107,** 683–692.

CHAPTER 50

In Vitro Polymerization of Tubulin from Echinoderm Sperm Flagellar Microtubules

Ryoko Kuriyama and Richard W. Linck
Department of Cell Biology and Neuroanatomy
University of Minnesota
Minneapolis, Minnesota 55455

I. Introduction

The purpose of this chapter is to outline two basic methods for the purification and *in vitro* polymerization of tubulin from echinoderm sperm flagellar microtubules and to summarize the uses and advantages of this system.

In vitro polymerization of tubulin into microtubules was first accomplished using soluble homogenates of mammalian brain (Weisenberg, 1972). This and subsequent studies (for reviews, see Goldman *et al.*, 1976; Roberts and Hyams, 1979; Soifer, 1986) established a number of important points: (1) Methods were developed for the purification and assembly of tubulin and microtubule-associated proteins (MAPs), by depolymerization at 0°C and repolymerization at 37°C, in Mg^{2+}/GTP-containing buffers. (2) Studies established the optimal conditions and kinetic parameters of tubulin polymerization *in vitro*. (3) Relative to the basal body (minus) and the distal (plus) ends of cilia and flagella, microtubules were defined kinetically as having plus and minus ends; the plus end corresponds to the fast-growing end with a higher on rate for tubulin assembly and it is distal relative to the cell center, whereas the minus end is a slow-growing end generally associated with the centrosomes/spindle poles in both interphase and mitotic cells.

Sperm flagella from marine invertebrates, such as echinoderms and, in particular, sea urchins, provide another source of tubulin and offer several advantages for the study of microtubule assembly, structure, and function. First, it is quite easy to obtain large amounts of sperm from these animals. Second, because of their simple structural organization, axonemal microtubules can be purified in a simple and reproducible manner. Finally, cilia and flagella are composed of microtubules of the most stable class, raising an important, but yet unanswered question regarding the stability of the microtubules.

The principal obstacle to the *in vitro* assembly of ciliary and flagellar tubulin was the inherent stability of these native microtubules. Methods had been developed to fractionate/solubilize ciliary and flagellar microtubule proteins (Stephens, 1968), but the protocols employed ionic detergent or other conditions which yielded tubulin incapable of Mg^{2+}/GTP/temperature-dependent polymerization into tubular structures. Kuriyama (1976) first demonstrated that flagellar doublet microtubules could be solubilized by sonication in Mg^{2+}/GTP-containing buffers at 0°C and repolymerized by elevating the temperature to 37°C; the conditions and kinetic parameters of flagellar tubulin assembly *in vitro* were essentially identical to those for mammalian brain tubulin, as shown by this and subsequent studies (Binder and Rosenbaum, 1978; Farrell *et al.*, 1978, 1979). The fact that sonication is rather nonspecific in its fractionation of the A- and B-tubules of flagellar doublet microtubules still left unanswered the question of how tubulin isoforms might influence tubulin assembly and microtubule structure. Linck and Langevin (1981) developed a method by combining the fractionation procedure used to selectively solubilize tubulin from the B-subfiber (Stephens, 1970) and the need to protect the GTP binding site during solubilization (Kuriyama, 1976). This method had the advantage that it provided a high yield source of assembly-competent tubulin isoforms specifically from the B-tubules of axonemal outer fibers.

II. Methods

A. Solutions

TED: 1 m*M* Tris, 0.1 m*M* ethylenediaminetetraacetic acid (EDTA), 0.1–0.5 m*M* dithiothreitol (DTT), pH 7.8–8.0

Buffer A: 5 m*M* 4-morpholineethanesulfonic acid (Mes), 0.5 m*M* $MgCl_2$, 1 m*M* ethylene glycol bis (β-aminoethyl ether)-*N,N'*-tetraacetic acid (EGTA), 0.5 m*M* DTT, pH 6.7–6.9

Buffer B: 10 mM Mes, 150 mM KCl, 10 mM $MgCl_2$, 1 mM EGTA, 1 mM DTT, pH 6.7

B. Solubilization of Doublet Tubules by Sonication

1. The starting material consist of axonemes from sea urchin sperm flagella (Fig. 1a) and protozoan/algal cilia, as described elsewhere in this book.
2. Resuspend axonemes to ~10 mg/ml in TED, note volume, and dialyze overnight at 0°C.

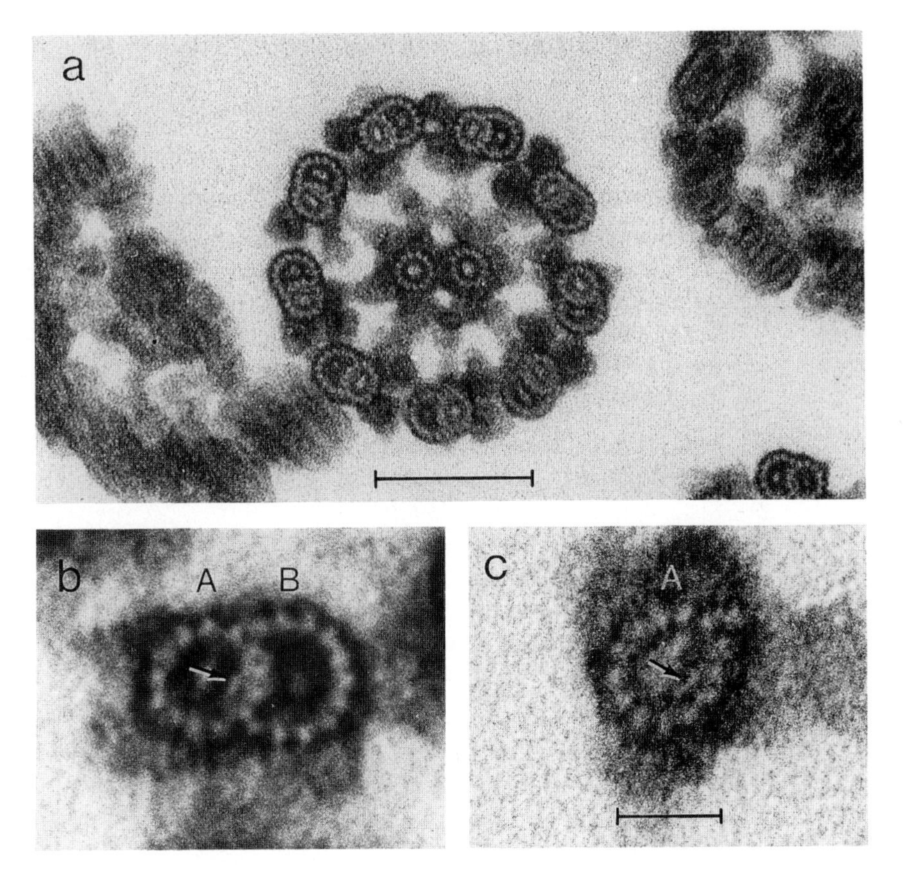

Fig. 1 Cross sections of sea urchin (*Strongylocentrotus purpuratus*) sperm flagellar axonemes (a), purified doublet microtubule (b), and thermally fractionated A-tubule (c) after fixation with tannic acid–glutaraldehyde. Each doublet tubule purified by low-ionic-strength dialysis in TED buffer is formed from a complete, cylindrical A-tubule (A) composed of 13 protofilaments and a "C-shaped" B-subfiber or B-tubule (B) composed of 10 protofilaments (b). An adluminal component (arrows in b and c) is bound to the inner wall of the A-tubule. Thermal treatment of doublet tubules at 40°C for 6 minutes selectively solubilizes the B($\alpha\beta$)-tubulin of the B-subfiber, leaving the more stable A-tubule and its adluminal component intact (c). Bar-0.1 μm in a, and 25 nm in b and c. [Reprinted with permission from R. W. Linck (1982) *Ann. NY Acad. Sci.* **383,** 98–121.]

3. Centrifuge dialysate at 40,000*g* for 30 minutes at 2°C; discard supernatant.
4. Resuspend pellets (Fig. 1b) in and dialyze overnight against buffer A; centrifuge dialysate at 40,000*g* for 30 minutes at 2°C; discard supernatant.
5. Resuspend pellet at 10–20 mg/ml (20–40% of axoneme volume from step 2) in buffer A plus 1 m*M* GTP.
6. Sonicate at 50–100 W for 2–3 minutes in a conical glass tube (3 × 9 cm) immersed in ice water.
7. Centrifuge at 100,000*g* for 1 hour at 2°C. Save supernatant for use in Section II,D,1. If necessary, the protein solution can be concentrated using an appropriate device. The solubilized tubulin can be stored at −80°C. Continue with Section II,D,1.

C. Thermal Fractionation of Doublet Tubules into Crude, Soluble B-Tubulin

The starting material consists of freshly prepared sea urchin sperm flagellar axonemes (Fig. 1a), as described in Chapter 8 of this volume. This method of tubulin assembly has been used successfully with *Strongylocentrotus purpuratus,* but for reasons that are not clear it does not work with *Lytechinus pictus*. (Approximate yields, starting from 100 mg of axonemes, are given in parentheses.)

1. Dilute sample of axonemes to ~12 mg/ml with TED at 0–4°C (note final volume); dialyze against 50–100 vol of TED for 18–20 hours; change dialysis after first 6–8 hours.
2. Centrifuge dialysate at 40,000*g* for 1 hour at 2–4°C; discard supernatant. Pellet = doublet tubules (approx yield = 70 mg).
3. Resuspend pellets of doublet tubules (Fig. 1b) in TED, 0–4°C, to ~20 mg/ml (~40% of original axoneme volume from step 1) and transfer to a graduated, conical glass tube.
4. Bring protein to 0.25 m*M* GTP, using a 0.1 *M* GTP stock neutralized to pH 7.
5. Incubate tube in water bath at 40°C; bring tube to 38.5°C (~2.5 minutes), stirring with thermometer; continue heating and mixing for an additional 6 minutes. Chill rapidly on ice.
6. Transfer protein solution to centrifuge tube on ice, using ~0.1 vol TED to wash glass tube; centrifuge at 100,000*g* for 45 minutes at 2°C. Save supernatant of crude B-tubulin (approx yield = 28 mg), transferring to graduated conical tube; measure volume. Pellet consists of thermally fractionated A-tubules (approx yield = 42 mg), which can be used for that purpose or glycerinated and stored (see Fig. 1c). Immediately continue with Section II,D,2.

D. Cycled Polymerization and Depolymerization of Soluble Doublet Tubulin

1. Tubulin Prepared by Ultrasonication

To induce microtubule polymerization, the final concentration of KCl in buffer A must be brought to 0.15 *M*. Moreover, addition of minute amounts of microtubule-nucleating agents, such as fragments of ciliary/flagellar axonemes, crude neuronal microtubules, or a purified MAP fraction from brain tissue, will markedly accelerate microtubule polymerization. Although the addition of such seeds is essential to obtain better yields in samples of low tubulin concentration, tubulin assembles into microtubules without addition of any seed fragments when a more concentrated tubulin fraction is incubated at 35°C. Continue with step 4, below.

2. B-Tubulin Prepared by Thermal Fractionation

1. To induce polymerization, the tubulin must be changed from TED to buffer B. This step must be conducted, based on the volume of tubulin from Section II,C,6, by adding in order each reagent below, at 0–4°C, and mixing after each addition:

ml reagent/ml tubulin	Final concentration
0.011 ml 1 *M* Mes, pH 6.7	10 m*M* Mes, pH 6.7
0.086 ml 2 *M* KCl	150 m*M* KCl
0.011 ml 0.1 *M* EGTA	1 m*M* EGTA
0.011 ml 0.1 *M* GTP	1 m*M* GTP
0.011 ml 1 *M* $MgCl_2$	10 m*M* $MgCl_2$
0.011 ml 0.1 *M* DTT	1 m*M* DTT

Final volume of protein in buffer B = 1.141 ml/ml of original protein.

2. Transfer to centrifuge tubes. Incubate on ice for 20 minutes. Centrifuge at 100,000*g* for 20 minutes at 0–4°C. Save supernatant for step 4 and on ice.

3. Transfer known volume of supernatant (essentially that from step 1) to clean centrifuge tube. Save ≤50 μl for protein determination in step 7 (approx yield of protein = 23 mg).

4. Incubate tube in water bath at 37°C for 45 minutes.

5. Centrifuge at 100,000*g* for 45 minutes at 37°C. Save pellet of reconstituted microtubules (for step 8) (approx yield = 20 mg) and supernatant (for step 7, then discard).

6. Use pellet immediately for step 8 or freeze in liquid nitrogen; store at −80°C.

7. Determine protein concentrations of samples and calculate yields as follows:

Crude B-tubulin before assembly (step 3):
_____ mg/ml × _____ ml = _____ mg
Supernatant after assembly (step 5):
_____ mg/ml × _____ ml = _____ mg
Difference = total B-tubulin/pellet/tube:
_____ mg

8. Resuspend pellet of reconstituted microtubules at 0–4°C, in approximately two pellet volumes of buffer A plus 150 m*M* KCl and 1 m*M* GTP (for sonicated protein) or in approximately two pellet volumes of buffer B (for thermal fractionated protein). Incubate on ice for 20 minutes. Centrifuge at 100,000*g* for 20 minutes at 20°C. Save supernatant (approx. yield = 4.2 mg).

9. Transfer supernatant to clean centrifuge tube and use for intended purpose; e.g., add 1 m*M* GTP and incubate in water bath at 37°C for 45 minutes; centrifuge at 100,000*g* for 45 minutes at 37°C. Discard supernatant; save pellet of twice-repolymerized outer-doublet tubulin/B-tubulin (see Fig. 2) (approx yield = 3.8 mg).

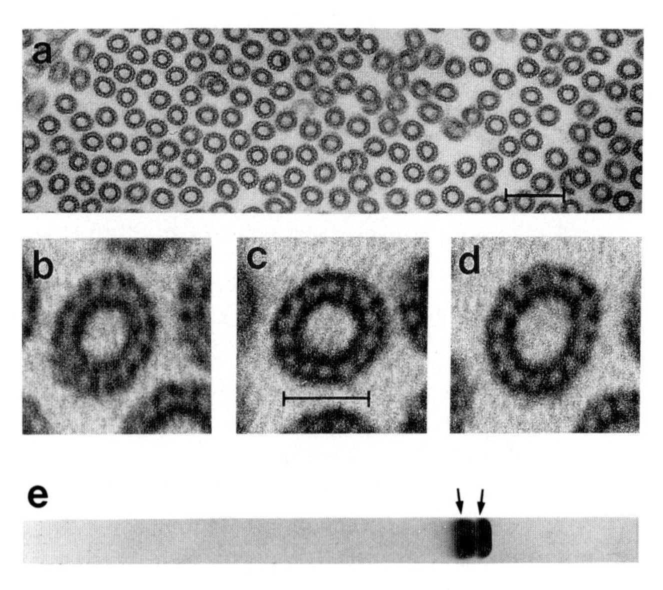

Fig. 2 Thermally fractionated flagellar B(aβ)-tubulin polymerizes *in vitro* into singlet microtubules (a–d). In cross section these synthetic B-microtubules are composed of varying numbers of protofilaments, e.g., 13 (b), 14 (c), and 15 (d); 73% are composed of 14 protofilaments. After two cycles of temperature-dependent polymerization/depolymerization, the B-microtubules are composed of more than 95% α- and β-tubulins (arrows), as judged by sodium dodecyl sulfate–polyacrylamide gel electrophoresis (e). Bar-0.1 μm in a, and 25 nm in b–d. [Reprinted with permission from R. W. Linck (1982) *Ann. NY Acad. Sci.* **383,** 98–121.]

III. Discussion

Stable flagellar doublet microtubules from echinoderms (e.g., sea urchins) can be solubilized into the native form of tubulin dimers by sonication (Kuriyama, 1976), the French pressure cell (Pfeffer *et al.,* 1978), and thermal fractionation (Linck and Langevin, 1981, after Stephens, 1970); futhermore, the latter procedure selectively solubilizes B-tubulin from the more stable A-microtubules. All of these methods have been successfully applied not only to sea urchin sperm flagella but also to cilia from various sources, including marine molluscan gill tissue, *Chlamydomonas,* and *Tetrahymena* (Johnson, 1986), although the later two sources provide limited yields. There are several advantages to using tubulin from sea urchin sperm compared with mammalian brain tissue: (1) With very little effort large quantities of sperm flagellar axonemes can be obtained, typically 1 g of axonemes (of which 600 mg is tubulin) from 25 male *S. purpuratus.* (2) Because the tubulin comes from a single organelle from a homogeneous cell type, the theoretical versus practical yields of protein can be meaningfully compared. (3) Known isoforms of tubulin can be obtained from specific kinds of microtubules. (4) Finally, sperm flagellar axonemes provide a means to investigate microtubule-stabilizing cofactors.

In contrast to the native outer-doublet microtubules, microtubules reconstituted from solubilized outer-doublet tubulin display two striking differences. Native doublet microtubules are among the most stable microtubules and they are complex structures consisting of an A-tubule with 13 protofilaments and an incomplete B-tubule with 10 protofilaments that are attached to the A-tubule in a specific pattern (Figs. 1a, b). On the other hand, soluble doublet tubulin polymerizes *in vitro* to give singlet microtubules with variable numbers of protofilaments (Figs. 2a–d) and these microtubules are cold labile. In these respects the structure and stability of repolymerized flagellar tubulin are indistinguishable from those of repolymerized mammalian brain tubulin (Soifer, 1986). Thus, it is likely that the stability and correct structure of doublet microtubules are due to specific microtubule-associated proteins rather than to any unique properties of flagellar tubulin isoforms. Ciliary and flagellar axonemes are composed of more than 150 polypeptides (Piperno *et al.,* 1977), but few besides dynein motor proteins (Porter and Johnson, 1989) have been extensively characterized. The only seemingly likely proteins to function in determining doublet microtubule structure and stability might be the tektins (see Chapter 51 of this volume). Tektins A (~53 kDa), B (51 kDa), and C (47 kDa) have been shown to form heteropolymeric filaments associated with a biochemically unique and highly stable ribbon of three to four protofilaments in the A-microtubule, near its inner attachment with the B-tubule (Steffen and Linck, 1990). Antitektin antibodies suggest that tektin-like proteins may also be present in cilia, flagella, basal bodies, centrioles, and midbodies, all of which contain stable microtubules. Finally, sequence analysis of the predicted tektin polypeptide structure (Chen *et al.,* 1993) suggests how linear tektin filaments might interact with and stabilize sets of tubulin protofilaments. Much has been learned from ciliary and flagellar systems regarding microtubule polarity, assembly, structure, and motility, and it is hoped that the methods presented here will provide for continued advances in our understanding of microtubules.

Acknowledgments

This work was supported by CTR No. 3157 (R.K.) and USPHS GM21527/GM35648 (R.L.).

References

Binder, L. I., and Rosenbaum, J. L. (1978). The in vitro assembly of flagellar outer doublet tubulin. *J. Cell Biol.* **79,** 500–515.

Chen, R., Perrone, C. A., Amos, L. A., and Linck, R. W. (1993). Tektin B1 from ciliary microtubules, primary structure as deduced from the cDNA sequence and comparison with tektin A1. *J. Cell Sci.* **106,** 909–918.

Farrell, K. W., and Wilson, L. (1978). Microtubule reassembly *in vitro* of *Strongylocentrotus purpuratus* sperm tail outer doublet tubulin. *J. Mol. Biol.* **121,** 393–410.

Farrell, K. W., Morse, A., and Wilson, L. (1979). Characterization of the *in vitro* reassembly of tubulin derived from stable *Strongylocentrotus purpuratus* outer doublet microtubules. *Biochemistry* **18,** 905–911.

Goldman, R., Pollard, T., and Rosenbaum, J. (1976). "Cell Motility". Cold Spring Harbor Laboratory Press, New York.

Johnson, K. A. (1986). Preparation and properties of dynein from *Tetrahymena* cilia. *In* "Methods in Enzymology" (R. B. Vallee, ed.), Vol. 134, pp, 306–317. Academic Press, San Diego.

Kuriyama, R. (1976). *In vitro* polymerization of flagellar and ciliary outer fiber tubulin into microtubules. *J. Biochem.* **80,** 153–165.

Linck, R. W., and Langevin, G. L. (1981). Reassembly of flagellar B($a\beta$) tubulin into singlet microtubules: consequences for cytoplasmic microtubule structure and assembly. *J. Cell Biol.* **89,** 323–337.

Pfeffer, T. A., Asnes, C. F., and Wilson, L. (1978). Polymerization and colchicine-binding properties of outer doublet microtubules solubilized by the French Pressure Cell. *Cytobiologie* **16,** 367–372.

Piperno, G., Huang, B., and Luck, D. J. L. (1977). Two-dimensional analysis of isolated flagellar proteins from wild-type and paralyzed mutants of *Chlamydomonas reinhardtii*. *Proc. Natl. Acad. Sci. U.S.A.* **74,** 1600–1604.

Porter, M. E., and Johnson, K. A. (1989). Dynein structure and function. *Annu. Rev. Cell Biol.* **5,** 119–151.

Roberts, K., and Hyams, J. S. (1979). "Microtubules". Academic Press, London and New York.

Soifer, D. (1986). Dynamic aspects of microtubule biology. *Ann. N.Y. Acad. Sci.* **466,** 1–978.

Steffen, W., and Linck, R. W. (1989). Tektins in ciliary and flagellar microtubules and their association with other cytoskeletal systems. *In* "Cell Movement" (J. R. McIntosh and F. D. Warner, eds.), Vol. 2, pp. 67–81. A. R. Liss, New York.

Stephens, R. E. (1968). Reassociation of microtubule protein. *J. Mol. Biol.* **33,** 517–519.

Stephens, R. E. (1970). Thermal fractionation of outer doublet microtubules into A- and B-subfiber components: A- and B-tubulin. *J. Mol. Biol.* **47,** 353–363.

Weisenberg, R. C. (1972). Microtubule formation *in vitro* in solutions containing low calcium concentrations. *Science* **177,** 1104–1105.

CHAPTER 51

Methods for the Isolation of Tektins and Sarkosyl-Insoluble Protofilament Ribbons

Mark A. Pirner and Richard W. Linck

Department of Cell Biology and Neuroanatomy
University of Minnesota
Minneapolis, Minnesota 55455

I. Introduction

This chapter describes methods to fractionate ciliary and flagellar microtubules into a stable subset of protofilaments (known as pf-ribbons) and to subfractionate these pf-ribbons into filaments composed of the proteins tektins. Witman and colleagues first reported that flagellar doublet microtubules from *Chlamydomonas reinhardtii* could be fractionated by Sarkosyl detergent into stable ribbons of three protofilaments (Witman, 1969; Witman *et al.*, 1972a,b), and similar observations were reported by Meza *et al.* (1972) for sea urchin sperm flagellar microtubules. Subsequently, Linck (1976) reported that protofilament (pf) ribbons could be isolated from sperm flagella from a variety of echinoderm and molluscan species and that the pf-ribbons were composed of tubulin and several other unique polypeptides (Figs. 1, 2). Although the precise origin of the pf-

METHODS IN CELL BIOLOGY, VOL. 47

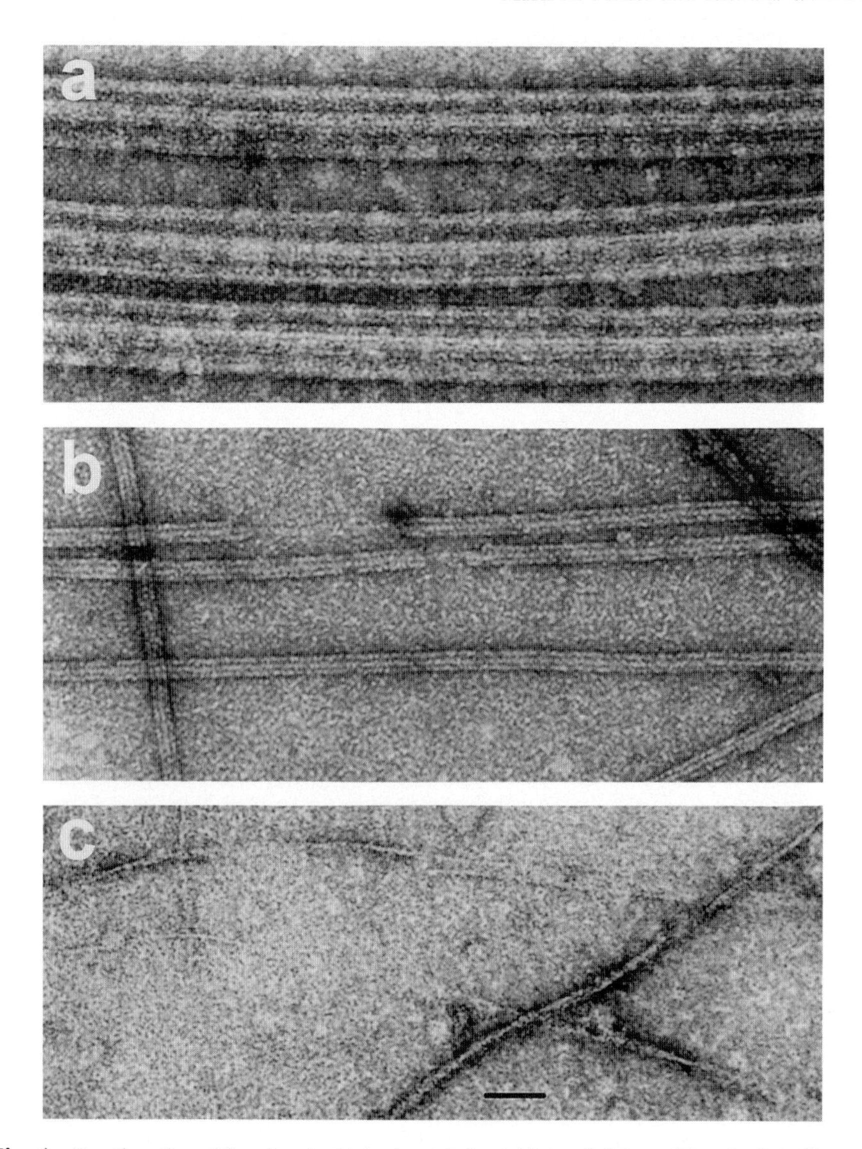

Fig. 1 Fractionation of flagellar doublet microtubules (a) into pf-ribbons (b) and tektin filaments (c), as observed by negative-stain electron microscopy. Bar = 50 nm for a–c.

ribbons in ciliary and flagellar microtubules is still not known, their approximate location in doublet microtubules has been indicated (Fig. 3).

As the stability of the pf-ribbons is presumably due to their unique polypeptide composition, efforts were made to isolate and study the relevant proteins. The first attempts to solubilize pf-ribbons with KI, KSCN, and urea led to a significant reduction of tubulin and yielded a filamentous residue enriched in nontu-

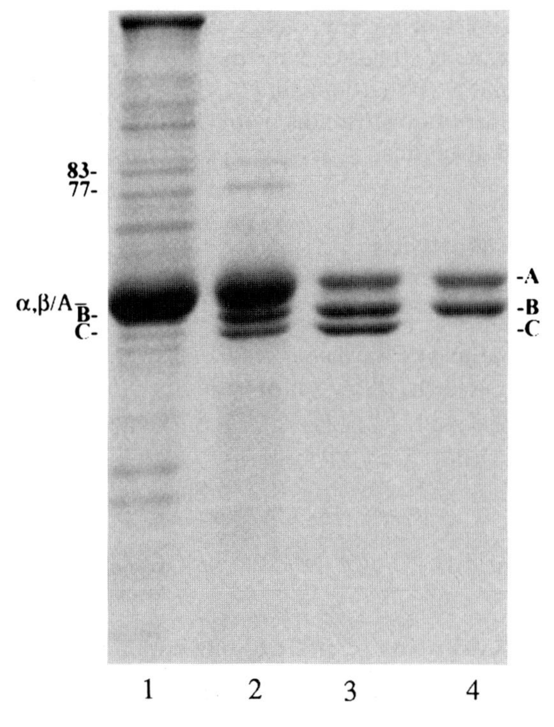

Fig. 2 Fractionation of flagellar axonemes (lane 1) into pf-ribbons (2), tektin ABC filaments (3), and tektin AB filaments (4), analyzed by sodium dodecyl sulfate–polyacrylamide gel electrophoresis.

bulin polypeptides (Linck, 1976; Linck *et al.,* 1982; Linck and Langevin, 1982). Based on the synergistic effects of detergents and urea to disrupt protein–protein interactions (Zweidler, 1978), a Sarkosyl–urea extraction was devised that provided a remarkably clean fractionation of the pf-ribbons (Linck and Langevin, 1982; Linck *et al.,* 1985; Linck and Stephens, 1987). Under optimal conditions of 0.5% Sarkosyl and 2 *M* urea, flagellar microtubules could be fractionated into 2- to 3-nm-diameter filaments composed almost exclusively of equimolar

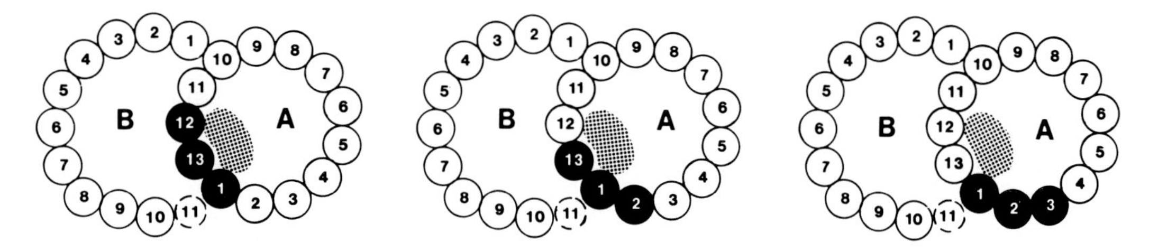

Fig. 3 Models showing the approximate location (in black) of a pf-ribbon in a doublet microtubule. [Reprinted with permission from Linck (1976, 1990)].

amounts of three proteins named tektins A (~55 kDa), B (~51 kDa), and C (~47 kDa). It has become possible to isolate filaments composed of only tektins A and B (Pirner and Linck, 1994). Methods for isolating pf-ribbons and tektin filaments are presented below, followed by a brief discussion of the characterization of tektins.

II. Methods

A. Solutions

Sarkosyl[1]: 0.5% Sarkosyl, 10 m*M* Tris, 1 m*M* ethylenediaminetetraacetic acid (EDTA), 1 m*M* dithiothreitol (DTT), pH 8.0

Sarkosyl–2 *M* urea: 0.5% Sarkosyl, 2 *M* urea, 50 m*M* Tris, 1 m*M* EDTA, 1 m*M* DTT, pH 8.0

Sarkosyl–4 *M* urea: 0.5% Sarkosyl, 4 *M* urea, 50 m*M* Tris, 1 m*M* EDTA, 1 m*M* DTT, pH 8.0

50TED: 50 m*M* Tris, 0.1 m*M* EDTA, 1 m*M* DTT, pH 8.0

B. Isolation and Purification of pf-Ribbons and Tektin Filaments

The methods described here were developed for sperm flagellar axonemes from the sea urchins *Lytechinus pictus* and *Strongylocentrotus purpuratus* (Chapter 8); isolation from other sources may require modification. It is important to use fresh material prepared and stored at 0–4°C and less than a week old; although it is possible to use glycerinated axonemes, the fractionation may be imperfect. If glycerination is desired, it is best to fractionate fresh axonemes and then glycerinate the pf-ribbons and/or tektin filaments. One hundred milligrams of sea urchin axonemes yields ~8.7 mg of pf-ribbons and ~2.9 mg of filaments composed of tektins A, B, and C. All steps are conducted at 0–4°C, and all resuspensions are to be done with a siliconized Pasteur pipet or a micropipet to avoid protein loss. Depending on the desired goal, the method below can be used in several ways:

To isolate pf-ribbons from axonemes, use the Sarkosyl solution only; these pf-ribbons can be glycerinated and stored for subsequent fractionation into tektins as indicated.

To isolate tektin filaments from pf-ribbons or directly from axonemes, use Sarkosyl–2 *M* urea solution for tektin ABC filaments or Sarkosyl–4 *M* urea solution for tektin AB filaments. At different points the material can be either stored overnight on ice or glycerinated, as indicated.

[1] Sarkosyl (sodium lauroyl sarcosinate) can be obtained from W. R. Grace (Nashua, NH) as Hamposyl L-95 (95% powder). It can be prepared as a 5% stock solution, filtered with glass filter paper, and should be stored frozen.

1. Place ~30 mg of axonemes at ~15 mg/ml in a polycarbonate ultracentrifuge tube (e.g., for a Beckman 50.2 Ti rotor). Fill the tube to capacity with the appropriate solution, adding at least 10 vol of solution for effective extraction; mix and extract for 1 hour. Centrifuge at 100,000*g* for 90 minutes. Discard supernatant and retain pellet.

2. For the second extraction of each pellet, add 5 ml of the solution used in step 1, let stand for ~5 minutes to loosen the pellet, then resuspend thoroughly. Fill tube to capacity (~23 ml), mix, and extract for 1 hour. Centrifuge at 100,000*g* for 90 minutes. Discard supernatant and retain pellet.

3. Drain and remove excess supernatant. Use ~2 ml 50TED to quickly wash the tube wall, avoiding the pellet, then discard wash.

4. At this point the pellet consists of either pf-ribbons (~2.6 mg; see Fig. 1b and Fig. 2, lane 2) or tektin filaments (~0.9 mg; see Fig. 1c and Fig. 2, lanes 3 and 4). The pf-ribbons or tektin filaments can be overlaid with 50TED to soften the pellets as described below for glycerination. Glycerinated pf-ribbons can be converted into tektin filaments (see below).

C. Isolation and Purification of Tektin Filaments from pf-Ribbons

To prepare tektin filaments from pf-ribbons, proceed from step 4 (or step 10 below), as follows:

5. To a centrifuge tube containing a pellet of ~2.6 mg of pf-ribbons, add ~2.5 ml of the appropriate Sarkosyl–urea solution to loosen the pellet, then resuspend thoroughly. Fill tube to capacity, mix, and extract for 1 hour. Centrifuge at 100,000*g* for 90 minutes. Discard supernatant and retain pellet.

6. Drain and remove excess supernatant. Use ~2 ml 50TED solution to quickly wash the tube wall, avoiding the pellet, then discard wash.

7. At this point the pellet consisting of ~0.9 mg of tektin filaments (Fig. 1c and Fig. 2, lanes 3 and 4) can be used for experimentation, stored at −80°C, or glycerinated as described below. For immediate use, the pellet can be overlaid with a few drops of the desired buffer and left overnight on ice.

D. Glycerination

8. To one pellet of pf-ribbons or tektin filaments, add 0.75 ml of 50TED and let stand overnight on ice to soften the pellet, then resuspend. Transfer protein to a graduated plastic tube; use a small amount (e.g., 0.25 ml) of 50TED to wash centrifuge tube and pipet/tip, and add wash to tube.

9. To each 1.0 ml of protein from step 8, add in order the following, mixing after each addition: 59 μl 1 *M* Tris, pH 8.0, 12 μl 0.1 *M* EDTA, 24 μl 0.5 *M* DTT, and 1.09 ml glycerol; final concentrations are 25 m*M* Tris, 0.5 m*M* EDTA, 5 m*M* DTT, pH 8.0, and 50% glycerol. Let stand with occasional mixing for 1

hour. Centrifuge out the air bubbles, using a clinical, table-top centrifuge at low speed. Flush with N_2 gas. Store at $-20°C$.

10. To recover pf-ribbons or tektin filaments after glycerination, dilute a desired volume of sample with at least 4 vol of 50TED, mix, and centrifuge at 100,000*g* for 90 minutes. Discard supernatant and retain pellet.

III. Discussion

Tektins from *S. purpuratus* sperm flagella have been partially characterized biochemically (Linck and Langevin, 1982; Linck and Stephens, 1987), structurally (Amos *et al.,* 1986; Linck and Langevin, 1982; Nojima *et al.,* 1995; Pirner and Linck, 1994), and immunologically (Chang and Piperno, 1987; Linck *et al.,* 1987; Steffen and Linck, 1988; Steffen *et al.,* 1994), and the cDNAs for *S. purpuratus* embryonic ciliary tektins have been cloned and sequenced (Chen *et al.,* 1993; Norrander *et al.,* 1992). From these studies the individual tektin polypeptide chains are predicted to be highly α-helical, and evidence indicates that tektins exist as longitudinal, heterodimeric protofilaments in the pf-ribbon and thus in flagellar A-tubules (Nojima *et al.,* 1995; Pirner and Linck, 1994). Tektin heterodimers are predicted to be linear, rodlike molecules, measuring ~48 nm long, giving the tektin filament the potential to interact with and stabilize adjacent tubulin protofilaments, as well as providing longitudinal binding sites for axonemal components with periodicities that are multiples of the 8-nm tubulin dimer and the 48-nm tektin spacing. This role for tektins and the predicted location of the pf-ribbon in the doublet microtubule (Fig. 3) would fit with the observation that Sarkosyl-extracted cilia consist of pf-ribbons still held in a ninefold pattern, presumably by nexin filaments (Stephens *et al.,* 1989) associated with the underlying tektin filaments. Furthermore, the intense staining of sea urchin sperm basal bodies and even human centrioles (Steffen and Linck, 1988) suggests that tektins are also present in triplet microtubules. While pf-ribbons and tektins certainly derive from the ciliary, flagellar, and probably basal body A-tubules, central pair singlet microtubules also appear to have a relatively stable subset of protofilaments (Linck, 1990), and the possibility has to be considered that tektins are not specific to doublet tubules alone. Indeed, by immunomicroscopy and immunoblotting with antitektin antibodies, cross-reactions have been observed with centrioles, centrosomes, mitotic spindles, and midbodies of a variety of species (see Steffen and Linck, 1989, 1992; Steffen *et al.,* 1994).

Studies of cilia and flagella have contributed much to our understanding of such important principles as microtubule motility, structure, polarity, and assembly. It is hoped that the methods presented here will encourage a continued investigation of tektins and their associated stable subset of protofilaments, leading to a greater understanding of the general microtubule cytoskeleton.

Acknowledgments

This work was supported by USPHS Grants GM21527 and GM35648 and NSF Grant BIR-9113444.

References

Amos, W. B., Amos, L. A., and Linck, R. W. (1986). Studies of tektin filaments from flagellar microtubules by immunoelectron microscopy. *J. Cell Sci.* **5**(Suppl), 55–68.

Chang, X. J., and Piperno, G. (1987). Cross-reactivity of antibodies for flagellar tektins and intermediate filament subunits. *J. Cell Biol.* **104,** 1563–1568.

Chen, R., Perrone, C. A., Amos, L. A., and Linck, R. W. (1993). Tektin B1 from ciliary microtubules: primary structure as deduced from the cDNA sequence and comparison with tektin A1. *J. Cell Sci.* **106,** 909–918.

Linck, R. W. (1976). Flagellar doublet microtubules: Fractionation of minor components and ‡-tubulin from specific regions of the A-tubule. *J. Cell Sci.* **20,** 405–439.

Linck, R. W. (1990). Tektins and microtubules. *Adv. Cell Biol.* **3,** 35–63.

Linck, R. W., and Langevin, G. L. (1982). Structure and chemical composition of insoluble filamentous components of sperm flagellar microtubules. *J. Cell Sci.* **58,** 1–22.

Linck, R. W., and Stephens, R. E. (1987). Biochemical characterization of tektins from sperm flagellar doublet microtubules. *J. Cell Biol.* **104,** 1069–1075.

Linck, R. W., Albertini, D. F., Kenney, D. M., and Langevin, G. L. (1982). Tektin filaments: chemically unique filaments of sperm flagellar microtubules. *Cell Motil.* **1**(Suppl), 127–132.

Linck, R. W., Amos, L. A., and Amos, W. B. (1985). Localization of tektin filaments in microtubules of sea urchin sperm flagella by immunoelectron microscopy. *J. Cell Biol.* **100,** 126–135.

Linck, R. W., Goggin, M. J., Norrander, J. M., and Steffen, W. (1987). Characterization of antibodies as probes for structural and biochemical studies of tektins from ciliary and flagellar microtubules. *J. Cell Sci.* **88,** 453–466.

Meza, I., Huang, B., and Bryan, J. (1972). Chemical heterogeneity of protofilaments forming the outer doublets from sea urchin flagella. *Exp. Cell Res.* **74,** 535–540.

Nojima, D., Linck, R. W., and Egelman, E. H. (1995). At least one of the protofilaments in flagellar microtubules is not composed of tubulin. *Current Biol.* **5,** 158–167.

Norrander, J. M., Amos, L. A., and Linck, R. W. (1992). Primary structure of tektin A1: comparison with intermediate-filament proteins and a model for its association with tubulin. *Proc. Natl. Acad. Sci. U.S.A.* **89,** 8567–8571.

Pirner, M. A., and Linck, R. W. (1994). Tektins are heterodimeric polymers in flagellar microtubules with axial periodicities matching the tubulin lattice. *J. Biol. Chem.* **269,** 31800–31806.

Steffen, W., and Linck, R. W. (1988). Evidence for tektinfrelated components in axonemal microtubules and centrioles. *Proc. Natl. Acad. Sci. U.S.A.* **85,** 2643–2647.

Steffen, W., and Linck, R. W. (1989). Tektins in ciliary and flagellar microtubules and their association with other cytoskeletal systems. *In* "Cell Movement, Vol. 2: Kinesin, Dynein and Microtubule Dynamics, Chapter 2" (F. D. Warner and J. R. McIntosh, eds.), pp. 67–81. Alan R. Liss, New York.

Steffen, W., and Linck, R. W. (1992). Evidence for a non-tubulin spindle matrix and for spindle components immunologically related to tektin filaments. *J. Cell Sci.* **101,** 809–822.

Steffen, W., Fajer, E. A., and Linck, R. W. (1994). Centrosomal components immunologically related to tektins from ciliary and flagellar microtubules. *J. Cell Sci.* **107,** 2095–2105.

Stephens, R. E., Oleszko-Szuts, S., and Linck, R. W. (1989). Retention of ciliary ninefold structure after removal of microtubules. *J. Cell Sci.* **92,** 391–402.

Witman, G. B. (1969). Fractionation and biochemical characterization of the flagella of *Chlamydomonas reinhardtii*. *J. Cell Biol.* **47,** 229a.

Witman, G. B., Carlson, K., and Rosenbaum, J. L. (1972a). *Chlamydomonas* flagella. I. The distribution of tubulins 1 and 2 in the outer doublet microtubules. *J. Cell Biol.* **54,** 540–555.

Witman, G. B., Carlson, K., Berliner, J., and Rosenbaum, J. L. (1972b). *Chlamydomonas* flagella. I. Isolation and electrophoretic analysis of microtubules, matrix, membranes, and mastigonemes. *J. Cell Biol.* **54,** 507–539.

Zweidler, A. (1978). Resolution of histones by polyacrylamide gel electrophoresis in presence of nonionic detergents. *Methods Cell Biol.* **17,** 223–233.

CHAPTER 52

Isolation of Radial Spoke Heads from *Chlamydomonas* Axonemes

Gianni Piperno

Department of Cell Biology and Anatomy
Mount Sinai School of Medicine
New York, New York 10029

I. Introduction

The radial spoke heads of *Chlamydomonas* axonemes have three binding sites. They are bound to the central microtubule complex, paired by a thin fiber, and connected through a stalk to the A-tubule of the outer-doublet microtubules (Huang *et al.,* 1981; Piperno *et al.,* 1981). The interactions occurring at these binding sites result in a modification of axonemal waveforms that is necessary for efficient swimming and displacement of the cell body (Brokaw *et al.,* 1982).

The radial spoke heads of *Chlamydomonas* are considered as a model system for studying organelle assembly (Curry and Rosenbaum, 1993). They can be partially purified as a protein complex (Piperno *et al.,* 1981) and are composed of six distinct proteins with apparent molecular weights in the range 24,000–123,000 (Piperno *et al.,* 1981). Two of these subunits have been isolated and sequenced (Curry *et al.,* 1992).

Five components of the radial spoke head complex are missing from flagella of distinct *Chlamydomonas* mutants (Huang *et al.,* 1981). These mutants are paralyzed and have radial spoke components as defective gene products (Huang *et al.,* 1981). Rescue of radial spoke heads in axonemes of dikaryons between a radial spoke mutant and a wild-type strain requires coassembly of radial spoke proteins from both strains (Luck *et al.*, 1977).

The following procedure was developed for partially purifying radial spoke heads from *Chlamydomonas* axonemes (Piperno *et al.*, 1981). It should be applicable to the isolation of radial spoke heads from axonemes of other systems, assuming that in these systems radial spoke head components form a complex and can be separated from radial spoke stalks as in *Chlamydomonas* axonemes.

II. Method

The procedure for isolation of radial spoke heads from *Chlamydomonas* axonemes is based on the preferential solubilization of radial spoke heads. That solubilization occurs under conditions of low ionic strength. The axonemes are previously extracted with 0.5 *M* NaCl and 1 m*M* ATP–Mg to remove the outer and inner dynein arms. The final step of the procedure consists of a sedimentation of solubilized radial spoke heads in a sucrose gradient. This step does not separate the radial spoke heads from the tubulin subunits; a further step, such as column chromatography, would be necessary to purify the spoke heads completely.

A. Solutions

Resuspension buffer: 50 m*M* NaCl, 4 m*M* $MgCl_2$, 2.5 m*M* (4-(2-hydroxyethyl)-1-piperazineethanesulfonic acid (Hepes), pH 7.

3× Dynein extraction solution: 2 *M* NaCl, 4 m*M* ATP, 16 m*M* $MgCl_2$, 4 m*M* dithiothreitol, 40 m*M* Hepes (pH 7.2 at 0°C).

Low-ionic-strength solution: 0.2 m*M* ethylenediaminetetraacetic acid (EDTA), 0.1% 2-mercaptoethanol, 0.1 m*M* phenylmethylsulfonyl fluoride, 5 m*M* Tris–chloride, pH 8.3.

Sucrose gradient solutions: 5 and 20% sucrose in 0.2 m*M* EDTA, 0.1% 2-mercaptoethanol, 0.1 m*M* phenylmethylsulfonyl fluoride, 5 m*M* Tris–chloride pH 8.3. Solutions are filtered through a 0.45-μm filter before use.

B. Procedure

All operations are performed at 4°C.

1. Axonemes are isolated (see King, this volume) and suspended at a concentration of 2 mg/ml in 50 m*M* NaCl, 4 m*M* $MgCl_2$, 2.5 m*M* Hepes, pH 7.

2. Dynein is solubilized by addition of 3× dynein extraction solution at 0°C to yield 0.5 *M* NaCl, 1 m*M* ATP, 4 m*M* $MgCl_2$, 1 m*M* dithiothreitol, 10 m*M* Hepes, pH 7.2 at 0°C.

3. After 10 minutes the suspension is centrifuged for 40 minutes at 30,000 rpm in a SW65 Beckman rotor at 4°C.

4. The pellet is suspended at a protein concentration of 1–2 mg/ml in 0.2 m*M* EDTA, 0.1% 2-mercaptoethanol, 0.1 m*M* phenylmethylsulfonyl fluoride, 5 m*M* Tris–chloride pH 8.3, and then dialyzed against the same solution for 2 hours at 4°C, followed by centrifugation as above. The supernatant containing the radial spoke heads is collected.

5. The spoke heads are partially purified by sedimentation in a 4-ml 5-20% sucrose gradient also containing 0.2 m*M* EDTA, 0.1% 2-mercaptoethanol, 0.1 m*M* phenylmethylsulfonyl fluoride, 5 m*M* Tris–chloride, pH 8.3. The centrifugation is performed in an IEC SB405 rotor at 42,000 rpm for 10 hours at 5°C. The radial spoke heads sediment as a peak in the lightest quarter of the gradient. They may be identified by their "signature" in two-dimensional gels (see Fig. 6 of Piperno *et al.,* 1981).

References

Brokaw, C. J., Luck, D. J., and Huang, B. (1982). Analysis of the movement of *Chlamydomonas* flagella: the function of the radial-spoke system is revealed by comparison of wild-type and mutant flagella. *J. Cell Biol.* **92,** 722–732.

Curry, A. M., and Rosenbaum, J. L. (1993). Flagellar radial spoke: a model molecular genetic system for studying organelle assembly. *Cell Motil. Cytoskel.* **24,** 224–232.

Curry, A. M., Williams, B. D., and Rosenbaum, J. L. (1992). Sequence analysis reveals homology between two proteins of the flagellar radial spoke. *Mol. Cell Biol.* **12,** 3967–3977.

Huang, B., Piperno, G., Ramanis, Z., and Luck, D. J. (1981). Radial spokes of *Chlamydomonas* flagella: genetic analysis of assembly and function. *J. Cell Biol.* **88,** 80–88.

Luck, D., Piperno, G., Ramanis, Z., and Huang, B. (1977). Flagellar mutants of *Chlamydomonas:* studies of radial spoke-defective strains by dikaryon and revertant analysis. *Proc. Natl. Acad. Sci. U.S.A.* **74,** 3456–3460.

Piperno, G., Huang, B., Ramanis, Z., and Luck, D. J. (1981). Radial spokes of *Chlamydomonas* flagella: polypeptide composition and phosphorylation of stalk components. *J. Cell Biol.* **88,** 73–79.

CHAPTER 53

Isolation of the Dense Fibers of Mammalian Sperm Flagella

Monica Brito and Luis O. Burzio

Instituto de Bioquímica
Facultad de Ciencias
Universidad Austral de Chile
Valdivia, Chile

I. Introduction

In mammalian sperm most of the length of the axoneme is surrounded by nine outer dense fibers generating the classic 9 + 9 + 2 cross-sectional pattern (Fawcett and Phillips, 1969; Fawcett, 1975). These fibers are joined anteriorly to the connecting piece, and differ from one another in cross-sectional shape and size. Fibers 1, 2, 4, 5, 6, 7, and 9 extend longitudinally from the connecting piece to almost the end of the principal piece (Fawcett, 1975). On the other hand, fibers 3 and 8 terminate abruptly at the limit between the midpiece and the principal piece (Fawcett, 1975).

The function of the dense fibers in sperm motility is not well understood. The early suggestion that the dense fibers possess ATPase activity and contain proteins similar to actin and myosin (Nelson, 1962) has been challenged by evidence that the dense fiber polypeptides are unrelated to contractile proteins

(Baccetti *et al.*, 1973; Olson and Sammons, 1980; Vera *et al.*, 1984; Brito *et al.*, 1986; Oko, 1988).

In rat sperm the dense fibers account for about 40% of the total sperm proteins (Vera *et al.*, 1984), indicating that during spermiogenesis, the synthesis and assembly of these fibrillar structures should be one of the most important events (Vera *et al.*, 1987; Oko and Clermont, 1989; Clermont *et al.*, 1990).

To learn more about the function and morphogenesis of the dense fibers, a method to isolate these structures from sperm was developed.

II. Methods

A. Solutions

Most of the chemicals used in our experiments were either from Sigma or from E. Merck. The following buffers and solutions are used for the isolation of the dense fibers:

PBS: 10 m*M* sodium phosphate plus 150 m*M* NaCl adjusted to pH 7.0. To this solution, 0.2 m*M* ethylene glycol bis(β-aminoethyl ether)-*N,N'*-tetraacetic acid (EGTA) and 0.5 m*M* phenylmethylsulfonyl fluoride (PMSF) were usually added before use.

Solution A: 1.6 *M* sucrose, 0.1% cetyltrimethylammonium bromide (CTAB), 10 m*M* Tris–HCl (pH 8.0), 60 m*M* 2-mercaptoethanol, and 0.2 m*M* PMSF.

Solution B: 2% sodium dodecyl sulfate (SDS), 5% 2-mercaptoethanol, 50 m*M* Tris–HCl (pH 8.0), and 0.5 m*M* PMSF.

B. Sperm Isolation

In our experiments, rat epididymal sperm were obtained from up to 30 Holtzman rats, and mouse epididymal sperm from a Rockefeller strain. After sacrifice by cervical dislocation, the epididymides are removed and stripped of fatty tissue, reduced to small pieces with scissors, and suspended in cold PBS containing 0.5 m*M* PMSF and 0.2 m*M* EGTA. The sperm are allowed to diffuse into the solution and the suspension is filtered through a nylon screen with a mesh size of 100 μm. The cell suspension is centrifuged at 3000*g* for 10 minutes at 4°C and the pellet is washed two more times with the PBS solution. The final sperm pellet is resuspended in 10 m*M* Tris–HCl (pH 8.0) containing 0.5 m*M* PMSF at a final concentration of about 2×10^8 sperm/ml, and stored at −20°C.

Fresh human ejaculates were obtained from healthy donors and were selected for a large proportion of normal cells and more than 30 million sperm. The ejaculate is diluted with 15 ml of PBS plus 1 m*M* PMSF and 0.5 m*M* EGTA, and centrifuged at 3000*g* for 10 minutes at 4°C. The pelleted cells are washed three times with the same solution, resuspended in 10 m*M* Tris–HCl (pH 8.0) containing 0.5 m*M* PMSF to a final concentration of 10^8 sperm/ml, and stored at −20°C. Bull semen was obtained from the Artificial Insemination Center of the University. The samples are transported to the laboratory in ice, diluted with 5 vol of PBS, and centrifuged at 3000*g* for 10 minutes at 4°C. The sedimented cells are washed three times with PBS, resuspended in 10 m*M* Tris–HCl (pH 8.0) plus 0.5 m*M* PMSF at a concentration of 10^8 sperm/ml, and stored at −20°C. About 1.5×10^8 bull sperm are used for isolation of outer dense fibers.

C. Isolation of the Outer Dense Fibers

About 2×10^8 rat sperm are diluted to 75 ml with 10 m*M* Tris–HCl (pH 8.0) containing 0.2 m*M* PMSF. The suspension then is mixed with an equal volume of solution A and the mixture incubated at room temperature (about 18°C) for about 60 minutes with occasional shaking. The dissociation of the sperm structures is followed by phase microscopy until the only visible structures are the sperm heads and the complex of outer dense fibers. Between 20 and 25 ml of this suspension is layered over 10 ml of solution A and centrifuged at 9000*g* for 30 minutes at 15°C (rotor HB-4, Sorvall Centrifuge RCB-2, or rotor SW 402, International Ultracentrifuge IEC B-60).

After centrifugation the outer dense fibers–connecting piece complexes form a layer at the interface between the 0.8 and 1.6 *M* sucrose layers; the heads form a compact pellet at the bottom of the tube. The dense fibers can be removed with a disposable pipet, diluted with 3 to 5 vol of 10 m*M* Tris–HCl (pH 8.0), centrifuged at 10,000*g* for 15 minutes, and resuspended in the same buffer solution. Under Nomarski microscopy the dense fibers appear as a tangled mat of filaments. To avoid aggregation of the fibers, the suspension obtained from the 0.8–1.6 *M* sucrose interface can be dialyzed overnight at 4°C against 100 vol of a solution containing 0.1% CTAB, 30 m*M* 2-mercaptoethanol, 10 m*M* Tris–HCl (pH 8.0), and 0.5 m*M* PMSF. The purity of the isolated dense fibers can be determined by Nomarski microscopy or by electron microscopy (Vera *et al.*, 1984).

D. Purification of the Dense Fiber Polypeptides

The isolated dense fibers are dissolved in 3% SDS, 5% 2-mercaptoethanol, 50 m*M* Tris–HCl (pH 8.0), and 0.5 m*M* PMSF. After centrifugation at 10,000*g* for 15 minutes to eliminate any insoluble material, the solubilized polypeptides are separated by chromatography on a Sephacryl S-200 (Pharmacia) column (2×150 cm) equilibrated in 2% SDS, 70 m*M* 2-mercaptoethanol, 25 m*M* Tris–HCl (pH 8.0), and 0.2 m*M* PMSF. It is recommended that the flow rate be adjusted to a maximum of 15 ml/h. Under these conditions the six major polypeptides (87, 30.4, 26, 18.4, 13, and 11.5 kDa) of the dense fibers are well resolved (Vera *et al.*, 1984). Further purification of each polypeptide can be achieved by rechromatography on the same column or by preparative SDS–gel electrophoresis (Vera *et al.,* 1984; Brito *et al.,* 1986). Alternatively, and prior to chromatography, the cysteine residues of the proteins can be modified by iodoacetate or iodoacetamide as described (Means and Feeney, 1971).

III. Comments

The procedure described has been used with success for the isolation of outer dense fibers from rat caput and caudal epididymal sperm (Vera *et al.,* 1984, 1987) and ejaculated bull sperm (Brito *et al.,* 1986). Small differences, however, were observed depending on the origin of the sperm. The fibers of rat sperm frayed apart from the connecting piece, forming a suspension of single fibers. In contrast, the fibers from bull sperm remained attached to the connecting piece, even after long treatment with solution A.

Treatment of human and mouse sperm with solution A also produced the complex of dense fibers (data not shown); however, phase microscopy at high magnification (1000 ×) was necessary to observe the small isolated structures. Also, after centrifugation on the discontinuous sucrose gradient, it is difficult to see the layer of fibers at the 0.8–1.6 *M* sucrose interface. It is recommended that about 2 cm be collected from the top of the 1.6 *M* sucrose cushion, diluted with Tris buffer, and centrifuged at high speed (10,000*g* for 20 minutes) to collect the fibers.

Another important consideration is partial degradation of the dense fiber polypeptides, especially with human sperm. With rat, mouse, and bull sperm the presence of 0.5 m*M* PMSF and 0.2 m*M* EGTA in the solutions seems to prevent proteolysis (for discussion see Vera *et al.,* 1984). During the isolation of human sperm dense fibers, however, a marked degradation of the polypeptides was observed. The addition of soybean trypsin inhibitor (25 μg/ml), L-1-*p*-tosylamino-2-phenylethyl chloromethyl ketone (TPCK, 0.2 m*M*), or aprotinin (20 μg/ml) was found to eliminate the problem.

Acknowledgments

This work was supported by Grant 1930365 from the Fondo Nacional de Ciencias y Tecnología, FONDECYT, Chile, and Grant I/65 516 from the Stiftung Volkswagenwerk, Germany.

References

Baccetti, B., Pallini, V., and Burrini, A. G. (1973). The accessory fibers of the sperm tail. I. Structure and chemical composition of the bull coarse fibers. *J. Submicrosc. Cytol.* **5,** 237–256.

Brito, M., Figueroa, J., Vera, J. C., Cortés, P., Hott, R., and Burzio, L. O. (1986). Phosphoproteins are structural components of bull sperm outer dense fibers. *Gamete Res.* **15,** 327–336.

Clermont, Y., Oko, R., and Hermo, L. (1990). Immunocytochemical localization of proteins utilized in the formation of outer dense fibers and fibrous sheath in rat spermatids: An electron microscope study. *Anat. Rec.* **227,** 447–457.

Fawcett, D. W. (1975). The mammalian spermatozoon. *Dev. Biol.* **44,** 394–436.

Fawcett, D. W., and Phillips, D. M. (1969). The structure and development of the neck region of the mammalian spermatozoon. *Anat. Rec.* **165,** 153–184.

Means, G. E., and Feeney, R. E. (1971). "Chemical Modification of Proteins." Holden-Day, Inc., San Francisco.

Nelson, L. (1962). Actin localization in sperm. *Biol. Bull.* **123,** 468–473.

Oko, R. (1988). Comparative analysis of proteins from the fibrous sheath and outer dense fibers of rat spermatozoa. *Biol. Reprod.* **39,** 169–182.

Oko, R., and Clermont, Y. (1989). Light microscopic immunocytochemical study of the fibrous sheath and outer dense fiber formation in the rat spermatid. *Anat. Rec.* **225,** 46–55.

Olson, G. E., and Sammons, D. W. (1980). Structural chemistry of outer dense fibers of rat sperm. *Biol. Reprod.* **22,** 319–332.

Vera, J. C., Brito, M., and Burzio, L. O. (1987). Biosynthesis of rat sperm outer dense fibers during spermiogenesis. In Vivo incorporation of [^{3}H] leucine into the fibrillar complex. *Biol. Reprod.* **36,** 193–202.

Vera, J. C., Brito, M., Zuvic, T., and Burzio, L. O. (1984). Polypeptide composition of rat sperm outer dense fibers. A simple procedure to isolate the fibrillar complex. *J. Biol. Chem.* **259,** 5970–5977.

CHAPTER 54

Isolation of the Fibrous Sheath of Mammalian Sperm Flagella

Monica Brito and Luis O. Burzio

Instituto de Bioquímica
Facultas de Ciencias
Universidad Austral de Chile
Valdivia, Chile

I. Introduction

The axoneme, which runs along the axis of the mammalian sperm tail, is surrounded by the outer dense fibers, the mitochondrial sheath, and the fibrous sheath. Depending on the characteristics of these structures, three segments of the sperm tail have been defined. The cross-sectional pattern of the middle segment contains nine outer dense fibers in close proximity with the central axoneme (Fawcett and Phillips, 1969; Fawcett, 1975). On the outside, this complex is encircled by the mitochondrial sheath; the annulus marks the caudal end of this segment. In most species, the main segment or principal piece is the longer flagellar segment. Here the axoneme is in contact with seven outer dense fibers, and on the outside the complex is surrounded by the fibrous sheath (Fawcett, 1975).

METHODS IN CELL BIOLOGY, VOL. 47

The fibrous sheath is probably one of the most fascinating biological structures. It begins immediately at the caudal margin of the annulus and extends for most of the flagellar length (Fawcett, 1975). From the annulus two longitudinal columns run along opposite sides of the tail, lying in the same plane as the central pair of microtubules of the axoneme. Perpendicular to the axis of the columns are a series of circumferentially oriented ribs that pass halfway around the principal segment and fuse at their ends with the column, so that the whole complex forms a cagelike structure, resembling the ribs of a snake (Fawcett, 1975). This structure, together with the dense fibers, poses a challenging problem in reproductive biology with regard to both their function in sperm motility and the synthesis and assembly of their corresponding polypeptides during spermiogenesis (Iron and Clermont, 1982; Sakai *et al.*, 1986; Oko and Clermont, 1989; Clermont *et al.*, 1990).

There have been several attempts to isolate the fibrous sheath. Incubation of sperm with a solution containing 6 *M* urea and 2-mercaptoethanol or dithiothreitol dissolves most of the structure, leaving the fibrous sheath and the sperm heads (Olson *et al.*, 1976; Oko, 1988; Brito *et al.*, 1989). The isolated fibrous sheath has a less complex polypeptide composition than the outer dense fibers. The most abundant polypeptide has a molecular weight of 80,000; less prominent components of 24,000 and 11,500 are also present (Olson *et al.*, 1976; Oko, 1988; Brito *et al.*, 1989). The 80,000 polypeptide is a phosphoprotein that contains about 1.5 mole of *o*-phosphoserine per mole of protein (Brito *et al.*, 1989).

II. Methods

A. Solutions

The chemicals used in our study were either from Sigma or from E. Merck. The following solutions are used:

PBS: 10 m*M* sodium phosphate plus 150 m*M* NaCl adjusted to pH 7.0. To this solution 0.2 m*M* ethylene glycol bis (β-aminoethyl ether)-*N*,*N*′-tetraacetic acid (EGTA) and 0.5 m*M* phenylmethylsulfonyl fluoride (PMSF) are usually added. To be effective, the PMSF has to be added just before using the solution. We recommend having a stock solution 100 m*M* in methanol and kept at −20°C.

Solution A: 50 m*M* Tris–HCl (pH 8.0), 1 m*M* PMSF, 0.5 m*M* EGTA, 2 m*M* dithiothreitol (DTT), and 1% Triton X-100.

Solution B: 6 *M* urea, 50 m*M* Tris–HCl (pH 8.0), 2 m*M* DTT, 0.5 m*M* PMSF, and 0.5 m*M* EGTA. The urea (E. Merck or BRL) solution used to isolate the fibrous sheath should be freshly prepared and treated with Amberlite MB-2 (Mallinckrodt) resin for 2 hours at room temperature to remove cyanates. The presence of cyanate might modify some amino acids of the polypeptides of the fibrous sheath.

Solution C: 1.8 *M* sucrose, 1% Triton X-100, 50 m*M* Tris–HCl (pH 8.0), 0.5 m*M* PMSF and 0.5 m*M* EGTA.

B. Sperm Isolation

Rat sperm are obtained from the caput and cauda epididymis of adult Holtzman male rats. The animals are sacrificed by cervical dislocation and the epididymides freed of fatty tissue, minced with scissors, and suspended in cold PBS. After diffusion of the sperm into the solution, the suspension is filtered through a 100- to 150-μm-mesh Nitex screen and centrifuged at 3000*g* for 10 minutes at 4°C. The sperm are washed three times with the PBS solution and resuspended in the same solution at a concentration of $\sim 10^8$ cells/ml. In our laboratory, bull sperm are isolated from fresh ejaculates obtained from the Artificial Insemination Center of the University, as described in Chapter 53.

C. Isolation of the Fibrous Sheath

About 3×10^8 rat sperm are diluted with about 20 ml of 50 m*M* Tris–HCl (pH 8.0), 0.5 m*M* PMSF, and 0.5 m*M* EGTA, and centrifuged at 3000*g* for 10 minutes at 4°C. The sediment is resuspended in 50 ml of solution A and incubated for 15 minutes at room temperature, and the cells are recovered by centrifugation at 3000*g* for 10 minutes. The treated sperm are then resuspended in 50 ml of solution B containing 6 *M* urea and incubated for a minimum of 2 hours at 20°C. The progress of the release of the fibrous sheath should be followed by phase microscopy.

The above suspension is mixed with 1 vol of solution C, and 25 ml is layered on top of 10 ml of the same solution C. The tubes are centrifuged at 9000*g* to 10,000*g* for 15 minutes (Sorvall HB-4 rotor). The heads form a compact pellet at the bottom of the tube. The fibrous sheaths form an opalescent band at the interface between the 0.9 and 1.8 *M* sucrose layers; these are removed with a disposable pipet and diluted with 3 to 5 vol of a solution containing 50 m*M* Tris–HCl (pH 8.0), 0.5 m*M* PMSF, and 0.5 m*M* EGTA. The fibrillar material is recovered by centrifugation at 10,000*g* for 15 minutes and washed with the buffer described above. The purity should be checked by phase or Nomarski microscopy at 1000× or by electron microscopy as described previously (Brito *et al.*, 1989).

D. Purification of the Fibrous Sheath Polypeptides

The fibrillar material is suspended in a solution containing 2% sodium dodecyl sulfate (SDS), 5% 2-mercaptoethanol, 0.5 m*M* PMSF, 0.5 m*M* EGTA, and 50 m*M* Tris–HCl (pH 8.0), and the mixture incubated for 2 minutes at 100°C. The protein solution is then centrifuged at 10,000*g* for 20 minutes to remove any insoluble material. The amount of protein can be determined by turbidimetry using bovine albumin as standard (Vera *et al.*, 1984).

To separate the different components of the fibrous sheath, the solubilized fibrillar material is chromatographed on a Sephacryl S-200 (Pharmacia) column (2 × 150 cm) equilibrated in 2% SDS, 70 m*M* 2-mercaptoethanol, 25 m*M* Tris–HCl (pH 8.0), and 0.2 m*M* PMSF (Vera *et al.*, 1984). The flow rate should be adjusted to ~15 ml/h for optimum resolution. Under these conditions a major peak eluting close to the void volume contains the major polypeptides with a molecular weight of 80,000. A second chromatographic peak contains a mixture of the polypeptides of 24,000 and 11,500 plus other minor components. Alternately, these polypeptides can be purified by preparative SDS–gel electrophoresis (Brito *et al.*, 1989).

III. Comments

Treatment of rat sperm for 12 hours with solution B containing 6 *M* urea plus DTT induces the dissolution of most cell structures with the exception of the fibrous sheath and the sperm head (Olson *et al.*, 1976; Brito *et al.*, 1989). The polypeptide composition of the isolated fibrillar material, as determined by SDS–polyacrylamide gel electrophoresis, was the same after 4, 8, or 12 hours of incubation (Brito *et al.*, 1989). It is, however, important to periodically check the progress of the treatment by phase or Nomarski microscopy. The sucrose gradient centrifugation step yields a pure fraction of fibrous sheaths at the interface between the 0.9 and 1.8 *M* layers. Nevertheless, it is important to load in each tube (40-ml capacity) no more than 10^8 sperm. Otherwise, contamination with sperm heads will be observed.

This method also has been successfully used to isolate the fibrous sheath from bull and mouse sperm (unpublished observations); however, because in these species the fibrous sheath is small, the fibrillar material has to be checked by phase microscopy at high magnification (1000×). Moreover, after centrifugation in the discontinuous sucrose gradient no opalescent layer at the interface between the 0.9 and 1.8 *M* sucrose layers was noticed. Also, it is difficult to observe the fibrillar material by phase microscopy in the presence of ~ 1 *M* sucrose. Therefore, ~2 cm of solution over the interface has to be removed with a disposable pipet. After dilution with 4 vol of the Tris buffer solution and centrifugation at 10,000*g* for 15 minutes, the fibrillar material is recovered.

Acknowledgments

This work was supported by Grant 1930365 from the Fondo Nacional de Ciencias y Tecnología, FONDECYT, Chile, and Grant I/65 516 from the Stiftung Volkswagenwerk, Germany.

References

Brito, M., Figueroa, J., Maldonado, E., Vera, J. C., and Burzio, L. O. (1989). The major component of the rat sperm fibrous sheath is a phosphoprotein. *Gamete Res.* **22,** 205–217.

Clermont, Y., Oko, R., and Hermo, L. (1990). Immunocytochemical localization of proteins utilized in the formation of outer dense fibers and fibrous sheath in rat spermatids: An electron microscope study. *Anat. Rec.* **227,** 447–457.

Fawcett, D. W. (1975). The mammalian spermatozoon. *Dev. Biol.* **44,** 394–436.

Fawcett, D. W., and Phillips, D. M. (1969). The structure and development of the neck region of the mammalian spermatozoon. *Anat. Rec.* **165,** 153–184.

Iron, M. J., and Clermont, Y. (1982). Kinetics of fibrous sheath formation in the rat spermatid. *Am J. Anat.* **165,** 121–130.

Oko, R. (1988). Comparative analysis of proteins from the fibrous sheath and outer dense fibers of rat spermatozoa. *Biol. Reprod.* **39,** 169–182.

Oko, R., and Clermont, Y. (1989). Light microscopic immunocytochemical study of the fibrous sheath and outer dense fiber formation in the rat spermatid. *Anat. Rec.* **225,** 46–55.

Olson, G. E., Hamilton, D. W., and Fawcett, D. W. (1976). Isolation and characterization of the fibrous sheath of rat epididymal spermatozoa. *Biol. Reprod.* **14,** 517–530.

Sakai, Y., Koyama, Y. P., Fujimoto, T., Nakamoto, T., and Yamashina, S. (1986). Immunocytochemical study of fibrous sheath formation in mouse spermiogenesis using a monoclonal antibody. *Anat. Rec.* **215,** 119–126.

Vera, J. C., Brito, M., Zuvic, T., and Burzio, L. O. (1984). Polypeptide composition of rat sperm outer dense fibers. A simple procedure to isolate the fibrillar complex. *J. Biol. Chem.* **259,** 5970–5977.

CHAPTER 55

Isolation and Fractionation of Ciliary Membranes from *Tetrahymena*

William L. Dentler
Department of Physiology and Cell Biology
University of Kansas
Lawrence, Kansas 66045

I. Introduction

The isolation of ciliary membranes is relatively simple and the success of each preparation depends primarily on the purity of the cilia used as starting material. As cilia and flagella are principally insoluble cytoskeletal structures (the axoneme) surrounded by a membrane, one can study membrane protein composition either by isolating intact membrane vesicles or by extracting the cilia with nonionic detergent, which solubilizes most of the membrane and leaves the axoneme intact. The major difficulty with isolating intact membrane vesicles is that they must be released from the axonemes with some mechanical agitation, which can result in microtubule breakage and contamination of the membranes with axonemal fragments. Additionally, many of the membranes remain bound to the microtubules by microtubule–membrane bridges (Dentler, 1980, 1992), so relatively few pure membrane vesicles are released from the axonemes and isolated. The composition of membranes that remains bound to the axonemes and those released as vesicles may not be identical (Dentler, 1992). Solubilization of the membrane proteins with nonionic detergent produces a greater quantity of membrane protein and results in the release of a greater percentage of membranes from the axoneme, but these fractions might also contain all of the soluble proteins associated with the ciliary matrix. We have examined a variety of ciliary isolation procedures and have found that most of these methods result in the partial disruption of the ciliary membrane. After washing, most of the readily soluble matrix components are probably lost, so detergent-solubilized material comprises primarily membrane-associated protein (and lipid). Additional fractionation of the detergent-solubilized proteins into hydrophobic, integral membrane protein and more hydrophilic fractions can be accomplished using the nonionic detergent Triton X-114. When combined with labeling of the ciliary surface proteins with biotin (see Chapter 57), this method provides a simple and rapid method to identify membrane proteins.

II. Methods

A. Isolation of Cilia

Tetrahymena cells are cultured and cilia isolated as described in Chapter 3 of this volume.

B. Isolation of Membrane Vesicles

1. Solutions

HEEMS: 50 m*M* 4-(2-hydroxyethyl)-1-piperazineethanesulfonic acid, 1 m*M* ethylenediaminetetraacetic acid, 1 m*M* ethylene glycol bis(ß-aminoethyl ether)-*N,N'*-tetraacetic acid, 3 m*M* Mg^{2+} acetate, 250 m*M* sucrose, 0.1 m*M* dithiothreitol, pH 7.4

HEEM: HEEM without sucrose

Leupeptin stock: 40 m*M* leupeptin in water

Phenylmethylsulfonyl fluoride (PMSF) stock: 29 m*M* stock in propanol, store in freezer

2. Procedure

1. Suspend cilia isolated as above in 3–4 ml of HEEM. For all subsequent procedures add fresh PMSF and leupeptin to each solution immediately before adding cilia. Mechanically agitate cilia by repeated suspensions with a glass Pasteur pipet for 4 minutes while keeping the suspension on ice. Slightly better

release of the membranes can be accomplished by adding 0.02% Triton X-100 or Nonidet P-40 to the solution, although we generally avoid using detergent in this procedure.

2. Overlay the suspension on three sucrose gradients composed of 1 ml of 20% sucrose (w/v), 1 ml of 30% sucrose (w/v), and 1.2 ml of 40% sucrose (w/v), all in HEEM. Centrifuge in a SW65 rotor (Beckman Instruments) for 120 minutes at 200,000*g* at 4°C.

3. Collect membranes from each of the tubes with a Pasteur pipet. Membranes are present at the interfaces between the 30 and 40% sucrose solutions, although some small vesicles are found at the top of the 30% sucrose layer. Axonemes with attached membranes are found in the pellet.

4. Pool the membrane fractions from each tube, dilute with cold HEEM, and pellet the vesicles by centrifugation at 48,000*g* for 1 hour at 4°C.

C. Detergent Extraction of Cilia

Fractionation of the membrane proteins into hydrophobic and more hydrophilic fractions can be accomplished with Triton X-114 (Bordier, 1981; Dentler, 1992). In our hands, most of the hydrophobic membrane polypeptides isolated with Triton X-114 are less than 60 kDa and are integral membrane proteins exposed to the cell surface, as revealed by surface labeling with NHS-LC biotin (Dentler, 1992; see Chapter 57). Higher-molecular-weight polypeptides exposed to the membrane surface generally appear in the Triton X-114 aqueous phase (Dentler, 1992). All of these detergent extraction methods fail to damage the axonemes, which remain intact and most of which contain microtubule capping structures. To isolate membrane-free axonemes, it is best to suspend the cilia in a relatively large volume of detergent, as less mechanical damage to the cilia will occur during resuspension.

1. Triton X-100 or Nonidet P-40

Make a stock solution of 10% Triton X-100 or Nonidet P-40 in water and store at 4°C.

1. Suspend isolated cilia in 5–6 ml of HEEM**S**. Then add 0.5–0.6 ml of *cold* 10% Triton X-100 to a final concentration of 1% detergent. Incubate for 20 minutes on ice.

2. Centrifuge at 17,400*g* for 20 minutes at 4°C to pellet axonemes.

3. The detergent-extracted membrane + matrix fraction can be further fractionated (see Chapter 61) or can be fractionated by running on sodium dodecyl sulfate–polyacrylamide gel electrophoresis (SDS–PAGE). In our experience, the presence of up to 1% nonionic detergent does not interfere with separation of polypeptides by SDS–PAGE. If detergent is a problem, membrane proteins

can be precipitated by adding 10% perchloric acid (final concentration) to the suspension at room temperature. After 5–10 minutes, pellet the precipitate in a conical centrifuge tube using a clinical centrifuge. Wash the pellet with water and suspend in SDS–PAGE sample buffer. It may be necessary to add NaOH to raise the pH to that of normal sample buffer: simply add NaOH dropwise and judge the pH by the indicator dye in the sample buffer.

2. Triton X-114

Make a stock solution of 10% Triton X-114 in water and store at 4°C.

1. Suspend isolated cilia in 5–6 ml of HEEMS. Then add 0.5–0.6 ml of *ice cold* 10% Triton X-114 to a final concentration of 1% detergent. Incubate for 20 minutes on ice.
2. Centrifuge at 17,000*g* for 20 minutes at 4°C to pellet axonemes. Save the supernatant on ice in a conical glass centrifuge tube.
3. Repeat steps 1 and 2 one or two times to remove all soluble membrane proteins.
4. Warm the tube containing the supernatant in a water bath at 30°C for 5 minutes to induce cloud formation by the detergent.
5. Centrifuge the sample at 1800*g* for 10 minutes at room temperature.
6. Carefully remove the clear aqueous phase and place on ice. Store the small detergent phase, containing hydrophobic proteins, on ice.
7. To remove additional hydrophobic proteins from the aqueous phase, add Triton X-114 to the aqueous phase to a final concentration of 1%. Then repeat steps 4–6.
8. To resolve polypeptides by SDS–PAGE, dilute the Triton X-114 detergent phase to a volume equal to that of the aqueous phase. Add concentrated sample buffer to each of the fractions and load the gel as usual.

References

Bordier, C. (1991). Phase separation of integral membrane proteins in Triton X-114 solution. *J. Biol. Chem.* **256,** 1604–1607.

Dentler, W. L. (1980). Microtubule-membrane interactions in cilia. I. Isolation and characterization of ciliary membranes from *Tetrahymena pyriformis*. *J. Cell. Biol.* **84,** 364–380.

Dentler, W. L. (1992). Identification of *Tetrahymena* ciliary surface proteins labeled with sulfosuccinimidyl 6-(biotinamido)hexanoate and Concanavalin A and fractionated with Triton X-114. *J. Protozool.* **39,** 368–378.

CHAPTER 56

Measurement of Membrane Potential and Na^+ and H^+ Transport in Isolated Sea Urchin Sperm Flagella and Their Membrane Vesicles

Lee Hon Cheung
Department of Physiology
University of Minnesota
Minneapolis, Minnesota 55455

I. Introduction

The motility of sea urchin spermatozoa is regulated by the intracellular pH (Lee *et al.*, 1983; Christen *et al.*, 1982), which is in turn regulated by a Na^+/H^+ exchanger present in the flagellar membrane (Lee, 1984a, b, 1985). The flagellar exchanger is novel in that it is insensitive to amiloride but is inhibited by membrane depolarization (Lee, 1984b). The voltage sensitivity is not due to the electrogenicity of the exchange process but is due to the exchanger being turned off by depolarization (Lee, 1984b, 1985). Data from both the isolated flagella (Lee, 1984b) and the plasma membrane vesicles derived from them (Lee, 1985) support the idea that the exchanger is regulated by a voltage-sensitive gating mechanism (Lee, 1985). The physiological role of the voltage sensitivity of the exchanger appears to be related to the action of speract, a peptide purified from the extracellular matrix of sea urchin eggs. This peptide can modulate the flagellar Na^+/H^+ exchanger by hyperpolarizing the membrane potential (Lee and Garbers, 1986). This is apparently due to an increase in K^+ permeability of the flagellar membrane through activation of K^+ channels (Lee and Garbers, 1986; Babcock *et al.*, 1992; Reynaud *et al.*, 1993). The exact mechanism of how the K^+ channels are activated by the speract receptor remains to be determined but it appears to involve GTP (Lee, 1988; Cook and Babcock, 1993). It has been proposed that the process may be mediated either by GTP-binding proteins (Lee, 1988) or by cyclic GMP (Cook and Babcock, 1993).

Isolated flagella and flagellar membrane vesicles are single-compartment systems that allow unambiguous analysis and interpretation of ion transport measurements. An added advantage of the flagellar membrane preparation is that the ionic conditions in both the internal and external media can be controlled. These subcellular systems are thus of value for investigating the ionic mechanisms underlying sperm motility initiation. In Chapter 8 of this volume, procedures for isolating sea urchin sperm flagella and preparing membranes from them are described. In this chapter, methods for measuring membrane potential and Na^+ and H^+ transport in these preparations are described.

II. Methods

A. Media

Na^+-free seawater (NaFSW) contains 460 m*M* choline chloride, 27 m*M* $MgCl_2$, 28 m*M* $MgSO_4$, 8 m*M* KCl, 10 m*M* $CaCl_2$, 2 m*M* $KHCO_3$, 5 m*M* Tris, 5 m*M* 4-(2-hydroxyethyl)-1-piperazineethanesulfonic acid (Hepes), pH 7.9.

Regular seawater (NaSW) has the same composition as NaFSW except choline chloride is replaced with 460 m*M* NaCl.

The internal medium (IM) used for isolation of flagellar membranes contains 20 m*M* K_2SO_4, 2.5 m*M* 1,4-piperazinediethanesulfonic acid (Pipes), and 50 μM 8-hydroxypyrene-1,2,6-trisulfonic acid (pyranine), with pH adjusted to 6.5 with methylglucamine.

The external medium (EM) has the same composition as IM except without pyranine and with the addition of 10 m*M* $MgSO_4$.

B. Fluorometric Measurement of Membrane Potential

The membrane potential of intact sperm (Babcock *et al.*, 1992; Reynaud *et al.*, 1993), isolated flagella (Lee and Garbers, 1986), and flagellar vesicles (Lee, 1988) can be monitored with a lipophilic fluorescent probe, diS-C_3-(5) (Molecular Probes). The presence of an internally negative membrane potential results in uptake of the cationic probe and quenching of its fluorescence. When it is applied to the intact cell, the interpretation of the measurements is complicated by the fact that organelles such as mitochondria also possess a highly negatively charged membrane potential and can accumulate the probe. The membrane potential measured is thus a combination of both the plasma membrane potential and the mitochondrial potential. This necessitates the use of a mitochondrial uncoupler, such as bis(hexafluoroacetonyl)acetone (5 μM), to discharge the mitochondrial membrane potential (Babcock *et al.*, 1992). In the cases of isolated flagella and flagellar membrane vesicles, this complication does not arise as they are single-compartment systems devoid of any internal membranous organelles.

An alternative method for measuring membrane potential in intact sperm and isolated flagella is to determine the equilibrium distributions of lipophilic ions such as [^{3}H]tetraphenolphosphonium and [^{14}C]thiocyanide (Schackman *et al.*, 1981; Lee, 1984). The advantage of the fluorometric method over the equilibrium distribution method is that the former monitors the membrane potential continuously and is also a much more sensitive assay.

A concentrated stock solution (40-100 μM) of diS-C_3-(5) is prepared in dimethyl sulfoxide. The final concentration used is 0.2 μM and the fluorescence is measured at an emission wavelength of 670 nm with an excitation wavelength of 630 nm. Flagella (about 1.9 mg/ml) are isolated from spermatozoa of *Strongylocentrotus purpuratus* as described (see Chapter 8 of this volume) and diluted 100 times into 1 ml of NaSW containing the probe. The suspension is mixed by continuous stirring with a glass-coated magnetic bar while the temperature of the cuvette is maintained at 16–17°C by a circulating water jacket. Teflon-coated stirring bars should not be used as the probe has quite a high affinity for Teflon. The resultant fluorescence change is measured with a spectrofluorometer such as the Perkin–Elmer 650-10S.

The fluorescence change of diS-C_3-(5) can be used as a qualitative index of the membrane potential. An increase in fluorescence indicates depolarization, whereas a decrease signifies hyperpolarization. If quantitative measurements are desired, the fluorescence can be calibrated by systematically varying the equilibrium potential of K^+ (Lee and Garbers, 1986). To ensure that the membrane potential is determined solely by the K^+ gradient across the flagellar membrane, a K^+ ionophore, valinomycin (Sigma, St. Louis, MO), can be used to increase the membrane permeability of K^+ such that it becomes the dominant permeant ion. Isolated flagella (about 1.9 mg/ml) are diluted 100 times into 1 ml of NaSW containing 0.2 μM diS-C_3-(5) as described above. After the fluorescence reaches a steady value (2–4 minutes), 1–3 μl of concentrated valinomycin stock in methanol is added to give a final concentration of 2 μM. This should result in a further decrease in fluorescence as the increase in K^+ permeability by valinomycin induces hyperpolarization. Stepwise additions of K^+ depolarize the flagella and should reverse the fluorescence quenching.

To calculate the equilibrium potential from the K^+ gradient using the Nernst equation, the value of internal $[K^+]$ needs to be determined. This can be done by measuring the total amount of K^+ released by a detergent such as Nonidet P-40. Isolated flagella (3–4 mg/ml) are diluted 20-fold into a medium containing 560 m*M* choline chloride, 10 m*M* Tris, 10 m*M* Hepes, pH 7.9. Nonidet P-40 is added to a final concentration of 0.5% to demembranate the flagella and release the intracellular K^+. The resultant change in K^+ activity of the suspension is measured using a K^+-specific electrode (Corning) and quantified by adding a known amount of KCl. The intracellular $[K^+]$ is calculated by dividing the amount of K^+ released by the flagellar volume of 4.9 μl/mg (Lee, 1984a). In this way, a value of 219 ± 61 m*M* has been determined (Lee, 1984b).

With the value of intraflagellar $[K^+]$, the stepwise increases in the K^+ equilibrium potential corresponding to the additions of K^+ can be calculated. From a series of such measurements, a calibration curve can be constructed. The fluorescence of diS-C_3-(5) should be linear when plotted against the equilibrium potential (Lee and Garbers, 1986).

A similar procedure also can be used to measure membrane potential in flagellar vesicles. The vesicles are prepared by osmotic lysis of the isolated flagella in a medium containing 40 m*M* KCl, 10 m*M* Pipes, 1 m*M* $MgCl_2$, pH 6.7 (adjusted with methylglucamine) as described in Chapter 8 of this volume. The vesicles are washed once by centrifugation (130,000*g*, 15 minutes, 5°C) and resuspended in a medium with similar composition except with the addition of 20 m*M* $MgSO_4$. The vesicle stock (60–100 μg/ml) is diluted 30- to 100-fold into an assay medium containing 20 m*M* $MgSO_4$, 2 m*M* $Ca(OH)_2$, 0.4 μM diS-C_3-(5), 20 m*M* Hepes, pH 8.0. The fluorescence can be calibrated by varying the K^+ equilibrium potential in the presence of valinomycin (0.2 μM) as described above for the isolated flagella. The intravesicular $[K^+]$ used for calculation of the Nernst potential is known and is the same as that of the lysis medium.

C. Fluorometric Measurement of Internal pH

The intraflagellar pH can be monitored with a fluorescent amine, acridine orange (Lee, 1984a). The probe, being a weak base, is concentrated into acidic compartments, resulting in the formation of dye aggregates with spectral characteristics different from those of the monomer (for a review, see Lee *et al.*, 1982). The procedure is to monitor the change in the monomer fluorescence at 530 nm (excitation at 490 nm). Isolated flagella (0.45 mg/ml) are diluted 10-fold into 1 ml of NaFSW containing 5 μM acridine orange. This should result in quenching of the fluorescence at 530 nm as the probe is accumulated into the internally acidic flagella. A diagnostic test is to discharge the pH gradient with 5 m*M* NH_4Cl and the fluorescence should return to its original level before the addition of the flagella. Similarly, treatments that affect the integrity of the flagellar membrane, such as detergents (e.g., Triton X-100, 0.1 mg/ml, or lysolecithin, 20 μg/ml) or osmotic lysis should also discharge the pH gradient and reverse the quenching.

The fluorescence change can be calibrated by equilibrating the pH gradient with the K^+ gradient using the ionophore nigericin (Sigma) (Lee and Garbers, 1986). By changing the external $[K^+]$, the pH gradient can be systematically varied and the resultant change in fluorescence intensity measured. The pH_i can be calculated from the equation $pH_i = pH_o - \log [K^+]_i/[K^+]_o$, where the subscripts i and o denote the ion concentrations inside and outside the flagella, respectively. Flagella (3.6 mg/ml) are diluted 40-fold into 1 ml of NaFSW containing 5 μM acridine orange. After the fluorescence reaches a steady value (1–2 minutes), nigericin is added to a final concentration of 0.8 μM. This should result in a further decrease in fluorescence as the pH gradient is equilibrated with the K^+ gradient. After 1–2 minutes, various concentrations of KCl are added stepwise to raise $[K^+]_o$ and should result in corresponding increases in fluorescence. The pH_o is buffered at pH 8.0 and the $[K]_i$ has been measured to be 219 mM as described above. Using this procedure, the pH_i of the isolated flagella in NaFSW was determined to be 6.7 ± 0.02 before the addition of nigericin, which is identical to that measured using the equilibrium distribution of [^{14}C]methylamine (Schackman *et al.*, 1982; Lee, 1984,b). The advantage of the fluorometric method over the methylamine method is that the former monitors the pH_i continuously and is also a much more sensitive assay.

In the case of flagellar membrane vesicles, a more direct method is available for measuring internal pH, which is to entrap an impermeant pH indicator, pyranine, into the vesicles during their preparation (Lee, 1985). The 8-hydroxy group of pyranine has a pK_a of 7.2 and ionization of the group is associated with a pronounced red shift in the fluorescence excitation maximum, from 405 nm (acidic) to 450 nm (alkaline), while the 510-nm emission maximum remains unchanged. The fluorescence change can be calibrated by using the nigericin procedure as described above. Flagellar membrane vesicles are prepared by osmotic lysis in IM containing pyranine as described in Chapter 8 of this volume. After washing with EM to remove external pyranine, the vesicles (0.15-0.27 mg/ml) are diluted 40-fold into a thermostated (16°C) cuvette containing 1 ml of assay medium composed of 20 mM Hepes, 20 mM $MgSO_4$, 2 mM $Ca(OH)_2$, pH 7.9. Excitation and emission wavelengths used are 450 and 510 nm, respectively. The pyranine entrapment is about 927 pmole/mg of membrane protein (Lee, 1985), and 3–9 μg/ml flagella membrane vesicles should give sufficient fluorescence intensity for the measurements. To set the gain of the spectrofluorometer, 1 μl of 20% Nonidet P-40 (a detergent) is added to 1 ml of suspension to solubilize the vesicles and the total fluorescence measured is set to 95 fluorescence units by adjusting the gain. In this way, results from day to day can be compared directly.

D. Measurement of Na^+ Uptake

The Na^+ movements can be monitored directly using $^{22}Na^+$. Uptake measurements are initiated by the addition of 10 mM ^{22}NaCl (2 μCi/ml final concentration) to a suspension containing 0.6–1.0 mg protein/ml freshly isolated flagella in NaFSW. After various periods of incubation, an aliquot of the suspension (130–200 μl) is filtered through a Whatman GFC glass-fiber filter under suction and washed quickly three times with 5 ml of ice-cold wash medium, which has a composition similar to that of NaSW except with the 100 mM NaCl replaced by KCl. The flagellar Na^+/H^+ exchanger is voltage sensitive. The KCl included in the wash medium is designed to depolarize the flagella and prevent loss of $^{22}Na^+$ via the exchanger. The radioactivity on the filter is then determined with a gamma counter. The filtering and washing procedure is completed within 8 seconds. Nonspecific trapping of radioactivity on the filter was determined to be negligible by using [^{14}C]sorbitol as an external marker (Lee, 1984a). The flagellar Na^+ uptake is fast and reaches a constant value of 134 ± 36 nmole/mg within 1 minute.

Acknowledgments

I thank Richard Graeff for suggestions and critical reading of the manuscript. The research in my laboratory is supported by a grant from the National Institutes of Health.

References

Christen, R., Schackman, R. W., and Shapiro, B. M. (1982). Elevation of the intracellular pH activates respiration and motility of sperm of the sea urchin, *Strongylocentrotus purpuratus*. *J. Biol. Chem.* **257,** 14881–14890.

Cook, S. P., and Babcock, D. F. (1993). Selective modulation by cGMP of the K^+ channel activated by speract. *J. Biol. Chem.* **268,** 22402–22407.

Babcock, D. F., Bosma, M. M., Battaglia, D. E., and Darzon, A. (1992). Early persistent activation of sperm K^+ channel by the egg peptide speract. *Proc. Natl. Acad. Sci. U.S.A.* **89,** 6001–6005.

Lee, H. C., Johnson, C., and Epel, D. (1983). Changes in internal pH associated with initiation of motility and acrosome reaction of sea urchin sperm. *Dev. Biol.* **95,** 31–45.

Lee, H. C. (1984a). Sodium and proton transport in flagella isolated from sea urchin spermatozoa. *J. Biol. Chem.* **259,** 4957–4963.

Lee, H. C. (1984b). A membrane potential-sensitive Na^+/H^+ exchange system in flagella isolated from sea urchin spermatozoa. *J. Biol. Chem.* **259,** 15315–15319.

Lee, H. C. (1985). The voltage-sensitive Na^+/H^+ exchange in sea urchin spermatozoa flagellar membrane vesicles studied with an entrapped pH probe. *J. Biol. Chem.* **260,** 10794–10799.

Lee, H. C. (1988). Internal GTP stimulates the speract receptor mediated voltage changes in sea urchin spermatozoa membrane vesicles. *Dev. Biol.* **126,** 91–97.

Lee, H. C., and Garbers, D. (1986). Modulation of the voltage-sensitive Na^+/H^+ exchange in sea urchin spermatozoa through membrane potential changes induced by the egg peptide speract. *J. Biol. Chem.* **261,** 16026–16032.

Lee, H. C., Forte, J. G., and Epel, D. (1982). The use of fluorescent amines for the measurement of pH_i: Applications in liposomes, gastric microsomes, and sea urchin gametes. *In* "Intracellular pH: Its Measurement, Regulation and Utilization in Cellular Functions" (R. Nuccitelli and D. Deamer, eds.), pp. 135–160. Alan R. Liss, New York.

Reynaud, E., La Torre, D. D. L., Zapata, O., Lievano, A., and Darzon, A. (1993). Ionic bases of the membrane potential and intracellular pH changes induced by speract in swollen sea urchin sperm. *FEBS Lett.* **329,** 210–214.

Schackman, R. W., Christen, R., and Shapiro, B. M. (1981). Membrane potential depolarization and increased intracellular pH accompanying the acrosome reaction of sea urchin sperm. *Proc. Natl. Acad. Sci. U.S.A.* **78,** 6066–6070.

CHAPTER 57

Nonradioactive Methods for Labeling and Identifying Membrane Surface Proteins

William L. Dentler

Department of Physiology and Cell Biology
University of Kansas
Lawrence, Kansas 66045

I. Introduction

In addition to motility, cilia provide surfaces on which a variety of cell surface molecules are displayed for cell–cell interactions during mating, chemoreception, and a variety of cellular functions. For a typical biflagellate *Chlamydomonas* cell, the surface area of the flagellar membrane is approximately 15 μm^2, about 0.3% of the total surface area of the cell. For *Tetrahymena pyriformis,* which contains approximately 600 cilia per cell, the total surface area of the membrane can be greater than 40% of the total cell surface area. At present, we have little understanding of the function and identity of most of the ciliary surface proteins, other than the fact that many are present. In a study of *Tetrahymena thermophila* ciliary membrane proteins, we identified more than 40 ciliary membrane surface proteins, at least 16 of which also bind concanavalin A (Dentler, 1992). Other than 36- and 50-kDa polypeptides that appear to be associated with a microtubule–membrane bridge, the functions of these proteins remain unknown.

Putative membrane surface proteins can be identified by iodination (Bloodgood and Workman, 1984; Williams *et al.,* 1980) followed by resolution of proteins by sodium dodecyl sulfate–polyacrylamide gel electrophoresis (SDS–PAGE) and autoradiography. There is, however, concern about this method due to its potential to label cytoplasmic proteins (Thompson *et al.,* 1987) and the use of ^{125}I. In contrast to iodination, labeling with biotin provides a gentle and nonradioactive method to label surface proteins in a variety of cells (see Goodloe-Holland and Luna, 1987). Biotinylation involves the covalent coupling of biotin to a protein or other molecule and the subsequent binding of biotin to avidin or the more specific streptavidin, with which biotin forms an extremely high affinity bond (with a dissociation constant of $\sim 10^{-15}$ M). Avidin or streptavidin can be coupled to alkaline phosphatase or other reagents that produce a colorimetric reaction used to identify biotinylated proteins bound to nitrocellulose. Although streptavidin–gold particles should be useful in identifying biotinylated proteins by electron microscopy, we have had relatively little success in obtaining convincing data using this procedure.

A variety of compounds, some of which are reversible, can be used to link biotin to proteins, carbohydrates, and nucleic acids. One of the best sources for these compounds is Pierce Chemical, and the reader is encouraged to consult their catalog for new compounds and labeling procedures. We use sulfosuccinimidyl 6-(biotinamido) hexanoate (NHS-LC biotin), which reacts with primary amines (generally lysine epsilon groups) to form a stable covalent amide bond. Reactions are favored in slightly alkaline media, in which the primary amines are unprotonated. The NHS-LC biotin is water soluble and does not penetrate the ciliary membrane to label axonemal or cytoplasmic proteins (Dentler, 1992). The sensitivity of the reagent depends on the detection methods used after SDS–PAGE and blotting to nitrocellulose. Adequate labeling, comparable to iodination, can be achieved using horseradish peroxidase (HRP)–streptavidin to stain the biotin, but more than a 10-fold increase in sensitivity can be achieved if alkaline phosphatase-labeled streptavidin is used (Dentler, 1992).

In addition to labeling membrane proteins prior to SDS–PAGE, putative membrane proteins can be identified by staining blotted proteins with concanavalin A (Con A) and HRP (Pagliaro and Wolfe, 1987). This method is also described below. Although we find this to be an excellent method to detect concanavalin A, an alternative method uses biotinylated concanavalin A coupled with HRP–avidin (R. A. Bloodgood, personal communication).

These methods were developed for *Tetrahymena* cells, but similar methods have been used for *Chlamydomonas* (Reinhart and Bloodgood, 1988). The method is simple and exceptionally reproducible. Care should be taken to avoid labeling in buffers containing amines that will compete for the ligand (e.g., Tris, azide, glycine, or ammonia). As opposed to other biotinylated compounds, which have half-lives of hours in aqueous solutions, NHS-LC biotin has an extremely short half-life and virtually all labeling is completed within 5–10 minutes. For all labeling, dissolve the reagent in water or culture medium immediately before use or add solid compound to the medium containing cells.

II. Methods

A. Labeling *Tetrahymena* Cilia with NHS-LC Biotin

1. Solution and Material

HNMK: 50 mM 4-(2-hydroxyethyl)-1-piperazineethanesulfonic acid (Hepes), pH 7.4, 36 mM NaCl, 0.1 mM $MgSO_4$, 1 mM KCl

NHS-LC Biotin (Pierce)

2. Procedure

1. Harvest *Tetrahymena* and wash three times with HNMK. Use a clinical centrifuge with round-bottom tubes. Centrifuge at setting 5 for 5 minutes. Suspend the cells gently by squirting a jet of buffer from a Pasteur pipet into the pellet and then pulling cells up and down in the pipet as gently as possible. Check for cell breakage by examining the cells with the light microscope. For labeling, suspend cells to approximately 10^7 cells/ml.
2. Take one portion of cells and set aside as a control for nonspecific binding of streptavidin to the polypeptides.
3. Make 8 ml of cells 1 mg/ml with NHS-LC biotin. This can be done by adding solid biotin to the solution of cells or by dissolving biotin in a small volume of HNMK. Cells can be labeled in other than HNMK buffers but the buffer must not contain biotinylatable amine groups. Incubate cells in a small beaker on a shaker for 10 minutes and then dilute with 10 vol of HNMK.
4. Pellet cells and wash at least three times with HNMK.
5. Deciliate cells and purify cilia and membrane fractions as described in Chapters 3 and 55.

Separate proteins by SDS–PAGE. If desired, proteins can initially be separated by isoelectric focusing and then separated by SDS–PAGE. Then blot polypeptides to nitrocellulose following standard procedures (see Dentler, 1992). Stain the biotinylated proteins on the blot using the following procedure.

B. Alkaline Phosphatase–Streptavidin Staining of Biotinylated Protein

1. Solutions

TBS: 150 mM NaCl, 10 mM Tris, pH 7.5

TBST: TBS + 0.1% Tween 20

TBSAT: TBST + 0.1% bovine serum albumin (BSA)

Alkaline phosphatase-labeled streptavidin (Zymed Laboratories, San Francisco, CA)

Alkaline phosphatase developer: For 10 ml, mix 66 ml NBT + 33 ml BCIP (see below) + 9.9 ml alkaline phosphatase buffer

Alkaline phosphatase buffer: 100 m*M* Tris, pH 9.5, 100 m*M* NaCl, 5 m*M* $MgCl_2$

NBT (nitroblue tetrazolium, Sigma N6876): 50 mg/ml in 70% dimethyl formamide

BCIP (5-bromo-4-chloro-3 indoyl-phosphate, Sigma B8503): 50 mg/ml in 100% dimethyl formamide

Stop solution: 3% trichloracetic acid (or rinse well with water)

2. Procedure

1. Wash blot twice with TBSAT, 10 minutes per wash.
2. Incubate in alkaline phosphatase-labeled streptavidin in TBSAT for 1–2 hours at room temperature.
3. Wash three times with TBSAT, 10 minutes per wash.
4. Wash three times with TBST, 15 minutes per wash.
5. Develop with alkaline phosphatase developer.
6. Pour off developer and stop reaction with 3% trichloracetic acid (optional—can just wash with water).
7. Photograph on Kodak Technical Pan film with a yellow filter. Develop film in Kodak HC-110 developer, dilution B.

C. Labeling Blotted Proteins with Concanavalin A

The key to this procedure is the fact that HRP is a Con A-binding protein (Hawkes, 1982). As a control for nonspecific HRP binding, mix 0.1 *M* α-D-methylmanoside with Con A to competitively inhibit Con A binding.

1. Solutions

TPBS: PBS + 0.1% BSA (optional, many do not use BSA), 0.1% Tween 20, 1 m*M* $CaCl_2$, 0.1 m*M* $MnCl_2$, 1 m*M* $MgCl_2$

50 m*M* Tris–Cl, pH 6.9, with HCl

Developer: 20 ml of 0.3% 4-chloro-1-naphthol in 100% methanol

H_2O_2

HRP

2. Procedure

1. Blot proteins onto nitrocellulose.
2. Wash three times in TPBS.
3. Incubate with 50 μg/ml Con A in TPBS for 60 minutes.

4. Wash three times in TPBS.
5. Incubate with 50 μg/ml HRP in TPBS for 60 minutes.
6. Wash twice with TPBS.
7. Wash three times in 50 mM Tris–HCl, pH 6.9, ~3 minutes per wash.
8. Develop with 4-chloro-1-naphthol and H_2O_2, for 5–10 minutes.
9. Stop development by rinsing with water. Photograph immediately or store in a dark place.

References

Bloodgood, R. A., and Workman, L. J. (1984). A flagellar surface glycoprotein mediating cell-substrate interaction in *Chlamydomonas*. *Cell Motil.* **4,** 77–87.

Dentler, W. L. (1992). Identification of *Tetrahymena* ciliary surface proteins labeled with sulfosuccinimidyl 6-(biotinamido)hexanoate and Concanavalin A and fractionated with Triton X-114. *J. Protozool.* **39,** 368–378.

Goodloe-Holland, C. M., and Luna, E. J. (1987). Purification and characterization of Dictyocstelium discoideum plasma membranes. *Methods Cell Biol.* **28,** 103–128.

Hawkes, R. (1982). Identification of Concanavalin A-binding proteins after sodium dodecyl sulfate-gel electrophoresis and protein blotting. *Anal. Biochem.* **123,** 193–196.

Pagliaro, L., and Wolfe, J. (1987). Concanavalin A binding induces association of possible mating-type receptors with the cytoskeleton of *Tetrahymena*. *Exp. Cell Res.* **168,** 138–152.

Reinhart, F. D., and Bloodgood, R. A. (1988). Membrane-cytoskeleton interactions in the flagellum: a 240,000 Mr surface-exposed glycoprotein is tightly associated with the axoneme in *Chlamydomonas moewusii*. *J. Cell Sci.* **89,** 521–531.

Thompson, J. A., Law, A. L., and Cunningham, D. D. (1987). Selective radiolabeling of cell surface proteins to a high specific activity. *Biochemistry* **26,** 743–750.

Williams, N. E., Subbaiah, P. V., and Thompson, G. A., Jr. (1980). Studies of membrane formation in *Tetrahymena*. The identification of membrane proteins and turnover rates in nongrowing cells. *J. Biol. Chem.* **255,** 296–303.

CHAPTER 58

Fluorescence Labeling of Flagellar Membranes

Robin Wright
Department of Zoology
University of Washington
Seattle, Washington 98195

I. Introduction

Lipophilic fluorescent dyes are useful tools for examining general cell structure, as well as for investigating the structure and function of specific organelles (Haugland, 1989; Wang and Taylor, 1989; Taylor and Wang, 1989; for algal staining see: Klut *et al.*, 1988, 1989). One of these dyes, 3,3′-dihexyloxacarbocyanide iodide [$DiOC_6(3)$, Fig. 1], has been used to investigate the properties of the endoplasmic reticulum in both living and fixed cells (Terasaki *et al.*, 1984; Terasaki, 1990, 1993; Terasaki and Reese, 1992). The abbreviated name reflects the structure of the molecule: $DiOC_6(3)$ is a carbocyanine dye, consisting of two (''Di'') 6-carbon hydrocarbon chains (''C_6'') that each extend from an oxygen-containing (''O'') double-ring structure held together by a three-carbon chain (''3'').

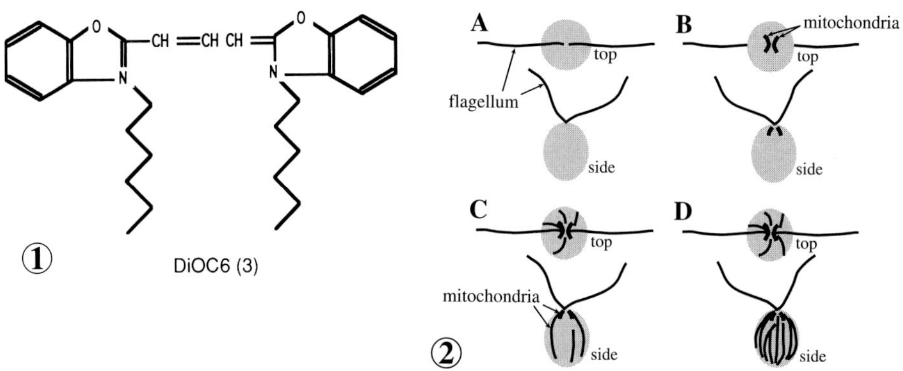

Fig. 1 Structure of DiOC$_6$(3)

Fig. 2 Patterns of DiOC$_6$(3) staining in living *Chlamydomonas* cells. The upper diagram in each panel shows the pattern of staining in a cell whose flagella have attached to the coverslip, with the cell body hanging below, out of the plane of focus. The lower diagram in each panel shows a cell whose flagella and cell body are both in the plane of focus. The position of the cell body is outlined for reference. (A) Incubation with 0.2 μg/ml of DiOC$_6$(3) stains flagella only. (B) Incubation with 0.3 μg/ml of DiOC$_6$(3) stains both flagella and the pair of mitochondria that lie near the basal body apparatus. (C) Incubation with 0.4 μg/ml of DiOC$_6$(3) stains both flagella, the mitochondria near the basal body apparatus, and begins to stain mitochondria in the cell body. (D) Incubation with 1 μg/ml of DiOC$_6$(3) stains the flagella and the mitochondria.

DiOC$_6$(3) diffuses through the plasma membrane and enters all cell membranes, but in living cells the dye becomes concentrated within specific organelles. The concentration of DiOC$_6$(3) determines which organelles appear stained. This selectivity of DiOC$_6$(3) staining probably reflects accumulation of the positively charged dye molecules in subcellular membranes that have negative membrane potential. For example, in mammalian cells (Terasaki, 1989) and in yeast (Koning *et al.*, 1993), low concentrations of DiOC$_6$(3) specifically stain mitochondria. As DiOC$_6$(3) concentrations are raised, the dye continues to accumulate in the mitochondria until the positively charged DiOC$_6$(3) molecules reduce or neutralize the potential difference across the inner mitochondrial membrane. The dye then moves into the cellular membranes with the next highest membrane potential, which in both mammalian cells (Terasaki, 1989) and in yeast (Koning *et al.*, 1993) is apparently the nuclear envelope/endoplasmic reticulum. The sensitivity of DiOC$_6$(3) accumulation to membrane potential has been used to measure plasma membrane potential in a variety of mammalian cells (e.g., see Hasmann *et al.*, 1989; Nishiyama *et al.*, 1992). The general characteristics of DiOC$_6$(3) and its use in labeling the endoplasmic reticulum are described in an excellent review by Terasaki (1989).

I recently began a series of experiments aimed at determining whether the flagellar membrane arises as a simple extension of the plasma membrane during flagellar elongation. I discovered that, when *Chlamydomonas reinhardtii* cells are exposed to very low concentrations of DiOC$_6$(3), flagella rather than mitochondria are specifically stained. Staining of mitochondria is observed only when the DiOC$_6$(3) concentration is increased, and even then, the flagella continue to stain. At the highest $DiOC_6(3)$ concentrations tested, other cellular membranes, including the chloroplast membrane, also became labeled.

If the staining characteristics of $DiOC_6(3)$ reflect its membrane potential-sensitive partitioning into membranes, flagellar ion pumps must maintain a higher negative membrane potential across the flagellar membranes than that of the rest of the plasma membrane, or even that of mitochondria. It is conceivable that this membrane potential may have functional roles in controlling flagellar length, autotomy, or aspects of flagellar beating. Thus, use of $DiOC_6(3)$ raises interesting questions and may also provide the means to answer some of them.

II. Cell Culture and Staining

A. Medium and Culture Conditions

Chlamydomonas reinhardtii strains NO^+ and CW-92, obtained from the *Chlamydomonas* Genetics Center (Duke University, Durham, NC), were grown in TAP or Sager and Granick medium with or without addition of 0.1% yeast extract (Harris, 1989). For solid medium, extensively washed Difco Bacto-agar was added to 1.5%. For cultures in liquid medium, a long-necked Erlenmeyer flask was filled one-tenth full with medium and inoculated with a single colony that had been pregrown for 5–7 days in 2 ml of TAP medium contained within a test tube. Liquid cultures were not stirred or shaken, but the large surface-to-volume ratio of the liquid within the flask allowed vigorous growth even without additional aeration. Cultures were grown under warm fluorescent lights on a 14-hour light: 10-hour dark cycle. Most experiments used cells that had reached densities of approximately 1 to 5 $\times$ 10^4 cells/ml.

B. Preparation and Storage of $DiOC_6(3)$

$DiOC_6(3)$ was purchased from either Kodak Laboratory and Research Products or Molecular Probes. We noticed that best staining was obtained with dye lots that produced a pronounced orange rather than yellow color when dissolved in ethanol (Koning *et al.*, 1993). For use, a 10 mg/ml stock solution was prepared in 100% ethanol. This stock solution was kept in the dark at room temperature, conveniently by placing it in a screw-capped microcentrifuge tube in a sealed, black plastic film canister. The dye was used for approximately a month, at which time staining quality began to deteriorate.

C. Staining of *Chlamydomonas* with $DiOC_6(3)$

The 10 mg/ml $DiOC_6(3)$ ethanolic stock solution was diluted 1 : 1000 into TAP medium, producing a final concentration of 10 μg/ml $DiOC_6(3)$. To determine ideal staining concentrations, the 10 μg/ml $DiOC_6(3)$ solution was added in increasing amounts to 50 μl of cell culture. Under our growth conditions, specific staining of flagella was usually best achieved by addition of 10 μl of the diluted $DiOC_6(3)$ to 50 μl of cell culture at a density of 10^4 cells/ml (see Fig. 2). A critical issue is that the cells actually titrate the dye. Consequently, to achieve the desired staining, one cannot merely suspend any number of cells in a given concentration of dye. Instead, specific and reproducible staining can be obtained only by adding a specific amount of dye to a specific number of cells. Most easily, the amount of dye needed is empirically determined, by adding small aliquots of diluted dye to an aliquot of cells and observing the staining characteristics until sufficient dye is added so that the flagella (or other desired structures) are well stained. This approach also allows use of the smallest possible amount of dye, an important consideration for mimizing phototoxicity.

High concentrations of the dye (approximately 50 μg/ml/10^4 cell) are toxic to cells, even before exposure to light of the excitation wavelength (484 nm). However, at the concentrations needed to stain flagella (approximately 0.2 μg/ml/10^4 cells), there is no noticeable toxicity: the cells continue to divide and maintain apparently normal motility, including phototaxis.

D. Microscopy

Immediately after addition of the dye, cells are observed with epifluorescence or confocal microscopy. $DiOC_6(3)$ has an absorbance maximum at 484 nm and emission maximum at 501 nm (Haugland, 1989), so fluorescence is readily observed using standard filters for fluorescein. Our filter set has an excitation maximum at 485 nm and absorbance maximum at 530 nm. For short-term observations, 10 μl of the stained cell culture is placed on an unwashed slide and covered with an unwashed coverslip. Many cells attach to the coverslip via their flagella under these conditions, allowing easy viewing of the flagella as the cells glide along the underside of the coverslip. To immobilize the cells or for long-term viewing, stained cells are mounted in 0.5% low-melting-point agarose dissolved in medium. To prepare these slides, 20 μl of stained cell suspension is placed on a microscope slide, rapidly mixed with 20 μl of melted 1% agarose that has been cooled to 55°C, and immediately sealed with a coverslip. Observations were made using a 60× objective lens, 1.4 numerical aperature.

Cells with $DiOC_6(3)$ staining limited to their flagella appear very robust during observations and can be observed until the dye slowly bleaches; however, cells that have accumulated sufficient dye to stain their internal membranes are very sensitive to exposure to excitation light and often drop their flagella and begin to show signs of distress during observation. This phototoxicity can be lessened by attenuating the excitation light with neutral density filters. The possibility that phototoxicity is occurring as the cells are being observed should, however, be kept in mind when designing experiments and interpreting results.

Images obtained on the confocal microscope are reasonably clear, lacking any noticeable chlorophyll autofluorescence. In contrast, although the $DiOC_6(3)$ staining patterns are readily visible by eye in a standard fluorescence microscope, chlorophyll autofluorescence makes photography impossible with standard fluorescein filter sets. One's eyes see both the green $DiOC_6(3)$ and the red chloroplast, at approximately equal intensities, so that the flagella appear a vivid green against the red background of the cell body. Apparently, the film is more sensitive to red or is capturing light that is not visible by eye. Consequently, photographs of cells visualized with standard fluorescein filter sets show only the red chlorophyll autofluorescence. Use of an emission filter that blocks light from 555 to 1.2 nm should eliminate chlorophyll autofluorescence, an absolute necessity for obtaining photographs on a standard light microscope.

E. Staining Patterns

Figure 2 summarizes the staining patterns of $DiOC_6(3)$ in *Chlamydomonas* treated with concentrations ranging from 0.1 to 1 μg/ml/10^4 cells. Addition of 0.1–0.2 μg/ml $DiOC_6(3)$ to cultures of approximately 10^4 cells/ml produces specific staining of the flagella. The entire length of the flagellum from the distal tip to the base appears uniformly fluorescent. In addition, rather than gradually attenuating, the staining appears to stop abruptly at the junction where the flagellar shaft joins the cell body. The abrupt loss of fluorescence is particularly apparent when observing cells that are attached to the coverslip via their flagella. In these cases, there is a pronounced absence of fluorescence between the two flagellar bases. When the dye concentration is increased to 0.3 μg/ml, in addition to flagellar staining, staining of a subset of mitochondria is observed. Specifically, the pair of mitochondria that extend along the centrin-containing fibers that link the basal bodies to the nucleus are stained, producing a cruciform structure at the flagellar bases. When the concentration of $DiOC_6$ is increased to 1 μg/ml, the entire network of mitochondria, in addition to the flagella, is observed. Interestingly, the mitochondrial patterns are strikingly similar to that of cytoplasmic microtubules visualized by immunofluorescence.

To eliminate the possibility that flagellar staining merely reflects limited accessibility of the dye to the plasma membrane because of the cell wall, I examined the staining properties of CW-92, a strain that lacks a cell wall. The flagellar staining patterns seen in CW-92 cells were identical to that of wild-type cells. Thus, the specific staining of flagella cannot be explained by limitations of dye access to the cell surface. The pattern of mitochondria appeared slightly different in the two strains. In the CW-92 strain, the mitochondrial profiles were straighter than in the wild-type strain.

Stained flagella that had detached from the cell body continued to fluoresce brightly. Thus, complete loss of $DiOC_6(3)$ staining was neither a prerequisite to nor a consequence of flagellar autotomy; however, transient changes in membrane potential would not have been observed under these conditions. In contrast to the continued staining of flagella that had been mechanically detached from the cell, little $DiOC_6(3)$ staining was observed in flagella that had been released as a consequence of lowered pH. In addition, treatments that killed the cells, such as fixation with formaldehyde or glutaraldehyde, also markedly decreased flagellar fluorescence. These observations are consistent with the possibility that $DiOC_6(3)$ staining may be revealing physiologically relevant information concerning flagellar membrane potential.

III. Prospects

Although we successfully used $DiOC_6(3)$ to stain *Chlamydomonas* flagella and mitochondria, we have thus far failed to label *C. reinhardtii* cells with dyes that have longer hydrocarbon chains, such as $DiOC_{18}(3)$ (P. Y. Lum and R. Wright, unpublished attempts). The advantage of long-chain dyes is that, once incorporated into a membrane, the dye is retained within that membrane. Our initial failures undoubtedly represent technical difficulties that will be surmounted with due effort. Thus, the use of $DiOC_6(3)$ represents only the first foray into an apparently untapped approach to understanding flagellar structure and function. A casual examination of the Molecular Probes catalog should convince even the most skeptical of the great variety of fluorescent probes currently available. It is conceivable that at least some of these fluorescent probes can provide novel insights into a variety of questions concerning flagellar structure and function.

Acknowledgments

I appreciate the assistance of Pek Yee Leem and Kathy Geil in certain of the studies described here.

References

Harris, E. (1989). ''The *Chlamydomonas* Handbook''. Academic Press, San Diego.
Hasmann, M., Valet, G. K., Tapiero, H., Trevorrow, K., and Lampidis, T. (1989). *Biochem. Pharmacol.* **38,** 305–312.
Haugland, R. P. (1989). ''Handbook of Fluorescent Probes and Research Chemicals'' (Molecular Probes Catalog). Molecular Probes, Eugene, OR.
Klut, M. E., Bisalputra, T., and Antia, N. J. (1988). *Histochem. J.* **20,** 35–40.
Klut, M. E., Stockner, J., and Bisalputra, T. (1989). *Histochem. J.* **21,** 645–650.
Koning, A. J., Lum, P. Y., Williams, J. M., and Wright, R. (1993). *Cell Motil. Cytoskel.* **25,** 111–128.
Nishiyama, M., Aogi, K., Saeki, S., Kim, R., Kuroi, K., Yamaguchi, K. Y., and Toge, T. (1992). *Anticancer Res.* **12,** 849–852.
Taylor, D. L., and Wang, Y. (eds.) (1989). ''Fluorescence Microscopy of Living Cells in Culture, Part B: Quantitative Fluorescence Microscopy—Imaging and Spectroscopy''. Academic Press, San Diego.
Terasaki, M. (1989). *Methods Cell Biol.* **29,** 125–135.
Terasaki, M. (1990). *Cell Motil. Cytoskel.* **15,** 71–75.
Terasaki, M. (1993). *Methods Cell Biol.* **38,** 211–220.
Terasaki, M., and Reese, T. S. (1992). *J. Cell Sci.* **101,** 315–322.
Terasaki, M., Song, J., Wong, J. R., Weiss, M. J., and Chen, L. B. (1984). *Cell* **38,** 101–108.
Wang, Y., and Taylor, D. L. (ed) (1989). ''Fluorescence Microscopy of Living Cells in Culture, Part A: Fluorescent Analogs, Labeling Cells, and Basic Microscopy''. Academic Press, San Diego.

CHAPTER 59

Electrophysiology of Ciliates

Hans Machemer
Arbeitsgruppe Zelluläre Erregungsphysiologie
Fakultät für Biologie
Ruhr-Universität
D-44780 Bochum, Germany

I. Introduction

The rapid responses of protozoan and metazoan cells to a variety of stimuli require a rapid communication system. The system must not only distribute information but also must integrate or filter the information and respond in the presence of a persistent flow of extrinsic and intrinsic signals. One of the most common procedures used to study stimulus–response activities is to measure changes in the separation of electric charges or, for many behavioral responses, Ca^{2+} ions, across each cell and organelle membrane.

Ciliated protozoa provide excellent systems to study the roles of electrical currents in behavioral responses because their swimming behavior—whether they swim forward, backward, or are arrested in one place—depends directly on the concentration of Ca^{2+} ions within the axoneme. A typical ciliate can change its motility in response to a stimulus within 10^{-3} to 10^{-2} second; this involves changes in cellular currents of the order of 10^{-8} A and potential changes of $\sim 10^{-2}$ V. Within a single cilium, a few voltage-sensitive channels of the range of 10^{-11} S conductance each, together with Ca^{2+}-pumping molecules and/or membranous or intraciliary Ca^{2+} binding sites, shuffle Ca^{2+} ions between the axoneme and the extracellular environment with high precision. The primary role of the voltage-sensitive Ca^{2+} channels is to maintain a level of 10^1 to 10^3 free Ca^{2+} ions per cilium.

An advantage of electrophysiological studies of whole cells or single channels is that an experimenter can witness the rapid events of a living system. A disadvantage is that electrophysiology is difficult to explain or to teach in a cookbook style. An electrophysiologist must have an understanding of the theoretical aspects of DC and AC circuitry and of electrophysiological equipment coupled with practical experience. At an early stage of planning, it is useful to seek advice from a practitioner. Later, it may be possible to enter into a collaboration; this generally will involve substantial mutual commitment.

Prior to designing experiments, one should consult the "classic" chapter on intracellular recording from *Paramecium* by Naitoh and Eckert (1972) and a chapter on corresponding methods and results in *Stylonychia* (Machemer and Deitmer, 1987). Additional resources include Geddes (1972) for electrodes, Neher (1974) for basics in electrophysiology, Machemer (1987) for electrophysiology on dummy cells, Kung (1979) for the use of mutants in problem solving, Hamill *et al.* (1981) for general patch-clamp procedures, Martinac *et al.* (1988) for patch-clamp in *Paramecium,* Machemer (1988) for extracellular voltage-clamp of *Paramecium* in DC fields and behavioral responses (galvanotaxis), and Machemer (1989) for solution-mediated electric stimulation and adaptation in ciliates.

II. Methods

Electrophysiological techniques may be employed at different levels. Their adequacy depends on the research questions. As outlined above, practical experience and training in electrophysiology are not easily explained in a short article. Rather than providing a detailed list of experimental procedures, I have listed several research problems appropriate for electrophysiological studies and several experimental procedures to consider prior to undertaking electrophysiological studies of ciliary function in protozoans.

A. You want to determine if substance X or treatment Y has an effect on the membrane potential of a ciliate.

1. Use cells equilibrated for at least 3 hours in defined solution and keep other stimuli low and constant.

2. Minimize evaporation of solution by using an appropriate volume and surface of bath and/or using a moist chamber.

3. Avoid mechanical and other disturbances while you introduce X or Y to the cells.

4. Use film or video recording to document the rate of locomotion before and for about 30 minutes after introduction of X or Y.

5. Evaluate the velocity and direction of swimming and the frequency of reversals. If the majority of cells increase their rate of forward swimming, or become immobile, or start swimming backward, one might conclude that the

cells were hyperpolarized, weakly depolarized, or strongly depolarized, respectively. Why this happened requires careful analysis. Not all ciliates can modulate the rate of forward swimming.

6. Determine if the factor elicits nonsaturated responses within the physiological range.

7. Ignoring the time dependence of the phenomena may lead to grave misjudgments (see D below).

8. Differentiate between effects on surface potential, driving force on ions, and interference with membrane channel conductance.

B. You wish to induce a depolarization (hyperpolarization) for diagnostic reasons.

1. $[K^+]_o$-dependent polarization of the membrane is a common procedure: raising $[K^+]_o$ depolarizes the membrane, whereas lowering $[K^+]_o$ hyperpolarizes the membrane potential. It is important that cells are equilibrated in the starting solution of known composition for several hours.

2. Immediately after solution exchange, the membrane potential begins to readapt toward the previous value. Adaptation time in *Paramecium* can be determined crudely by measuring the duration of reversed swimming following depolarization (or duration of enhanced forward swimming rate following hyperpolarization).

3. A preparation with a DC field of ≤ 2 V/cm and an inert high-viscosity solution allows the observation of specimens that are depolarized in the cathodal part and hyperpolarized in the anodal part of the cell body. Only constant-current sources can establish reliable voltage gradients because they overcome irregularities of electrode resistance. Find the lowest possible effective voltage gradient.

4. Isolate electrolysis products from cells during DC application by embedding electrodes in agar of appropriate ionic composition.

5. Keep in mind adaptation to persistent voltage shifts.

C. You wish to apply Ca^{2+} to see Ca^{2+}-dependent responses.

1. External application of Ca^{2+} is usually inappropriate for modulation of $[Ca^{2+}]_i$ due to a number of variables, including external surface potential and competition of external ions near channels. Ca^{2+} equilibrium potential and Ca^{2+} conductance may change at the same time, generating the "Ca paradox" (see Machemer, 1989).

2. Ca^{2+} ionophores provide crude tools to supply excess Ca^{2+} with unphysiological consequences. Injection of a calibrated ethylene glycol bis (B-aminoethyl ether)-*N, N'-tetraacetic acid (EGTA)–Ca^{2+}* buffer by micropipet or similar techniques are comparatively safe (Tsien and Rink, 1980).

3. Intracellular Ca^{2+} buffering, pumps, and sequestering systems tend to quickly renormalize an artificially modified $[Ca^{2+}]_i$.

4. Very high $[Ca^{2+}]_i$ and very low $[Ca^{2+}]_o$ can raise a nonspecific conductance of the membrane. With weak membranes, the use of intracellular electrodes is prohibited, but noninvasive recording of behavior may still be possible.

D. You observe slow changes in behavior, while the stimulating substance X or treatment Y persists.

Behavioral adaptation is a universal phenomenon of great importance for response to persistent stimuli, orientation along stimulus gradients, and survival in dramatically changed environments. At the membrane level, the descriptive term is *accommodation*. Some electrophysiological mechanisms have been described in the literature (see Machemer, 1989). Intracellular analysis (e.g., input resistance, time-dependent shift in steady-state potential, early and late current/voltage relationships) may be useful.

E. You have identified a behavioral mutant and wish to know the nature of the defect.

1. Conscientious intracellular electrophysiological analysis may identify a defect in the functioning of one or several types of membrane channels. Consult the literature (e.g., Kung, 1979).

2. The mutation may concern a channel-regulating enzyme (not the channel structure), or the axonemal machinery may be defective. In this case reactivation of ciliary motility or axonemal sliding or analysis of axonemal structure by transmission electron microscopy may be appropriate.

F. You have characterized a particular behavior of cells or cellular components in the course of the ciliate life cycle. You suspect a crucial role of ion channels and wish to identify it.

1. Motor behavior is under electrophysiological control, but complex behavioral sequences (e.g., conjugation) may follow from controlled electrophysiological parameters such as leakage conductance or changes in the concentration of internal Ca^{2+} or other cations. A rule of thumb is that slow events (that occur in the time range of seconds and minutes) are unlikely to be controlled exclusively in the electric domain. An anteroposterior intracellular organization might be indirectly guided by polar distribution of channel conductance(s) establishing subcortical gradients in ions and transcellular current.

2. Current per se is physiologically effective only as far as it can establish a voltage drop across a resistive element.

G. You wish to know the point of attack of substance X or treatment Y in the sensorimotor signaling pathway.

1. Are strongly charged agents that might alter surface potentials involved? If the answer is negative, consider intracellular recording.

2. Check passive and active electric properties of the cell in the presence and absence of the test agent or treatment.

3. If possible, test orthodox receptor pathway and Ca^{2+} channel excitability.

4. Use voltage-clamp and video recording to screen the electromotor coupling sequence.

5. Do simple tests first. Seek advice from experienced laboratories.

H. You have documented a stimulus-oriented behavior (taxis) and wish to analyze the regulation.

1. Determine the stimulus specificity and sensitivity of the response.

2. Can you experimentally manipulate the stimulus (e.g., intensity, direction, time characteristics) and the cellular responses? Electrophysiological approaches for identification and distribution of receptor channels and a search for anatomical specialties may be considered.

3. A cellular taxis is not necessarily regulated along a physiological pathway.

4. Cell size may be limiting for experimental approaches; even large cells may resist attempts to record electrophysiologically.

I. You wish to study sensory transduction (following photic, chemical, or mechanical stimuli).

Quantitative stimulation and behavioral screening can help to guide the experimenter to ask the proper questions for a committed electrophysiological analysis. Patch-clamp techniques in ciliates are, to date, not sufficiently advanced for routine investigations of receptor channels *in vivo*. Ciliary membranes of protists have not been patched at all. Seek advice from and training in an experienced laboratory.

Acknowledgments

This work was supported by the Deutsche Forschungsgemeinschaft, the Minister für Wissenschaft und Forschung of the State of Northrhine-Westphalia, and the Deutsche Agentur für Raumfahrtangelegenheiten, Grant 50QV8857-5.

References

Geddes, L. A. (1972). "Electrodes and the measurement of bioelectric events" Wiley-Interscience, New York.

Hamill, O. P., Marty, E., Neher, E., Sakmann, B., and Sigworth, F. J. (1981). Improved patch-clamp techniques for high-resolution current recording from cells and cell-free membrane patches. *Pflügers Arch.* **391,** 85–100.

Kung, C. (1979). Biology and genetics of *Paramecium* behavior. *In* "Topics in Neurogenetics" (C Brakefield, ed.), pp. 1–26. Elsevier, New York.

Machemer, H. (1987). "Übungen zur Elektrophysiologie tierischer Zellen und Gewebe", pp. 1–93, I–XI. Edition Medizin, VCH Weinheim.

Machemer, H. (1988). Galvanotaxis: Grundlagen der elektromechanischen Kopplung und Orientierung bei *Paramecium. In* "Praktische Verhaltensbiologie" (G. H. K. Zupanc, ed.), pp. 60–82. Paul Parey, Berlin.

Machemer, H. (1989). Cellular behavior modulated by ions: electrophysiological implications. *J. Protozool.*, **36**, 463–487.
Machemer, H., and Deitmer, J. W. (1987). From structure to behavior: *Stylonychia* as a model system for cellular physiology. *In* "Progress in Protistology" (J. O. Corliss and D. J. Patterson, eds.), Vol 2, pp. 213–330. Biopress Ltd., Bristol.
Martinac, B., Saimi, Y., Gustin, M. C., and Kung, C. (1988). Ion channels of three microbes: *Paramecium,* Yeast and *Escherichia coli. In* "Calcium and Ion Channel Modulation" (A. D. Grinnell, D. Armstrong, and M. B. Jackson, eds.), pp. 415–430. Plenum, New York and London.
Naitoh, Y., and Eckert, R. (1972). Electrophysiology of the ciliate protozoa. *In* "Experiments in Physiology and Biochemistry" (G. A. Kerkut, ed.), Vol V, pp. 17–38. Academic Press, New York and London.
Neher, E. (1974). "Elektronische Meβtechnik in der Physiologie" Springer, Berlin and Heidelberg.
Tsien, R. Y., and Rink, T. J. (1980). Neutral carrier ion-selective microelectrodes for measurement of intracellular free calcium. *Biochem. Biophys. Acta* **599**, 623–638.

CHAPTER 60

Isolation of *Chlamydomonas* Mastigonemes

Mitchell Bernstein
Department of Biology
Yale University
New Haven, Connecticut 06520

I. Introduction

Mastigonemes are hairlike structures attached to the flagellar membrane of free-living protista. The morphologies of mastigonemes range from elaborate branched structures, composed of up to a dozen different proteins, to simple unbranched structures composed primarily of a single type of polypeptide. [See Moestrup (1982) for a discussion of the different types of mastigonemes and Andersen *et al.* (1991) for a discussion of mastigoneme nomenclature.] An organism may contain multiple types of mastigonemes on one flagellum; for example, the *Euglena* flagellum contains both branched tubular mastigonemes and simple unbranched mastigonemes (Bouck *et al.,* 1978).

Chlamydomonas mastigonemes are simple structures that are uniformly 0.9 μm long and 16 nm wide (Witman *et al.,* 1972). Electron microscopy of isolated and negatively stained mastigonemes reveals a repeating pattern with a periodicity of 20 nm along the length of the mastigoneme. The major constituent of *Chlamydomonas* mastigonemes is a glycoprotein of approximately 200 kDa (Bernstein and Rosenbaum, 1993; Monk *et al.*, 1983; Witman *et al.*, 1972), and it is possible that the repeat unit of the *Chlamydomonas* mastigoneme stems from the ability of this protein to form filaments. The function of *Chlamydomonas* mastigonemes is unknown. In other protista, opposing rows of rigid mastigonemes may effectively reverse the thrust formed by the sinusoidal flagellar waveform of flagellar beating (Sleigh, 1991), allowing flagella that beat with the flagellar waveform to pull a cell through the medium, as observed for *Ochromonas*. *Chlamydomonas* mastigonemes, however, are not held at a defined angle relative to the flagellar surface and so are presumed to have little hydrodynamic effect. Several investigators have also examined, and ruled out, a possible role of mastigonemes in flagellar adhesion during *Chlamydomonas* mating (Bergman *et al.*, 1975; McLean *et al.*, 1974; Snell, 1976).

II. Starting Material

A. Cells

Mastigonemes have been described on the flagella of *Chlamydomonas reinhardtii* and *Chlamydomonas moewusii*. I have isolated mastigonemes from *C. reinhardtii* wild-type strains 21*gr* and 125^+ (*nit1*$^-$*nit2*$^-$) with equal success. Cultures are grown to a density of $2–6 \times 10^6$ cells/ml in acetate-supplemented medium (R/2 medium, Kindle *et al.*, 1989). To ensure that all cells in a culture contain flagella, cells are synchronized by growth on a cycle of 14 hours light : 10 hours dark and flagella are isolated during the light cycle. Cells can be concentrated by tangential-flow filtration with the Millipore Pellicon filtration system (see Witman, 1986) with no loss or shearing of mastigonemes.

B. Source of Mastigonemes

1. Flagella

Mastigonemes are present only on the flagella and are not found on the plasma membrane, making isolated flagella (see below) the starting material of choice for mastigoneme isolation. Flagella should be detached by pH shock (Witman *et al.*, 1972) as flagellar removal with dibucaine (Witman, 1986) removes large portions of the flagellar membrane and, presumably, extracts mastigonemes as well. A starting volume of 32 liters of cells (5×10^6 cells/ml) will yield 5–10 μg of mastigonemes, an amount sufficient for electron microscopy, but insufficient for certain types of biochemical analysis. For isolation of larger quantities of mastigonemes, isolated flagella may be resuspended in 10 m*M* Tris–HCl, pH 8.3 (TB), frozen in liquid nitrogen, and stored at −70°C. When sufficient numbers of flagella have been accumulated, samples can be thawed and pooled for isolation of mastigonemes. The freeze–thaw cycle may cause breakdown of up to 50% of the axonemes during extraction with nonionic detergents (see Sections III and IV, B, step 2), but this does not affect the purity or yield of the final mastigoneme fraction. In general, purity of the flagellar fraction is not critical and contamination of flagella by cell bodies will not affect the purity or yield of the final mastigoneme fraction.

2. Culture Medium

Mastigonemes can also be isolated from culture medium, where they are attached to flagellar membrane vesicles that are continually shed from the flagellar surface (Bergman *et al.*, 1975; Bloodgood, 1984; McLean *et al.*, 1974; Snell, 1976). The quantity of mastigonemes that can be isolated from the medium of a culture grown to 5×10^6 cells/ml is approximately equal to that which can be isolated from the flagella of the cells in the culture. Mastigoneme-containing vesicles in the culture medium are concentrated along with cells by the Pellicon filtration system (Witman, 1986). Even so, relatively large volumes of medium concentrate must be centrifuged at a high speed to obtain a crude mastigoneme-containing pellet, and I have found it easiest to use cells and flagella as starting material, rather than isolate mastigonemes from culture medium.

III. Extraction of Flagellar Membranes

Mastigonemes are attached to the flagellar membrane and are released by either nonionic or ionic detergents. Because the mastigonemes are very stable to ionic detergent, axonemes can be dissociated with 0.7% Sarkosyl prior to obtaining an insoluble crude mastigoneme pellet (Witman *et al.*, 1972). Dissolution of flagella and axonemes with 0.7% Sarkosyl yields the cleanest crude mastigoneme pellet; however, the final mastigoneme fraction is not purer than that obtained when flagellar membranes are solubilized with nonionic detergents. Extraction of the flagellar membrane with nonionic detergents, e.g., 1% Nonidet P-40, will yield a soluble membrane/matrix fraction (Witman *et al.*, 1972) and an insoluble axonemal fraction. Axonemes are removed by a low-speed centrifugation and mastigonemes remain quantitatively in the low-speed supernatant. A crude mastigoneme pellet is then obtained by high-speed centrifugation. The resulting supernatant (soluble membrane/matrix fraction) can be used as starting material for isolation of other proteins.

IV. Materials and Methods

A. Flagellar Isolation

1. Grow 32 liters of *C. reinhardtii* strain 21*gr* to 5×10^6 cells/ml in R/2 medium (Kindle *et al.*, 1989) on a cycle of 14 hours light : 10 hours dark.

2. Harvest cells during the light cycle by concentrating cells to a volume of 2 liters with a Millipore Pellicon filtration unit (Witman, 1986) followed by centrifugation in conical polypropylene bottles (Corning No. 25350-250) at 1100*g* for 6 minutes at room temperature (RT). Decant supernatant.

3. Resuspend cells in 600 ml of 10 m*M* Tris–HCl, pH 8.3 (TB) at RT. If many cells have lost flagella, the concentrated cells in TB can be stirred under fluorescent light for up to 90 minutes until flagella have regenerated.

4. Pellet cells as in step 2 and decant supernatant.

5. Resuspend cells in 250 ml ice-cold TB + 6% sucrose. All subsequent steps are at 4°C and all solutions are chilled on ice.

6. While stirring vigorously, deflagellate cells by dropwise addition of 0.5 *N* acetic acid to pH 4.3.

7. Wait 1 minute, then add 0.5 *N* KOH to pH 7.0. Examine cells by phase microscopy: the efficiency of flagellar removal should be greater than 95%.

8. Place 15 ml of the cell–flagellum mixture into 50-ml conical polycarbonate centrifuge tubes (Nalgene No. 3105) and underlay with 25 ml TB + 25% sucrose.

9. Centrifuge in a swinging bucket rotor at 2000*g* for 10 minutes at 4°C. Cell bodies will pellet and flagella will be visible at the 5%–25% sucrose interface and in the 5% sucrose layer.

10. Collect the 5% sucrose layer and the 5%–25% sucrose interface by aspiration or with a pipet.

11. Aliquot 35 ml of the pooled flagellum-containing fractions into 50-ml conical centrifuge tubes and underlay with 5 ml of TB + 25% sucrose.

12. Repeat steps 9 and 10.

13. Pellet flagella by centrifugation at 30,000*g* for 20 minutes at 4°C. The flagellar pellet varies in color from white for uncontaminated preparations to pale green for flagellar preparations that are contaminated by cell bodies.

14. Resuspend the flagellar pellet in 5 ml of cold TB. Resuspended flagella can be frozen in liquid nitrogen and stored at −70°C. The yield from a starting culture volume of 32 liters is 10–20 mg of flagellar protein.

B. Mastigoneme Isolation

1. To 5 ml of concentrated flagella (either freshly isolated or thawed), add

0.55 ml of 10% Nonidet P-40 (1% final concentration) to extract flagellar membranes.

2. Rotate the sample gently on a wheel, end-over-end (10 rpm), for 1 hour at 4°C.

3. Centrifuge 16,000*g* for 30 minutes at 4°C to remove axonemes.

4. Centrifuge supernatant in an SW55 Ti rotor 125,000*g* (36,000 rpm) for 1 hour at 4°C to obtain the crude mastigoneme pellet.

5. Resuspend the crude pellet in 4 ml of TB and then add 2.16 g of CsCl (2.8 *M* CsCl final; final density, 1.34 g/ml).

6. Centrifuge in ultraclear tubes (Beckman No. 344057) in an SW55 Ti rotor at 144,000*g* (39,000 rpm) for 24 hours at 4°C.

7. At equilibrium, mastigonemes will band in the middle of the gradient. Other protein contamination is present at the top of the gradient, well separated from the mastigonemes. The mastigoneme band is best visualized in a darkened room by placing the gradient in front of a black background while shining a light on it. Under these conditions the mastigonemes appear as a thin white line against the black backdrop. The mastigonemes are collected with a 21-gauge needle attached to a 1-ml syringe that is inserted through the wall of the centrifuge tube slightly below the level of the band.

8. Dialyze the collected fraction against TB for 24 hours at 4°C to remove sucrose. If necessary, the mastigonemes can be concentrated by centrifugation as described in step 4; however, this often leads to a loss of material. A starting culture volume of 32 liters yields approximately 5–10 μg of pure material, whereas a scaled-up procedure starting with the frozen flagella from 550 liters of cells yields approximately 250 μg of mastigoneme protein. Purified mastigonemes are stable in TB at 4°C for at least 2 months.

V. Analysis of Purified Mastigonemes

Analysis of purified mastigonemes by sodium dodecyl sulfate–polyacrylamide gel electrophoresis (Bernstein and Rosenbaum, 1993; Monk *et al.*, 1983; Witman *et al.*, 1972) showed that the major protein of mastigonemes is a 220-kDa glycoprotein. Removal of carbohydrate by acid treatment yields a 200-kDa polypeptide (Bernstein and Rosenbaum, 1993). Native mastigonemes can be examined in the electron microscope by negatively staining with 1–2% uranyl acetate (Bergman *et al.*, 1975; Snell, 1976) or 4% phosphotungstate acid (Witman *et al.*, 1972). These staining procedures also allow the visualization of mastigonemes *in situ,* though care must be taken not to overstain specimens to the point where stain accumulates around the flagella and obscures the mastigonemes.

Acknowledgments

This work was supported by National Institutes of Health Grant GM14642 and National Science Foundation Grant CB45147 to Joel L. Rosenbaum.

References

Andersen, R. A., Barr, D. J. S., Lynn, D. H., Melkonian, M., Moestrup, O., and Sleigh, M. A. (1991). Terminology and nomenclature of the cytoskeletal elements associated with the flagellar/ciliary apparatus in protists. *Protoplasma* **164,** 1–8.

Bergman, K., Goodenough, U. W., Goodenough, D. A., Jawitz, J., and Martin, H. (1975). Gametic differentiation in *Chlamydomonas reinhardtii.* II. Flagellar membranes and the agglutination reaction. *J. Cell Biol.* **667,** 606–622.

Bernstein, M., and Rosenbaum, J. L. (1993). Transport to the *Chlamydomonas* cell surface: mastigonemes as a marker for the flagellar membrane. *In* "Molecular Mechanisms of Membrane Traffic" (D. J. Morre, R. E. Howell, and J. J. M. Bergeron, eds.), Vol 74 NATO ASI Series H, pp. 179–180. Springer-Verlag, Heidelberg, Germany.

Bloodgood, R. A. (1984). Preferential turnover of membrane proteins in the intact *Chlamydomonas* flagellum. *Exp. Cell Res.* **150,** 488–493.

Bouck, G. B., Rosiere, T. K., and Valaitis, A. (1978). Surface organization and composition of Euglena. II. Flagellar mastigonemes. *J. Cell Biol.* **77,** 805–826.

Kindle, K. L., Schnell, R. A., Fernandez, E., and Lefebvre, P. A. (1989). Stable nuclear transformation of *Chlamydomonas* using a gene for nitrate reductase. *J. Cell Biol.* **109,** 2589–2601.

McLean, R. J., Laurendi, C. J., and Brown, R. M., Jr. (1974). The relationship of gamone to the mating reaction in *Chlamydomonas moewusii. Proc. Natl. Acad. Sci. U.S.A.* **71,** 2610–2613.

Moestrup, O. (1982). Flagellar structure in algae: a review, with new observations particularly on the Chrysophyceae, Phaeophyceae (Fucophyceae), Euglenophyceae, and *Reckertia. Phycologia* **21,** 427–528.

Monk, B. C., Adair, W. S., Cohen, R. A., and Goodenough, U. W. (1983). Topography of *Chlamydomonas:* fine structure and polypeptide components of the gametic flagellar membrane surface and the cell wall. *Planta* **158,** 517–533.

Sleigh, M. A. (1991). Mechanisms of flagellar propulsion a biologist's view of the relation between structure, motion, and fluid mechanics. *Protoplasma* **164,** 45–53.

Snell, W. J. (1976). Mating in *Chlamydomonas:* a system for the study of specific cell adhesion. I. Ultrastructural and electrophoretic analysis of flagellar surface components involved in adhesion. *J. Cell Biol.* **68,** 48–69.

Witman, G. B. (1986). Isolation of *Chlamydomonas* flagella and flagellar axonemes. *In* "Methods in Enzymology" (R. B. Vallee, ed.), Vol 134, pp. 280–290. Academic Press, San Diego.

Witman, G. B., Carlson, K., Berliner, J., and Rosenbaum, J. L. (1972). *Chlamydomonas* flagella I. Isolation and electrophoretic analysis of microtubules, matrix, membranes, and mastigonemes. *J. Cell Biol.* **54,** 507–539.

CHAPTER 61

Ciliary Membrane Tubulin: Isolation and Fractionation

Raymond E. Stephens
Department of Physiology
Boston University School of Medicine
Boston, Massachusetts 02118

I. Introduction

Ciliary membrane-associated tubulin was first described in molluscan gill cilia wherein this protein was by far the major protein constituent of the detergent-solubilized membrane/matrix fraction. This was in striking contrast to the composition of the sperm flagellar membrane where an equivalent amount of protein was found but, in this case, it was composed mainly of a distinctive high-molecular-weight glycoprotein that was only a minor constituent in cilia (Stephens, 1977). Later studies showed that membrane tubulin (but not brain tubulin) could be reincorporated into membranes of uniform buoyant density when the solubilizing detergent was removed (Stephens, 1983). Furthermore, it was demonstrated by a number of approaches that the tubulin and associated minor proteins formed a large micellar complex with lipids and detergent on partial delipidation (Stephens, 1985a,b). More recent evidence, based on gentle delipidation with low concentrations of detergent, suggests that membrane tubulin and associated proteins form a discrete "membrane skeleton" surrounding the axoneme (Stephens *et al.*, 1987). Tubulin has also been described as a prominent constituent of frog olfactory but not respiratory cilia (Chen and Lancet, 1984) and of bovine retinal rod outer segment plasma but not disk membranes (Matesic *et al.*, 1992). Such observations, coupled with the consistent appearance of tubulin in mechanically sensitive cilia (Stommel, 1984) and in various neuronal tissues, has led to suggestions that membrane tubulin may serve some structural or functional role in sensory membranes (Chen and Lancet, 1984; Stephens, 1986a, 1990).

More recent evidence supports a very different role for membrane tubulin. Tubulin is the major constituent of embryonic cilia in sea urchin embryos, where, as in scallop cilia, it is lipid associated and can be recycled *in vitro* into membranes (Stephens, 1991). Pulse–chase studies, however, have demonstrated that this tubulin most likely is the precursor to the tubulin of the assembling axoneme (Stephens, 1992). Interestingly, most other newly synthesized axonemal protein are not found associated with the tubulin-rich detergent-solubilized membrane/matrix fraction but, rather, are associated with the axoneme proper, suggesting two separate or separable pathways for tubulin and lipids, on one hand, and the various structural proteins of the axoneme, on the other (Stephens, 1994). Sea urchin embryonic cilia turn over quite rapidly (Stephens, 1994), as do retinal rod, neuronal, and presumably olfactory ciliary membranes. Earlier work demonstrated that terminally differentiated gill cilia also turn over rapidly (Stephens, 1988) but the implication of this observation was not obvious at the time. It may be that the presence of "membrane" tubulin is a measure of tubulin and lipid on the move, with the mechanism and motors of conveyance yet to be determined.

II. Preparation of Membrane Tubulin by Reconstitution

This procedure is based on Stephens (1983, 1991) and uses *in vitro* membrane reconstitution to separate membrane tubulin and its associated proteins from soluble proteins of the axonemal matrix. A variation on this procedure has also been used to reconstitute ciliary membranes not containing membrane tubulin, for the purpose of studying surface components (Hastie *et al.*, 1990).

A. Solutions

Nonidet–Tris–magnesium–EDTA (NP-40/TME): 0.25% Nonidet P-40 (NP-40) in 30 m*M* Tris–HCl, pH 8, 3 m*M* $MgCl_2$, 0.1 m*M* ethylenediaminetetraacetic acid (EDTA), 0.25 m*M* phenylmethylsulfonyl fluoride (PMSF). Add NP-40 and PMSF just before use.

Nonidet–low-salt solution (NP-40/LSS): 0.25% NP-40 in 0.1 *M* NaCl, 4 m*M*

$MgSO_4$, 0.1 m*M* EDTA, 1 m*M* dithiothreitol, 5 m*M* imidazole, pH 7.0, 0.25 m*M* PMSF. Add NP-40 and PMSF just before use.

NP-40 stock: 10% P-40 sealed under nitrogen (Surfact-Amps NP-40, Pierce No. 28324).

100 m*M* PMSF stock: 17 mg/ml in isopropanol. Use 2.5 μl/ml of solution.

Bio-Bead slurry: 33% (v/v) SM-2 styrene–divinylbenzene copolymer beads (Bio-Rad No. 1523920, 20–50 mesh) in water, prepared by washing with methanol according to Holloway (1973). Store in refrigerator.

Condensed Triton X-114: dense, detergent-rich phase (13–15% detergent) resulting from warming Triton X-114 above its cloud point, prepared by the method of Bordier (1981) or purchased as a purified 10% solution (Surfact-Amps X-114, Pierce No. 28332).

B. Procedure

1. Beginning with isolated cilia, extract the membrane with 10 vol of 0.25% NP-40 in either TME or LSS (see Chapter 7 of this volume) at least 10 minutes on ice.

2. In the meantime, take the equivalent of the above extraction volume of well-mixed Bio-Bead slurry and place it into a 15-ml conical plastic centrifuge tube, spin briefly in a clinical centrifuge to sediment the beads, and withdraw the supernatant with a Pasteur pipet, inverting the tube slightly so that any remaining interstitial fluid may be removed. This simple step gives the proper ratio of beads to extraction volume (cf. Bordier, 1981).

3. Spin the extracted cilia at 25,000*g* for 15 minutes. Carefully remove the supernatant, placing it into the tube containing the semidry Bio-Beads.

4. Cap the tube and place it on a sample tube rotator or roller, adjusting so that the Bio-Beads remain suspended, without foaming, and mix for 2 hours at 0–4°C. This batch method removes essentially all of the detergent without diluting the sample or adsorbing significant protein. A flow column (Holloway, 1973) or a spin column (Horigome and Sugano, 1983) of Bio-Beads also may be used. The former, taking about 15 minutes, results in some dilution; the latter, involving a 2-minute centrifugation, is limited to relatively small sample volumes.

5. Remove the Bio-Beads from the extract. Because of density differences and surface effects, the Bio-Beads are typically difficult to remove by centrifugation at this stage. A simple and quick way to achieve a clean separation is to filter the slurry through a small filter cone made of a very porous paper (e.g., Miracloth or a piece of coffee filter).

6. The opalescent filtrate contains small vesicles or leaflets of reconstituted membranes which can be sedimented with some difficulty at 100,000*g* for 60 minutes. To obtain larger vesicles that can be sedimented more conveniently

into a very dense pellet, the preparation should be subjected to a freeze–thaw step. Slow freezing at −20°c produces large, tubular vesicles which break up on resuspension; quick freezing in liquid nitrogen produces uniform but small vesicles and leaflets. In either case, sedimentation at 25,000*g* for 15 minutes produces a pellet containing virtually all of the tubulin and a supernatant containing soluble matrix components, e.g., calmodulin (cf. Fig. 1 in Stephens, 1983).

7. For further purification by recycling, the pellet of membranes can be resolubilized in 10 vol of 0.25% NP-40 in TME or LSS and the above steps repeated. At each cycle, there is a nonspecific loss of proteins due to denaturation (cf. Stephens, 1985a).

Note: An alternative detergent adsorbent, XAD-2 (Cheetham, 1979), causes considerably more protein adsorption and denaturation than SM-2 Bio-Beads and hence its use is not recommended.

III. Additional Purification

These methods are based on Stephens (1983) and Stephens (1985b), respectively.

A. Isopycnic Centrifugation

1. To obtain the reconstituted membranes as a monodisperse population of vesicles, resuspend the pellet in a minimal volume of buffer (TME or LSS) without detergent and layer atop a linear 20–60% (w/w) sucrose gradient made up with the respective buffer. A convenient way to prepare reproducible gradients is by layering equal 5% step increments at room temperature and allowing the steps to diffuse to uniformity overnight at 0–4°C.

2. Spin at 40,000 rpm in a Beckman SW40 rotor at 0–4°C for 3–6 hours. Higher speeds or longer times do not improve the results, except with very small vesicles or leaflets.

3. The vesicles should equilibrate about halfway down the gradient. Using intense side lighting, place a needle or a drawn-out Pasteur pipet just above the scattering layer of vesicles and "vacuum" the layer into the syringe or pipet, minimizing the uptake of surrounding sucrose. To recover the vesicles as a pellet, dilute the sucrose with buffer at least fivefold before centrifuging.

B. Triton X-114 Delipidation

1. Dissolve the reconstituted vesicles in 1% Triton X-114 in TME or LSS buffer by extracting on ice for 30 minutes with 5–10 pellet volumes. Spin at 25,000*g* for 15 minutes to pellet any denatured protein.

2. Warm the extract to 25–30°C for 5 minutes. Triton X-114 will come out of solution and condense into a dense layer containing much of the lipid and any small or uncomplexed hydrophobic proteins. Spin briefly in a clinical centrifuge at 25–30°C (or use a prewarmed tube adapter or sleeve) to compact the condensate. Withdraw the supernatant containing the partially delipidated membrane tubulin/associated protein–detergent complex.

3. Add sufficient Triton X-114 stock to make the supernatant 1% in detergent and repeat the condensation (step 2) to further remove any residual, loosely bound lipids.

4. The final product consists of a water-soluble but micellar complex of membrane tubulin and associated proteins, certain tightly bound lipids, and some residual Triton X-114. When most of the detergent is removed with Bio-Beads, a complex of constant protein composition, still containing some tightly bound detergent and lipid, may be subjected to a freeze–thaw cycle and further purified on a sucrose density gradient (Stephens, 1985b).

Note: Triton X-114 condensation work is very conveniently accomplished in clinical centrifuge tubes designed for urine sedimentation analysis (10-ml Kolmer tubes, Pyrex brand, No. 8360). These tubes have a large-diameter upper section tapering to a much narrower, uniformly cylindrical bottom section. The detergent-rich phase sediments into the latter, thus allowing thorough removal of the upper phase without mixing; graduations simplify volume estimations.

References

Bordier, C. (1981). Phase separation of integral membrane proteins in Triton X-114 solution. *J. Biol. Chem.* **256,** 1604–1607.

Cheetham, P. S. (1979). Removal of Triton X-100 from aqueous solution using Amberlite XAD-2. *Anal. Biochem.* **92,** 447–452.

Chen, Z., and Lancet, D. (1984). Membrane proteins unique to vertebrate olfactory cilia: candidates for sensory receptor molecules. *Proc. Natl. Acad. Sci. U.S.A.* **81,** 1859–1863.

Hastie, A. T., Krantz, M. J., and Colizzo, F. P. (1990). Identification of surface components of mammalian respiratory tract cilia. *Cell Motil. Cytoskel.* **17,** 317–328.

Holloway, P. W. (1973). A simple procedure for removal of Triton X-100 from protein samples. *Anal. Biochem.* **53,** 304–308.

Horigome, T., and Sugano, H. (1993). A rapid method for removal of detergents from protein solution. *Anal. Biochem.* **130,** 393–396.

Matesic, D. F., Philp, N. J., Murray, J. M., and Liebman, P. A. (1992). Tubulin in bovine retinal rod outer segments. *J. Cell Sci.* **103,** 157–166.

Stephens, R. E. (1977). Major membrane protein differences in cilia and flagella: evidence for a membrane associated tubulin. *Biochemistry* **20,** 4716–4723.

Stephens, R. E. (1983). Reconstitution of ciliary membranes containing tubulin. *J. Cell Biol.* **96,** 68–75.

Stephens, R. E. (1985a). Evidence for a tubulin-containing lipid-protein structural complex in ciliary membranes. *J. Cell Biol.* **100,** 1082–1090.

Stephens, R. E. (1985b). Ciliary membrane tubulin and associated proteins: a complex stable to Triton X-100 dissociation. *Biochim. Biophys. Acta* **821,** 413–419.

Stephens, R. E. (1986). Membrane tubulin. *Biol. Cell* **57,** 95–110.

Stephens, R. E. (1988). Rapid incorporation of architectural proteins into terminally differentiated molluscan gill cilia. *J. Cell Biol.* **107,** 20a (abstract).

Stephens, R. E. (1990). Ciliary membrane tubulin. *In* "Ciliary and Flagellar Membranes" (R. A. Bloodgood, ed.), pp. 217–240. Plenum Publishing, New York.

Stephens, R. E. (1991). Tubulin in sea urchin embryonic cilia: characterization of the membrane-periaxonemal matrix. *J. Cell Sci.* **100,** 521–531.

Stephens, R. E. (1992). Tubulin in sea urchin embryonic cilia: post-translational modifications during regeneration. *J. Cell Sci.* **101,** 837–845.

Stephens, R. E. (1994). Tubulin and tektin in sea urchin embryonic cilia: pathways of protein incorporation during turnover and regeneration. *J. Cell Sci.* **107,** 683–692.

Stephens, R. E., Oleszko-Szuts, S., and Good, M. J. (1987). Evidence that tubulin forms an integral membrane skeleton in molluscan gill cilia. *J. Cell Sci.* **88,** 527–535.

Stommel, E. W. (1984). Calcium activation of mussel gill abfrontal cilia. *J. Comp. Physiol.* **155A,** 457–469.

CHAPTER 62

Peptide Receptors in Sea Urchin Sperm Flagella

Lawrence J. Dangott
Department of Neurobiology
Harvard Medical School
Boston, Massachusetts 02115

I. Introduction

Sea urchin eggs possess small peptides that activate sperm metabolism at very low concentrations (~30 p*M*) and that have been implicated in sea urchin sperm chemoattraction and the induction of the acrosome reaction (Garbers, 1989; Domino and Garbers, 1991). Egg peptides have been isolated from a number of echinoderm species and grouped into families based on their primary structure (Suzuki and Yoshino, 1992; Suzuki, 1990). Two peptides that have been extensively characterized are speract (Gly–Phe–Asp–Leu–Asn–Gly–Gly–Gly–Val–Gly) and resact (Cys–Val–Thr–Gly–Ala–Pro–Gly–Cys–Val–Gly–Gly–Gly–Arg–Leu-NH_2), isolated from the jellycoats of *Strongylocentrotus purpuratus* and *Arbacia punctulata,* respectively. Although the two peptides do not cross-react in any detectable manner with sperm from the other species, they appear to possess nearly identical biological properties and potencies in homologous species (Garbers, 1989; Domino and Garbers, 1991).

METHODS IN CELL BIOLOGY, VOL. 47

Receptors for the peptides were identified using classic chemical crosslinking techniques. Radiolabeled peptide analogs were synthesized and characterized and were subsequently used to label putative receptor proteins. The speract receptor was initially reported to be an acidic integral membrane glycoprotein of approximately 80-kDa (Dangott and Garbers, 1984; Dangott *et al.*, 1989) and the receptor for resact was determined to be the 150-kDa membrane form of guanylyl cyclase (Shimomura *et al.*, 1986; Singh *et al.*, 1988). Subsequent studies have strongly suggested that the receptors for each peptide contain both polypeptide chains and may exist as oligomeric structures on the sperm surface.

Successful identification of receptors is affected by several factors including (1) the choice of ligand, (2) the choice of crosslinking agent, and (3) the conditions (pH, salt, ligand concentration, temperature) used for ligand/receptor binding and crosslinking.

A critical component is the development of crosslinkable ligands that resemble closely the natural product. For peptide ligands this is accomplished by chemically synthesizing a variety of structural analogs and characterizing their potencies in the appropriate bioassay and in competition binding experiments. Alterations to the primary structure of the peptide may be required to introduce reactive groups or radioiodination sites which may alter the binding kinetics and specificity of the ligand. The importance of the ligand selection cannot be overstated. It has been demonstrated that different ligands for the same receptor can lead to different patterns of crosslinked proteins (Changeux *et al.*, 1992; Pearson and Miller, 1987). This suggests that the researcher may want to test more than one ligand under a variety of conditions to ensure that all the receptors or receptor subunits are identified.

The choice of crosslinker may be dictated by the known characteristics of both the ligand and the receptor. Several questions need to be addressed: (1) What is the chemical nature of the receptor? (2) What are the binding constants for the particular ligand for the receptor? (3) How much structural alteration of the ligand can be tolerated and what are the reactive sites on the proteins that will be exploited in the labeling?

Finally, the conditions under which the compounds are premixed and crosslinked and the length of the crosslinker can have profound effects on the final outcome of the experiment. pH and choice of crosslinking agent can greatly alter the pattern of tagged proteins when the same radiolabeled ligand is crosslinked to intact spermatozoa. Figure 1 illustrates the labeling patterns obtained when the same radiolabeled speract analog is conjugated onto intact *S. purpuratus* spermatozoa at two different pHs. The different patterns may reflect conformational changes in the ligand or receptor that alter the accessability of the crosslinker to reactive sites. Similarly, long or short crosslinking agents may be more or less able to span the distance between reactive groups on the ligand and the receptor. Therefore, the researcher may wish to try reagents of different length.

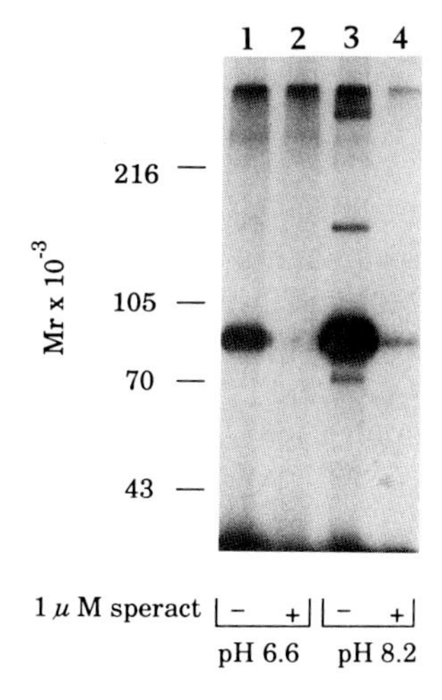

Fig. 1 Effect of pH on crosslinking radiolabeled speract to isolated sperm flagella. Lanes 1 and 2 were labeled with ^{125}I-GGG[Y^2]-speract-I at pH 6.6 and lanes 3 and 4 were labeled at pH 8.2. All samples were crosslinked with 5 m*M* disuccinimidyl suberate for 15 minutes at 15°C. Labeled bands were detected by autoradiography at approximately 65, 80, 150, and >216 kDa at pH 8.2, but only at 80 kDa at pH 6.6 (lanes 1 and 3). The specificity of the labeling is indicated by the ability of 1 μ*M* unlabeled speract to greatly reduce or abolish the labeling (lanes 2 and 4). The label at the bottom of the figure represents noncovalently associated radioligand that is migrating with the dye front.

The procedures described here are meant to act solely as a guide. Each ligand/receptor system needs to be tested individually and the appropriate crosslinking conditions determined empirically. The procedure also assumes that the researcher has a biologically active ligand and some knowledge of receptor/ligand interactions, including the results of equilibrium binding and binding competition studies (Dangott and Garbers, 1984; Shimomura *et al.*, 1986).

II. Methods

A. Collection and Washing of Sea Urchin Spermatozoa

1. Background

Egg peptide receptors can be identified by crosslinking ligands to intact sperm cells, isolated flagella, or spermatozoan plasma membrane vesicles. Preliminary experiments can be most quickly and easily performed on intact spermatozoa with a minimum of equipment and expertise. Detailed procedures are available elsewhere that describe methods used to isolate sperm heads and flagella (Stephens, 1986) and sperm plasma membranes (Bentley and Garbers, 1986) and, therefore, are not described here.

2. Solution

Several formulations for artificial seawater are published. The recipe we follow is buffered with 4-(2-hydroxyethyl)-1-piperazineethanesulfonic acid (Hepes, pH 7.8–8.2) or 2-[(2-amino-2-oxoethyl)amino]ethanesulfonic acid (ACES, pH 6.6) and is presented in Table I. The seawater is stored at 4°C and filtered through a 0.45-μm Millipore filter before use.

3. Procedure

Gravid *Strongylocentrotus purpuratus, Arbacia punctulata,* or *Lytechinus pictus* are spawned by injecting 0.5 *M* KCl with a 25-gauge syringe needle inserted through the peristomal membrane. The sperm are then pipetted directly from the gonopores on the dorsal surface of the animal with a glass Pasteur pipet and placed undiluted into 1.5-ml polypropylene microcentrifuge tubes on ice. Stored in this manner the sperm will remain capable of fertilization for several days.

A small population of contaminating pigmented cells can be removed from the dry semen by suspending the cells in 50 vol of artificial seawater and centrifuging in a swinging bucket rotor at 200*g* for 15 minutes at 4°C. The pigmented cells will form a tight pellet at the bottom of the tube, and the sperm can be decanted to a clean tube and concentrated by centrifugation at 1500*g* for

Table I
Composition of Artificial Seawater[a]

Final concentration	g/liter
454 m*M* NaCl	26.54
9.7 m*M* KCl	0.723
24.9 m*M* $MgCl_2$	5.06
9.6 m*M* $CaCl_2 \cdot 2H_2O$	1.411
27.1 m*M* $MgSO_4 \cdot 7H_2O$	6.679
4.4 m*M* $NaHCO_3$	0.369
Seawater can be buffered with:	
10 m*M* 4-Morpholineethanesulfonic acid	1.95
10 m*M* Hepes	2.4
10 m*M* Aces	1.82
10 m*M* Tris	1.211

[a] Add all salts in the order shown and titrate immediately.

15 minutes as described above. The washing step can be repeated as necessary.

The "wet weight" of the washed sperm pellet can be determined quickly by carefully pipetting 100 μl of the cells into preweighed 1.5-ml microcentrifuge tubes and centrifuging the cells at 15,000*g* for 10 minutes. After the supernatant fluids are decanted, the tube is weighed again and the weight of the cells is determined by subtracting the weight of the empty tube. The weight is determined in duplicate samples and an average obtained.

B. Radioiodination and Purification of Ligand

1. Background

Peptide ligands can be radioiodinated by several methods including enzymatic reaction with lactoperoxidase or chemical reaction using chloramine-T. We have used both but prefer the latter method for simplicity and reproducibility (Cardullo *et al.*, 1994). The procedure can be performed in solution or by having the chloramine-T reagent associated with a solid support such as glass beads (Iodobeads, Pierce Chemical) or coating the reaction vessel with the material (Iodogen, Pierce Chemical). Regardless of the labeling technique used, the labeled peptide can be separated from unincorporated iodine by gel filtration or high-pressure liquid chromatography (HPLC). Although the latter method has the additional benefit of resolving monoiodinated from diiodinated peptides, a method using gel filtration that requires a minimum of equipment and expertise is described.

2. Solutions and Materials

Appropriate safeguards for handling, shielding, and disposal of radioactive materials should be followed.

50 μM peptide
$Na^{125}I$ (Amersham, Catalog No. IMS 30, 100 mCi/ml)
0.5 *M* sodium phosphate, pH 7.5
Chloramine-T (Sigma, Catalog No. C-9877, 0.38 mg/ml in water)
Sodium metabisulfite (Sigma, Catalog No. S-1516, 0.36 mg/ml in water)
Column buffer: 10 m*M* sodium phosphate, pH 7.2, 150 m*M* NaCl, 0.1% gelatin
Bio-Gel P-2 (Bio-Rad)
0.7 × 30-cm column (Bio-Rad Econo-column)

3. Procedure

1. Mix together the reagents in the following order:

Solution	Volume	Final concentration
50 μM peptide	10 μl	0.5 nmol
$Na^{125}I$	20 μl	2 mCi
0.5 *M* sodium phosphate, pH 7.5	40 μl	0.25 *M*
Chloramine-T	10 μl	19 nmol

2. Allow the reaction to proceed for 5 minutes at room temperature.

3. Add 10 μl of the sodium metabisulfite solution and react for 5 minutes at room temperature.

4. Apply the sample to the top of a Bio-Gel P-2 column (0.7 × 25 cm) that was previously equilibrated in column buffer. Rinse the reaction tube with 100 μl of column buffer and combine with the first sample. Allow the sample to drain into the column, overlay the column bed with buffer, and allow the column to flow. A pump may be used if desired.

5. Collect 1-ml fractions and transfer 5 μl of each fraction to a separate tube along with the pipet tip. Including the tip in the tube will help to reduce the variability due to residual counts that are stuck to the pipet tip. Count the radioactivity present with a gamma counter. Using the data, construct a plot of cpm versus fraction number to identify the radioactive fractions. Bio-Gel P-2 was selected based on its fractionation range to resolve the labeled peptides from unincorporated radioiodine; however, the appropriate gel matrix must be selected based on the molecular weight of the ligand being purified. Combine the fractions that contain the labeled ligand, mix, and prepare small aliquots for storage at −20°C. The labeled peptides are relatively stable under these conditions for periods up to approximately 1 month, after which time we have noticed unacceptable amounts of nonspecific binding, presumably due to the breakdown of the labeled peptides.

C. Crosslinkage of Iodinated Peptides to Intact Sperm and Analysis of Labeled Proteins

1. Background

The labeling reaction is performed by mixing the labeled ligand with the intact, washed cells until equilibrium conditions are obtained [as determined in separate experiments (Dangott and Garbers, 1984)]. The crosslinking reagent is then added to the mixture and allowed to react. The labeled proteins are analyzed by electrophoresis on polyacrylamide slab gels containing sodium dodecyl sulfate (SDS) and autoradiography.

2. Solutions and Materials

Washed spermatozoa

Radiolabeled ligand

100 mM disuccinimidyl suberate (Pierce Chemical, Catalog No. 21555) freshly prepared in dimethyl sulfoxide (DMSO, Sigma, Catalog No. D-8779)

DNase I (Worthington, code DPFF, 1 mg/ml in 0.5 M Tris, pH 7.0, 50 mM $MgCl_2$)

3. Procedure

1. React washed, intact sperm (25 mg) with radiolabeled peptide in a microcentrifuge tube at 15°C until equilibrium is reached. Reaction volume is 100 μl. Include a control incubation to which an excess of unlabeled ligand has been added. The concentration of unlabeled ligand added should be at least 100-fold higher than the apparent K_d of the ligand.
2. Add the crosslinking agent directly to the cell suspension and allow it to react for 15 minutes at 15°C. Care should be taken to maintain the final DMSO concentration below 10%. A range of crosslinker concentrations (0.05–10 mM) and reaction times should be tested. Include a control that receives DMSO only.
3. Add 10 μl of a 1 M glycine solution to the mixture and allow it to react for an additional 5 minutes at 15°C. This step will block any unused crosslinking reagent.
4. Collect the labeled cells by centrifugation at 15,000g for 2 minutes at 4°C in a microcentrifuge.
5. Decant and discard the radioactive supernatant fluids appropriately. Place the labeled cells on ice. The cells may be washed with artificial seawater and centrifuged as described in step 4.
6. Add 100–200 μl of SDS sample buffer that contains reducing agent and 10 mM $MgCl_2$ to the cells and solubilize the cells for 5 minutes on ice. Add 5 μl of a 1 mg/ml solution of DNase I and react for 5 minutes. Heat the samples at 95°C for 5 minutes and load onto an SDS–polyacrylamide slab gel. The DNase will reduce the viscosity of the sample by degrading the genomic DNA which could interfere with the electrophoresis.
7. Stain the gel with Coomassie brilliant blue or silver. Dry the gel and expose to Kodak X-Omat film at −80°C. An intensifying screen will shorten the exposure time but it is not necessary to detect the radiolabeled bands. Exposure times will range from one to several days.

III. Comments

The procedures outlined here have been used by the author to identify peptide receptors on a variety of echinoderm sperm using a variety of ligands; however, each researcher will need to determine empirically the conditions and variations required for each particular system tested. The examples have used certain crosslinkers which, of course, need to be selected on an individual basis. Additionally, the amount of radiolabeled ligand added is determined by the binding characteristics of the ligand. To ensure labeling of low-affinity sites, the ligand must be added to a concentration in excess of the K_d. If the ligand/receptor interaction is of very low affinity, several strategies using photoactivatable crosslinking agents should be explored.

References

Bentley, J. K., and Garbers, D. L. (1986). Retention of the speract receptor by isolated plasma membranes of sea urchin spermatozoa. *Biol. Reprod.* **34,** 413–421.

Cardullo, R. A., Herrick, S. B., Peterson, M. J., and Dangott, L. J. (1994). Speract receptors are localized on sea urchin sperm flagella using a fluorescent peptide analog. *Dev. Biol.* **162,** 600–607.

Changeux, J.-P., Galzi, J.-L., Devillers-Thiery, A., and Bertrand, D. (1992). The functional architecture of the acetylcholine nicotinic receptor explored by affinity labelling and site-directed mutagenesis. *Qu. Rev. Biophys.* **25,** 395–432.

Dangott, L. J., and Garbers, D. L. (1984). Identification and partial characterization of the receptor for speract. *J. Biol. Chem.* **259,** 13712–13716.

Dangott, L. J., Jordan, J. E., Bellet, R. A., and Garbers, D. L. (1989). Cloning of the mRNA for the protein that crosslinks to the egg peptide speract. *Proc. Natl. Acad. Sci. U.S.A.* **86,** 2128–2132.

Domino, S. E., and Garbers, D. L. (1991). Mode of action of egg peptides. *In* "Controls of Sperm Motility: Biological and Clinical Aspects" (C. Gagnon, ed.), pp. 91–101. CRC Press, Boca Raton, FL.

Garbers, D. L. (1989). Molecular basis of signalling in the spermatozoon. *J. Androl.* **10,** 99.

Pearson, R. K., and Miller, L. J. (1987). Affinity labeling of a novel cholecystokinin-binding protein in rat pancreatic plasmalemma using new short probes for the receptor. *J. Biol. Chem.* **262,** 869–876.

Shimomura, H., Dangott, L. J., and Garbers, D. L. (1986). Covalent coupling of a resact analogue to guanylate cyclase. *J. Biol. Chem.* **261,** 15778–15782.

Singh, S., Lowe, D. G., Thorpe, D. S., Rodriguez, H., Kuang, W.-J., Dangott, L. J., Chinkers, M., Goeddel, D. V., and Garbers, D. L. (1988). Membrane guanylate cyclase is a cell-surface receptor with homology to protein kinases. *Nature* **334,** 708–712.

Stephens, R. E. (1986). Isolation of embryonic cilia and sperm flagella. *In* "Methods in Cell Biology" (T. E. Schroeder, ed.), Vol 27, pp. 217–228. Academic Press, New York.

Suzuki, N. (1990). Structure and function of egg-associated peptides of sea urchins. *In* "Mechanism of Fertilization: Plants to Humans" (B. Dale, ed.), Springer-Verlag.

Suzuki, N., and Yoshino, K.-I. (1992). The relationship between amino acid sequences of sperm-activating peptides and the taxonomy of echinoids. *Comp. Biochem. Physiol.* **102B,** 679–690.

PART VIII

Assay and Isolation of Flagellar Enzymes

CHAPTER 63

Assaying Protein Phosphatases in Sperm and Flagella

Gerácimo E. Bracho and Joseph S. Tash

Department of Physiology
University of Kansas Medical Center
Kansas City, Kansas 66160

I. Introduction

Increasing evidence suggests that protein phosphorylation is a major mechanism for regulation of motility in sperm and flagella. Dynein has been indicated as a potential site of convergence for regulation through phosphorylation/dephosphorylation by protein kinases and phosphatases (Takahashi *et al.*, 1985; Murofushi *et al.*, 1986; Dey and Brokaw, 1991; Hamasaki *et al.*, 1991). Phosphorylation of dynein has been correlated with increased ATPase activity and an increase in the velocity of dynein-mediated gliding of microtubules *in vitro* (Tash and Means, 1989; Bonini *et al.*, 1991; Hamasaki *et al.*, 1991). An important role for protein phosphatases in motility regulation has been suggested by adding protein phosphatases or their inhibitors to flagellar models (Takahashi *et al.*, 1985; Brokaw, 1987; Tash *et al.*, 1988; Klumpp *et al.*, 1990; Ahmad *et al.*, 1995). In this regard, it should be noted that a significant decrease in the gliding velocity of microtubules *in vitro* by dynein has been observed by preincubation of dynein with a novel serine/threonine protein phosphatase, SPP, recently purified from sea urchin sperm (Bracho *et al.*, 1995). The enzyme specifically dephosphorylates a single dynein-associated subunit phosphorylated by cAMP-dependent protein kinase.

Almost all of the four catalytic subunits of serine/threonine protein phosphatases that have been identified in the cytoplasm of eukaryotic cells can be divided into four groups, termed types 1, 2A, 2B, and 2C on the basis of their substrate specificities and their mode of activation and inhibition *in vitro* (Ingebritsen and Cohen, 1983a,b).

Because the identification, purification, and characterization of protein phosphatases from sperm and flagella have been limited, it would be convenient to choose a substrate that could be easy to prepare in pure form and that could be used in a general assay for all types of serine/threonine protein phosphatases. Such a substrate is a synthetic peptide corresponding to residues 81 to 99 of rat skeletal muscle cAMP-dependent protein kinase type II regulatory subunit (RII, Scott *et al.*, 1987). We have used the RII peptide to purify protein phosphatase 2B from bovine brain and the novel SPP protein phosphatase from sea urchin sperm and to assay protein phosphatases 1, 2A, and 2C from rabbit skeletal muscle (Bracho *et al.*, 1995). In this chapter, we describe the assay for serine/threonine protein phosphatases using the dephosphorylation of RII peptide phosphorylated by cAMP-dependent protein kinase as substrate. In addition, we discuss special considerations when designing experiments to examine the effect of protein phosphatases and kinases on the gliding velocity of microtubules *in vitro* by dynein.

II. Methods

A. Purification of the Catalytic Subunit of cAMP-Dependent Protein Kinase

1. Solutions and Materials

Homogenization buffer: 45 m*M* KH_2PO_4, 7 m*M* β-mercaptoethanol, 1 m*M* ethylenediaminetetraacetic acid (EDTA), pH 6.8

Homogenization buffer + 100 μ*M* cAMP

Ammonium sulfate

Ammonium hydroxide

DEAE-cellulose (~1 kg Whatman DE-52) equilibrated against homogenization buffer containing 0.4 *M* NaCl

Phosphocellulose (~200 g Whatman P11 in 1 liter) equilibrated with homogenization buffer

DEAE-cellulose 3-liter column equilibrated with column buffer

Column buffer: 55 m*M* KH_2PO_4, 7 m*M* β-mercaptoethanol, 1 m*M* EDTA, pH 6.8

For column gradient: 100 mM KH_2PO_4(pH 6.8) and 300 mM KH_2PO_4 (pH 6.8)

2. Procedure

The catalytic subunit of cAMP-dependent protein kinase (PKA) is purified from fresh bovine heart obtained at a local slaughterhouse using a modification of the method of Peters *et al.* (1977).

1. Grind 10 kg of fresh beef heart in a meat grinder and then homogenize in a 4-liter Waring blender in 25 liters of homogenization buffer in 4-liter batches for three 1-minute intervals at high speed.
2. Centrifuge the homogenate at 14,300*g* for 30 minutes at 2°C.
3. Bring the supernatant to 50% saturation with ammonium sulfate and maintain the pH near 6.8 by the addition of ammonium hydroxide.
4. Pellet the precipitate as in step 2.
5. Dissolve the pellet in homogenization buffer diluted twofold with water and dialyze against this buffer until the ionic strength (as measured by conductivity) of the redissolved protein reaches the equivalent of 0.4 M NaCl.
6. Centrifuge as in step 2 to remove nondissolved material.
7. Mix the supernatant with 2 liters of DEAE-cellulose and pellet the DEAE.
8. Dialyze the supernatant against homogenization buffer (without NaCl) and then mix with 1 liter of phosphocellulose. Centrifuge to remove phosphocellulose. This step removes proteins that have charge characteristics similar to those of the catalytic subunit, but because the catalytic subunit is still bound to holoenzyme, it is not removed.
9. Apply the supernatant to a 3-liter column of DEAE-cellulose. Wash with approximately 20 liters of column buffer until the absorbance at 280 nm is 0.05 or less. Wash the column with 1 liter of homogenization buffer and then with homogenization buffer containing 100 μM cAMP. Collect fractions of 25 ml in siliconized glass tubes. Monitor the 280 nm/260 nm absorbance to determine when the column has been saturated with cAMP.
10. Pool the column fractions from the point where the 45 mM phosphate buffer elutes (as monitored by a drop in conductivity) to where the cAMP starts to come through the column (as monitored by an increase in absorbance at 260 nm over that at 280 nm).
11. Apply the pooled fractions to a 20-ml siliconized column of phosphocellulose equilibrated with column buffer. After the entire sample has been applied, wash the column with 1 vol of column buffer.
12. Elute the catalytic subunit with a 100 to 300 mM KH_2PO_4(pH 6.8) gradient and collect fractions in siliconized tubes. The catalytic subunit is eluted as a

single protein peak between 200 and 250 mM phosphate. The overall yield from 10 kg of heart is about 120 mg (approximately 80% recovery), and the enzyme normally has a specific activity of about 2×10^6 pmole min^{-1} mg^{-1} as assayed by the procedure of Fakunding and Means (1977).

B. Phosphorylation and Purification of RII Substrate

The RII peptide, representing residues 81 to 99 of the phosphorylation site of type II regulatory subunit of rat skeletal muscle PKA (DLEVPIPAKFTRRV-SVCAE, Scott *et al.*, 1987) is synthesized and purified, and its sequence is confirmed by standard Edman degradation and amino acid analysis. Phosphorylation of RII by the catalytic subunit of PKA at the serine residue (Takio *et al.*, 1982, 1984) is performed essentially as described by Blumenthal *et al.* (1986).

1. Solutions and Materials

Reaction buffer: 20 mM 4-morpholinepropanesulfonic acid (Mops, pH 7.0), 250 mM NaCl, 2 mM magnesium acetate, 15 mM β-mercaptoethanol, 0.5 mM [^{32}P]ATP (200–400 cpm/pmole), 0.25 mM RII, ~2 μM PKA

0.5 mM [^{32}P]ATP

0.25 mM RII

2 μM PKA

75 mM H_3PO_4

P81 phosphocellulose filter paper (Whatman)

2.5 × 15-cm Bio-Gel P-2 column (Bio-Rad) preequilibrated and run with 0.2 M NH_4HCO_3/5% 2-propanol.

2. Procedure

1. Prepare 1-ml reaction mixture containing reaction buffer and 0.5 mM [^{32}P]ATP (200–400 cpm/pmole), 0.25 mM RII, and ~2 μM PKA.
2. Start the reaction by adding catalytic subunit and incubate at 30°C for 30 minutes. Stop the reaction by boiling for 2–3 minutes.
3. Determine the time course of the reaction (Fig. 1) by spotting a 3-μl aliquot onto P81 phosphocellulose filter paper, wash three times for 5 minutes each in 75 mM H_3PO_4, dry, and then count with scintillation cocktail as described by Roskoski (1983).
4. Purify the phosphorylated RII substrate by gel filtration chromatography on the 2.5 × 15-cm Bio-Gel P-2 column. The phosphorylated peptide, as detected by A_{280} and radioactivity counting (Fig. 2), is pooled, lyophilized, and dissolved in water to a final concentration of 80 μM RII.

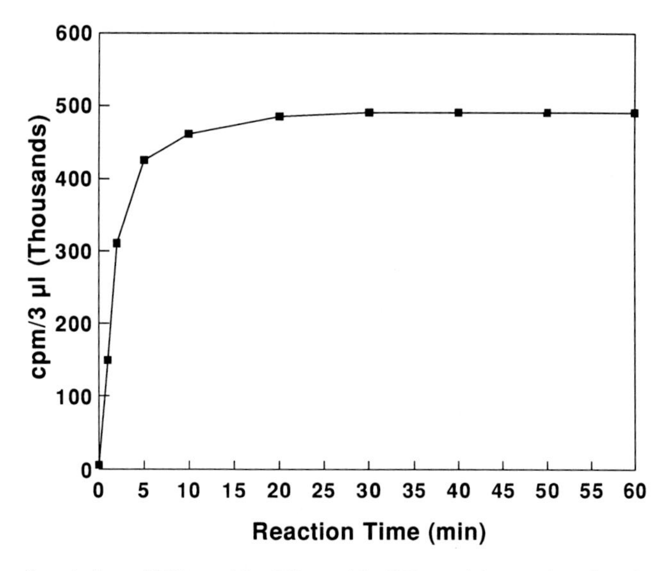

Fig. 1 Phosphorylation of RII peptide. RII peptide (250 nmole) was phosphorylated with about 2 nmole of catalytic subunit of cAMP-dependent protein kinase at 30°C in the presence of 500 nmole of [^{32}P]ATP. Three microliter aliquots were removed and the reaction was terminated by 75 mM H_3PO_4 at the times indicated for determination of the extent of phosphorylation by liquid scintillation counting. Other details are described in the text.

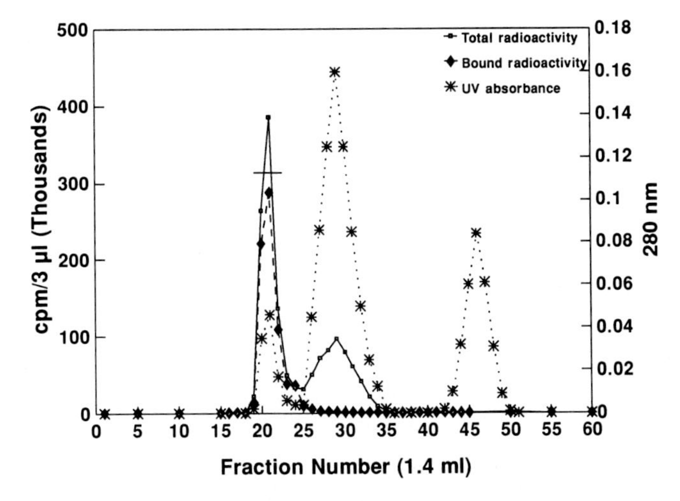

Fig. 2 Gel filtration chromatography of phosphorylated RII peptide. Elution profile of phosphorylated RII peptide on a Bio-Gel P-2 column at 4°C. The flow rate was about 25 ml/h. The pooled fractions are indicated by the horizontal bar. Other details are described in the text.

5. Store the solution in small aliquots at −20°C until needed.

Under these phosphorylation conditions, the specific activity of the radiolabeled RII peptide is normally about 500,000 cpm/nmole. A stoichiometry of 1 mole of phosphate/mole of substrate is verified by comparing the amount of phosphate incorporated with that expected based on the concentration of substrate used and the specific activity of the radioisotope (see Concluding Remarks).

C. Protein Phosphatase Assay

Dephosphorylation of ^{32}P-RII is determined in duplicate assays using the P81 phosphocellulose paper technique described previously (Roskoski, 1983).

1. Solutions and Materials

Assay mixture: 50 m*M* Mops (pH 7.0), 18.5 m*M* β-mercaptoethanol, 0.1% Triton X-100, 1 mg/ml bovine serum albumin, 2 m*M* Mg acetate, 0.1 m*M* $CaCl_2$, 16 μ*M* ^{32}P-RII, 10 μl of enzyme

75 m*M* H_3PO_4

2. Procedure

1. Prepare a 50-μl assay mixture.
2. Determine the substrate concentration by liquid scintillation counting using the specific radioactivity of the incorporated ^{32}P label.
3. Adjust the concentration of protein phosphatase to maintain linear rates of ^{32}P label removal with respect to time.
4. Preincubate all reactions at 30°C for 3 minutes before starting reaction by addition of substrate.
5. Remove 4-μl aliquots at 0 time and at 5-minute intervals up to 30 minutes.
6. Terminate reactions with 75 m*M* H_3PO_4 as described above, and quantitate paper-bound radioactivity by liquid scintillation counting. Calculate protein phosphatase activity from the slope of a linear regression analysis of the filter paper-bound ^{32}P data and the specific activity of the substrate.

A typical result of this assay is shown in Fig. 3. One unit of protein phosphatase activity is defined as the amount of enzyme that removes 1 nmole of phosphate per minute at 30°C under these assay conditions. ^{32}P released from ^{32}P-RII peptide needs to be confirmed to be inorganic phosphate, rather than radioactive peptide by the action of proteinases, by electrophoresis of ^{32}P-RII with a high-resolution sodium dodecyl phosphate–polyacrylamide gel electrophoresis (SDS–PAGE) peptide system (Schagger and von Jagow, 1987) after controlled incubation with protein phosphatase as described by Bracho *et al.*

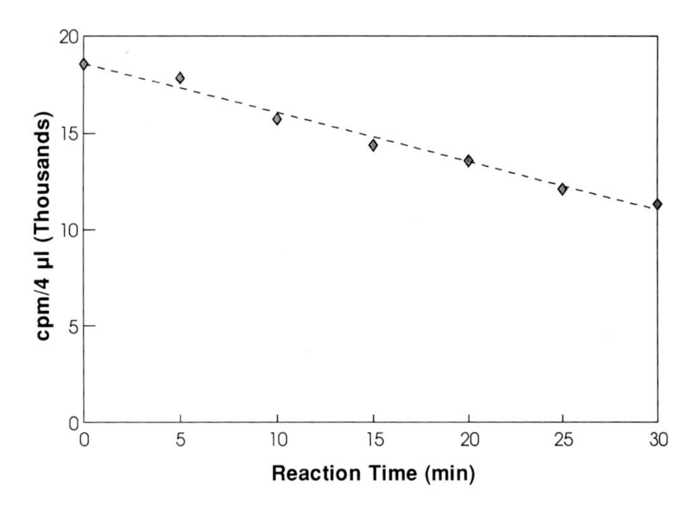

Fig. 3 Dephosphorylation of ^{32}P-RII peptide. ^{32}P-RII (0.8 nmole) was incubated with 10 μl of diluted SPP sea urchin sperm protein phosphatase in a total volume of 50 μl at 30°C. Four-microliter aliquots were removed at the times indicated. The reaction was terminated by 75 m*M* H_3PO_4, and the extent of dephosphorylation was estimated by liquid scintillation counting. Other details are described in the text.

(1995), or by the molybdate/isobutanol/benzene extraction procedure described by Foulkes *et al.* (1981). Figure 4 shows a typical determination of total protein phosphatase activity from a sea urchin sperm extraction using procedures described by Bell *et al.* (1982) and Fox and Sale (1987). It is important to point out that in many instances, the total activity in the starting extract from complex tissues could be underestimated. Dilution of the starting material before activity analysis is recommended.

The activity determined under these assay conditions is the total activity contributed by all the protein phosphatases present in the enzyme sample. In many cases, however, identification and quantitation of each activity are needed. In these situations, the standard assay described above is modified by the addition of okadaic acid and anticalmodulin agents or chelators as described by Cohen *et al.* (1990). The inclusion of okadaic acid in the assay buffer greatly facilitates the identification and quantitation of activities. Addition of about 1 n*M* okadaic acid is enough to completely inhibit protein phosphatase 2A (PP2A). At this concentration, PP1 is unaffected, but it is completely inhibited when the concentration is increased to about 20 n*M* (Cohen *et al.*, 1990). The amount of PP2A activity present in the enzyme sample is the fraction of activity inhibited by 1 n*M* okadaic acid, PP1 is the activity resistant to 1 n*M* okadaic acid but inhibited at 1 μM, and PP2C is the Mg^{2+}-dependent activity (~10 m*M* Mg^{2+}) resistant to 1 μM okadaic acid. The Ca^{2+}/calmodulin-dependent PP2B activity is the fraction of activity in the presence of 1 μM okadaic acid inhibited by about 150 μM trifluoroperazine or by ethylene glycol

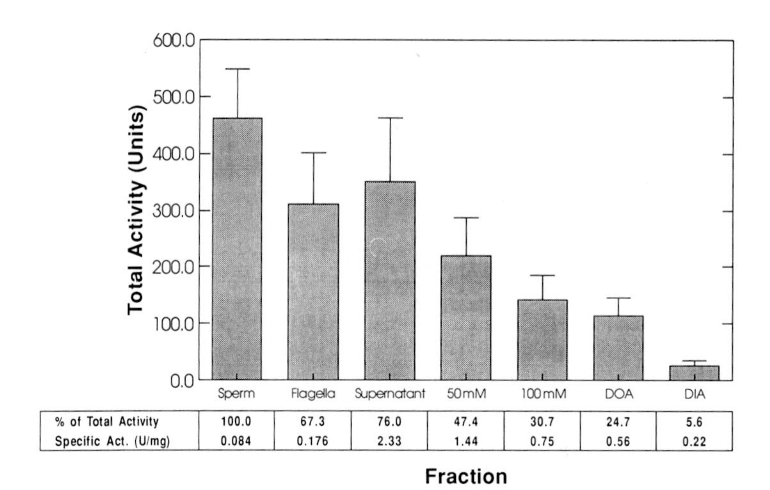

	Sperm	Flagella	Supernatant	50mM	100mM	DOA	DIA
% of Total Activity	100.0	67.3	76.0	47.4	30.7	24.7	5.6
Specific Act. (U/mg)	0.084	0.176	2.33	1.44	0.75	0.56	0.22

Fig. 4 Distribution of protein phosphatase activity in sea urchin sperm. Total activity in each fraction was determined using the standard assay described in the text, and the percentage of activity and the specific activity are also indicated. Fractions are sperm, starting activity in the whole sperm; flagella, activity associated with flagella after sperm heads were removed; supernatant, activity in the supernatant after flagella were diluted twofold with isolation buffer and pelleted by centrifugation; 50 m*M*, activity in 50 m*M* NaCl-containing buffer used to extract pelleted flagella; 100 m*M*, activity in 100 m*M* NaCl-containing buffer used to extract pelleted flagella; DOA, activity associated with dynein outer arm extract; and DIA, activity associated with dynein inner arm extract. Each bar represents the average of five sperm extractions.

bis(β-aminoethyl ether)-*N,N'*-tetraacetic acid (Cohen *et al.*, 1990); see also Concluding Remarks).

3. Considerations for Assaying the Effect of Protein Phosphatases and Kinases on Dynein-Driven Microtubule Gliding on Glass Surfaces

The ability to assay dynein function by examining the propulsion of taxol-stabilized microtubules on glass surfaces offers a powerful tool for examining the role of protein phosphorylation in this process. The basic microtubule gliding assay we use is essentially the same as described in Chapter 37 of this volume. To distinguish alterations in gliding produced by changes in phosphorylation as opposed to changes produced by the incubations and mechanical manipulations, careful experimental design is required. Two approaches can be taken regarding manipulation of dynein phosphorylation: (1) treat dynein prior to loading into the gliding chamber, or (2) treat dynein after loading into the chamber. Each of these techniques has inherent strengths and weakness. Dynein is more stable during extended incubation in solution than after binding to glass. We have found that dynein can be incubated in control buffers for up to 30 minutes at room temperature (but not at 30°C or higher) in solution with no apparent change in subsequent gliding function, whereas after binding to the gliding chamber, incubations longer than about 6 minutes in control solutions produce a marked decrease in gliding function. On the other hand, manipulations of the dynein in solution require parallel incubations with control compounds, as the added kinases or phosphatases would also be carried into the gliding chamber with the dynein during loading. Treatment of the dynein after binding offers the opportunity to observe gliding of microtubules by dynein in the same chamber before and after manipulation; however, we have observed that there is a limit of about four wash/incubation steps that can be done before the gliding function deteriorates. Whether this is due to a loss of dynein arms or to denaturation or loss of specific subunits has not yet been determined. In any case, to obtain interpretable data, incubations are limited in number (less than four after the initial wash step) and time (2 minutes or less), and always performed with replicate control incubations carried through with an identical number of manipulations. As a result of these observations, we always use dynein at a concentration that requires no more than two initial loadings to obtain optimal gliding and prepare the incubation additions such as pure catalytic subunit of cAMP-dependent protein kinase, SPP, or control additions such as PKI, so that the volume added is no more than 1 μl per 25 μl of dynein to minimize dilution. In addition, all experiments examining effects on gliding of manipulation in solution are conducted in parallel with incubations using dynein that has been treated under identical conditions with [^{32}P]ATP and analyzed by SDS–PAGE (Laemmli, 1970) and autoradiography, to confirm changes in protein phosphorylation that are associated with changes in gliding velocity.

III. Concluding Remarks

A peptide substrate radiolabeled to a high specific activity is essential for the high sensitivity and the good reproducibility of the protein phosphatase activity assay described above. Under the phosphorylation conditions described here, the specific activity of the labeled peptide is normally about 500,000 cpm/nmole. To achieve this high specific activity some general guidelines should be followed: (1) The synthetic peptide must be of high purity (>95%), which is easily achievable with today's peptide technology. (2) [^{32}P]ATP and cold ATP must be of high purity. Purity can be determined by thin-layer chromatography on polyethyleneimine-coated cellulose plates for about 2.5 hours with 0.75 *M* potassium phosphate, pH 3.75, at room temperature, with about 100 m*M* cold ATP (detected by UV radiation) and about 0.5 μCi of [^{32}P] ATP (detected by autoradiography after 10–15 minutes X-ray film exposure). (3) The preparation of PKA catalytic subunit must be good. Purity shuold be checked by analytical SDS–PAGE and the enzyme should be active enough to complete the reaction at low micromolar concentrations in about 30 minutes.

The final concentration of the peptide substrate should be kept constant. This is especially important when comparing different protein phosphatases from the same tissue or the same protein phosphatases from different tissues. It is very likely that the final concentration of substrate in the reaction will be below K_m; therefore, the activity determination would change significantly with any change in the final substrate concentration.

Contaminating proteases can be kept in check by using general inhibitors for serine, acid, and thioproteases; however, metalloproteases could be a problem. As these enzymes are more active in the presence of Ca^{2+}, which also activates PP2B, special precautions should be taken. Most of the contaminating metalloproteases can be removed by chromatography on hemoglobin–Sepharose as described by Klee *et al.* (1983). In addition, the action of metalloproteases can be substantially decreased by running the PP2B assay for 1–5 minutes only and starting the reaction by adding enzyme instead of substrate (Klee *et al.,* 1983).

The classification of serine/threonine protein phosphatases into types 1, 2A, 2B, and 2C (Ingebritsen and Cohen, 1983a,b should be taken as a general guideline for characterization of protein phosphatase activity. There are examples of protein phosphatases that are not adequately classified by these guidelines. Four enzymes that do not fit into the four major groups of protein phosphatases are the phage phosphatase (Cohen *et al.,* 1988; Cohen and Cohen, 1989), a bovine brain protein recently characterized (Honkanen *et al.,* 1991), a *Drosophila* enzyme recently found by DNA sequencing (Steele *et al.,* 1992), and the SPP sea urchin sperm enzyme recently purified and characterized in our laboratory (Bracho *et al.,* 1995). The discovery of these novel enzymes should serve to indicate the progress that has been made in understanding the structure and regulation of protein phosphatases.

References

Ahmad, K., Bracho, G. E., Wolf, D. P., and Tash, J. S. (1995). Regulation of human sperm motility and hyperactivation component by Ca^{2+}, calmodulin, and phosphoprotein phosphatases. *Arch. Androl.* (in press).

Bell, C. W., Fraser, C. L., Sale, W. S., Tang, W. J., and Gibbons, I. R. (1982). Preparation and purification of dynein. *Methods Enzymol.* **85,** 450–474.

Blumenthal, D. K., Takio, K., Hansen, R. S., and Krebs, E. G. (1986). Dephosphorylation of cAMP-dependent protein kinase regulatory subunit (type II) by calmodulin-dependent protein phosphatase. Determinants of substrate specificity. *J. Biol. Chem.* **261,** 8140–8145.

Bonini, N. M., Evans,T. C., Miglietta, L. A., and Nelson, D. L. (1991). The regulation of ciliary motility in Paramecium by Ca2+ and cyclic nucleotides. *Adv. Second Messenger Phosphoprotein Res.* **23,** 227–272.

Bracho, G. E., Rodvelt, T. J., Hedge, A.-M., and Tash, J. S. (1995). Purification and characterization of a novel serine/threonine protein phosphatase from sea urchin sperm. submitted.

Brokaw, C. J. (1987). Regulation of sperm flagellar motility by calcium and cAMP-dependent phosphorylation. *J. Cell Biochem.* **35,** 175–184.

Cohen, P. T., and Cohen, P. (1989). Discovery of a protein phosphatase activity encoded in the genome of bacteriophage lambda. Probable identity with open reading frame 221. *Biochem. J.* **260,** 931–934.

Cohen, P., Holmes, C. F., and Tsukitani, Y. (1990). Okadaic acid: a new probe for the study of cellular regulation. *Trends Biochem. Sci.* **15,** 98–102.

Cohen, P. T., Collins, J. F., Coulson, A. F., Berndt, N., and da Cruz, e. S. O. B. (1988). Segments of bacteriophage lambda (orf 221) and phi 80 are homologous to genes coding for mammalian protein phosphatases. *Gene* **69,** 131–134.

Dey, C. S., and Brokaw, C. J. (1991). Activation of Ciona sperm motility: phosphorylation of dynein polypeptides and effects of a tyrosine kinase inhibitor. *J. Cell Sci.* **100,** 815–824.

Fakunding, J. L., and Means, A. R. (1977). Characterization and follicle stimulating hormone activation of Sertoli cell cyclic AMP-dependent protein kinases. *Endocrinology* **101,** 1358–1368.

Foulkes, J. G., Howard, R. F., and Ziemiecki, A. (1981). Detection of a novel mammalian protein phosphatase with activity for phosphotyrosine. *FEBS Lett.* **130,** 197–200.

Fox, L. A., and Sale, W. S. (1987). Direction of force generated by the inner row of dynein arms on flagellar microtubules. *J. Cell Biol.* **105,** 1781–1787.

Hamasaki, T., Barkalow, K., Richmond, J., and Satir, P. (1991). cAMP-stimulated phosphorylation of an axonemal polypeptide that copurifies with the 22S dynein arm regulates microtubule translocation velocity and swimming speed in *Paramecium*. *Proc. Natl. Acad. Sci. U.S.A.* **88,** 7918–7922.

Honkanen, R. E., Zwiller, J., Daily, S. L., Khatra, B. S., Dukelow, M., and Boynton, A. L. (1991). Identification, purification, and characterization of a novel serine/threonine protein phosphatase from bovine brain. *J. Biol. Chem.* **266,** 6614–6619.

Ingebritsen, T. S., and Cohen, P. (1983a). The protein phosphatases involved in cellular regulation. 1. Classification and substrate specificities. *Eur. J. Biochem.* **132,** 255–261.

Ingebritsen, T. S., and Cohen, P. (1983b). Protein phosphatases: properties and role in cellular regulation. *Science* **221,** 331–338.

Klee, C. B., Krinks, M. H., Manalan, A. S., Cohen, P., and Stewart, A. A. (1983). Isolation and characterization of bovine brain calcineurin: a calmodulin-stimulated protein phosphatase. *Methods Enzymol.* **102,** 227–244.

Klumpp, S., Cohen, P., and Schultz, J. E. (1990). Okadaic acid, an inhibitor of protein phosphatase 1 in Paramecium, causes sustained Ca2(+)-dependent backward swimming in response to depolarizing stimuli. *EMBO J.* **9,** 685–689.

Laemmli, U. K. (1970). Cleavage of structural proteins during the assembly of the head of bacteriophage T4. *Nature (London)* **227,** 680–685.

Murofushi, H., Ishiguro, K., Takahashi, D., Ikeda, J., and Sakai, H. (1986). Regulation of sperm flagellar movement by protein phosphorylation and dephosphorylation. *Cell Motil. Cytoskel.* **6,** 83–88.

Peters, K. A., Demaille, J. G., and Fischer, E. H. (1977). Adenosine 3′ : 5′-monophosphate dependent protein kinase from bovine heart. Characterization of the catalytic subunit. *Biochemistry* **16,** 5691–5697.

Roskoski, R., Jr. (1983). Assays of protein kinase. *Methods Enzymol.* **99,** 3–6.

Schagger, H., and von Jagow, G. (1987). Tricine-sodium dodecyl sulfate-polyacrylamide gel electrophoresis for the separation of proteins in the range from 1 to 100 kDa. *Anal. Biochem.* **166,** 368–379.

Scott, J. D., Glaccum, M. B., Zoller, M. J., Uhler, M. D., Helfman, D. M., McKnight, G. S., and Krebs, E. G. (1987). The molecular cloning of a type II regulatory subunit of the cAMP-dependent protein kinase from rat skeletal muscle and mouse brain. *Proc. Natl. Acad. Sci. U.S.A.* **84,** 5192–5196.

Steele, F. R., Washburn, T., Rieger, R., and O'Tousa, J. E. (1992). Drosophila retinal degeneration C (rdgC) encodes a novel serine/threonine protein phosphatase. *Cell* **69,** 669–676.

Takahashi, D., Murofushi, H., Ishiguro, K., Ikeda, J., and Sakai, H. (1985). Phosphoprotein phosphatase inhibits flagellar movement of Triton models of sea urchin spermatozoa. *Cell Struct. Funct.* **10,** 327–337.

Takio, K., Smith, S. B., Krebs, E. G., Walsh, K. A., and Titani, K. (1982). Primary structure of the regulatory subunit of type II cAMP-dependent protein kinase from bovine cardiac muscle. *Proc. Natl. Acad. Sci. U.S.A.* **79,** 2544–2548.

Takio, K., Smith, S. B., Krebs, E. G., Walsh, K. A., and Titani, K. (1984). Amino acid sequence of the regulatory subunit of bovine type II adenosine cyclic 3′,5′-phosphate dependent protein kinase. *Biochemistry* **23,** 4200–4206.

Tash, J. S., and Means, A. R. (1989). cAMP- and Ca2+-calmodulin (CaM)-dependent phosphorylation/dephosphorylation pathways regulate dynein. *J. Cell Biol.* **107,** 247a(abstract).

Tash, J. S., Krinks, M., Patel, J., Means, R. L., Klee, C. B., and Means, A. R. (1988). Identification, characterization, and functional correlation of calmodulin-dependent protein phosphatase in sperm. *J. Cell Biol.* **106,** 1625–1633.

CHAPTER 64

Flagellar Adenylyl Cyclases in *Chlamydomonas*

Yuhua Zhang and William J. Snell
Department of Cell Biology and Neuroscience
University of Texas Southwestern Medical School
Dallas, Texas 75235

I. Introduction

In *Chlamydomonas,* cAMP has been shown to be important in at least two flagellar functions: control of flagellar motility (Tash, 1989) and initiation of cellular responses during the mating reaction. This second messenger was first reported to function as a signaling molecule during fertilization by van den Ende's and Goodenough's laboratories (Pijst *et al.,* 1984; Pasquale and Goodenough, 1987). Since then it has been shown that flagellar adhesion both *in vivo* and *in vitro* between mt^+ and mt^- flagella activates the flagellar adenylyl cyclase severalfold in a mechanism that may not require G-proteins (Pasquale and Goodenough, 1987; Goodenough *et al.*, 1993; Saito *et al.*, 1993; Zhang *et al.*, 1991; Zhang and Snell, 1993, 1994). Our laboratory reported that the gametic flagellar enzyme, but not the vegetative enzyme, is regulated by phosphorylation and dephosphorylation (Zhang *et al.*, 1991; Zhang and Snell, 1993, 1994). Both the regulation by cell–cell interactions and the differences between the vegetative and gametic forms of the enzyme are interesting for several reasons. First, because cAMP is important for control of flagellar motility, gametes must carefully regulate low, basal levels of this molecule when they are not interacting with gametes of the opposite type. But, on flagellar adhesion with a gamete of the opposite mating type, cAMP levels rise severalfold as the cells prepare for fusion. While this adhesion-dependent regulation is critical for gametes, evidence is lacking that vegetative cells, which do not undergo flagellar adhesion, exhibit any dramatic regulation of the enzyme. Thus, learning more about flagellar adenylyl cyclases should reveal new features of ciliary and flagellar function and, at the same time, may contribute to our understanding of the regulatory repertoire of this widely used enzyme. In this chapter we summarize methods for studying *Chlamydomonas* adenylyl cyclase in both vegetative and gametic flagella.

II. Methods

A. Chemicals, Reagents, and Solutions

1. Chemicals and Reagents

Adenosine 5′-triphosphate (ATP, Sigma)
Adenosine 3′ : 5′-cyclic monophosphate (cAMP, Sigma)
[α-^{32}P]ATP (800 Ci/mmole, NEN)
[2,8-^{3}H]cAMP (30–50 Ci/mmole, ICN)
Imidazole (Sigma)
Dowex AG 50W-$\times$4, 200–400 mesh (Bio-Rad)
Neutral alumina WN-3 (Sigma)
Papaverine (Sigma)
RO20-1724 (Biomol, Plymouth, PA)
Pyruvate kinase (Boehringer-Mannheim)
Potassium phosphoenolpyruate (K_2PEP, Sigma)

Chemicals not listed are of reagent grade.

2. Stock Solutions

Flagella buffer: 20 m*M* sodium 4-(2-hydroxyethyl)-1-piperazineethanesulfonic acid (Na-Hepes), pH 7.2, 4% sucrose, 1 m*M* ethylenediaminetetraacetic acid (EDTA), 0.5% bovine serum albumin (BSA), 2.5 m*M* $MgCl_2$, 0.1 m*M* papaverine or 0.1 mM RO20-1724.

Protease inhibitors (1000×): 7m*M* leupeptin, 3.2 mg/ml trypsin inhibitor from lima bean in H_2O; 60 m*M* L-1-*p*-tosylamino-2-phenylethyl chloromethyl ketone (TPCK), 130 m*M* phenylmethylsulfonyl fluoride (PMSF) in isopropanol.

Assay buffer: 20 m*M* Na-Hepes, pH 7.2, 4% sucrose, 1 m*M* EDTA, 0.5% BSA, 2.5 m*M* $MgCl_2$, 0.1 m*M* papaverine or 0.1 m*M* RO20-1724, 30 μg/ml pyruvate kinase, 3 m*M* K_2PEP, 0.5 m*M* ATP, 10^7 cpm/ml [α-^{32}P]ATP. Assay buffer is made from the following stock solutions:

i. Flagella buffer (2.5×).
ii. 50 m*M* ATP in 20 m*M* Hepes, pH 7.2 (100×). (It is critical to adjust the final pH of the ATP solution to 7.2.)
iii. [α-^{32}P]ATP
iv. Pyruvate kinase (33×): The original suspension (10 mg/ml, Boehringer-Mannheim) is diluted 1 : 10 in 20 m*M* Hepes, 2 m*M* $MgCl_2$, 1 m*M* EDTA, pH 7.2, divided into aliquots, and frozen at −20°C.
v. 1 *M* K_2PEP in 20 m*M* Hepes, pH 7.2 (333×).

Stop solution: 2.5% sodium doderyl sulfate (SDS), 50 m*M* ATP, 1.75 m*M* cAMP, 10^4 cpm [^{3}H]cAMP.

Imidazole washing buffer: 20 m*M* imidazole, 200 m*M* NaCl, pH 7.45 (with HCl).

B. Flagellar Isolation

For two reasons most of the methods described here for assaying adenylyl cyclase are carried out with frozen and thawed whole flagella: First, large quantities of homogenous samples of flagella can be frozen in aliquots and used for several experiments. Second, freezing and thawing disrupt the flagellar membrane, thereby making the [α-^{32}P]ATP used as substrate accessible to the enzyme. To isolate flagella, gametic or vegetative cells are deflagellated by a modification of the pH shock, sucrose method of Witman *et al.* (1972) as described earlier (Zhang *et al.*, 1991). The sedimented flagella are resuspended to a final concentration of about 10–20 mg/ml flagellar protein in flagellar buffer. For best preservation of adenylyl cyclase activity, the flagella should be resuspended at 4°C, as rapidly as possible and without causing the sample to foam. After resuspension, a mixture of protease inhibitors (final concentrations:

7 μM leupeptin, 3.2 μg/ml trypsin inhibitor from lima bean, 60 μM TPCK, 60 μM 1-chloro-3-tosylamido-7-amino-2-heptanone hydrochloride (TLCK), and 0.13 mM PMSF) is added. The flagellar suspension then is divided into 100- to 400-μl aliquots, flash-frozen, and stored in liquid nitrogen.

C. Assaying Adenylyl Cyclase

Flagellar adenylyl cyclase is usually assayed in duplicate in 12 × 100-mm borosilicate glass test tubes coated with Aqua Sil Siliconizing Fluid (Pierce) according to the manufacturer's instructions. The reaction is started by addition of 5–10 μl of highly concentrated (final concentration in the assay 0.5–2 mg/ml), freshly thawed flagella into prewarmed assay buffer (final assay volume, 100 μl). The optimal pH for the assay is between 7.0 and 7.5. Linearity of the activity of gametic adenylyl cyclase with time can be achieved only at an assay temperature of about 37°C. At 30°C or lower, activity declines rapidly (within 3–5 minutes) after initiation of the assay. This effect is attributed to an inhibitor of the flagellar adenylyl cyclase as discussed below. In contrast, the adenylyl cyclase of vegetative flagella does not have this temperature-dependent inhibition and activity is linear with time at 30°C. The assay is stopped by addition of 0.9 ml of stop solution. The [^{3}H]cAMP is included as an internal control to determine recovery of cAMP in the subsequent steps of the assay.

To separate [^{32}P]ATP from [^{32}P]cAMP after the incubation, a double-column technique is used as described by Salomon (1979). All of the following steps are done at room temperature. Briefly, the reaction mixtures are loaded onto 175 × 7-mm plastic columns containing 1 ml Dowex AG 50W-X4. The columns are held in a Plexiglas rack with Plexiglas across the front to protect the user from radiation. The Plexiglas rack is constructed so that the column positions correspond to the positions of the rack used to hold scintillation vials. (We use commonly available cardboard trays.) The Plexiglas rack is suspended in a plywood frame. The flow-through is collected in a wash tray placed at the bottom of the assembly. The columns are washed twice with 1 ml of H_2O, and the flow-through is discarded. Then, a Plexiglas rack (identical to the first rack) containing plastic columns with 0.75 g of neutral alumina/column is slid into place in the plywood frame below the rack of Dowex columns and above the wash tray. The Dowex columns are washed twice with 2 ml of H_2O, and because the upper and lower racks are vertically aligned, the effluent from the Dowex columns drains into the alumina columns. Then the upper rack of Dowex columns is removed, and the wash tray is also removed and replaced with a box of scintillation vials. The cAMP is eluted from the alumina columns by addition of 3 ml of imidazole washing buffer, and the run-through is collected in the scintillation vials. Scintillation fluid (12 ml, Liquiscint, National Diagnostics, Manville, NJ) is added to the vials and the ^{32}P and ^{3}H contents of the samples are determined in a scintillation counter. Recovery of [^{3}H]cAMP is usually about 50%, but it varies from column to column and recoveries and corresponding counts per minute must be determined for each sample. Duplicate samples should vary by less than 10%.

III. Evaluating Parameters of the Assay Conditions

A. Divalent Cations

Both gametic and vegetative adenylyl cyclases have an absolute requirement for Mg^{2+}, the optimal concentration of Mg^{2+} for both enzymes being about 2.5 m*M* (Zhang and Snell, 1993; Pasquale and Goodenough, 1987). Compared with no added Mn^{2+}, the gametic (but not the vegetative) enzyme is activated about fivefold at 0.5 m*M* free Mn^{2+}. Both the gametic and vegetative enzymes are inhibited by free Ca^{2+} in the millimolar range. The gametic adenylyl cyclase is more sensitive to Ca^{2+} than the vegetative enzyme and is inhibited by micromolar concentrations of free Ca^{2+}. To test this an ethylene glycol bis (β-aminoethyl ether)-*N,N'*-tetraacetic acid (EGTA)/Ca^{2+} buffer is used, with 0.5 m*M* EGTA and varying amounts of Ca^{2+}. It is important to check the pH of these solutions as addition of Ca^{2+} to EGTA can cause dramatic changes in pH. In our hands the Ca^{2+}-induced inhibition seems to be independent of calmodulin (unpublished data).

B. Detergents

In unpublished experiments we have found that the following detergents inhibit about 50% of enzyme activity when present in the assay solution at the indicated concentrations: octylglucoside, 8 m*M*; cholate, 0.15%; digitonin, 0.02%; Lubrol PX, 0.06%; 3-[(3-cholamidopropyl)dimethylammonio]propane sulfonate (CHAPS), 5 m*M*. This property of the *Chlamydomonas* adenylyl cyclase has made purification of the enzyme particularly difficult. The vegetative enzyme seems to be less sensitive to detergents than the gametic enzyme.

IV. Regulation of Gametic and Vegetative Adenylyl Cyclase

A. ATP-Dependent Inhibition and Heat-Induced Activation of Gametic Adenylyl Cyclase

Flagellar adenylyl cyclase from gametes, but not from vegetative cells, can be stimulated up to 10-fold by prior treatment of flagella for 10 minutes at 45°C (Zhang *et al.*, 1991; Zhang and Snell, 1993). This heat-induced activation of gametic adenylyl cyclase activity can be blocked by ATP, but not by ATPγS and AMPPNP. To test the effects of temperature, flagella are heated to 45°C for varying times and then diluted and assayed as described above. In contrast to the heat-induced activation, preincubation of gametic flagella with ATP at 30°C leads to a 50% reduction of the activity and this ATP-dependent inhibition can be blocked by the protein kinase inhibitor, staurosporine (1 μM). These results suggest that gametic adenylyl cyclase is regulated by phosphorylation and dephosphorylation. During gametogenesis the appearance of the gametic form of the enzyme is coincident with the acquisition of the ability of the cells to undergo sexual signaling and cell fusion. These observations in combination with the result that both cell types must maintain low levels of cAMP for proper flagellar motility suggest that there may be multiple adenylyl cyclases in *Chlamydomonas*.

B. Adhesion-Induced Activation of Adenylyl Cyclase in Gametic Flagella

Gametic flagellar adenylyl cyclase can be activated two- to threefold by flagellar adhesion. This has been shown both in experiments in which freshly isolated flagella from gametes of opposite mating types are mixed together *in vitro* (Zhang and Snell, 1994) and in experiments in which flagella were isolated from adhering mt^+ and mt^- gametes (Saito *et al.*, 1993). Here we describe the method we developed for studying regulation of flagellar adenylyl cyclase during adhesion in a cell-free system.

1. Mixing mt^+ and mt^- Flagella *in Vitro*

Flagella are isolated as described in Section II,B and resuspended in flagella buffer (see above). Equal volumes of freshly isolated mt^+ and mt^- flagella are mixed together on ice in flagellar buffer for 15 seconds and immediately frozen in liquid N_2. (The adhesiveness of isolated gametic flagella should be monitored by microscopic examination.) The samples are thawed and assayed as described above. If frozen and thawed flagella are mixed together, no activation of the adenylyl cyclase can be detected, implying that the coupling between adhesion molecules (agglutinins) and adenylyl cyclase is interrupted by freezing and thawing.

2. Testing the Effects of Protein Kinase and Protein Phosphatase Inhibitors

To study the coupling between agglutinins and adenylyl cyclase, the effect of several inhibitors of protein kinases and protein phosphatases on adenylyl cyclase activation can be tested (Zhang and Snell, 1994). To do this the inhibitors are added to freshly isolated mt^+ and mt^- flagella before they are mixed. We have tested okadaic acid, calyculin A, NaF, vanadate, H-8, H-7, and staurosporine. Although most of these inhibitors do not have a significant effect on the activation of adenylyl cyclase, 50 nM staurosporine blocks about 70% of the adhesion-induced activation of adenylyl cyclase *in vitro*. The adenylyl cyclase activity in nonadhering flagella should be unaffected by the same concentration of staurosporine.

References

Goodenough, U. W., Shames, B., Small, L., Saito, T., Crain, R. C., Sanders, M. A., and Salisbury, J. L. (1993). The role of calcium in the *Chlamydomonas reinhardtii* mating reaction. *J. Cell Biol.* **121,** 365–374.

Pasquale, S. M., and Goodenough, U. W. (1987). Cyclic AMP functions as a primary sexual signal in gametes of *Chlamydomonas reinhardtii. J. Cell Biol.* **105,** 2279–2292.

Pijst, H. L. A., van Driel, R., Janssens, P. M. W., Musgrave, A., and van den Ende, H. (1984). Cyclic AMP is involved in sexual reproduction of *Chlamydomonas eugametos. FEBS Lett.* **174,** 132–136.

Saito, T., Small, L., and Goodenough, U. W. (1993). Activation of adenylyl cyclase in *Chlamydomonas reinhardtii* by adhesion and by heat. *J. Cell Biol.* **122,** 137–147.

Salomon, Y. (1979). Adenylate Cyclase Assay. *In "Advances in Cyclic Nucleotide Research",* Vol 10, pp. 35–55, Academic Press, New York.

Tash, J. S. (1989). Protein phosphorylation: The second messenger signal transducer of flagellar motility. *Cell Motil. Cytoskel.* **14,** 322–339.

Witman, G. B., Carlson, K., Berliner, J., and Rosenbaum, J. L. (1972). *Chlamydomonas* flagella. I. Isolation and electrophoretic analysis of microtubules, matrix, membranes, and mastigonemes. *J. Cell Biol.* **54,** 507–539.

Zhang, Y., and Snell, W. J. (1994). Flagellar adhesion-dependent regulation of *Chlamydomonas* adenylyl cyclase *in vitro:* A possible role for protein kinases in sexual signalling. *J. Cell Biol.* **125,** 617–624.

Zhang, Y., Ross, E. M., and Snell, W. J. (1991). ATP-dependent regulation of flagellar adenylylcyclase in gametes of *Chlamydomonas reinhardtii. J. Biol. Chem.* **266,** 22954–22959.

Zhang, Y., and Snell, W. J. (1993). Differential regulation of adenylylcyclases in vegetative and gametic flagella of *Chlamydomonas. J. Biol. Chem.* **268,** 1786–1791.

CHAPTER 65

Isolation and Characterization of Sea Urchin Flagellar Creatine Kinase

Robert M. Tombes
Massey Cancer Center
Division of Hematology and Oncology
Medical College of Virginia
Richmond, Virginia 23298

I. Introduction

Two isozymes of creatine kinase (CrK) exist in sperm of the sea urchin *Strongylocentrotus purpuratus* (Tombes and Shapiro, 1985). They reside at each end of a metabolic shuttle pathway to transport high-energy phosphate from the sole site of production in the mitochondrion at the base of the sperm head to the main site of energy consumption at dynein ATPases situated along the entire length of the flagellum (Tombes *et al.*, 1987). Both isozymes have been purified and have been shown to be associated with the mitochondrion and the axoneme (Tombes and Shapiro, 1987; Tombes *et al.*, 1988).

Comparable CrK enzymatic activities have been found in sperm heads and tails from diverse species and in other ciliated or flagellated cells, such as photoreceptor cells and ciliated oviduct epithelial cells, which have polarized structure and function (Tombes and Shapiro, 1989). In species whose sperm have alternative sources of energy production and less stringent energetic demands, CrK is greatly diminished (Tombes and Shapiro, 1989).

II. Methods

Phosphagen kinase detection is based on straightforward, yet highly specific spectrophotometric assays. To purify CrK, an initial nucleotide-mimic affinity chromatography matrix is used. Unique labeling and peptide mapping approaches to analyze the properties of these enzymes are straightforward; results obtained using those methods have been confirmed by molecular cloning (Wothe *et al.*, 1990).

A. Separation of Sperm Heads (Mitochondrion and Nucleus) from Tails (Flagellum)

A first simple and worthwhile step in the isolation of any mitochondrial or flagellar protein from sea urchin sperm is to separate heads from tails. Sperm heads and tails are separated from each other by shearing via homogenization followed by differential centrifugation (Vacquier, 1983).

Collect dry sperm by spawning sea urchins with 0.5 *M* KCl, injected intracoelomically. Dilute approximately 1 ml of dry sperm with 20 ml of ice-cold artificial Ca^{2+}-free seawater to prevent activation. Filter this mixture through cheesecloth, to remove any spines or other debris, into a 25-ml-capacity Dounce homogenizer (Wheaton) on ice. Sever the tails from the heads with 15–20 strokes using a low-clearance homogenizer head. This homogenizer should be reserved for embryological work by never exposing it to detergent. Heads are centrifuged away from the tails at 750*g* for 10 minutes (Sorvall SS34 fixed-angle rotor at 3000 rpm). Tails can also be severed from heads using a Waring blender; however, this method generates small tail fragments rather than intact tails. In either case, a small tail "stub" remains embedded in the sperm heads, making it impossible to completely separate the entire axoneme from the head. Resuspend the pellet in artificial Ca^{2+} -free seawater, recentrifuge at 3000 rpm, and reserve the supernatant. Repeat a final third time to ensure that heads are as free of severed tails as possible. Pool all supernatants and concentrate sperm tails by centrifuging at 10,000*g* for 30 minutes (Sorvall SS34 rotor at 10,000 rpm).

B. Efficient Extraction of Creatine Kinase Isoforms from Sperm Heads and Tails

As CrK enzymatic activities are labile, once the enzyme has been extracted, continue the procedure until purified enzyme is concentrated into glycerol in the freezer, where the enzymes are stable indefinitely.

Resuspend sperm tails at 10 mg protein/ml in 20 m*M* 4-(2-hydroxyethyl)-1-piperazineethanesulfonic acid (Hepes), 5 m*M* Tris, pH 7.0, 0.05% Nonidet P-40, 10% glycerol, 5 m*M* $MgCl_2$, 100 m*M* KCl, 1 m*M* ethylene glycol bis(β-aminoethyl ether)-*N,N'*-tetraacetic acid (EGTA), 1 m*M* dithiothreitol (DTT), 0.5 m*M* ADP, 0.2 m*M* phenylmethylsulfonyl fluoride (PMSF), 10 μg/ml soybean trypsin inhibitor (SBTI). Homogenize this suspension using a Dounce homogenizer for 10 strokes. Stir on ice for 30 minutes and homogenize for 10 more strokes. Centrifuge at 100,000*g* for 60 minutes (28,000 rpm, Type 30 Beckman rotor).

Sperm head CrK also can be efficiently extracted using the exact same conditions except for 0.5% instead of 0.05% Nonidet P-40.

The KCl is important for efficient extraction and the ADP and glycerol are also critical for enzyme stability. Other nucleotides also stabilize activity, but ADP is used because it is one of the substrates for the CrK enzyme assay. Unlike muscle enzymes, sea urchin sperm CrK is not efficiently extracted by organic solvents (Dawson and Eppenberger, 1970).

C. Procion Red Column Chromatography

Dilute the 100,000*g* supernatant in 4 vol of 10% glycerol, 20 m*M* Hepes, 10 m*M* Tris, pH 7.0, 1 m*M* EGTA, 0.5 m*M* DTT, 0.2 m*M* ADP, 2 μg/ml SBTI so that the final concentration of KCl is 25 m*M* and the Nonidet P-40 is also decreased fivefold. Up to 200 ml of this diluted tail extract can be applied directly to a 50-ml bed volume of Procion red–agarose (Sigma) with efficient binding. Preequilibrate the column in 10% glycerol, 1 m*M* EGTA, 0.5 m*M* DTT, 0.2 m*M* ADP, 20 m*M* Hepes, 10 m*M* Tris, pH 7.0. Wash the column with this column buffer until the OD_{280} reaches baseline (about 6 column volumes). Elute flagellar CrK with a 0 to 1.0 *M* KCl gradient in column buffer. The flagellar isoform elutes at around 0.3 *M* KCl. Elute the head (mitochondrial) isoform with a gradient of 0 to 2.0 *M* KCl in column buffer. This isoform elutes at 0.6 *M* KCl.

D. Selective Precipitation of the Head Isoform by Dialysis

Pool the activity peaks and dialyze into Procion red column buffer lacking the glycerol, but containing 10 m*M* KCl. The head sample becomes visibly cloudy as the head isoform completely precipitates. Solubility of the head isozyme at any point of purity requires a minimum of 75 m*M* KCl and 10%

glycerol. The insoluble CrK activity can be centrifuged at 10,000*g* for 10 minutes and then resuspended and purified by DEAE-Sepharose column chromatography. Head CrK remains fully active under these conditions and flagellar CrK remains soluble. This approach can be used to separate the isozymes from each other. These solubility properties are consistent with the mitochondrial membrane localization of the head isozyme and the axonemal localization of the flagellar isozyme (Tombes and Shapiro, 1987; Tombes *et al.,* 1988).

E. DEAE Column Chromatography

Dialyze the pooled or resuspended CrK into DEAE column buffer which consists of 10% glycerol, 15 m*M* Hepes, 15 m*M* Tris, pH 8.0, 0.1 m*M* EGTA, 0.2 m*M* DTT, and 0.5 m*M* ADP. Load activities onto the DEAE-agarose column, wash until baseline is reached, and elute with a 0 to 0.5 *M* KCl gradient in DEAE column buffer. Mitochondrial CrK elutes at 0.15 *M* KCl and flagellar CrK at 0.11 *M* KCl. Pool the activity peaks and dialyze into 50% glycerol, 10 m*M* Hepes, 10 m*M* Tris, pH 8.0, 0.1 m*M* ADP, 0.1 m*M* DTT, 0.1 m*M* EGTA. Store frozen at −70°C. Purified sea urchin sperm CrK has been characterized extensively and although it shows immunological and enzymatic similarity to mammalian isozymes, it has an unusual size (Tombes and Shapiro, 1987). The flagellar isozyme has a mass of 145 kDa, which is the result of an apparent triplication of the normal 45-kDa catalytic domain (Wothe *et al.,* 1990).

F. Enzyme Activity

CrK enzymatic activity is measured using a spectrophotometric, coupled enzyme assay. Enzymatic activity can be measured in either direction using either phosphocreatine (PCr) or creatine (Cr) as selective substrate for this enzymatic activity. The "reverse" reaction (using PCr) is recommended because it is more sensitive. The assay is carried out in 115 m*M* imidazole acetate, pH 6.5, 11.5 m*M* magnesium acetate, 2.3 m*M* EGTA, 2.0 m*M* nicotinamide adenine dinucleotide phosphate (NADP), 20 m*M* *N*-acetyl cysteine, 20 m*M* dextrose, 5 m*M* AMP, 10 μM diadenosine pentaphosphate, 2.0 m*M* ADP, 2 U/ml glucose 6-phosphate dehydrogenase, 4 U/ml hexokinase, and 45 m*M* PCr. The assay is conducted in 0.5- to 1.0-ml volume at room temperature, and initial rates, detected by an increase in absorbance at 340 nm, should be linear with respect to sample amount, which is typically 10 μl. The reaction is begun by PCr addition.

The "forward" reaction is also carried out in 0.5 ml and uses 115 m*M* imidazole acetate, 11.5 m*M* Mg acetate, 1 m*M* potassium acetate, 1 m*M* EGTA, 0.50 m*M* nicotinamide adenine dinucleotide, reduced form (NADH), 1.5 m*M* phosphoenolpyruvate, 16 U/ml pyruvate kinase, 22 U/ml lactate dehydrogenase, 20 m*M* *N*-acetyl cysteine, 5 m*M* ATP, and 100 m*M* creatine. All coupling enzymes are obtained in their lyophilized, salt-free form to avoid inhibitory ion contamination. Rates are determined by measuring the decrease in 340-nm absorbance.

G. *In Vivo* Labeling

A unique and useful property of CrK is its sensitivity to modification at neutral pH by Sanger's reagent, 1-fluoro-2,4-dinitrobenzene (FDNB). Labeling occurs at low concentrations of FDNB on a reactive thiol in the active site (Mahowald *et al.,* 1962), which not only helps identify CrK isoforms (Tombes and Shapiro, 1985, 1987, 1989), but also inhibits activity and is therefore a reasonable tool to analyze function in many cell types (Infante and Davis, 1965; Yang and Dubick, 1977; Carpenter *et al.,* 1983; Tombes and Shapiro, 1985; Tombes *et al.,* 1987).

Dilute pure FDNB to approximately 10 m*M* in isopropanol. Add small volumes of this stock solution to sperm in seawater at 4×10^9 sperm/ml (a 10% solution) and incubate at 10°C for 2 hours. Tritiated FDNB (1-fluoro-2,4-dinitro[3,5-^3H]benzene) (Amersham) can also be used. Typically, 10 μM FDNB under these conditions yields relatively selective incorporation into CrK isoforms, with nonspecific labeling at much higher concentrations (Tombes and Shapiro, 1985). [^{3}H]FDNB-labeled sperm can be separated into heads and tails and processed for fluorography, either before or after limited proteolysis in gels (Cleveland *et al.,* 1977). Such analyses can yield information on the similarity to other isozymes and on the number of active sites (Tombes and Shapiro, 1987).

III. Summary

Phosphagen kinases play a role in metabolic function in diverse species (Bessman and Carpenter, 1985). Enzyme assays and inhibitors have been used successfully to demonstrate the role of one of these enzyme families in sea urchins. Although arginine kinase was not purified, its enzymatic activity has been detected in sea urchin eggs and embryos and many other species by simply substituting phosphoarginine (PArg) for PCr (Fujimaki and Yanagisawa, 1978; Tombes and Shapiro, 1989). ArgK does not show sensitivity to FDNB, but CrK proteins of the appropriate molecular weight were detected with this reagent in cells from many species (Tombes and Shapiro, 1989). The use of these techniques to identify and study function of phosphagen kinases in diverse species has much potential.

References

Bessman, S. P., and Carpenter, C. L. (1985). The creatine-creatine phosphate energy shuttle. *Annu. Rev. Biochem.* **54,** 831–862.

Carpenter, C. L., Mohan, C., and Bessman, S. P. (1983). Inhibition of protein and lipid synthesis

presence of PKC in cilia (Kim, 1994). In addition, a family of casein kinases, not dependent on a second messenger for activation, has been isolated from cilia (Walczak *et al.*, 1993).

PKG (Ann, 1991), two CaPKs (Gundersen and Nelson, 1987; Son *et al.*, 1993), and one form of PKA (Hochstrasser and Nelson, 1989), all found in deciliated cell bodies as well as cilia, have been purified to homogeneity from whole cells. The amount of each kinase obtained when purified from whole cells is much higher than from cilia, which contain only 2% of the total cell protein. One form of PKA is found only in cilia (Hochstrasser, 1989), however, and we do not yet know which forms of casein kinase in cilia are the same as those detected in the cell body (Walczak *et al.*, 1993). Cilia are therefore an essential starting fraction for the isolation of certain kinases.

Here we describe the isolation and properties of several protein kinases from *Paramecium* cilia. Using kinases isolated from *Paramecium* rather than heterologous kinase preparations for the study of the behavior of *Paramecium* is advisable; *in vitro* phosphorylation of *Paramecium* proteins by *Paramecium* and bovine PKAs is distinctly different (Hochstrasser, 1989). We have not included the isolation of the two CaPKs detected in cilia because they have only been isolated from whole cells.

PKA, PKG, and a messenger-independent kinase have been characterized from *Tetrahymena* cilia as well (Murofushi, 1973, 1974) and have some similar chromatographic properties.

II. Materials

Solutions for the isolation and fractionation of *Paramecium* cilia are described in Chapter 4 of this volume.

A. Kinase Assay Buffers

PKA: 50 mM 4-morpholinepropanesulfonic acid (Mops), pH 6.5, 10 mM $MgCl_2$, 1 mg/ml mixed histones, 100 μM [γ-^{32}P]ATP at 50 Ci/mole (use 100 μM ATP + 0.5 μC_i [γ-^{32}P]ATP per reaction), 0.2 μM cAMP

PKG: 50 mM 4-morpholineethanesulfonic acid (Mes)–NaOH, pH 6.0, 10 mM $MgCl_2$, 1 mM ethylene glycol bis(β-aminoethyl ether)-*N*,*N*′-tetraacetic acid (EGTA), 1 mg/ml mixed histones, 100 μM [γ-^{32}P]ATP, 0.2 μM 8-Br-cGMP

CaPK: 20 mM 4-(2-hydroxyethyl)-1-piperazineethanesulfonic acid (Hepes), pH 7.2, 5 mM Mg-acetate, 0.5 mM EGTA, 1 mM dithiothreitol (DTT), 1 mg/ml total casein, 100 μM [γ-^{32}P]ATP, 0.6 mM $CaCl_2$

PKC: same as for CaPK except use histone as substrate and add 200 μg/ml phosphatidylserine

Casein kinase: same as for PKA except use casein as substrate and omit the cAMP

B. Chromatographic Reagents

Reactive Red 120–agarose, cAMP–agarose (use only Catalog No. A7775), α-casein agarose (all from Sigma)

DEAE-cellulose, Sephacryl S-200, Sephadex G-150, or a similar sizing resin (generally available)

MELP: 10 m*M* Mops, pH 6.8, 0.2 m*M* ethylenediamine tetraacetic acid (EDTA), 0.2 μg/ml leupeptin, 0.03 m*M* phenylmethylsulfonyl fluoride (PMSF) 0.02% NaN_3

METP1: 5 m*M* Mops, pH 6.8, 0.5 m*M* EDTA, 0.03 m*M* *N*-*p*-tosyl-L-arginine methyl ester (TAME), 0.03 m*M* PMSF, 0.02% NaN_3;

METP2: 20 m*M* Mops, pH 7.5, 0.1 m*M* EDTA, 0.03 m*M* TAME, 0.03 m*M* PMSF, 0.02% NaN_3

MMKETP: 20 m*M* Mops, pH 7.5, 1 m*M* $MgCl_2$, 0.5 *M* KCl, 0.1 m*M* EDTA, 0.03 m*M* TAME, 0.03 m*M* PMSF, 0.02% NaN_3

III. Assays

The *in vitro* kinase assay is based on the procedure of Corbin and Reiman (1974). Assay kinase activity at room temperature (use 37°C for CaPKs) in a 96-well microtiter plate or in microcentrifuge tubes. Start the reaction by adding 10 μl enzyme to 90 μl assay mix or vice versa. After 10 minutes, terminate each reaction by spotting 90μl onto a Whatman GF/A filter (2.4-cm diameter) and dropping the filter into cold 10% trichloroacetic acid (TCA). Wash the filters extensively in 10% TCA, and measure ^{32}P for each filter (in 5 ml double-distilled water). Units are defined as picomoles of P_i transferred per minute.

Certain protein kinases can also be detected by means of an in-gel kinase assay using the method of Kameshita and Fujisawa (1989) (see Chapter 20 in this volume).

Protein concentration is assayed in a 96-well microtiter plate with 10 μl protein sample and 100 μl Bradford reagent (Bradford, 1976). Protein standards are 0.1–1 mg/ml bovine serum albumin. Absorbance at 570 nm is measured in an automated plate reader.

IV. Properties of Protein Kinases

Each ciliary kinase can be detected selectively in a fraction that contains several kinases by using unique conditions for optimal activation. Some fractionation may be required to detect individual kinases in crude cilia. Activation of histone-phosphorylating kinase activity in whole cilia by cAMP and 8-Br-cGMP is 2.7- and 1.6-fold, respectively, above background (Miglietta, 1987). The properties and subciliary location of each kinase are summarized in Table I.

Table I
Properties of Ciliary Protein Kinases

	PKA[a]		PKG	CKS1	CKS2	CKA	CaPK1	CaPK2	PKC
Molecular mass by gel filtration (kDa)	220	70	88	36	36	28	58	51	?
Molecular mass by SDS–PAGE (kDa)	R = 48 C = 40	R = 44 C = 40	77	45	45	33–42	52	50	80–85
Subciliary localization	Soluble	Soluble > membrane > axoneme	Soluble > membrane	Soluble	Soluble	Axoneme	?	?	?
Specific activator	0.2 μM cAMP	0.2 μM cAMP	0.2 μM 8-Br-cGMP[b]	None	None	None	1 μM free Ca^{2+}	1 μM free Ca^{2+}	Ca^{2+} + PS or 100 μg/ml PS[c]
In vitro substrate	Histone	Histone	Histone	β-Casein	β-Casein α-Casein	α-Casein	Casein	Casein Histone	Histone
Autophosphorylated?	Yes (R)	Yes (R)	Yes	?	?	?	Yes	Yes	Yes

[a] Abbreviations: PKA, cAMP-dependent protein kinase; PKG, cGMP-dependent protein kinase; CKS1 and CKS2, soluble casein kinases 1 and 2; CKA, axonemal casein kinase; CaPK1 and CaPK2, Ca^{2+}-dependent protein kinases 1 and 2; PKC, protein kinase C; R, regulatory subunit of PKA; C, catalytic subunit of PKA; PS, phosphatidylserine.

[b] 8-Br-cGMP is used because it is less effective than cGMP at activating PKA as well (Mason, 1989b).

[c] Free $[Ca^{2+}]$ for maximal PKC activation has not been determined.

No potent inhibitors specific for each kinase have been identified. Quercetin at 100 μM and Gpp(NH)p at 4 mM inhibit PKG completely and PKA by approximately 50% (Miglietta and Nelson, 1988). Heparin at 10 μg/ml inhibits 75% of the activity of the casein kinase CKS2 (Walczak *et al.*, 1993). Surprisingly, the PKA-specific mammalian inhibitor PKI is not an effective inhibitor of PKA activity in *Paramecium* (Mason and Nelson, 1989b; Carlson, 1994).

V. Isolation of Protein Kinases from Cilia

Protocols to isolate and fractionate cilia are described in Chapter 4 in this volume. Perform all chromatographic steps at 4°C. Collect column fractions in an 80-well microcentrifuge tube rack that has been rinsed thoroughly in concentrated acid. Isolate all kinases from fresh preparations of cilia. Each liter of cell culture harvested yields approximately 1 mg of ciliary protein. Approximately 25% of this protein fractionates into the soluble ME extract, the source of all but one of the kinases below.

A. Protein Kinase A

The two forms of ciliary PKA can be separated on the basis of size (steps 1–4 below) (Mason and Nelson, 1989a). The large PKA (M_r 220,000) is separated from all other known ciliary kinases. The small PKA (M_r 70,000) is not fully resolved from PKG (M_r approx 80,000), but PKA in a mixture of both can be selectively activated by 0.2 μM cAMP. The small ciliary PKA is immunologically identical to the cell body PKA (Hochstrasser, 1989).

The catalytic (C) subunit of both PKAs can be purified by cAMP–agarose chromatography (steps 4–6 below) (Carlson, 1994). Note that only one type of cAMP–agarose resin is effective (see Section II). Once purified, the C subunit from either form of PKA is constitutively active and, thus, can be characterized in the absence of cAMP. Its instability in this fraction (see step 6 below), however, makes partially purified fractions a more practical source of PKA. Only a trace amount of PKG is present in the pure C-subunit fractions of the small form of PKA, as judged by immunoblotting, and cGMP-dependent activity is undetectable.

1. Prepare at least 5 mg of ME extract of cilia in a volume under 2 ml. Adjust Mops to 10 mM and NaCl to 0.1 M.
2. Run a large (at least 150-ml) Sephacryl S-200 column in MELP buffer (see Section II) + 0.1 M NaCl at 10 ml/h. A similar sizing resin such as Sephadex G-150 can be substituted. Collect at least 1 column volume as 1- to 2-ml fractions. The cAMP-dependent activity resolves into two peaks; pool fractions within each peak.

3. Store both PKA pools at 4°C. They are stable at this stage for at least 1 month without significant loss of cAMP-dependent activity. Do not freeze. Each of the two PKA peaks contains 30–40% of the starting activity and has been purified approximately threefold. Stimulation of PKA activity by cAMP in each fraction is six- to eightfold over basal activity.

4. For further purification, run a cAMP–agarose column in MELP + 0.1 *M* NaCl at 10 ml/h. Load pooled gel filtration fractions directly onto this column. Collect 1- to 2-ml fractions.

5. Wash column with 5 vol each of MELP + 0.1 *M* NaCl, MELP + 1 *M* NaCl, and MELP + 2 m*M* cAMP. Pure C subunit elutes in 1 *M* NaCl and is active without cAMP. The 44-kDa R subunit elutes in 2 m*M* cAMP and is detected by sodium dodecyl sulfate–polyacrylamide gel electrophoresis (SDS–PAGE) (Laemmli, 1970). The 48-kDa R subunit of the large PKA is not significantly recovered in this step.

6. Store pure C subunit at 4°C; use immediately as it loses all of its activity within 2 days. Do not freeze. Pure C-subunit fractions desalted by dialysis have no activity. The overall yield of kinase activity is 10–15% of the starting material. C is purified 60- and 30-fold, respectively, to a specific activity of approximately 360,000 and 170,000 units/mg, respectively, for the small and large forms of PKA.

B. Protein Kinase G

PKG is enriched in cilia, and has been partially purified and efficiently separated from the ciliary PKAs and casein kinases (Miglietta and Nelson, 1988; Walczak *et al.*, 1993). The ciliary form, a monomer with an approximate M_r of 80,000, is identical to PKG found in deciliated cell bodies.

1. Prepare ME extract of cilia. [The isolation of PKG has been described using 144 mg of extract; less starting material can be used. For PKG release, the repeated freeze–thaw cycles used in the original purification by Miglietta and Nelson (1988) are not essential.] Dialyze against 1000 vol METP1 buffer (see Section II).

2. Centrifuge dialysate at 150,000*g* for 1 hour at 4°C.

3. Run a DEAE-cellulose column in METP1 at 15–20 ml/h. Most of the PKG activity flows through the column. Concentrate flow-through fractions to a volume of 1 ml by ultrafiltration using Centricon-10 microconcentrators (Amicon). Wash column in 5–10 vol METP1 + 1 *M* NaCl to elute all proteins retained by the resin.

4. Apply the concentrated flow-through to a large (at least 150-ml) Sephadex G-150 gel filtration column, and run column in METP1 + 0.1 M NaCl at 10 ml/h. A comparable sizing resin can be substituted. Collect at least 1 column volume as 1- to 2-ml fractions. PKG elutes at M_r approximately 80,000.

5. Store PKG fractions at 4°C. Less than 10% of the kinase activity is lost

after 3 months. Do not freeze. The overall yield of PKG is 12%, and it is purified at least 25-fold to a specific activity of 50,000 units/mg. Stimulation of kinase activity by cGMP is 29-fold.

C. Soluble Casein Kinases

Two soluble casein kinases from cilia, CKS1 and CKS2, can be partially purified and efficiently separated from PKA, PKG, and CaPKs (Walczak *et al.*, 1993).

1. Prepare at least 10 mg of ME extract of cilia in a volume under 2 ml. Adjust Mops to 20 m*M* and NaCl to 0.1 *M*.
2. Run a large (at least 150-ml) Sephacryl S-200 column in METP2 (see Section II) + 0.1 *M* NaCl at 10 ml/h. A comparable sizing resin can be substituted. Collect at least 1 column volume as 1- to 2-ml fractions. CKS1 and CKS2 elute at M_r 36,000.
3. Run α-casein–agarose column in METP2 + 0.1 *M* NaCl at 10 ml/h. Load pooled gel filtration fractions directly onto this column. Collect 1- to 2-ml fractions. Wash the column in 5 vol each of METP + 0.1 *M* NaCl, METP2 + 0.5 *M* NaCl, and METP2 + 1 *M* NaCl. CKS1 flows through the α-casein–agarose column; CKS2 elutes in 0.5 *M* NaCl.
4. Store both CKs at 4°C. The activities of CKS1 and CKS2 are lost rapidly (50% within one week). Do not freeze. Combined, CKS1 and CKS2 are purified eightfold from the ME extract with a 29% yield of activity; specific activities are 24,000 and 43,000 units/mg, respectively.

D. Axonemal Casein Kinase

An axonemal casein kinase, CKA, also can be separated from PKA, PKG, and CaPKs (Walczak *et al.*, 1993). It cannot be purified on the α-casein–agarose column, however, because it loses activity.

1. Prepare the high-salt extract of axonemes.
2. Run Sephacryl S-200 column as for CKS1 and CKS2 (step 2 above) in MMKETP buffer (see Section II). CKA elutes at M_r 28,000.
3. Store CKA at 4°C. Most of its activity is also lost within 1 week. Do not freeze. The apparent yield of CKA is 500% (presumably due to the removal during purification of an inhibitor present in the starting fraction), and CKA is purified 66-fold from the high-salt extract of axonemes to a specific activity of 11,000 units/mg.

Acknowledgment

The work described in this chapter was supported by National Institutes of Health Grants GM32514, GM34906, and T32-GM07215.

References

Ann, K. (1991). Cyclic GMP-dependent protein kinase and nucleoside diphosphatekinase from *Paramecium tetraurelia:* Biochemical and immunological studies. Ph.D thesis. University of Wisconsin at Madison.

Bradford, M. M. (1976). A rapid and sensitive method for the quantitation of microgram quantities of protein utilizing the principle of protein-dye binding. *Anal. Biochem.* **72,** 248–256.

Carlson, G. L. (1994). Cyclic AMP-dependent protein kinases of *Paramecium tetraurelia:* subunit structures and substrate recognition. Ph.D. thesis. University of Wisconsin at Madison.

Corbin, J. D., and Reiman, E. M. (1974). Assay of cAMP-dependent protein kinases. *Methods Enzymol.* **38,** 287–299.

Gundersen, R. E., and Nelson, D. L. (1987). A novel Ca^{2+}-dependent protein kinase from *Paramecium tetraurelia. J. Biol. Chem.* **262,** 4602–4609.

Hochstrasser, M. (1989). Tools for the study of cAMP-mediated events in *Paramecium tetraurelia:* Purification of a cAMP-dependent protein kinase and production of monoclonal antibodies against it. Ph.D thesis. University of Wisconsin at Madison.

Hochstrasser, M., and Nelson, D. L. (1989). Cyclic AMP-dependent protein kinase in *Paramecium tetraurelia:* Its purification and the production of monoclonal antibodies against both subunits. *J. Biol. Chem.* **264,** 14510–14518.

Kameshita, I., and Fujisawa, H. (1989). A sensitive method for detection of calmodulin-dependent protein kinase II activity in sodium dodecyl sulfate-polyacrylamide gel. *Anal. Biochem.* **183,** 139–143.

Kim, K.-H. (1994). Cloning and characterization of calcium-dependent protein kinases from *Paramecium tetraurelia.* Ph.D. thesis. University of Wisconsin at Madison.

Laemmli, U. K. (1970). Cleavage of structural proteins during the assembly of the head of bacteriophage T4. *Nature (London)* **227,** 680–685.

Mason, P. A., and Nelson, D. L. (1989a). Cyclic AMP-dependent protein kinases of *Paramecium.* I. Chromatographic and physical properties of enzymes from cilia. *Biochim. Biophys. Acta* **1010,** 108–115.

Mason, P. A., and Nelson, D. L. (1989b). Cyclic AMP-dependent protein kinases of *Paramecium.* II. Catalytic and regulatory properties of type II kinase from cilia. *Biochim. Biophys. Acta* **1010,** 116–121.

Miglietta, L. P. (1987). Purification and characterization of cGMP-dependent protein kinase from *Paramecium.* Ph.D thesis. University of Wisconsin at Madison.

Miglietta, L. A. P. and Nelson, D. L. (1988). A novel cGMP-dependent protein kinase from *Paramecium. J. Biol. Chem.* **263,** 16096–16105.

Murofushi, H. (1973). Purification and characterization of a protein kinase in *Tetrahymena* cilia. *Biochim. Biophys. Acta* **327,** 354–364.

Murofushi, H. (1974). Protein kinases in *Tetrahymena* cilia II. Partial purification and characterization of adenosine 3′,5′-monophosphate-dependent and guanosine 3′,5′-monophosphate-dependent protein kinases. *Biochim. Biophys. Acta* **370,** 130–139.

Satir, P., Barkalow, K., and Hamasaki, T. (1993). The control of ciliary beat frequency. *Trends Cell Biol.* **3,** 409–412.

Son, M. (1991). Biochemical and immunological studies of two novel calcium-dependent protein kinases from *Paramecium tetraurelia.* Ph.D thesis. University of Wisconsin at Madison.

Son, M., Gundersen, R. E., and Nelson, D. L. (1993). A second novel Ca^{2+}-dependent protein kinase in *Paramecium:* Purification and biochemical characterization. *J. Biol. Chem.* **268,** 5940–5948.

Walczak, C. E., and Nelson, D. L. (1994). Regulation of dynein-driven motility in cilia and flagella. *Cell Motil. Cytoskel.* **27,** 101–107.

Walczak, C. E., Anderson, R. A., and Nelson, D. L. (1993). Identification of a family of casein kinases in *Paramecium:* Biochemical characterization and cellular localization. *Biochem. J.* **296,** 729–735.

CHAPTER 67

Isolation of Inner- and Outer-Arm Dyneins

David R. Howard and Winfield S. Sale

Department of Anatomy and Cell Biology
Emory University School of Medicine
Atlanta, Georgia 30322

I. Introduction

Dyneins are the main structural and functional components of the inner and outer arms of eukaryotic flagella and cilia. The composition of axonemal dyneins varies from species to species and within a single axoneme. Within an axoneme, outer-arm dyneins are thought to be homogenous in form (Witman, 1989), whereas the inner-arm dyneins consist of at least three subforms (Piperno *et al.*, 1990; Muto *et al.*, 1991; Mastronarde *et al.*, 1992). Dyneins were first separated from other axonemal components by selective extraction in low-ionic-strength solution containing ethylenediaminetetraacetic acid (EDTA) (Gibbons, 1963); however, this procedure causes dynein molecules to dissociate into subunits (Tang *et al.*, 1982; Pfister and Witman, 1984; King *et al.*, 1990). It is now most common to use high-ionic-strength (elevated NaCl or KCl) solution to extract dynein from axonemes.

Biochemical quantities of dynein can be obtained from (at least) five different sources. For this chapter, we describe procedures for isolation of axonemes and outer-arm dynein from the flagella of the sea urchin *Strongylocentrotus purpuratus* and for the isolation of inner-arm dynein from the flagella of the green alga *Chlamydomonas reinhardtii*. Similar procedures are used to isolate dynein from a number of sources. Previously published protocols for the isolation of dynein from specific sources include those for *Tetrahymena* cilia (Johnson, 1986); *Paramecium* cilia (Travis and Nelson, 1988; Hamasaki *et al.*, 1991); trout sperm flagella (Moss *et al.*, 1991); sea urchin sperm flagella (Bell *et al.*, 1982; Sale *et al.*, 1989); and *Chlamydomonas* flagella (King *et al.*, 1986).

II. Isolation of Sea Urchin Sperm Axonemes and Dynein

A. Collection of Sperm

1. Obtain semen by injecting a small amount of 0.53 *M* KCl into the body cavities of sea urchins, and dilute the product approximately four-fold into seawater containing 0.1 m*M* EDTA. Centrifuge the sperm suspension at 30*g* for 5 minutes to remove debris.

2. Carefully remove the supernatant and centrifuge at 3000*g* for 8 minutes to pellet the sperm.

B. Isolation and Demembranation of Flagella

Two alternative methods to demembranate flagella, osmotic shock (method 1) and detergent (method 2), are presented here. The choice of demembranation agent can be crucial, depending on the intended use of the dynein. Because exposure of sea urchin dynein to detergent nonphysiologically activates the dynein ATPase activity (Bell *et al.*, 1982), it is preferable to demembranate by osmotic shock with high sucrose for investigations into the regulation of axonemes and dynein.

1. Demembranation by Osmotic Shock

This procedure was developed by Bell *et al.*, (1982).

1. Resuspend the flagellar pellet into 20% sucrose in isolation buffer [0.1 *M* NaCl, 5 m*M* imidazole/HCl, pH 7.0, 4 m*M* $MgSO_4$, 1 m*M* dithiothreitol (DTT), 5 m*M* 2-mercaptoethanol (2-ME), 0.2 m*M* phenylmethylsulfonyl fluoride (PMSF)] with eight strokes of a Dounce homogenizer with a tight pestle. This and all further steps are carried out at 4°C.
2. Centrifuge the homogenate twice at 3000*g* for 7 minutes to pellet the sperm heads.

3. Pellet the axonemes at 27,000*g* for 15 minutes.
4. Wash the axonemes twice in half the original volume of isolation buffer lacking sucrose or Triton.

2. Demembranation with Detergent

This procedure was described by Sale *et al.* (1985).

1. Resuspend the flagellar pellet into 0.5% Triton X-100 in isolation buffer by a homogenization at 4°C similar to that described in Section II, B, 1, step 1.
2. Centrifuge the homogenate twice at 2000*g* for 5 minutes to pellet the sperm heads.
3. Pellet the axonemes at 12,000*g* for 8 minutes.
4. Wash the axonemes as described in Section II, B, 1, step 4.

C. Isolation of Outer-Arm Dynein from Sea Urchin Sperm Flagella

Large quantities of outer-arm dynein can be isolated from sea urchin flagella (Bell *et al.*, 1982). This purification is facilitated by the resistance of the inner-arm dyneins to extraction by high-ionic-strength solution (Gibbons and Gibbons, 1973; Fox and Sale, 1987). Following high-salt extraction of the outer arms, inner-arm dyneins can be partially extracted by low-ionic-strength dialysis, but this inner-arm preparation contains about 15–20% outer arms (for procedure, see Sale *et al.*, 1989). Furthermore, the structural and functional properties of inner-arm dynein prepared in this manner are not well characterized. Therefore, we present only the procedure for extraction of the outer-arm dynein (from Bell *et al.*, 1982; Sale *et al.*, 1985).

1. Measure the protein concentration of the axonemal solution.
2. Pellet the axonemes at 12,000*g* for 8 minutes and resuspend to a concentration of 2–8 mg/ml in high-salt buffer (isolation buffer with NaCl adjusted to 0.6 *M* and 0.1 m*M* EDTA).
3. Incubate 15 minutes on ice.
4. Centrifuge extracted axonemes at 30,000*g* for 20 minutes.
5. Carefully remove the dynein-containing supernatant and save for further purification or experimentation.

III. Isolation of *Chlamydomonas* Inner-Arm Dynein

The existence of numerous outer-arm and inner-arm mutants (Kamiya *et al.*, 1989; Piperno *et al.*, 1990) and the ability to extract both inner and outer arms with high-salt solutions (Piperno and Luck, 1979; Huang *et al.*, 1979; Pfister *et al.,* 1982) make *Chlamydomonas* an excellent system for purification of inner-arm dyneins. To isolate inner-arm dyneins from *Chlamydomonas,* a cell line genetically lacking the entire outer-arm structure is used [e.g., *pf*28 (Mitchell and Rosenbaum, 1985); *oda*1-10 (Kamiya, 1988)]. Likewise, when purifying outer-arm dynein from *Chlamydomonas,* mutants missing certain inner arms [e.g., *pf*9 (*pf*30, *ida*1, Porter *et al.,* 1992) or *pf*23 (Huang *et al.,* 1979] can reduce initial contamination by inner arms. The following procedure is based on the methods of Piperno *et al.* (1990) and Smith and Sale (1991).

1. Collect flagella from an outer armless mutant as described in Chapter 2 of this volume. Resuspend flagellar pellets in buffer A [30 m*M* NaCl, 10 m*M* 4-(2-hydroxyethyl)-1-piperazineethanesulfonic acid (Hepes), pH 7.4, 5 m*M* $MgSO_4$, 1 m*M* DTT, 0.5 m*M* EDTA, 0.06 TIU/ml aprotinin, 0.1 m*M* PMSF), ~2 ml for every liter of cell culture used or ~1 ml/ml of pelleted cells.
2. Demembranate flagella by adding Nonidet P-40 to 0.5%. Gently mix, taking care to minimize the formation of bubbles because proteins denature at the air/solution interface.
3. Centrifuge the axonemes at 27,000*g* for 15 minutes.
4. Resuspend the axonemes in buffer A to one-fourth the volume used in step 1. (To improve removal of membrane and matrix components, the axonemes can be washed once by centrifugation and resuspension.)
5. Measure the protein concentration of the axonemal solution.
6. Pellet the axonemes at 27,000*g* for 15 minutes at 4°C.
7. Resuspend the axonemes to 6–10 mg/ml in 0.6 *M* NaCl in buffer A to extract dynein, and incubate 15 minutes on ice.
8. Pellet extracted axonemes at 27,000*g* for 20 minutes at 4°C.
9. Carefully remove the dynein-containing supernatant and save on ice.

IV. Fractionation of Dyneins by Sucrose Density Gradient Zonal Centrifugation

1. Prepare pairs of 5–20% (w/v) continuous sucrose gradients in the appropriate buffer. The salt concentration will affect the conformation and consequently the mobility of proteins in the extract (Tang *et al.,* 1982).
2. From the top of the tube, remove a volume of the gradient equal to the volume of the extract (0.5–0.7 ml on a 12-ml gradient), and gently overlay the extract solution. Each bucket must be weighed and balanced according to the manufacturer's specifications to prevent rotor or centrifuge damage.
3. Centrifuge at 28,000 rpm (Beckman, SW50.1) or 35,000 rpm (Beckman, SW41) for 16 hours at 4°C.
4. Gradients can be fractionated in a number of ways. We usually collect gradient fractions by attaching a 50-μl capillary pipet to peristaltic pump tubing

and gently inserting the pipet to the bottom of the gradient. The solution is then drawn at a pump flow rate of 4–6 ml/min, collecting from bottom to top. We generally collect 20–21 equivalent fractions (~250 μl for SW50.1, ~550 μl for SW41) into 1.5-ml microcentrifuge tubes.

Peak fractions of dynein are identified by assay of ATPase activity of individual fractions (Chapter 21 or 22 of this volume) or by sodium dodecyl sulfate–polyacrylamide gel electrophoresis (see Chapter 15 in this volume). Sea urchin outer-arm and *Chlamydomonas* inner-arm I1 sediment at 21 S and are recognized in 2.5–5 or 3–7% gels by the presence of two prominent bands (heavy chains α and β) in the region of M_r 350K–500K (see Bell *et al.,* 1982, and Smith and Sale, 1991, respectively). Cosedimenting, diagnostic intermediate chains of M_r 122K, 90K, and 76K in sea urchin outer-arm dynein and 140K and 97K in *Chlamydomonas* I1 dynein also can be resolved in these gels.

For isolation of sea urchin outer-arm or *Chlamydomonas* inner-arm I1, we find that purification by zonal centrifugation produces dynein of sufficient purity for most uses. If further purification of individual dynein species is required (e.g., *Chlamydomonas* outer-arm purification), selected sucrose gradient fractions are pooled and purified by hydroxylapatite column chromatography (King et al., 1986) or high-pressure liquid chromatography (see Chapter 68 of this volume).

V. Other Considerations

To minimize contamination of the dynein extracts by axonemal tubulin, it is possible to include 10 μM taxol in the high-salt solution (Sale *et al.,* 1985). Taxol reduces the disassembly of axonemal microtubules.

Either NaCl or KCl can be used as the salt for these experiments; however, the presence of >0.2 M KCl will cause precipitates to form in the gel sample buffer.

Although we have not fully characterized this observation, the addition of ATP (≥1 mM) to the high-salt solution may enhance release of certain inner-arm dyneins in *Chlamydomonas*.

References

Bell, C. W., Fraser, C. L., Sale, W. S., Tang, W-J. Y., and Gibbons, I. R. (1982). Preparation and purification of dynein. *Methods Enzymol.* **85,** 450–474.

Fox, L. A., and Sale, W. S. (1987). Direction of force generated by the inner row of dynein arms on flagellar microtubules. *J. Cell Biol.* **105,** 1781–1787.

Gibbons, I. R. (1963). Studies on the protein components of cilia from *Tetrahymena pyriformis*. *Proc. Natl. Acad. Sci. U. S. A.* **50,** 1002–1010.

Gibbons, B. H., and Gibbons, I. R. (1973). The effect of partial extraction of dynein arms on the movement of reactivated sea urchin sperm. *J. Cell Sci.* **13,** 337–357.

Hamasaki, T., Barkalow, K., Richmond, J., and Satir, P. (1991). cAMP-stimulated phosphorylation of an axonemal polypeptide that copurifies with the 22S dynein arm regulates microtubule translocation velocity and swimming speed in *Paramecium*. *Proc. Natl. Acad. Sci. U. S. A.* **88,** 7918–7922.

Huang, B. Piperno, G., and Luck, D. J. L. (1979). Paralyzed flagella mutants of *Chlamydomonas reinhardtii*. *J. Biol. Chem.* **254,** 3091–3099.

Johnson, K. A. (1986). Preparation and properties of dynein from *Tetrahymena* cilia. *Methods Enzymol.* **134,** 306–317.

Kamiya, R. (1988). Mutations at twelve independent loci result in absence of outer dynein arms in *Chlamydomonas reinhardtii*. *J. Cell Biol.* **107,** 2253–2258.

Kamiya, R., Kurimoto, E., Sakakibara, H., and Okagaki, T. (1989). A genetic approach to the function of inner and outer arm dynein. *In* "Cell Movement. Vol 1: The Dynein ATPases," pp. 209–218. (F. D. Warner, P. Satir, and I. R. Gibbons, eds.), Alan R. Liss, New York.

King, S. M., Otter, T., and Witman, G. B. (1986). Purification and characterization of *Chlamydomonas* flagellar dyneins. *Methods Enzymol.* **134,** 291–306.

King, S. M., Gatti, J.-L., Moss, A. G., and Witman, G. B. (1990). Outer-arm dynein from trout spermatozoa: substructural organization. *Cell Motil. Cytoskel.* **16,** 266–278.

Mastronarde, D. N., O'Toole, E. T., McDonald, K. L., McIntosh, J. R., and Porter, M. E. (1992). Arrangement of inner dynein arms in wild-type and mutant flagella of *Chlamydomonas*. *J. Cell Biol.* **118,** 1145–1162.

Mitchell, D. R., and Rosenbaum, J. L. (1985). A motile *Chlamydomonas* flagellar mutant that lacks outer dynein arms. *J. Cell. Biol.* **100,** 1228–1234.

Moss, A. G., Gatti, J.-L., King, S. M., and Witman, G. B. (1991). Purification and characterization of *Salmo gairdneri* outer arm dynein. *Methods Enzymol.* **196,** 201–221.

Muto, E., Kamiya, R., and Tsukita, S. (1991). Double-rowed organization of inner dynein arms in *Chlamydomonas* flagella revealed by tilt-series thin-section electron microscopy. *J. Cell Sci.* **99,** 57–66.

Pfister, K. K., and Witman, G. B. (1984). Subfractionation of *Chlamydomonas* 18S dynein into two unique subunits containing ATPase activity. *J. Biol. Chem.* **259,** 12072–12080.

Pfister, K. K., Fay, R. B., and Witman, G. B. (1982). Purification and polypeptide composition of dynein ATPases from *Chlamydomonas* flagella. *Cell Motil.* **2,** 525–547.

Piperno, G., and Luck, D. J. L. (1979). Axonemal adenosine triphosphatases from flagella of *Chlamydomonas reinhardtii*. *J. Biol. Chem.* **254,** 3084–3090.

Piperno, G., Ramanis, Z., Smith, E. F., and Sale, W. S. (1990). Three distinct inner dynein arms in *Chlamydomonas* flagella: Molecular composition and location in the axoneme. *J. Cell Biol.* **110,** 379–389.

Porter, M. E., Power, J., and Dutcher, S. K. (1992). Extragenic suppressors of paralyzed flagellar mutations in *Chlamydomonas reinhardtii* identify loci that alter the inner dynein arms. *J. Cell Biol.* **118,** 1163–1176.

Sale, W. S., Goodenough, U. W., and Heuser, J. E. (1985). The substructure of isolated and in situ outer dynein arms of sea urchin sperm flagella. *J. Cell Biol.* **101,** 1400–1412.

Sale, W. S., Fox, L. A., and Milgram, S. A. (1989). Composition and organization of the inner row dynein arms. *In* "Cell Movement. Vol 1: The Dynein ATPases," pp. 89–102. (F. D. Warner, P. Satir, and I. R. Gibbons, eds.), Alan R. Liss, New York.

Smith, E. F., and Sale, W. S. (1991). Microtubule binding and translocation by inner dynein arm subtype I1. *Cell Motil. Cytoskel.* **18,** 258–268.

Tang, W.-J. Y., Bell, C. W., Sale, W. S., and Gibbons, I. R. (1982). Structure of the dynein-1 outer arm in sea urchin sperm flagella. *J. Biol. Chem.* **257,** 508–515.

Travis, S. M., and Nelson, D. L. (1988). Purification and properties of dyneins from Paramecium cilia. *Biochem. Biophys. Acta.* **966,** 73–83.

Witman, G. B. (1989). Composition and molecular organization of the dyneins. *In* "Cell Movement. Vol 1: The Dynein ATPases," pp. 25–35. (F. D. Warner, P. Satir, and I. R. Gibbons, eds.), Alan R. Liss, New York.

CHAPTER 68

Separation of Dynein Species by High-Pressure Liquid Chromatography

Osamu Kagami and Ritsu Kamiya
Zoological Institute
Graduate School of Science
Univeristy of Tokyo
Tokyo 113, Japan

I. Introduction

Traditionally, most studies on isolated dynein have used a purification step with sucrose density gradient centrifugation. Although this method can separate dynein from contaminants having low sedimentation coefficients, it has a significant disadvantage in that the centrifugation process takes as long as 5–8 hours. As an alternative to this method, high-pressure liquid chromatography (HPLC) on a Mono Q column has been used to separate various axonemal dyneins (Goodenough *et al.*, 1987). This procedure takes only 1 to 2 hours. It has the further advantage that it is able to separate multiple inner-arm dyneins, which sucrose gradient centrifugation is unable to separate (Kagami and Kamiya, 1992); however, the sample obtained by Mono Q chromatography may be contaminated by proteins that have similar charges. Therefore, it is recommended that sucrose gradient centrifugation be used in combination with Mono Q chromatography when very pure samples are desired. The following proce-

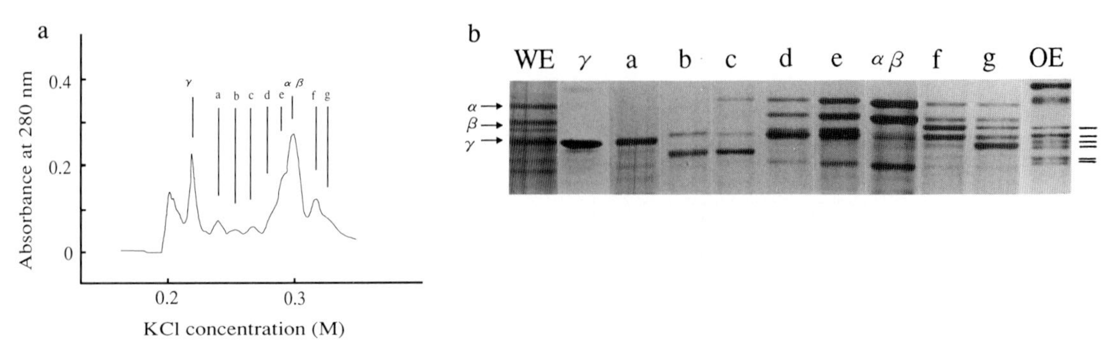

Fig. 1 Elution profile (a) and sodium dodecyl sulfate–polyacrylamide gel electrophoresis (SDS–PAGE) pattern of peak fractions (b) of crude dynein extract obtained from 30 mg of wild-type axonemes. Dynein is separated into nine peak fractions containing distinct subspecies. Peaks a–g originate from inner-arm dynein, whereas peaks γ and $\alpha\beta$ are from outer-arm dynein (Kagami and Kamiya, 1992). In the SDS–PAGE pattern, only the high-molecular-weight region containing the dynein heavy-chain bands (M_r 400–550K) is shown. The arrows indicate the outer-arm heavy chains; short lines on the right indicate the inner-arm heavy chains. Note that each peak fraction has discrete sets of heavy-chain bands. WE, crude extract from wild-type axonemes; OE, crude extract from *oda*1 axonemes lacking the outer-arm dynein. The two bands at the top of the lane are membrane proteins.

dure is based on that of Goodenough *et al.* (1987) and has been used to separate seven discrete subspecies of inner-arm dyneins from *Chlamydomonas* axonemal extract (Fig. 1) (Kagami and Kamiya, 1992).

II. Solutions

HMDE: 30 m*M* 4-(2-hydroxyethyl)-1-piperazineethanesulfonic acid (Hepes), 5 m*M* $MgSO_4$, 1 m*M* dithiothreitol, 1 m*M* ethylene glycol bis(β-aminoethyl ether)-*N*,*N*′-tetraacetic acid (EGTA), pH 7.4. This solution should be made in highly purified water, such as that obtained with a Milli-Q system (Millipore), and filtered through a 0.2-μm cellulose–acetate filter to remove any remaining small particles.

50 m*M* phenylmethylsulfonyl fluoride (PMSF) in ethanol: This solution should be stored at −20°C and used within 1 week after being made up.

1 mg/ml leupeptin.

HMDEI: HMDE containing 0.1 m*M* PMSF and 5 μg/ml leupeptin. Because aqueous PMSF solution is unstable, this solution should be made from the above stock solutions within 30 minutes of use.

III. High-Pressure Liquid Chromatography System

Our HPLC system (FPLC, Pharmacia) consists of a 2249 gradient pump, a 2510 UV monitor, a 2210 potentiometric recorder, and a Radifrac fraction collector. Any similar HPLC system should be suitable. A Mono Q HR5/5 column (Pharmacia) is used. The entire system should be kept at 4°C.

IV. Procedure

A. Dynein Extraction

1. Isolate a total of 10–30 mg flagellar axonemal proteins as described in Chapter 2 of this volume.
2. Resuspend axonemes in 0.5 ml of HMDEI containing 0.6 *M* KCl and leave on ice for 15 minutes. Centrifuge the suspension at 15,000*g* for 15 minutes. Retain the supernatant.
3. Resuspend the pellet with 0.5 ml of HMDEI containing 0.6 *M* KCl, centrifuge the suspension, and retain the supernatant, as above.
4. Combine the two supernatant fractions and dilute with HMDEI to 60 m*M* final KCl concentration to yield 10 ml of the sample.
5. Centrifuge the extract at 20,000*g* for 15 minutes. Save the supernatant.

B. Mono Q Chromatography

1. Equilibrate a Mono Q HR5/5 analytical anion-exchange column with HMDEI.
2. Adjust the sensitivity of the UV monitor.
3. Load the diluted extract on the column at a rate of 0.5 ml/min using a 10-ml Superloop (sample reservoir).
4. After the proteins that do not bind to the resin have flowed through completely, apply a linear 0.2 to 0.5 *M* KCl gradient for 100 minutes at a flow rate of 0.5 ml/min. Collect 0.5-ml fractions.

References

Goodenough, U. W., Gebhart, B., Mermall, V., Mitchell, D. R., and Heuser, J. E. (1987). High-pressure liquid chromatography fractionation of *Chlamydomonas* dynein extracts and characterization of inner-arm dynein subunits. *J. Mol. Biol.* **194,** 481–494.

Kagami, O., and Kamiya, R. (1992). Translocation and rotation of microtubules caused by multiple species of *Chlamydomonas* inner-arm dynein. *J. Cell Sci.* **103,** 653–664.

CHAPTER 69

Reconstitution of Dynein Arms *in Vitro*

Elizabeth F. Smith
Department of Genetics and Cell Biology
University of Minnesota
St. Paul, Minnesota 55108

I. Introduction

Reconstitution experiments often provide the most convincing evidence to support models of organelle function and assembly. These experiments can also be the most technically challenging as there are many unknowns in even the most extensively studied systems. Nonetheless, several laboratories have been successful in using this approach to begin elucidating mechanisms of flagellar dynein arm function and assembly. Gibbons and Gibbons (1976, 1979) were the first investigators to reconstitute structurally and functionally dynein arms *in vitro*. They found that sea urchin sperm whose outer dynein arms had been selectively extracted beat at half the original frequency (Gibbons and Gibbons, 1973). After the outer dynein arms were restored, beat frequency was restored to preextraction levels. Since these initial efforts, reconstitution experiments have also been used in studies of *Tetrahymena* cilia (Takahashi and Tonomura, 1978; Mitchell and Warner, 1980; Warner *et al.*, 1985) and

METHODS IN CELL BIOLOGY, VOL. 47

Chlamydomonas flagella (Sakakibara and Kamiya, 1989; Smith and Sale, 1992a,b) to begin answering questions of dynein function and regulation, as well as dynein assembly onto doublet microtubules. For these two organisms reconstitution is somewhat more difficult as both inner and outer dynein arms are extracted in high-salt buffers. For *Chlamydomonas* flagella, this difficulty is overcome to a certain extent with the aid of mutants lacking subsets of inner and/or outer dynein arms. As methods for obtaining axonemes and purifying dynein arms are presented in other chapters of this book, in this chapter I only briefly describe strategies for designing reconstitution experiments and focus on methods of assessing experimental success.

II. Methods for Isolating Axonemes, Extracted Axonemes, and Dynein

The buffers sited in the example below are those most commonly used for *Chlamydomonas reinhardtii*. As mentioned above, although both outer and inner dynein arms are extracted in high-salt-containing buffers, using mutants that are lacking outer dynein arms and/or subsets of inner dynein arms allows for greater diversity in experimental design. Similar protocols can be adapted for isolating axonemes and dyneins from other organisms for use in reconstitution.

A. Solutions

Buffer A: 10 m*M* 4-(2-hydroxyethyl)-1-piperazineethanesulfonic acid (Hepes), 5 m*M* $MgSO_4$, 1 m*M* dithiothreitol (DTT), 0.5 m*M* ethylenediaminetetraacetic acid (EDTA), 0.1 m*M* phenylmethylsulfonyl fluoride (PMSF), and 0.5 TIU aprotinin, pH 7.4

Buffer A + 600 m*M* NaCl

Buffer A + 30 m*M* NaCl

5–20% sucrose gradient in buffer A + 30 m*M* NaCl

Nonidet P-40 (NP-40)

B. Procedure

1. Flagella are severed from cell bodies using the dibucaine method (Witman, 1986) and isolated by differential centrifugation in 30 m*M* NaCl buffer A. Axonemes are isolated using 0.5% NP-40 to remove flagellar membranes.

2. To obtain crude dynein extracts, axonemes are resuspended to a concentration of 10 mg/ml in buffer A containing 600 m*M* NaCl and extracted for 15 minutes to 1 hour on ice. Axonemes are then pelleted and resuspended in 30 m*M* NaCl buffer A and the extract is prepared for reconstitution. If axonemes

completely devoid of outer and inner dynein arms are required, it may be necessary to extract the axonemes a second time under the same conditions to remove any residual dyneins.

3. If whole extract is to be used in reconstitution, then the extract is dialyzed into buffer A + 30 m*M* NaCl. If purified arms are used, the extract is loaded onto a 5–20% sucrose gradient made with buffer A and run for 16 hours at 35,000 rpm in an SW41 rotor (Beckman Instruments). Twenty equal fractions are collected and dynein peaks determined by rapid gel electrophoresis on 7% polyacrylamide minigels. Alternatively, dynein peaks may be determined using ATPase assays.

4. For reconstitution, dyneins and axonemes can be combined based on volume and the original stoichiometry with a final protein concentration of 7–8 mg/ml for visualization on Coomassie-stained gels. Reconstitutions are done for 15 minutes at room temperature.

III. Assessing Reconstitution

A. Biochemical Methods

After reconstitution is complete, axonemes are pelleted (15,000 rpm for 20 minutes, SS34 rotor, Sorvall Instruments), supernatants recovered, and axonemes resuspended in the appropriate volume of 30 m*M* NaCl buffer A. Samples are run on polyacrylamide gels composed of a 2.5–5% acrylamide resolving gel with both a urea (0–8 *M*) and a sucrose gradient (Sale *et al.*, 1985).

To quantitate stoichiometric binding, reconstitutions are generally performed using extracted or mutant axonemes and whole dialyzed extracts. This allows dynein and axonemes to be recombined in varying ratios based on the original protein concentration and stoichiometry. The resulting pellets are then run on polyacrylamide gels, and band intensities of dynein arm heavy chains are compared using a densitometer. We have used the Image-1 system gel scanning function to measure band intensity (Universal Imaging) (Smith and Sale, 1992a). Heavy chains were scanned horizontally across the gel to compare intensities of a single band for all conditions of reconstitutions. Final intensity values result from three subtractions. First, this program automatically subtracts the baseline for each lane. Second, the amount of dynein present in axonemes or extracted axonemes is subtracted from the amount pelleting after the addition of extract. And third, nonspecifically sedimenting dynein is subtracted. This value is determined by measuring dynein intensities in control experiments where extracts are added to axonemes containing all dynein arms. Electron microscopy has indicated that for *Chlamydomonas,* dynein arms do not bind to extraneous positions and, therefore, dynein pelleting with control axonemes does not represent dynein bound to axonemes.

B. Structural Analysis

The most obvious assessment of reconstitution is structural analysis using electron microscopy (Gibbons and Gibbons, 1979). Reconstituted axonemes are prepared for thin-section electron microscopy as described in Chapter 28 of this volume. Structural analysis of reconstitution experiments involving completely extracted axonemes can be complicated, as after the two extractions sometimes required to remove all dyneins, axonemes do not usually retain their original 9 + 2 arrangement. Doublet microtubules can dissociate from one another, forming sheets or small groups of doublets. Fortunately, the radial spokes have proven highly inextractable and can be used as structural markers in these cases (see Smith and Sale, 1992a).

C. Functional Analysis

As mentioned above, the first dynein reconstitution experiments of Gibbons and Gibbons (1979) assessed functional restoration by using methods for reactivating axonemes from sea urchin sperm. Sakakibara and Kamiya (1989) performed similar experiments reconstituting outer dynein arms onto mutant *Chlamydomonas* axonemes. In both of these experiments, reconstitution resulted in restoration of beat frequency to original levels. Interestingly, there are no published reports of reactivating completely extracted axonemes by the addition of dynein or of reactivating paralyzed mutant axonemes lacking dynein arms by reconstituting the missing arms. In the future, these experiments may be possible as we begin to gain more understanding of flagellar motility and assembly. In the meantime, the sliding disintegration assay can be employed to assess functional reconstitution in these axonemes (Yano and Miki-Noumura, 1981).

In this assay, axonemes are briefly exposed to a buffer (30 m*M* NaCl buffer A minus the protease inhibitors for *Chlamydomonas*) containing protease and ATP. The protease weakens the interdoublet links and the ATP allows the dynein arms to generate force, inducing the microtubule doublets to slide past one another. Sliding can be recorded using video light microscopy and sliding velocities measured directly. This method has been used to assess inner-dynein-arm binding to mutant or extracted axonemes (Smith and Sale, 1992a) and to determine that dynein is regulated by the radial spokes (Smith and Sale, 1992b). One important note: One adaptation of the sliding disintegration assay for *Chlamydomonas* (Okagaki and Kamiya, 1986) required that flagella be sonicated prior to demembranation, yielding transversely fractured flagellar bits one-half to one-third their original length. As sonication of axonemes results in complete disruption of axoneme structure, I have omitted sonication when performing sliding disintegration on reconstituted axonemes. The only result of this alteration is that fewer axonemes undergo sliding per field of view.

IV. Precautions

Stoichiometry of dynein to axonemes may be important. So far, we have found that for *Chlamydomonas,* inner dynein arms return only to inner-dynein-arm positions even in excess of inner dynein arms (Smith and Sale, 1992a). Takahashi and Tonomura (1978), however, reported that excess dynein added back to extracted axonemes from *Tetrahymena* binds to both the A- and B-subfibers of doublet microtubules as well as the central pair of microtubules (see also Warner and McIlvain, 1982).

Inner dynein arms and outer dynein arms of *Chlamydomonas* return to the original positions, but the inner dynein arms will not rebind if the outer dynein arms are present on the axoneme. Apparently, the outer dynein arms structurally prevent the inner arms from locating their appropriate positions on axonemal doublets. For this reason, when reconstituting inner dynein arms (or other structures in the vicinity), it is necessary to extract the outer dynein arms or make use of outer-dynein-arm-less mutants.

Reconstitution seems to work best using whole extracts that are fresh and dialyzed. Reconstitutions will work with sucrose gradient fractions (Warner *et al.,* 1985; Smith and Sale, 1992a) but it is more difficult to determine binding stoichiometry as the dynein becomes somewhat diluted. In addition, factors or adapter molecules important for binding or at least correct positioning of the dyneins may be lost in fractions other than those containing the dynein heavy chains. This problem might also be true for dyneins separated and isolated by FPLC although there are no published reports of reconstitutions with FPLC-purified dyneins.

References

Gibbons, B. H., and Gibbons, I. R. (1973). The effect of partial extraction of dynein arms on the movement of reactivated sea urchin sperm. *J. Cell Sci.* **13,** 337–357.

Gibbons, B. H., and Gibbons, I. R. (1976). Functional recombination of dynein 1 with demembranated sea urchin sperm partially extracted with KCl. *Biochem. Biophys. Res. Commun.* **73,** 1–6.

Gibbons, B. H., and Gibbons, I. R. (1979). Relationship between the latent ATPase state of dynein I and its ability to recombine functionally with KCl-extracted sea urchin sperm flagella. *J. Biol. Chem.* **254,** 197–201.

Mitchell, D. R., and Warner, F. D. (1980). Interactions of dynein arms with ß-subfibers of *Tetrahymena* cilia. Quantitation of the effects of magnesium and adenosine triphosphate. *J. Cell. Biol.* **87,** 84–97.

Okagaki, T., and Kamiya, R. (1986). Microtubule sliding in mutant *Chlamydomonas* axonemes devoid of outer or inner dynein arms. *J. Cell Biol.* **103,** 1895–1902.

Sakakibara, H., and Kamiya, R. (1989). Functional recombination of outer dynein arms with outer arm-missing flagellar axonemes of a *Chlamydomonas* mutant. *J. Cell Sci.* **92,** 77–78.

Sale, W. S., Goodenough, U. W., and Heuser, J. E. (1985). The substructure of isolated and in situ outer dynein arms of sea urchin sperm flagella. *J. Cell Biol.* **101,** 1400–1412.

Smith, E. F., and Sale, W. S. (1992a). Structural and functional reconstitution of inner dynein arms in *Chlamydomonas* flagellar axonemes. *J. Cell Biol.* **117,** 573–581.

precautions in doing experiments and in interpreting results with the knowledge of such a limitation enables numerous meaningful investigations.

With regard to inhibitors of the dynein ATPase, we have to consider their specificity for dynein and ciliary or flagellar motility relative to their inhibitory effects on other motile systems and motor enzymes.

II. Inhibition Kinetics

In discussing the inhibitors of interest, it is important to keep in mind certain aspects of inhibition kinetics. In some cases, these parameters are significant for diagnosis of an effect.

The first parameter is the half-maximal concentration of the inhibitor for the inhibition. The half-maximal inhibition varies depending on the assay conditions. It would be natural to assume that an inhibitor is good (useful) if the inhibition is most potent under physiological conditions.

The second parameter is the mode of inhibition. With enzymes, inhibition usually is classified into three categories: competitive, noncompetitive, and uncompetitive inhibition. These are based on the effects of the inhibitor on the Michaelis constant of an enzyme (cf. Segel, 1993). Ciliary or flagellar motility or *in vitro* motility by dynein or other motors are not true enzyme reactions, but usually they can be analyzed in an analogous manner.

The above applies to inhibitors that interact noncovalently with their targets. Chemical modifiers also are sometimes fairly specific and effective inhibitors of physiological events, especially for enzyme reactions. In this case, the second parameter above does not exist. In addition, the inhibition depends on the progression of chemical modification, so that it is influenced by the concentration of the inhibitor, as well as on the duration of modification.

III. Vanadate

Vanadate was originally identified as the membrane ATPase inhibitor in commercial ATP (Cantley *et al.*, 1977), but now many phosphate-metabolizing enzymes and processes are known to be inhibited, although the sensitivity varies. The inhibition is due to the physicochemical nature of vanadate; vanadate acts as a phosphate analog.

Dynein ATPase was found to be inhibited by vanadate as potently as membrane ATPases (Kobayashi *et al.*, 1978). Ciliary or flagellar motility also was found to be inhibited by vanadate but less potently. This inhibition is brought about by the formation of the dynein–ADP–vanadate dead-end kinetic block, which mimics the dynein–ADP–phosphate kinetic intermediate (Shimizu and Johnson, 1983).

Among motor enzymes, dynein is most sensitively inhibited by vanadate. This is especially true for the 22 S (outer arm) dynein from *Tetrahymena* cilia, which is inhibited in a noncompetitive manner and half-maximally at less than 100 n*M* (Shimizu, 1981). Other dyneins, including cytoplasmic dynein, are reported to be less sensitive, but still are inhibited in the micromolar range or less. Myosin ATPase inhibition by vanadate is potent, but the vanadate binding to myosin is a very slow process so that the inhibition is not apparent if the assay is performed without preincubation of myosin with vanadate (Goodno and Taylor, 1982). Kinesin motility is also inhibited by vanadate but less sensitively (10 μM or more), and this inhibition is even less at higher concentrations of ATP (e.g., 1 m*M*) (Cohen *et al.*, 1989).

Vanadate can be nonenzymatically reduced by natural substances such as norepinephrine, and loses inhibitory activity (Omoto and Moody, 1988). This reduction releases the vanadate inhibition of most enzymes and events. Norepinephrine must be made fresh, e.g., at 0.25 *M* in 0.25 *M* HCl. It can be added to a final concentration of 2 m*M* in the assay solution to reverse the effect of 0.1 m*M* vanadate in a few minutes.

There are a couple of precautions in using vanadate. Two forms of vanadate can be used: sodium orthovanadate and sodium metavanadate. In solution, both are the same. Sodium orthovanadate is readily soluble in water, but because it contains varying amounts of water, one must determine the concentration by atomic absorption spectroscopy or by a chemical method. A chemical method for doing this is to make the solution 1 to 2 *N* in sulfuric acid, add 0.25 ml of 3% hydrogen peroxide per 10 ml of test solution, and compare the absorbance at 450 nm. Standard vanadate solution can be made by dissolving ammonium vanadate in water or vanadium pentoxide in alkali solution. Sodium metavanadate (0.1 *M*) must be dissolved in 0.1 *N* NaOH, and the solution must be titrated to pH 10 and boiled. In either case, if the solution is yellow, it should not be used because the coloration indicates formation of oligomeric or polymeric species.

The other precaution is that an aqueous solution of vanadate contains not only monomeric vanadate but oligomeric forms. At pH 7, monomeric vanadate is the major species up to 1 m*M*. At acidic pH, oligomeric forms are more likely to exist. Thus it is advised to keep the stock solution alkaline in the dark, and dilute into the assay solution directly. Some effects of vanadate are assumed to be due to those oligomeric forms. For the inhibition of dynein ATPase and ciliary or flagellar motility, the monomeric form is believed to be responsible.

IV. Aluminum Fluoride

Aluminum fluoride and beryllium fluoride are other phosphate analogs inhibiting a wide range of phosphate-metabolizing enzymes. Investigation of the effect of these inhibitors on dynein ATPase or ciliary and flagellar motility is a future project. To make AlF_4^-, just include 5 to 10 m*M* NaF and various concentrations of $AlCl_3$. Note that Al^{3+} may be an inhibitor itself, possibly due to the formation of an AlATP inhibitory complex. Commercially available ATP or NTP may contain trace amounts of Al^{3+}. Beryllium is a cancer-causing agent, so pay special attention to safe handling of this compound.

V. *erythro*-9-[3-(2-Hydroxynonyl)]adenine

erythro-9-[3-(2-Hydroxynonyl)]adenine (EHNA), a potent inhibitor of adenosine deaminase, was shown to inhibit dynein ATPase as well as flagellar motility (Bouchard *et al.*, 1981; Penningroth, 1986). Half-maximal inhibition of the dynein was observed at 0.23 m*M* and not in a competitive manner, and that of sperm motility was in the same range. This inhibition is not potent but may be specific to dynein among molecular motors; myosin and actomyosin are not inhibited at millimolar concentrations, while the effect of EHNA on kinesin has not been investigated to our knowledge. Nonetheless, EHNA is known to affect actin assembly and actin-dependent motility so that its effects on complex biological systems should be interpreted with caution.

VI. Other Inhibitors

A. Ni^{2+}

This cation at 0.5 m*M* was reported to be inhibitory for motility of ciliary or flagellar detergent models, but was also shown to be ineffective on outer-doublet microtubule sliding (Lindemann *et al.*, 1980). Ni^{2+} appears to affect the motile axoneme directly but not the ATP-dependent dynein–microtubule interaction.

B. K-26

K-26 is an acylpeptide from *Bacillus* sp. 503, soluble in dimethyl sulfoxide, and inhibits motility of both live sea urchin spermatozoa and their Triton X-100 models in the micromolar range. Its target is not dynein ATPase, but apparently cAMP-dependent protein kinase (PKA) in flagella (Yokota *et al.*, 1989). K-26 is unavailable commercially.

C. W-7

W-7, *N*-(6-aminohexyl)-5-chloronaphthalenesulfonamide, is a calmodulin antagonist that inhibits live sperm motility at 10 to 50 μM (Iwasa *et al.*, 1987); however, it does not inhibit Triton X-100 models. Therefore, its target is unlikely to be the dynein–microtubule interaction. W-7 is commercially available and soluble in H_2O.

D. Palytoxin

This deadly toxin from a marine animal was reported to be inhibitory for sperm motility at as little as 10^{-13} *M* (Morton *et al.*, 1982). Triton X-100 models were reported to be far less sensitive, with micromolar amounts of this toxin having little effect on their motility. This toxin appears to affect the membrane of the spermatozoa.

Acknowledgments

The author thanks Dr. Charlotte K. Omoto (Washington State University) for critical reading and valuable suggestions. This work was supported by a grant-in-aid from the Agency of Industrial Science and Technology. The secretarial assistance of Ms. Iseko Akui is acknowledged.

References

Bouchard, P., Penningroth, S. M., Cheung, A., Gagnon, C., and Bardin, C. W. (1981). *erythro*-9-[3-(2-hydroxynonyl)]adenine is an inhibitor of sperm motility that blocks dynein ATPase and protein carboxylmethylase activities. *Proc. Natl. Acad. Sci. U.S.A.* **78,** 1033–1036.

Cantley, L. C., Jr., Josephson, L., Warner, R., Yanagisawa, M., Lechene, C., and Guidotti, G. (1977). Vanadate is a potent (Na,K)-ATPase inhibitor found in ATP derived from muscle. *J. Biol. Chem.* **252,** 7421–7423.

Cohn, S. A., Ingold, A. L., and Scholey, J. M. (1989). Quantitative analysis of sea urchin egg kinesin-driven microtubule motility. *J. Biol. Chem.* **264,** 4290–4297.

Goodno, C. C., and Taylor, E. W. (1982). Inhibition of actomyosin ATPase by vanadate. *Proc. Natl. Acad. Sci. U.S.A.* **79,** 21–25.

Iwasa, F., Hasegawa, Y., Ishijima, S., Okuno, M., Mohri, T., and Mohri, H. (1987). Effects of calmodulin antagonists on motility and acrosome reaction of sea urchin sperm. *Zool. Sci.* **4,** 61–72.

Kobayashi, T., Martensen, T., Nath, J., and Flavin, M. (1978). Inhibition of dynein ATPase by vanadate, and its possible use as a probe for the role of dynein in cytoplasmic motility. *Biochem. Biophys. Res. Commun.* **81,** 1313–1318.

Lindemann, C. B., Fentie, I., and Rikmenspoel, R. (1980). A selective effect of Ni^{2+} on wave initiation in bull sperm flagella. *J. Cell Biol.* **87,** 420–426.

Morton, B. E., Fraser, C. F., Thenawidjaja, M., Albagli, L., and Rayner, M. D. (1982). Potent inhibition of sperm motility by palytoxin. *Exp. Cell Res.* **140,** 261–265.

Omoto, C. K., and Moody, M. E. (1988). Kinetics of vanadate dissociation: estimation of the rate by inhibitor inactivation. *Anal. Biochem.* **168,** 337–344.

Penningroth, S. M. (1986). erythro-9[3-(2-hydroxynonyl)]adenine and vanadate as probes for microtubule-based cytoskeletal mechanochemistry. *Methods Enzymol.* **134,** 477–487.

Segel, I. H. (1993). "Enzyme Kinetics". Wiley & Sons, New York.

Shimizu, T. (1981). Steady-state kinetic study of vanadate-induced inhibition of ciliary dynein adenosinetriphosphatase activity from *Tetrahymena*. *Biochemistry* **20,** 4347–4354.

Shimizu, T., and Johnson, K. A. (1983). Presteady state kinetic analysis of vanadate-induced inhibition of the dynein ATPase. *J. Biol. Chem.* **258,** 13833–13840.

Yokota, E., Mabuchi, I., Kobayashi, A., and Sato, H. (1989). Effects of acylpeptide K-26 on the motility of sea urchin sperm model. *Cell Struct. Funct.* **14,** 299–310.

CHAPTER 71

Vanadate-Mediated Photolysis of Dynein Heavy Chains

Stephen M. King
Department of Biochemistry
University of Connecticut Health Center
Farmington, Connecticut 06030

I. Introduction

Dynein heavy chains (DHCs) may be cleaved at two discrete sites, termed V1 and V2, by UV irradiation in the presence of vanadate; these reactions were orginally described by Lee-Eiford *et al.* (1986) and Tang and Gibbons (1987), respectively. Although both reactions involve vanadate as the chromophore, the solution requirements for the two are quite different. In the first, cleavage occurs in the presence of ATP (or ADP) and low to submicromolar levels of vanadate. At these concentrations, vanadate remains monomeric and acts as a phosphate analog in an enzyme–ADP–vanadate complex. This reaction is supported by a variety of cations including Mg^{2+}, Ca^{2+}, and Zn^{2+}, but is inhibited by transition metals such as Mn^{2+}, Fe^{2+}, and Co^{2+} (Gibbons *et al.*, 1987). Interestingly, the requirement for nucleotide is not absolute, as in several cases V1 photocleavage has been obtained in the absence of ATP (Beckwith and Asai, 1993; King and Witman, 1987, 1988; Marchese-Ragona *et al.*, 1989). V1 cleavage is thought to occur within the active site of the enzyme, at or near the region that coordinates to the γ phosphate of the nucleotide. Several lines of evidence support this hypothesis. First. V1 photocleavage abolishes the

ATPase activity of the enzyme (Lee-Eiford *et al.*, 1986). Second, mapping studies indicate that V1 cleavage within the γ DHC from the *Chlamydomonas* outer arm occurs at or near the highly conserved P-loop motif that is thought to form part of the ATP hydrolytic site (King and Witman, 1988; Wilkerson *et al.*, 1994). Third, cleavage of both myosin (Cremo *et al.*, 1989) and adenylate kinase (Cremo *et al.*, 1992) under V1 conditions now has been reported. In both cases, the site of cleavage was demonstrated to occur within the glycine-rich P-loop involved in binding the triphosphate moiety of ATP.

Photocleavage at the V2 site requires the presence of higher concentrations of vanadate ($>\sim 100\ \mu M$) and the *absence* of nucleotide. Under these conditions, vanadate polymerizes and the chromophore for this reaction is probably a cyclic tetramer (see King and Witman, 1987, and Tang and Gibbons, 1987, for discussions of vanadate solution chemistry). The number of sites at which DHCs are cleaved under these conditions has been found to vary from one to three, depending on the DHC; in all cases, these sites are located C-terminal to the site of V1 cleavage. The precise nature of the V2 cleavage sites is uncertain at present; however, they appear to map at or close to the additional P-loop motifs found within each DHC (Gibbons *et al.*, 1991; Wilkerson *et al.*, 1994).

Vanadate-mediated photolysis has proved very useful in efforts to locate specific sites within DHCs (see, for example, King and Witman, 1987, 1994; Mocz *et al.*, 1988) and V1 cleavage also has been used as a diagnostic tool to identify presumptive cytoplasmic dyneins (e.g., Paschal *et al.*, 1987). In this chapter, I describe the basic techniques for obtaining photocleavage of dyneins at the V1 and V2 sites. These simple methods may be applied both to the purified enzyme and *in situ*.

II. Methods

A. Equipment and Reagents

1. UV Lamp

For V1 cleavage, lamps emitting at either 254 nm (UVG-11 lamp, UltraViolet Products, San Gabriel, CA; power output 580 μW cm^{-2} at 15 cm) or 365 nm (Spectroline EN-280L lamp, Spectronics, Westbury, NY; power output 1300 μW cm^{-2} at 15 cm) may be used. If the former is employed, it is essential to add free radical scavengers such as 1 mM dithiothreitol; extended irradiation times at 254 nm result in considerable nonspecific damage to the protein with a concomitant reduction in fragment yield. Therefore, 365-nm irradiation is highly recommended for V1 photolysis and is essential to obtain products from cleavage at V2 sites.

2. Dynein

The dynein sample should be prepared in 10 m*M* 4-(2-hydroxyethyl)-1-piperazineethanesulfonic acid (Hepes), pH 7.4, or 20 m*M* Tris–Cl, pH 7.5, containing 0.5 m*M* ethylenediaminetetraacetic acid (EDTA), appropriate salts (this will depend on the particular dynein under study), and the required cation (usually Mg^{2+}) at 1–5 m*M*. *Note:* For V2 cleavage, 1 m*M* Mn^{2+} should be added in place of Mg^{2+} to suppress photolysis at the V1 site.

3. Nucleotide

Stock solutions of nucleotides should be prepared in the same buffer as the dynein sample. Routinely, ATP at a final concentration of 10–100 μM is employed. Both ADP and 8-N_3ATP also give good results. At least for flagellar dyneins, CTP and UTP support a minor amount of cleavage but GTP and ITP do not.

4. Vanadate

A 10 m*M* stock solution of $NaVO_3 \cdot H_2O$ in water is prepared. *Note:* Metavanadate polymerizes over time to form decavanadate. This is evident as a yellow tinge to the solution, which then should be discarded. It probably is best to prepare the vanadate stock fresh for each experiment. For V1 cleavage, the final vanadate concentration may be less than 50 μM; V2 cleavage requires higher levels (100 μM to 1 m*M*).

B. Photolysis Reactions

1. Place a 20-μl aliquot of dynein in a 500-μl microfuge tube. For V1 cleavage, add vanadate and nucleotide and place the tube in a rack set in an ice bucket and packed around hard with ice. If required, also add free radical scavengers. For V2 cleavage, the samples should contain 1 m*M* Mn^{2+} and vanadate but *no* nucleotide.

2. Place the UV lamp directly on top of the tubes and irradiate. The time of irradiation required depends on the particular lamp used. Routinely, 30–60 minutes is sufficient to achieve a high degree of photolysis at the V1 site. V2 cleavage is slower and fragment yield may increase over several hours. *Note:* Some DHCs contain multiple V2 sites; cleavage at one of these sites does not necessarily preclude cleavage of the same molecule at a second or a third site. In such cases, the percentage yield of the various fragments will change as a function of the irradiation time.

3. Following photolysis, the samples may be prepared directly for gel electrophoresis or assayed to determine the effects on enzymatic activity. In the latter

case, it is necessary to reduce the added vanadate with arterenol (norepinephrine) prior to assay.

References

Beckwith, S. M., and Asai, D. J. (1993). Ciliary dynein of *Paramecium tetraurelia:* photolytic maps of the three heavy chains. *Cell Motil. Cytoskel.* **24,** 29–38.

Cremo, C. R., Grammer, J. C., and Yount, R. G. (1989). Direct chemical evidence that serine 180 in the glycine-rich loop of myosin binds ATP. *J. Biol. Chem.* **264,** 6608–6611.

Cremo, C. R., Loo, J. A., Edmonds, C. G., and Hatlelid, K. M. (1992). Vanadate catalyzes photocleavage of adenylate kinase at proline-17 in the phosphate-binding loop. *Biochemistry* **31,** 491–497.

Gibbons, I. R., Gibbons, B. H., Mocz, G., and Asai, D. J. (1991). Multiple nucleotide-binding sites in the sequence of dynein β heavy chain. *Nature* **352,** 640–643.

Gibbons, I. R., Lee-Eiford, A., Mocz, G., Phillipson, C. A., Tang, W. J. Y., and Gibbons, B. H. (1987). Photosensitized cleavage of dynein heavy chains: Cleavage at the "V1" site by irradiation of 365 nm in the presence of ATP and vanadate. *J. Biol. Chem.* **262,** 2780–2786.

King, S. M., and Witman, G. B. (1987). Structure of the α and β heavy chains of the outer arm dynein from *Chlamydomonas* flagella: masses of chains and sites of ultraviolet induced vanadate-dependent cleavage. *J. Biol. Chem.* **262,** 17596–17604.

King, S. M., and Witman, G. B. (1988). Structure of the γ heavy chain of the outer arm dynein from *Chlamydomonas* flagella. *J. Cell Biol.* **107,** 1799–1808.

King, S. M., and Witman, G. B. (1994). Multiple sites of phosphorylation within the α heavy chain of *Chlamydomonas* outer arm dynein. *J. Biol. Chem.* **269,** 5452–5457.

Lee-Eiford, A., Ow, R. A., and Gibbons, I. R. (1986). Specific cleavage of dynein heavy chains by ultraviolet irradiation in the presence of ATP and vanadate. *J. Biol. Chem.* **261,** 2337–2342.

Marchese-Ragona, S. P., Facemyer, K. C., and Johnson, K. A. (1989). Structure of the α, β, and γ heavy chains of 22 S outer arm dynein obtained from *Tetrahymena* cilia. *J. Biol. Chem.* **264,** 21361–21368.

Mocz, G., Tang, W.-J. Y., and Gibbons, I. R. (1988). A map of photolytic and tryptic cleavage sites on the β heavy chain of dynein ATPase from sea urchin sperm flagella. *J. Cell Biol.* **106,** 1607–1614.

Paschal, B. M., Shpetner, H. S., and Vallee, R. B. (1987). MAP 1C is a microtubule-activated ATPase which translocates microtubules *in vitro* and has dynein-like properties. *J. Cell Biol.* **105,** 1273–1282.

Tang, W.-J. Y., and Gibbons, I. R. (1987). Photosensitized cleavage of dynein heavy chains: cleavage at the V2 site by irradiation at 365 nm in the presence of oligovanadate. *J. Biol. Chem.* **262,** 17728–17734.

Wilkerson, C. G., King, S. M., and Witman, G. B. (1994). Molecular analysis of the γ heavy chain of *Chlamydomonas* flagellar outer-arm dynein. *J. Cell Sci.* **107,** 497–506.

CHAPTER 72

Modified Nucleotides as Probes for Dynein

Charlotte K. Omoto

Department of Genetics and Cell Biology
Washington State University
Pullman, Washington 99164

I. Introduction

Analogs of ATP have been used for a variety of purposes in the study of axonemal dyneins. Axonemal dyneins have very high affinity and relatively high turnover with ATP as compared with other nucleoside triphosphates (Shimizu, 1987), and there are relatively few analogs that support axonemal motion. Among these are those with modifications on the 2-position of adenine (Omoto and Brokaw, 1989; Omoto and Nakamaye, 1989), 2′- and 3′-positions on the ribose (Inaba *et al.*, 1989), and substitution of S for the α phosphate (Shimizu *et al.*, 1990).

This chapter focuses on the ribose-modified analogs (Fig. 1). Unlike many other analogs, these ribose-modified analogs have a high affinity for axonemal dyneins and support axonemal movement. Benzoyl benzoyl ATP (Bz_2-ATP) is a photoaffinity label; it has been used to identify the ribose-binding site of myosin

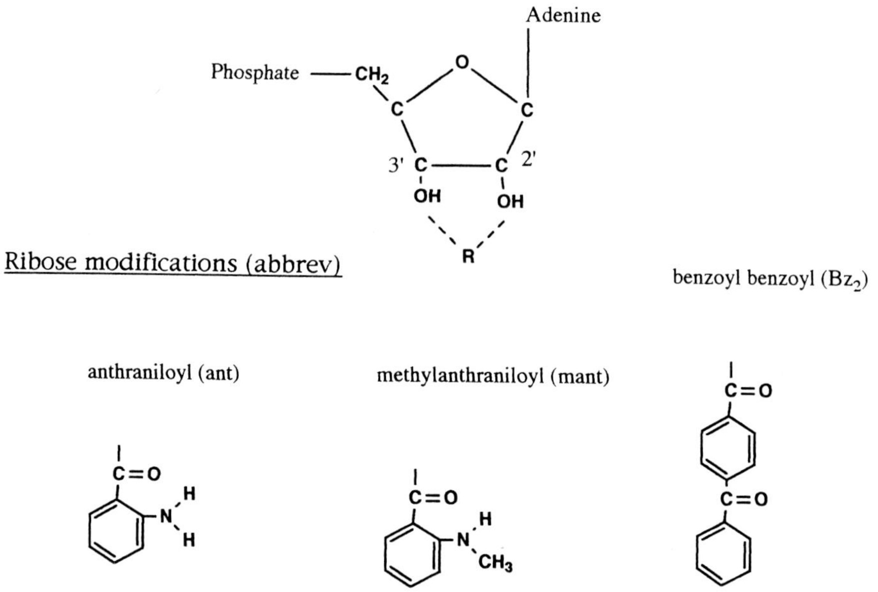

Adenine

Phosphate — CH_2

3' 2'

OH OH

R

Ribose modifications (abbrev)

benzoyl benzoyl (Bz_2)

anthraniloyl (ant)

methylanthraniloyl (mant)

Fig. 1 Diagram of ribose-modified ATP analogs that support axonemal movement. The substituents attach where indicated by R. They exchange between the 2′- and 3′-positions, indicated by the dotted line.

(Mahmood *et al.*, 1987) and may be useful for labeling dynein. Anthraniloyl ATP (antATP) and methylanthraniloyl ATP (mantATP) are fluorescent analogs that are easy to synthesize and have a large Stoke's shift ($\approx$100-nm separation between excitation and emission maxima). For these reasons, ribose-modified analogs appear to be useful probes for axonemal dynein structure and function.

Before discussing the use of these analogs in investigating axonemal dyneins, I remind the readers of a couple of key differences between axonemal dyneins and cytoplasmic dyneins or other motor molecules. First, an axoneme is a very complicated structure that requires chemomechanical coupling of many dyneins that are functionally distinct. Second, even a single axonemal dynein complex is composed of more than one heavy chain, each of which potentially has multiple nucleotide binding sites. This complexity may be dissected by the judicious use of ribose-modified analogs.

II. Synthesis and Purification

A. Benzoyl benzoyl ATP

Bz_2ATP is available from Sigma. It may be further purified by DEAE-Sephadex chromatography using a linear gradient of water: 1 *M* triethylammonium bicarbonate (TEAB), pH 7.6 (Mahmood *et al.*, 1987).

B. Anthraniloyl ATP and Methylanthraniloyl ATP

The synthesis of antATP and mantATP is detailed by Hiratsuka (1983). Molecular Probes is a good source of isatoic and methylisatoic anhydride. Vanadate-free ATP should be used, as axonemal dynein is particularly sensitive to vanadate inhibition.

1. Dissolve 1 mmole ATP in 15 ml of distilled water at 38°C.
2. Adjust pH to 9.6 with 2 *N* NaOH.
3. Add 1.5 mmole isatoic (antATP) or methylisatoic (mantATP) anhydride.
4. Maintain pH at 9.6 with 2 *N* NaOH for 2 hours with continuous stirring.
5. Bring pH to 7 with 1 *N* HCl; cool on ice.
6. Place on Sephadex LH-20 column (2.4 × 5.6 cm) and elute with water at 40 ml/h. Collect 5-ml fractions.
7. Assay fractions by silica gel thin-layer chromatography (TLC) with fluorescent indicator and develop in 1-propanol : NH_4OH : water [6 : 3 : 1, v/v, with 0.5 g/liter ethylenediaminetetraacetic acid (EDTA)].
8. Store the peak fractions as aqueous aliquots at neutral pH at −20°C.
9. Determine concentration by absorption at 252 nm using $\varepsilon = 20{,}200\ M^{-1}\ cm^{-1}$ for antATP and at 255 nm using $\varepsilon = 23{,}300\ M^{-1}\ cm^{-1}$ for mantATP.

C. Cautions

Note that the solution is not buffered so care should be taken not to overshoot the desired pH in step 2.

The reaction proceeds quickly in the beginning and the yield is greatly dependent on rapid dissolution of the anhydride. If some precipitate remains, filter out prior to column chromatography.

The elution of the analog can be followed by a UV lamp (366 nm). The analog has blue fluorescence, whereas the anthranilic acid and methylanthranilic acid are violet; however, ATP, the contaminant of greatest concern, elutes just before the analog and is not visible by this method so TLC with fluorescent indicator is essential.

III. Using Anthraniloyl ATP and Methylanthraniloyl ATP

A. Advantages

Unlike other nucleotide analogs, antATP and mantATP have higher K_m and/or K_i with axonemal dyneins than does ATP, so very low levels of ATP contamination are not as great a concern as with other analogs (Omoto, 1992; Lark and Omoto, 1994).

AntATP and mantATP are used very readily by many other enzymes; thus, ATP-regenerating systems with either pyruvate kinase or creatine kinase can be used (Hiratsuka, 1983).

Fluorescence of antATP and mantATP increases on binding to nucleotide binding sites (Cremo *et al.,* 1990). This may be useful for probing potential multiple nucleotide binding sites of dynein.

B. Cautions

For light microscopic observations, use a UV filter to prevent glow from the analogs and photodynamic damage.

Because of the ester linkage of the fluorophore, there is a potential for transfer of the fluorophore to protein. A UV filter may reduce this. This problem also may be avoided by using monodeoxy-ATP as the starting material; however, this is quite expensive and the cost may not be warranted.

The best storage is at neutral pH at −20°C. The ester linkage is readily hydrolyzed by base to yield ATP and anthranilic or methylanthranilic acid, and acid will hydrolyze the phosphate to yield antADP or mantADP.

Although sea urchin sperm axonemes and some *Chlamydomonas* mutant axonemes move with antATP and mantATP, wild-type *Chlamydomonas* axonemes do not, suggesting complex interactions between *Chlamydomonas* axonemal dyneins and the analogs (Lark and Omoto, 1994). Thus, the analogs may be useful for dissecting the function of individual dynein isoforms in some species.

References

Cremo, C. R., Neuron, J. M., and Yount, R. G. (1990). Interaction of myosin subfragment 1 with fluorescent ribose-modified nucleotides. A comparison of vanadate trapping and SH1-SH2 cross-linking. *Biochemistry* **29,** 3309–3319.

Hiratsuka, T. (1983). New ribose-modified fluorescent analogs of adenine and guanine nucleotides available as substrates for various enzymes. *Biochim. Biophys. Acta* **742,** 496–508.

Inaba, K., Okuno, M., and Mohri, H. (1989). Anthraniloyl ATP, a fluorescent analog of ATP, as a substrate for dynein ATPase and flagellar motility. *Arch. Biochim. Biophys.* **274,** 209–215.

Lark, E., and Omoto, C. K. (1994). Paralysis of axonemes due to the presence of dynein arms unable to use ribose-modified ATP. *Cell Motil. Cytoskel.* **27,** 161–168.

Mahmood, R., Cremo, C., Nakamaye, K. L., and Yount, R. G. (1987). The interaction and photolabeling of myosin subfragment 1 with 3′(2′)-O-(4-benzoyl)benzoyladenosine 5′-triphosphate. *J. Biol. Chem.* **262,** 14479–14486.

Omoto, C. K. (1992). Sea urchin axonemal motion supported by fluorescent, ribose-modified analogs of ATP. *J. Muscle Res. Cell Motil.* **13,** 635–639.

Omoto, C. K., and Brokaw, C. J. (1989). 2-Cl ATP as substrate for sea urchin axonemal movement. *Cell Motil Cytoskel.* **13,** 239–244.

Omoto, C. K., and Nakamaye, K. L. (1989). ATP analogs substituted on the 2-position as substrates for dynein ATPase activity. *Biochim. Biophys. Acta* **999,** 221–224.

Shimizu, T. (1987). The substrate specificity of dynein from Tetrahymena cilia. *J. Biochem.* **102,** 1159–1165.

Shimizu, T., Okuno, M., Marchese-Ragona, S. P., and Johnson, K. A. (1990). Phosphorothioate analogs of ATP as substrates of dynein and ciliary or flagellar movement. *Eur. J. Biochem.* **191,** 543–550.

PART IX

Genetics and Molecular Tools for the Analysis of Cilia and Flagella

CHAPTER 73

Transformation of *Chlamydomonas reinhardtii*

Julie A. E. Nelson and Paul A. Lefebvre

Department of Genetics and Cell Biology and Plant Molecular Genetics Institute
University of Minnesota
St. Paul, Minnesota 55108

I. Introduction

The development of reliable procedures for nuclear transformation of *Chlamydomonas reinhardtii* (Debuchy *et al.,* 1989; Kindle *et al.,* 1989; Kindle, 1990; Mayfield and Kindle, 1990) has led to greater usefulness of the organism in molecular studies of flagellar assembly and function. Transformation of *C. reinhardtii* is essential for confirming the relationship between DNA clones and genetic loci identified by mutation, and is very useful in generating tagged mutants for cloning genes of interest (Tam and Lefebvre, 1993). Exogenous DNA is stably integrated into the *C. reinhardtii* genome, although this integration is often associated with rearrangements of transformed DNA, as well as of the host DNA. Multiple copies of the transgenes can be integrated, either at separate sites or in linked arrays (Kindle *et al.,* 1989). Some of these copies are deleted or rearranged prior to insertion, which usually makes them nonfunctional. Also, deletions of the host genome at the point of insertion spanning 3–30 kb are common (Tam and Lefebvre, 1993).

The number of genes available for use as selectable markers is limited, because heterologous genes are very poorly expressed when transformed into *C.*

reinhardtii. The selectable marker genes used to date include genes for nitrate reductase (*NIT1*, Fernández *et al.*, 1989), argininosuccinate lyase (*ARG7*, Debuchy *et al.*, 1989), and oxygen-evolving enhancer protein 1 (*OEE1*, Mayfield and Kindle, 1990). Each of these markers is used to complement mutations in the cognate gene. There is currently only one dominant selectable marker gene, *CRY1-1*, which gives efficient transformation of all strains. *CRY1-1* encodes a mutant form of ribosomal protein S14 that confers resistance to the eukaryotic translational inhibitors cryptopleurine and emetine (Nelson *et al.*, 1994). Use of this gene requires extra manipulation of cells after transformation before selection, as discussed below.

The most common problem in making transformation work with *C. reinhardtii* is the variability in transformation efficiency between strains. The best remedy for a poor strain is to cross it to wild-type strains (e.g., strains 137c or 21gr) and use the progeny of the mating for further transformations. Cell-wall-less strains (i.e., *cw15*) are often used to eliminate the need for autolysin treatment to remove cell walls, but we have found that an incubation of nitrogen-starved cells under crowded conditions (see step 3 below) removes the cell walls efficiently (Nelson *et al.*, 1994). For arg^- strains, however, autolysin treatment is necessary for highest transformation efficiencies.

II. Materials

M (Sager–Granick), TAP, and SGII (modified Sager–Granick) media (Harris, 1989).

R medium (Schnell and Lefebvre, 1993).

MNO_3, $SGIINO_3$, and RNO_3 media: NH_4NO_3 is replaced by KNO_3 (Harris, 1989).

M-N and SGII-N media: NH_4NO_3 is omitted and the concentration of K_2HPO_4 is doubled (Harris, 1989).

Agar for solid medium: Wash agar several times with water (at least one overnight wash) to remove nitrogen when making NO_3 plates.

20% Polyethylene glycol (PEG) 8000: Filter-sterilize and store in the dark.

Emetine (Sigma): Make fresh 50 mg/ml stock in ethanol; add to autoclaved agar cooled to below 50°C.

15-ml disposable conical tubes, with one containing 0.3 g sterile, acid-washed glass beads (710–1180 μm in diameter, Sigma).

III. Procedure

1. Inoculate a bubbler flask containing 250 ml growth medium with 10 ml cells grown to about 10^7 cells/ml. Grow cells with air bubbled through the medium (see Harris, 1989, for bubbler setup) in bright light for 3 days or until density

is about 10^7 cells/ml. Cells should be grown in SGII, although some arg$^-$ strains do not grow well in SGII and should be grown in TAP. Cells may be counted before pelleting or just before transforming (Harris, 1989). If most of the cells drop to the bottom of the flask, the cells may be too dense or too old, or the pH of the medium may be wrong. Cells from such cultures transform poorly.

2. Pellet cells at 1700*g* for 10 minutes at 10°C; decant medium immediately.

3. Add 1.5 ml medium to cells, resuspend well with a pipet, measure the final volume, and transfer cells to 50-ml flask. Use SGIINO_3 if strain is Nit$^-$ and SGII-N if strain is Nit$^+$. Place the flask on a shaker under bright light and swirl cells gently for 4 hours (Sodeinde and Kindle, 1993). The transformation efficiency increases up to 4 hours, then decreases. This step not only increases transformation efficiency for most strains, it also eliminates the need to use autolysin for walled strains. If using an arg$^-$ strain, skip this step and treat with autolysin.

4. Resuspend aliquots of 5×10^7 cells in autolysin (Harris, 1989) to remove the cell walls. Slowly shake cells for 40 minutes. If cell/autolysin volume is 0.3 ml or less, cells may be transformed without further treatment. If the volume is greater than 0.3 ml, pellet cells and resuspend in 0.3 ml M-N. Cells begin to regenerate their cell walls immediately after the autolysin is removed, so do not pellet the cells until you are ready to transform them. For this reason, it is helpful to stagger autolysin treatment of cell aliquots at 5- to 10-minute intervals.

5. Adjust the concentration of the cells to 1.7×10^8 cells/ml with the medium used in step 3. Transfer 0.3 ml cells to each tube containing beads (5×10^7 cells).

6. Add plasmid to all but one of the tubes containing cells (always include a control transformation without DNA or with nonspecific DNA). Use between 0.5 and 2 μg of linearized plasmid purified by centrifugation on cesium chloride gradients prior to linearization. Linearized plasmid shows a two- to threefold increase in efficiency over supercoiled plasmid and decreases the rearrangement of plasmid sequences which can occur before integration into genomic DNA. This is especially important when using transformation for insertional mutagenesis because it increases the chance of rescuing the plasmid with flanking genomic DNA. The reaction mixture can be used directly after linearization of the plasmid without further purification.

7. Add 100 μl 20% PEG (5% final concentration) to one tube and vortex for 45 seconds at the highest setting. Immediately add 10 ml of the medium used in step 3, invert the tube to mix, and pour the cells without the beads into the empty conical tube. PEG is toxic to the cells and should not be left undiluted longer than necessary. After dilution, the cells may be set aside while additional transformations are performed.

8. Pellet cells in clinical centrifuge for 2 minutes and immediately decant medium.

9. Resuspend each transformation mixture in 0.5 ml medium used in step 3 and spread onto selective plates: RNO_3 (for acetate-requiring mutants) or MNO_3 for *NIT1*, R or M for *ARG7*. Plates should be sufficiently dry that the 0.5 ml of resuspension medium soaks into the plate within 15 minutes. Place plates under light and colonies will appear to the naked eye after approximately 1 week. Pick colonies into selective media, then isolate single colonies by restreaking the cells onto selective plates. If using the *CRY1-1* gene as the marker gene (Nelson *et al.*, 1994), the cells must be further manipulated to allow the marker gene to be expressed before selection on emetine. Skip step 9 and continue as follows.

10. Resuspend cells in 10 ml SGII (nonselective medium). Lay tubes on shaker and slowly shake under light for 3.5 hours.

11. Pellet cells and wash once with 10 ml SGII-N. Pellet again and transfer to bubbler flask containing 100 ml SGII-N. Bubble air through flask with good light for 4 days. The cells will differentiate into gametes during this period. They will also become yellow and accumulate on the bottom of the flask, but they are still viable.

12. Pellet cells and spread onto R + emetine plates. Because emetine potency is variable between batches, and because different strains show different emetine sensitivity, the concentration of emetine used for selection should be determined experimentally. First test the strain on plates with emetine concentrations ranging from 30 to 60 μg/ml to determine the effective killing concentration for nontransformed cells. Then do some test transformations starting at double the killing concentration in the previous test and ranging up to 40 μg/ml higher. These test transformations should compare a control transformation without DNA to a transformation with 1 μg of the *CRY1-1* DNA (pJN4) at each emetine concentration tested. Use the concentration at which the control has no more than 5 colonies and the pJN4 transformation has at least 50 colonies. Emetine is light sensitive, so plates should be made shortly before use and stored in the dark.

Acknowledgments

We thank Karen Kindle and Ola Sodeinde for helpful suggestions for improving transformation efficiency.

References

Debuchy, R., Purton, S., and Rochaix, J.-D. (1989). The argininosuccinate lyase gene of *Chlamydomonas reinhardtii:* an important tool for nuclear transformation and for correlating the genetic and molecular maps of the *ARG7* locus. *EMBO J.* **8,** 2803–2809.

Fernández, E., Schnell, R., Ranum, L. P. W., Hussey, S. C., Silflow, C. D., and Lefebvre, P. A. (1989). Isolation and characterization of the nitrate reductase structural gene of *Chlamydomonas reinhardtii*. *Proc. Natl. Acad. Sci. U.S.A.* **86,** 6449–6453.

Harris, E. H. (1989). "The *Chlamydomonas* Sourcebook: A Comprehensive Guide to Biology and Laboratory Use." Academic Press, San Diego.

Kindle, K. L. (1990). High-frequency nuclear transformation of *Chlamydomonas reinhardtii*. *Proc. Natl. Acad. Sci. U.S.A.* **87,** 1228–1232.

Kindle, K. L., Schnell, R., Fernández, A. E., and Lefebvre, P. A. (1989). Stable nuclear transformation of *Chlamydomonas* using the *Chlamydomonas* gene for nitrate reductase. *J. Cell Biol.* **109,** 2589–2601.

Mayfield, S. P., and Kindle, K. L. (1990). Stable nuclear transformation of *Chlamydomonas reinhardtii* by using a *C. reinhardtii* gene as the selectable marker. *Proc. Natl. Acad. Sci. U.S.A.* **87,** 2087–2091.

Nelson, J. A. E., Savereide, P. B., and Lefebvre, P. A. (1994). The *CRY1* gene in *Chlamydomonas reinhardtii:* structure and use as a dominant selectable marker for nuclear transformation. *Mol. Cell. Biol.* **14,** 4011–4019.

Schnell, R. A., and Lefebvre, P. A. (1993). Isolation of the *Chlamydomonas* regulatory gene *NIT2* by transposon tagging. *Genetics* **134,** 737–747.

Sodeinde, O. A., and Kindle, K. L. (1993). Homologous recombination in the nuclear genome of *Chlamydomonas reinhardtii*. *Proc. Natl. Acad. Sci. U.S.A.* **90,** 9199–9203.

Tam, L.-W., and Lefebvre, P. A. (1993). Cloning of flagellar genes in *Chlamydomonas reinhardtii* by DNA insertional mutagenesis. *Genetics* **135,** 375–384.

CHAPTER 74

Insertional Mutagenesis and Isolation of Tagged Genes in *Chlamydomonas*

Lai-Wa Tam and Paul A. Lefebvre

Department of Genetics and Cell Biology
University of Minnesota
St. Paul, Minnesota 55108

I. Introduction

Chlamydomonas reinhardtii has long been a favorable model system for genetic analysis of a wide range of cellular processes. Recently, its popularity for molecular characterization of cellular mechanisms has greatly increased due to the development of DNA insertional mutagenesis for tagging and cloning genes (Tam and Lefebvre, 1993). The success of this approach derives from the development of techniques for efficient introduction of DNA into *Chlamydomonas* cells (Kindle, 1990) and the almost random integration of the transforming DNA into nonhomologous sites in the nuclear genome (Kindle *et al.*, 1989). When the transforming DNA integrates into and disrupts a functional gene, it serves as a tag for the isolation of the mutated gene. Using this new approach, a large number of mutants with defects in different cellular processes such as cell motility, phototaxis, chloroplast RNA splicing, sulfur and acetate metabolism, nitrogen assimilation, mating, and cell division have been identified. Moreover, this approach has proven to be particularly useful for isolating genes involved in flagellar function and assembly because these genes are nonessential, and mutant phenotypes can easily be scored (Tam and Lefebvre,

1993). More than a hundred new flagellar mutations have been isolated by insertional mutagenesis and over half a dozen of these genes have been cloned in our laboratory in the last 2 years.

There are many advantages to using transforming DNA to disrupt genes in *Chlamydomonas*. First, integration of the transforming DNA into the nuclear genome appears to be effective in producing random mutations (Tam and Lefebvre, 1993). Second, gene tagging efficiency is very high. More than 75% of the mutants recovered by insertional mutagenesis appear to be caused by integration of the transforming DNA (Tam and Lefebvre, 1993). As tagging genes is so efficient, even tedious or difficult screens for rare mutants affecting a particular gene may be justified using insertional mutagenesis. Third, with some precautions in generating mutants (see below), the copy number of the integrated plasmid can be kept low so that many of the tagged genes can readily be cloned. Fourth, cloned fragments from mutants of interest can readily be placed on the molecular map of the *Chlamydomonas* genome (Ranum *et al.*, 1988; see Chapter 75 in this volume), so that insertional mutations can be mapped without the extensive strain construction and crosses required for standard genetic mapping.

One complication frequently encountered using insertional mutagenesis is the deletion of a large region (up to 23 kb) at the site of insertion (Tam and Lefebvre, 1993). Such deletions make it necessary in some cases to perform a short chromosome walk to isolate the gene of interest. In some mutants, we also found that the deletion covered more than one gene (unpublished results). Because the insertion of the plasmid DNA and the accompanying sequence rearrangements will almost certainly produce the null phenotype for any affected gene, insertional mutagenesis is not useful for studying essential genes in this haploid organism. Genes adjacent to essential genes may also be difficult to obtain because of the effect of sequence deletion.

When planning experiments to generate insertional mutants, a number of factors should be considered. First is the choice of the transforming plasmids. A number of selectable markers, such as nitrate reductase (Fernandez *et al.*, 1989; Kindle *et al.*, 1989), argininosuccinate lyase (Debuchy *et al.*, 1989), and a mutated ribosomal protein S14 gene that confers resistance to emetine (Nelson *et al.*, 1994), have all been shown to be effective in creating insertional mutations. In principle any selectable marker gene should be usable as an insertional tag in transformation experiments. When choosing a plasmid for experiment, however, one should consider what methods will subsequently be used to clone the sequence disrupted by the insertion of the plasmid. A plasmid with little or no repetitive sequence but with many suitable restriction sites for DNA analysis will be good for cloning by simple methods such as plasmid rescue and "bookshelf" library construction. On the other hand, if one chooses to make a genomic library to clone the sequence of interest, a small plasmid will be easier to manipulate. The second factor to consider is the recipient *Chlamydomonas* strains for transformation. It is important to choose a strain that crosses well with existing laboratory strains because linkage of the inserted plasmid and the mutation should be tested by genetic crosses to determine whether the mutated gene is tagged and therefore clonable. If available strains give poor mating efficiency, it will be advisable to backcross the strains to other laboratory strains until the mating efficiency and meiotic viability are satisfactory. The third factor to consider is the form of the transforming DNA. Although any of the available transformation protocols should be effective in generating insertional mutants, it is crucial that the transforming DNA be linearized to obtain high tagging efficiency. A very poor tagging efficiency has been observed when supercoiled plasmids were used for transformation (Pazour and Witman, personal communication). It is also advisable to establish the lowest amount of linearized plasmid DNA that gives an acceptable number of transformants in pilot experiments to minimize the copy number of plasmids in the transformants.

II. Methods

A. Isolating and Analyzing Insertional Mutants

1. Digest 50 μg of the plasmid DNA used for transformation with a restriction enzyme that linearizes the plasmid at the polylinker site; limit digestion to 4–5 hours (prolonged digestion may decrease the transformability of the DNA, probably due to some nuclease activity). Extract the reaction twice with an equal volume of 1 : 1 phenol/chloroform and precipitate the DNA in ethanol. Resuspend the DNA to about 0.5 μg/μl. Use varying amounts of the DNA (e.g., 0.5, 1, 1.5, 2 μg) to transform a fixed number of cells (we usually use 4×10^7 cells) following any of the standard protocols (see Chapter 73). Plot the number of transformants against the amount of DNA used. An amount of DNA in the linear range of the curve should be used for future mutagenesis experiments.

2. Once mutants of interest are identified among transformants, backcross them to a strain of the same genotype as the parent (in the opposite mating type) to determine whether the mutations are closely linked to the inserted plasmid. For mutants that show cosegregation of the two traits, it is very likely that the mutation of interest was caused by insertion of the transforming plasmid.

3. Analyze the integrated DNA by DNA blot analysis to determine both the number and orientation of the inserted plasmids. This information will be very useful for designing a strategy to clone the tagged gene. Digest mutant and wild-type DNA with a number of restriction enzymes which will be useful for diagnosis of the number and orientation of the integrated plasmid and hybridize with probes derived from different parts of the transforming plasmid. If a complete vector sequence is present and juxtaposed or close to an integration junction, the junction sequence may be cloned by plasmid rescue. If the vector sequence is incomplete, a "bookshelf" library can be created to clone the

particular integration junction. In cases where multiple copies of plasmids are present, it may be necessary to construct a genomic library using the mutant DNA in order to clone sequences flanking the integrated plasmids.

B. Cloning the Affected Genes

1. Plasmid rescue is the easiest method to isolate the sequence flanking an integration junction. In general, a junction fragment of less than 15 kb will be clonable by this method. Digest 5 μg of genomic DNA from the mutant of interest with an appropriate enzyme (one that generates sticky ends) which will liberate a fragment containing the vector and adjacent genomic sequence. After 4–5 hours of incubation, stop the reaction by heat denaturation of the enzyme at 65°C for 15–20 minutes. Add 45 μl of 10 × ligase buffer and bring final volume to 450 μl. Add 2 units of T4 DNA ligase and incubate the reaction at 16°C overnight. Such conditions favor unimolecular ligation and therefore favor circularization of DNA fragments. Next day, extract the reaction twice with an equal volume of 1 : 1 phenol/chloroform, precipitate DNA in ethanol, and resuspend in 10 μl of TE. Transform 2 μl of the DNA into an appropriate *Escherichia coli* strain (DH5α works well) by electroporation following the manufacturer's recommended procedures. Select colonies harboring the rescued plasmid on agar plates with the appropriate drug. As the amount of the circularized junction fragment is very low, plasmid rescue will work only if bacterial transformation efficiencies of $10^8/\mu$g DNA or higher are achieved.

2. Alternatively, the junction fragment from an insertional mutant can be cloned by constructing a "bookshelf" library, which is a partial genomic library constructed by using size-selected restriction fragments. Thorough restriction analysis of the site of insertion will be necessary to identify the appropriate junction fragment to be cloned. To construct a "bookshelf" library, digest 20 μg of the mutant DNA with an appropriate restriction enzyme (enzymes that generate sticky ends should be used) for 4–5 hours and fractionate on a 1% low-melting agarose gel. After electrophoresis, excise several 1-mm slices from the gel corresponding to the size of the junction fragment Melt the gel slices at 65°C for 10–15 minutes and electrophorese one-tenth of each fraction on an agarose gel. Transfer DNA from the gel to nylon filters and hybridize with a labeled probe to determine which fractions contain the junction fragment. Melt the fractions containing the junction fragments at 65°C; extract three times with phenol and three times with ether. Precipitate DNA with ethanol and resuspend in 10 μl of TE. Ligate 5 μl of DNA to 0.1 μg of vector DNA that has been digested with the same restriction enzyme and dephosphorylated with calf intestinal alkaline phosphatase in 50 m*M* Tris–HCl (pH 9.0), 1 m*M* $MgCl_2$, 0.1 m*M* $ZnCl_2$, 1 m*M* spermidine (Maniatis *et al.*, 1982). Transform the ligation mixture into *E. coli* cells by electroporation as described above. Transfer 2000–3000 ampicillin-resistant colonies to nitrocellulose membranes and screen

colonies with an appropriate probe derived from the transforming plasmid to identify colonies containing the junction fragment.

3. Procedures for constructing a genomic library are not covered here; readers should refer to Sambrook *et al.* (1989) for details.

4. To determine whether the cloned sequence corresponds to the site of integration, use Southern analysis to search for restriction fragments that are polymorphic between wildtype and the mutant strain. In some instances the cloned sequence may contain repetitive elements that obscure such an experiment. In these cases, a smaller unique sequence fragment can often be found among the larger clone.

5. Once a unique genomic fragment flanking the integration site is available, it can be used to screen a genomic library of wild-type DNA to obtain overlapping clones. To test whether any of these clones contain the gene of interest, they are introduced into the mutant by cotransformation using a second selectable marker to assay rescue of the mutant phenotype. In our experience, even 1–2 μg of a crude preparation of lambda DNA can give a cotransformation efficiency of a few percent, using the argininosuccinate lyase gene as the selectable marker for cotransformation. Therefore, it is often only necessary to examine 100–200 transformants to determine whether a particular DNA clone can rescue the phenotype.

References

Debuchy, R., Purton, S., and Rochaix, J.-D. (1989). The argininosuccinate lyase gene of *Chlamydomonas reinhardtii:* an important tool for nuclear transformation and for correlating the genetic and molecular maps of the ARG7 locus. *EMBO J.* **8,** 2803–2809.

Fernandez, E., Schnell, R., Ranum, L. P. W., Hussey, S. C., Silflow, C. D., and Lefebvre, P. A. (1989). Isolation and characterization of the nitrate reductase structural gene of *Chlamydomonas reinhardtii. Proc. Natl. Acad. Sci. U.S.A.* **86,** 6449–6453.

Kindle, K. L. (1990). High-frequency nuclear transformation of *Chlamydomonas reinhardtii. Proc. Natl. Acad. Sci. U.S.A.* **87,** 1228–1232.

Kindle, K. L., Schnell, R. A., Fernandez, E., and Lefebvre, P. A. (1989). Stable nuclear transformation of *Chlamydomonas* using the *Chlamydomonas* gene for nitrate reductase. *J. Cell Biol.* **109,** 2589–2601.

Maniatis, T., Fritsch, E. F., and Sambrook, J. (1982). "Molecular Cloning. A Laboratory Manual." pp. 133–134. Cold Spring Harbor Laboratory Press, Cold Spring Harbor, New York.

Nelson, J. A. E., Savereide, P. B., and Lefebvre, P. A. (1994). The *CRY1* gene in *Chlamydomonas reinhardtii:* Structure and use as a dominant selectable marker for nuclear transformation. *Mol. Cell Biol.* **14,** 4011–4019.

Ranum, L. P. W., Thompson, M. D., Schloss, J. S., Lefebvre, P. A., and Silflow, C. D. (1988). Mapping flagellar genes in *Chlamydomonas* using restriction fragment length polymorphism. *Genetics* **120,** 109–122.

Sambrook, J., Fritsch, E. F., and Maniatis, T. (1989). "Molecular Cloning. A Laboratory Manual. Second Edition." Cold Spring Harbor Press, Cold Spring Harbor, New York.

Tam, L.-W., and Lefebvre, P. A. (1993). Cloning of flagellar genes in *Chlamydomonas reinhardtii* by DNA insertional mutagenesis. *Genetics* **135,** 375–384.

CHAPTER 75

Molecular Mapping of Genes for Flagellar Proteins in *Chlamydomonas*

Carolyn D. Silflow,[*,†] Pushpa Kathir,[*] and Paul A. Lefebvre[*]

[*]Department of Genetics and Cell Biology
[†]Department of Plant Biology
University of Minnesota
St. Paul, Minnesota 55108

I. Introduction

The construction of linkage maps from molecular markers as originally proposed by Botstein *et al.* (1980) has led to remarkable progress in the cloning of genes identified by mutation and in the development of breeding strategies for agriculturally important species. The use of molecular markers is based on the observation that nucleotide sequence changes accumulate gradually over time in DNA, resulting in sequence differences between closely related, interfertile strains or individuals. These sequence changes generate detectable polymorphisms between strains, such as in the patterns of restriction enzyme recognition sites in genomic DNA. Restriction fragment length polymorphism (RFLP) markers corresponding to specific cloned DNA fragments can be obtained by hybridizing the labeled fragment to genomic DNA that has been isolated from the two strains, digested with a restriction enzyme, separated by gel electrophoresis, and blotted to a membrane. A resulting difference in the size of the hybridizing genomic fragments provides an RFLP marker and identifies segregational alleles at the locus represented by the hybridizing fragment. The advent of polymerase chain reaction (PCR) technology has provided powerful new methods to detect nucleotide sequence polymorphisms between strains and generate molecular markers termed sequence-tagged sites (STSs) (Olson *et al.,* 1989). These methods rely on specific PCR oligonucleotide primers that can be used to amplify a unique 200- to 500-bp fragment from total genomic DNA. A key difference between molecular markers and phenotypic markers (usually mutations) is that molecular markers are not necessarily associated with a detectable phenotype. As a result, one is not limited by the number of different phenotypic markers that may be included in a cross. The segregation of an unlimited number of different molecular markers can be analyzed in the progeny of a single cross.

Chlamydomonas reinhardtii is a valuable system for molecular genetic studies of flagellar assembly and motility and of basal body function (Luck, 1984; Huang, 1986; Curry and Rosenbaum, 1993). Mutations in more than 80 genes affecting the function of these organelles have been mapped (Harris, 1993). A growing number of genes encoding flagellar proteins have been isolated using a variety of methods including differential screening of cDNA libraries (Schloss *et al.,* 1984) and screening of expression libraries with antibodies (Williams *et al.,* 1989). Insertional mutagenesis methods developed recently rely on transposable elements (Schnell and Lefebvre, 1993) or plasmid DNA integrated into the genome by transformation (Tam and Lefebvre, 1993) to generate a molecular tag associated with the mutagenized gene which facilitates its isolation.

A molecular map anchored on the genetic map is currently available for the *Chlamydomonas* nuclear genome (Ranum *et al.,* 1988). The molecular map allows one to rapidly establish the genetic map position of cloned genes and to correlate the map positions of these clones with previously mapped mutations. In addition, the molecular map may be useful for positional cloning of genes by genomic walking from a molecular marker to a nearby gene of interest. The molecular map may also facilitate the future construction of a physical map of the *Chlamydomonas* genome based on large cloned overlapping DNA fragments. The existing molecular map has been established primarily with RFLP mapping technology and contains genetic and molecular markers that anchor the map on all 17 known linkage groups (Ranum *et al.,* 1988; unpublished data). To establish the map, crosses were carried out between strains of *C. reinhardtii* carrying various genetic markers and a polymorphic strain, *C. smithii.* Tetrad progeny were scored for segregation of the genetic markers; genomic DNA was isolated from the same progeny and then scored for the segregation of RFLP markers. Scoring all four members of a tetrad provides valuable information on linkage both to other markers and to the centromere. The majority of the existing map, however, has been obtained by scoring only one member of a tetrad (76 total random progeny). Although this method generates no information on centromere linkage, it permits one to maximize the number of meiotic events analyzed to determine linkage relationships using a minimum number of DNA samples. Crosses of *C. reinhardtii* strains to another wild isolate, S1-D2, that appears to be slightly more polymorphic than the *C.*

smithii strain for RFLP markers have also yielded tetrad progeny for mapping studies (Gross *et al.*, 1988). Data for the linkage map were analyzed using MAPMAKER/EXP 3.0 (Lander *et al.*, 1987; Lincoln *et al.*, 1992) software. Although MAPMAKER was written for diploid organisms, it can be used for crosses with haploid organisms using the F2 backcross subroutine, assigning one genotype as "homozygous" and the other as "heterozygous." At this writing, the segregation of more than 200 markers has been analyzed, resulting in an estimated average spacing of 4–5 centiMorgans between adjacent markers. These studies have provided molecular markers that map to 16 of the 17 known linkage groups. Currently, the map is being expanded with additional markers including RFLP markers and STS markers as described below. The molecular map and information on the molecular markers are available on the *Chlamydomonas* electronic database where it can be continually updated as new markers are added (Dr. E. Harris, *Chlamydomonas* Genetics Center, Duke University, Durham, NC)

For the investigation of genes encoding flagellar proteins, the most common use for molecular mapping information is to correlate cloned DNA sequences with known mutant loci. The cloned genomic DNA or cDNA is obtained using insertional mutagenesis or a biochemical approach. Mapping the cloned DNA allows one to determine whether it is linked to previously mapped mutations affecting flagellar function. The protocols below outline the steps for creating molecular markers and for analyzing their segregation in progeny of a cross between polymorphic *Chlamydomonas* strains. Section II,A describes the RFLP mapping approach based on nucleic acid hybridization methods. No DNA sequence for the isolated DNA fragment is necessary. Section II,B describes the identification of one type of STS markers that requires the availability of DNA sequence from the 3′ noncoding portion of a cDNA clone. This method is based on the observation that differences in nucleotide sequence accumulate more rapidly in evolution in the noncoding portion of genes than in the coding portion. Takahashi and Ko (1993) introduced the term *biallelic Expressed Sequence Tag* (bEST) to describe the molecular markers generated by PCR amplification of the 3′ noncoding portion of a cDNA clone combined with restriction enzyme digestion of the resulting amplified fragments. An advantage of this method is that specific PCR primers can be designed so that individual members of multigene families can be distinguished in mapping experiments.

II. Methods

A. Generating and Mapping RFLP Markers

1. Digest genomic DNA from *C. reinhardtii* (strain 137C, *Chlamydomonas* Genetics Center No. CC-124) and a polymorphic strain *C. smithii* (CC-1373) or the S1-D2 strain (CC-2290) with several restriction enzymes. Use 1 μg DNA

for hybridization with nonradioactive probes and 2 μg DNA for hybridization with radioactive probes. Because of the high G + C content of nuclear DNA, we use enzymes with a high G + C content in the recognition site including *Bam*HI, *Eco*RI + *Xho*I, *Pst*I, *Pvu*II, *Sac*I, *Sal*I, *Sma*I, and *Sph*I,

2. Fractionate the DNA on a 1.0% agarose gel with adjacent lanes containing DNA from the different strains digested with a common enzyme. Stain, photograph, and blot the DNA to a nylon membrane using standard techniques.

3. Hybridize the DNA with a labeled probe that hybridizes to one or a few DNA fragments. Probes may be labeled with radioactive nucleotides or with nonradioactive detection systems such as digoxygenin (Genius System, Boehringer-Mannheim). Examine the resulting autoradiographs to determine the restriction enzyme(s) useful for generating a clear RFLP for the probe.

4. The segregation of the molecular marker may be examined in progeny of a cross between the two polymorphic strains. Either complete tetrads or random progeny (one member of each tetrad) may be used. The random progeny used to generate the molecular map described above are available from the *Chlamydomonas* Genetics Center (strains designated ***.1 from among the tetrad strains CC-2125 through CC-2227 and CC-2537 through CC-2644). Isolate DNA from each strain and digest it with a restriction enzyme as determined in step 3. Prepare DNA blots for hybridization with the labeled probe. Examine the hybridization results and score each progeny for the presence of the *C. reinhardtii* allele or the *C. smithii* allele.

5. Examine the linkage of the unknown RFLP marker to other genetic and molecular markers that have been mapped previously on the same progeny using the MAPMAKER/EXP 3.0 software for random progeny.

B. Mapping "bEST" Markers

1. Examine the DNA sequence from a noncoding portion of a genomic clone or the 3′ noncoding portion of a cDNA clone and design two primers for amplification of a DNA fragment 200–500 bp long. We use the Primer Designer Version 1.01 program (Scientific Education Software) to design 20-mer oligonucleotide primers with a target melting temperature of 70°C.

2. Amplify DNA fragments from genomic DNA of *C. reinhardtii* and a polymorphic strain (*C. smithii* or strain S1-D2). The 20-μl reaction mixture contains 70 ng genomic DNA, 10 pmole of each primer, 50 m*M* KCl, 10 m*M* Tris–HCl (pH 9.0), 0.1% Triton X-100, 1 m*M* $MgCl_2$, 20 μM dNTP, and 2.5 U Taq DNA polymerase. The amplification is performed in a PTC-100 Programable Thermal Controller (MJ Research, Watertown, MA) using the following program: 94°C denaturation for 5 minutes, followed by 30 cycles of denaturation at 94°C for 1 minute, annealing at 60°C for 45 seconds, and extension at 72°C for 2 minutes, followed by a final cycle of incubation at 94°C for 1 minute, 60°C for 45 seconds, and extension at 72°C for 10 minutes. The reaction mixture is run on a 1.0%

agarose gel. If the amplified fragments differ in size between the two strains, this difference can serve as a polymorphism for analyzing the segregation of the marker in the progeny strains.

3. If the amplified fragments generated in step 2 are of similar size, the DNA sequence of the *C. smithii* (or S1-D2) fragment can be obtained to detect sequence polymorphisms and identify restriction enzyme sites that differ between the two sequences. This can be done most easily by unequal amplification of one strand of the amplified fragment. The amplified band from step 2 is sliced out of the gel, and the DNA is recovered by phenol extraction and ethanol precipitation. Single-strand DNA is amplified by asymmetric PCR (McCabe, 1990) using a 100-μl reaction volume containing the same reaction components as in step 2 except 1–5 ng amplified DNA fragment is substituted for the genomic DNA and 50 pmole of the first primer is used in combination with 0.5 pmole of the second primer. The reaction conditions are the same as described above. After running 10 μl of the reaction mixture on an agarose gel to check the amplification efficiency, the remaining product is precipitated in a final volume of 2 *M* ammonium acetate and 50% isopropanol to remove primers and is then resuspended in 10 μl TE. A 3-μl aliquot of the product and 0.5 pmole of the second primer are used in a DNA sequencing reaction using the Sequenase 2.0 protocol (United States Biochemical). The DNA sequences of the amplified fragments from the two *Chlamydomonas* strains are compared to find restriction enzyme site differences.

4. If a restriction site difference between the amplified fragments is detected in step 3, it can be used to score each progeny strain for the *reinhardtii* allele or the *smithii* allele. DNA from each progeny strain is amplified by PCR and the amplified fragments are digested with the informative enzyme.

5. If no restriction site difference is detected between the two strains, single base pair differences will likely exist, and these can be used to design allele-specific PCR primers (Wu *et al.*, 1989). Design an allele-specific primer that is complementary to the DNA sequence of one strain but not complementary to the sequence of the other strain in the 3′-terminal nucleotide. The allele-specific primer, in combination with one of the original primers, can be used to amplify DNA from strains carrying the complementary allele but not from strains carrying the nucleotide mismatch. To increase the accuracy of scoring alleles, an allele-specific primer should be designed for each allele at each locus.

6. Use the allele-specific PCR primers to amplify DNA from each progeny strain and score the allele present in each strain. Analyze the segregation of the marker as in Section II,A, step 5.

References

Botstein, D., White, R. L., Skolnick, M., and Davis, R. W. (1980). Construction of a genetic linkage map in man using restriction fragment length polymorphisms. *Am. J. Hum. Genet.* **32,** 314–331.

Curry, A. M., and Rosenbaum, J. L. (1993). Flagellar radial spoke: a model molecular genetic system for studying organelle assembly. *Cell Motil. Cytoskel.* **24,** 224–232.

Harris, E. H. (1993). *Chlamydomonas reinhardtii. In* "Genetic Maps of a Compilation of Linkage and Restriction Maps of Genetically Studied Organisms" (S. J. O'Brien, ed.), Vol 6, pp. 2.157–2.169. Cold Spring Harbor Laboratory, New York.

Huang, B. P.-H. (1986). *Chlamydomonas reinhardtii:* A model system for the genetic analysis of flagellar structure and motility. *Int. Rev. Cytol.* **99,** 181–215.

Lander, E., Green, P., Abrahamson, J., Barlow, A., Daly, M., Lincoln, S., and Newburg, L. (1987). MAPMAKER: an interactive computer package for constructing primary genetic linkage maps of experimental and natural populations. *Genomics* **1,** 174–181.

Lincoln, S., Daly, M., and Lander, E. (1992). Constructing genetic maps with MAPMAKER/EXP 3.0. Whitehead Institute Technical Report. 3rd Ed.

Luck, D. J. (1984). Genetic and biochemical dissection of the eukaryotic flagellum. *J. Cell Biol.* **98,** 789–794.

McCabe, P. C. (1990). Production of a single-stranded DNA by asymmetric PCR. *In* "PCR Protocols: A Guide to Methods and Applications" (M. A. Innis, D. H. Gelfand, J. J. Sninsky, and T. J. White, eds.), pp. 76–83. Academic Press, San Diego.

Olson, M., Hood, L., Cantor, C., and Botstein, D. (1989). A common language for physical mapping of the human genome. *Science* **245,** 1434–1435.

Ranum, L. P., Thompson, M. D., Schloss, J. A., Lefebvre, P. A., and Silflow, C. D. (1988). Mapping flagellar genes in *Chlamydomonas* using restriction fragment length polymorphisms. *Genetics* **120,** 109–122.

Schloss, J. A., Silflow, C. D., and Rosenbaum, J. L. (1984). mRNA abundance changes during flagellar regeneration in *Chlamydomonas reinhardtii. Mol. Cell Biol.* **4,** 424–434.

Schnell, R. A., and Lefebvre, P. A. (1993). Isolation of the *Chlamydomonas* regulatory gene NIT2 by transposon tagging. *Genetics* **134,** 737–747.

Takahashi, N. and Ko, M. S. H. (1993). The short 3′-end region of complementary DNAs as PCR-based polymorphic markers for an expression map of the mouse genome. *Genomics* **16,** 161–168.

Tam, L.-W. and Lefebvre, P. A. (1993). The use of DNA insertional mutagenesis to clone genes in *Chlamydomonas. Genetics* **135,** 375–384.

Williams, B. D., Velleca, M. A., Curry, A. M., and Rosenbaum, J. L. (1989). Molecular cloning and sequence analysis of the *Chlamydomonas* gene coding for radial spoke protein 3: flagellar mutation pf-14 is an ochre allele. *J. Cell Biol.* **109,** 235–245.

Wu, D. Y., Ugozzoli, L., Pal, B. K., and Bruce Wallace, R. (1989). Allele-specific enzymatic amplification of β-globin genomic DNA for diagnosis of sickle cell anemia. *Proc. Natl. Acad. Sci. U.S.A.* **86,** 2757–2760.

CHAPTER 76

Mating and Tetrad Analysis in *Chlamydomonas reinhardtii*

Susan K. Dutcher
Department of Molecular, Cellular, and Developmental Biology
University of Colorado
Boulder, Colorado 80309

I. Introduction

Chlamydomonas reinhardtii is a haploid organism. Haploid organisms have several advantages over diploid organisms when dissecting a process by a mutational approach. The ease of detecting mutations in haploid organisms is straightforward; any newly arising recessive mutation is apparent because there is no wild-type allele to mask the mutant phenotype. Thus, large-scale screens and/or selections can be performed for the isolation of mutations affecting the flagella of *Chlamydomonas*. Two different mating types exist in *C. reinhardtii:* mating type plus (mt^+) and mating type minus (mt^-). These mating types behave as alleles at a genetic locus on linkage group VI (Sager and Granick, 1953; Harris, 1989). The events of mating and meiosis involve several steps. Mating involves the cytoplasmic fusion of a mt^+ cell and a mt^- cell. Mating is initiated when the cells are starved for nutrients and specifically when they are starved for nitrogen. Recognition occurs between cells of opposite mating types at two levels. The initial recognition occurs by interactions between hydroxyproline-rich glycoproteins, agglutinins, in the membrane of the flagella of the two cell types (Adair, 1985). This process is known as tipping. It leads to an increase in the intracellular levels of cyclic AMP, which stimulates the formation of a fertilization tube on the surface of the mt^+ cell. This actin-filled projection interacts with a structure on the mt^- cell, and an enzyme that causes the cells to lose the cell walls is released (Goodenough and Weiss, 1975). After the walls are shed, the cells fuse in a specific orientation. The mt^- cell will fuse along the *cis* (eyespot-containing) side and the mt^+ cell will fuse along the *trans* side (Holmes and Dutcher, 1988). Recognition, formation of the fertilization tube, and cytoplasmic fusion can occur within 3–5 minutes of mixing populations of the two cell types. Within 1.5 to 3 hours of cytoplasmic fusion, the nuclei and chloroplasts fuse. The flagella are lost and a thick, impermeable, zygotic cell wall begins to form (Minami and Goodenough, 1978). Zygotes can remain dormant for months if maintained in the dark or in the absence of a nitrogen source. Meiosis is initiated in zygotes by reintroduction of light and nutrients and is completed within 18–24 hours to produce four meiotic zoospores.

II. Mating of Flagellated Strains

Mutations that affect flagellar function, length, or number do not block the ability of *C. reinhardtii* strains to mate. Short flagella or paralyzed flagella may reduce the efficiency of mating, but do not block mating. Flagella are required for conventional mating procedures.

1. To prepare cells for mating, 1×10^7 cells are used to inoculate a 100-mm Petri plate with gametic medium (see Table I for media recipes).
2. Cells are grown at 25°C in 17.5 μEinstein/m^2/s for 3 days; the cells are then moved to a lower-light-intensity environment (6 μEinstein/m^2/s) for 1–4 days.
3. For tetrad analysis, cells from about one-eighth of the plate are scraped with a wire loop or a sterile, flat toothpick from the solid medium into 0.3 ml of M-N/5 medium in 1.5-ml plastic tubes for 4 hours. The cell concentration should be approximately 1×10^8 cells/ml.
4. Cells of opposite mating types are mixed together. The mixture is monitored after 30 minutes for evidence of mating. An aliquot of 20 μl is mixed with 2 μl 1% glutaraldehyde in 0.02 *M* phosphate-buffered saline (pH 7.5) and observed at a magnification of 640× for cells that have four flagella. These cells are known as dikaryons and most (>99.9%) will become zygotes.

Table I
Recipes for Solid Media

Solution (w/v)	Modified Sagar and Granick medium I (ml/liter water)	M-N/5 medium (ml/liter water)
Rich plates		
10% Sodium acetate	10.0	
10% Sodium citrate	5.0	1.0
4% Calcium chloride	1.0	0.25
1% Ferrous chloride	1.0	0.25
Magnesium sulfate · $7H_2O$	3.0	0.75
Potassium phosphate, dibasic	1.7	3.0
Potassium phosphate, monobasic	1.0	
10% Ammonium nitrate	3.0	
Trace metals	1.0	0.25
Gametic plates[a]		
Magnesium sulfate · $7H_2O$	0.2	
Magnesium chloride	2.8	
Washed agar[b]	15 g/liter water	
Zygote plates[a]		
Washed agar[b]	40 g/liter water	
Dissection plates[a]		
Washed agar[b]	20 g/liter water	

[a] Use rich plate recipe with these changes.

[b] Washed agar is important to remove contaminants in the agar that decrease the viability of single cells. Five changes of deionized water are used and then the agar is dried before use. Washing takes 2 days and drying takes 3–5 days. Modified Sager and Granick medium I (1953) is used for routine culturing of cells. The trace metals of Hutner are used as described by Harris (1989).

5. Mating efficiencies are variable and can range between 0 and 100%. Efficiencies as low as 1% are sufficient for performing tetrad analysis.

6. Mating mixtures are placed onto zygote medium, which has 4% agar. The large percentage of agar facilitates the recovery of the zygotes at the time of dissection. The plates are allowed to remain in the presence of 17.5 μEinstein/m^2/s light for 18 hours and then placed in the dark by wrapping in aluminum foil.

7. Four days after being placed in the dark, the plate is exposed to chloroform vapors by inverting the plate over a glass Petri plate containing chloroform for 30 seconds. This exposure kills the unmated cells as well as the small percentage of dikaryons that become mitotically dividing diploid cells. The plates are returned to the dark for at least 8 hours.

8. To isolate the zygotes, the zygote plate is scraped with a sterile, dull razor blade. The zygotes stick to the surface of the agar and the dead vegetative cells come off onto the razor blade (Fig. 1). The zygotes can be loosened from the surface of the agar with a flattened platinum wire under a dissection microscope at 32× magnification.

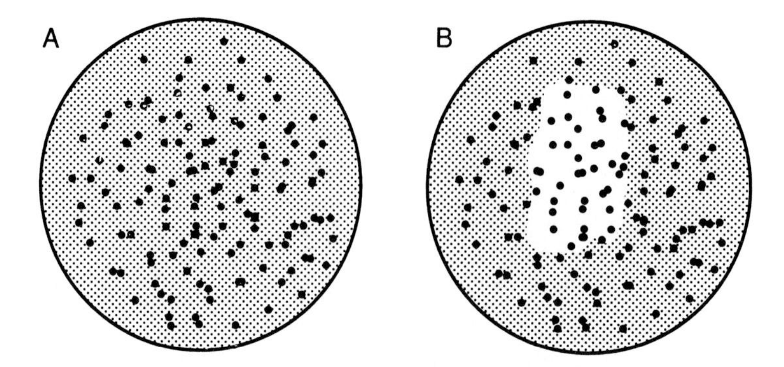

Fig. 1 Diagrammatic representation of a zygote plate. The larger black spheres are zygotic cells. The stippled area represents dead vegetative cells. (A). Appearance of the plate with 64× magnification before it is scraped with a dull razor blade. (B) The center area, which appears white, has been scraped to remove the dead vegetative cells and leave an enriched population of zygotic cells. These cells can be collected into a pile and transferred to a dissection plate.

9. To induce the meiotic cycle, the zygotic cells are transferred onto fresh dissection medium. The agar is lightly scored with a razor blade to make a grid for orientation purposes and the zygotes are spread at the two ends of the grid (Fig. 2). A single zygote is pulled into each lane using a glass needle. This is generally performed at a magnification of 64×.

10. Meiosis is generally complete in 18–24 hours; the timing is dependent on both the temperature and the strains. Incubation of the zygotic cells at 16

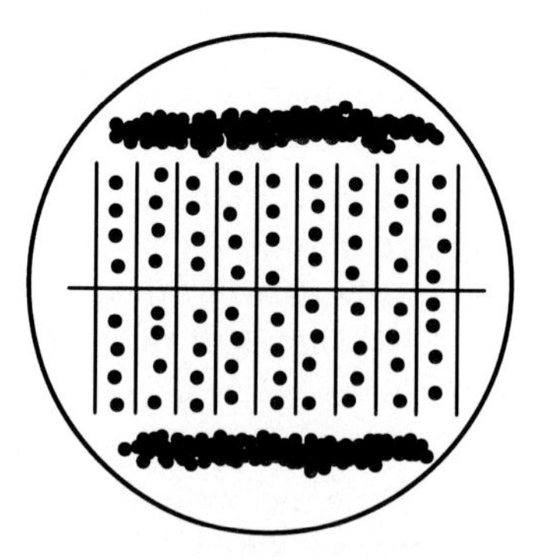

Fig. 2 Diagrammatic representation of a dissection plate 1 week after transfer and dissection of the zygotic cells. To prepare the plate a grid is marked onto the dissection plate with a sterile razor blade. The grid serves as an orientation to keep the zoospores from each meiotic zygote separate from those of other zygotes. The streaks across the top and bottom of the plate are the zygotes that were transferred from the zygote plate but not dissected. Each lane of the grid contains the four meiotic products of a single zygote that has completed meiosis and been separated using a glass needle. Each zoospore has grown into a colony that can be analyzed.

or 32°C will often produce eight products instead of four. Generally, each zygote left at 21 to 25°C will have four zoospores that can be separated with a glass needle by dragging the cells over the surface of the agar at a magnification of 64×. Separation of about 2–3 mm between the zoospores is sufficient to avoid mixing between colonies. It is advisable to have a location for the dissection microscope that is separated from traffic and air currents to avoid contamination by molds and yeasts. Zoospores grow up into colonies in 5–7 days.

Strains that are difficult to mate (mating efficiency is less than 1%) can be coaxed by several strategies.

11. Some strains show better mating frequencies when mated earlier or later in the starvation regime. With new strains, one can begin mating after 1 day in low light and continue until cells have been in reduced light for 5 or 6 days.

12. Better efficiencies of mating can be achieved for strains with paralyzed or short flagella by centrifuging the mating mixture for 10 seconds at 77*g* in a microfuge.

13. The techniques in Section III for aflagellated strains can be used.

III. Mating of Aflagellated Strains

Mutations that produce cells that lack flagella require additional steps to ensure mating. Because the first recognition events involve the tipping of flagella of the two mating types, this step must be bypassed. This can be accomplished by the addition of cyclic AMP (or analogs of cAMP) and inhibitors of phosphodiesterase, which cleaves the ring of cAMP (Pasquale and Goodenough, 1987). These treatments do not affect the specificity of the mating reaction, the ability of zygotes to complete meiosis, or the segregation of chromosomes.

1. After 3 hours in M-N/5 medium, add 20 μl of 300 m*M* dibutyryl cyclic AMP (dbcAMP, the highly purified grade is not necessary) and 20 μl of 10 m*M* isobutylmethylxanthine (IBMX) to 0.2 ml of each parental strain. The dbcAMP is freshly prepared for each mating in 10 m*M* 1,4-piperazinediethanesulfonic acid (Pipes) buffer, pH 7.2, and filter-sterilized. The IBMX is prepared in deionized water, filter-sterilized, and kept frozen in aliquots at −20°C. Cells are incubated for 1 hour at 21 to 25°C.

2. Mix parental strains, wait 1.5 hours, and spread 150 μl of the mating mixture onto zygote medium.

3. Mating is assayed the following day by the formation of pellicle in the plastic tube. Pellicle is the aggregation of zygotes and appears as large clumps and threads of zygotic cells.

IV. Tetrad Analysis

When meiosis is completed and the meiotic progeny have been collected, they can be analyzed as the progeny from individual meiotic events (a tetrad) or individually (random progeny). One advantage of tetrad analysis is the ability to map a gene with respect to its centromere as well as with respect to other genes. Only three classes of tetrads can be produced from crosses heterozygous at two genes: parental ditype (PD), nonparental ditype (NPD), and tetratype (T). If the two genes are linked then the frequency of PDs will be greater than the frequency of NPDs. In this case, the frequency of NPD and T tetrads will determine the frequency of recombination and the map distance. The equation to determine map distance in centimorgans (cM) is

$$\text{map distance} = T + 6\ NPD \times 100/2(PD + NPD + T),$$

where PD, NPD, and T represent the number of each class of tetrads.

A. Centromere Mapping

If two genes are on different chromosomes, then the frequencies of PDs and NPDs will be equal. The frequency of T tetrads provides a means to determine if the two genes are linked to their respective centromeres. If the frequency of T tetrads is less than 0.67, then the distance from gene 1 to its centromere plus the distance from gene 2 to its centromere is less than 33 cM. In general, centromere mapping is performed with a known gene that rarely shows recombination between itself and its centromere. In *C. reinhardtii,* this category includes the loci *ac17, maa9,* and *y1*. If one of these genes is used, then the frequency of T tetrads would give a distance from the unknown gene to its centromere described by the following equation:

$$\text{distance from unknown gene to its centromere} = T \times 100/2(PD + NPD + T).$$

B. Genetic Maps

At present the genetic map of *C. reinhardtii* is represented by 17 linkage groups (Dutcher *et al.,* 1991). The average number of genetic markers per linkage group is 12.2. The most saturated linkage groups (LG I and LG VI) have 26 mapped genetic loci and the least saturated linkage group (LG XVI/XVII) has only two loci. The patterns of recombination events have been studied extensively on linkage groups I and XIX (Ebersold and Levine, 1959; Holmes *et al.,* 1993). The analysis of crosses involving five to eight heterozygous loci has demonstrated the presence of chiasma interference. Chiasma interference is the observation that one recombination event lowers the probability of a second linked recombinational event. The presence of high levels of interference is advantageous for mapping new mutations. Almost all genes on a linkage group show linkage to a centromere-linked locus even if the map distance between a gene and its centromere is greater than 50 cM. Table II contains a list of strains and loci that are useful for mapping mutations.

1. In the worst case scenario, mapping new mutations requires six to seven different matings and the analysis of ~25 tetrads from each mating.

2. For a gene that is less than 33 cM from its centromere, the observation of any NPD tetrads with a given mapping locus is often sufficient to exclude the linkage group containing that locus.

3. For a gene that is more than 33 cM from its centromere, linkage can be established when PDs are greater than NPDs. In some cases, additional crosses may be necessary to exclude a linkage group.

4. Once the linkage group is established, additional crosses to known loci should be performed to determine linkage. In many cases, the ordering of loci

Table II
List of Available Mapping Strains

Strain	Locus	Centromere distance	Linkage group
MS-6	*arg7*	28	I
MS-1	*ery1*	60	I
MS-2; MS-3	*act1*	28	II
MS-5	*fsr1*	34	II
MS-2	*ac17*	0	III
MS-3	*pyr1*	5	IV
	ac31[a]	16	V
MS-6	*maa8*	4	VI
MS-3	*can1*	7	VII
MS-3; MS-5	*tun1*	5	VIII
MS-1	*sr1*	14	IX
MS-6	*nic13*	2	X
MS-1	*maa3*	10	X
MS-3	*maa9*	1	XI
MS-3	*nic15*	18	XII/XIII
MS-2	*maa4*	4	XIV
	nic1[a]	10	XV
MS-5	*y1*	0	XVI/XVII
MS-5	*spr1*	14	XVII
MS-5	*maa11*	22	XIX

[a] Indicates that the mutant allele is difficult to score or the strain acquires additional modifying mutations that do not give $2^+:2^-$ or Mendelian segregation. Strains are available from the *Chlamydomonas* Genetics Center at Duke University.

may require three-point crosses to be performed. Generally, 50–100 tetrads should be analyzed to ensure the map distances are accurate to 1 cM.

V. Constructing Diploid Strains

The ability to construct diploid strains allows the investigator to perform genetic analyses that are not possible in haploid strains. First, diploid strains provide a means to determine if mutations are dominant or recessive to the wild-type allele. Second, they offer the ability to perform complementation tests to determine the number of genes that produce a particular phenotype. Third, they provide a means to isolate additional alleles at a particular locus. For example, mutations at the β-dynein gene have a variety of phenotypes. Some alleles act as suppressors of central pair and radial spoke deficient mutant strains, whereas other alleles are missing the outer dynein arms and do not act as suppressors. Thus, diploid strains offer the ability to explore gene interactions and the range of functions of a particular gene product.

A small minority of dikaryons ($<0.1\%$) fail to enter the meiotic cycle; they become mitotically dividing diploid cells. These diploid cells can be maintained for extended periods although they show both mitotic recombination and chromosome loss at a rate of $\sim 5 \times 10^5$ loss events per cell division when monitored for LG VI (F. Lux and S. K. Dutcher, unpublished observations). Morphologically, diploid cells resemble haploid cells. They have two flagella at the anterior end and a single eyespot. One distinguishing feature is that they have about twice the volume of a haploid cell.

Several strategies are available for obtaining diploid strains. The strategies include the use of closely linked alleles at the *ARG7* locus that demonstrate interallelic complementation (Ebersold, 1967), complementing mutations at linked, auxotrophic loci, and the use of a variety of dominant mutations (Dutcher and Gibbons, 1988; Dutcher *et al.*, 1992) in combination with an auxotrophic mutation. For example, *arg7* mutant alleles require arginine to grow at 25°C. A diploid heterozygous for these alleles will be able to grow without the addition of arginine to the medium.

1. Mate cells as described in Section II or III. Plate onto the appropriate selective medium in the presence of light. Colonies will appear within 4–5 days. Pick and streak the colonies with a flat, sterile toothpick onto a new selective plate to ensure that the population of cells is not mixed. Picking colonies within the first 5 days is important to avoid zygotes that have completed meiosis and produced recombinant progeny with the appropriate genotype.

2. To test that diploid cells have been generated, mating type and viability in triploid crosses can be determined (see Section VI). Flow cytometry can be used to verify that the cells have a $2n$ DNA content.

VI. Meiotic Analysis of Stable Diploid Strains

Diploid strains cannot be induced to enter into the meiotic cycle directly; however, diploid mt^+/mt^- cells are able to mate as mt^-. Diploid cells homozygous for the mt^+ allele mate as mt^+ cells. If diploid strains are mated to haploid cells, a triploid zygote arises. Meiosis is initiated and meiotic progeny are produced. Less than 20% of the progeny are viable. These progeny are aneuploid and up to five additional backcrosses are needed to achieve more than 95% viable progeny. There are two approaches to analyzing diploid cells.

A. Inhibition of Nuclear Fusion

When a diploid strain is mated with a haploid strain in the presence of colchicine (0.5 *M*) or vinblastine (2.5 μM), nuclear fusion is inhibited in 25–50% of the zygotes. The products include four viable meiotic progeny from the diploid nucleus and two inviable progeny from the haploid nucleus (Dutcher, 1988).

B. Tetraploid Genetics

Two diploid strains can be mated together and the viable meiotic progeny can be generated. The four progeny will retain a diploid complement but segregation and behavior of traits can be followed. For a gene tightly linked to its centromere, parents that are heterozygous for a gene (say *Aa*) should produce ratios of 4 : 1 : 1 for the genotypes *Aa, AA,* and *aa*. For a gene that is unlinked to its centromere, ratios of 8 : 3 : 3 are expected. For intermediate map position, intermediate ratios are expected.

References

Adair, W. D. (1985). Characterization of *Chlamydomonas* sexual agglutinins. *J. Cell Sci.* **2,** Suppl. 233–260.

Dutcher, S. K. (1988). Nuclear fusion-defective phenocopies in *Chlamydomonas reinhardtii:* mating-type functions for meiosis act through the cytoplasm. *Proc. Natl. Acad. Sci. U.S.A.* **85,** 3946–3950.

Dutcher, S. K., and Gibbons, W. (1988). Isolation and characterization of dominant tunicamycin resistance mutation in *Chlamydomonas reinhardtii* (CHLOROPHYCEAE). *J. Phycol.* **24,** 230–236.

Dutcher, S. K., Galloway, R. E., Barclay, W. R., and Poortinga, G. (1992). Tryptophan analog resistance mutations in *Chlamydomonas reinhardtii*. *Genetics* **131,** 593–607.

Dutcher, S. K., Power, J., Galloway, R. E., and Porter, M. E. (1991). Reappraisal of the genetic map of *Chlamydomonas reinhardtii*. *J. Hered.* **82,** 295–301.

Ebersold, W. T. (1967). *Chlamydomonas reinhardtii:* Heterozygous diploid strains. *Science* **157,** 447–449.

Ebersold, W. T., and Levine, R. P. (1959). A genetic analysis of linkage group I of *Chlamydomonas reinhardtii*. *Z. Verebimgs*. **90,** 74–82.
Goodenough, U. W., and Weiss, R. L. (1975). Gametic differentiation in *Chlamydomonas reinhardtii*. III. Cell wall lysis and microfilament-associated mating structure activation in wild-type and mutant strains. *J. Cell Biol.* **76,** 430–438.
Harris, E. H. (1989). "The *Chlamydomonas* Sourcebook. A Comprehensive Guide to Biology and Laboratory Use". Academic Press, New York.
Holmes, J. A., and Dutcher, S. K. (1988). Cellular asymmetry in *Chlamydomonas reinhardtii*. *J. Cell Sci.* **94,** 273–285.
Holmes, J. A., Johnson, D. E., and Dutcher, S. K. (1993). Linkage group XIX of *Chlamydomonas reinhardtii* has a linear map. *Genetics* **133,** 865–874.
Minami, S., and Goodenough, U. W. (1978). Novel glycopolypeptide synthesis induced by gametic cell fusion in *Chlamydomonas reinhardtii*. *J. Cell Biol.* **77,** 165–180.
Pasquale, S. M., and Goodenough, U. W. (1987). Cyclic AMP functions as a primary sexual signal in gametes of *Chlamydomonas reinhardtii*. *J. Cell Biol.* **105,** 2279–2292.
Sager, R., and Granick, S. (1953). Nutritional control of sexuality in *Chlamydomonas reinhardtii*. *J. Gen. Physiol.* **37,** 729–742.

CHAPTER 77

Strategies for Isolation of Flagellar Motility and Assembly Mutants in *Chlamydomonas*

Ritsu Kamiya

Zoological Institute
Graduate School of Science
University of Tokyo
Tokyo 113, Japan

I. Introduction

Our understanding of the structure and function of cilia and flagella has benefited greatly from *Chlamydomonas* mutants that have abnormal flagella (see Luck, 1984). *Chlamydomonas* is uniquely suited for genetic studies because its flagellated vegetative cells are haploid and its mutants can be genetically analyzed by sexual crosses between different strains. More than 40 mutants are known that are deficient in internal structures of the flagella such as inner and outer dynein arms, central pair microtubules, or radial spokes (see Harris, 1989; Dutcher and Lux, 1989). There are also mutants with abnormal flagellar number, length, and waveform (Harris, 1989) and mutants deficient in the control of flagellar beating (Horst and Witman, 1993).

Given below is a simple method for the isolation of various motility mutants. It is based on enrichment for cells that grow on the bottom of test tube cultures after mutagenesis with UV light. As first demonstrated by Lewin (1954), it is effective in isolating mutants that have no or impaired motility. With this method, I have isolated a number of mutants deficient in outer dynein arms that swim slowly (Kamiya, 1988). The enrichment methods can similarly be used for the selection of paralyzed or slow-swimming mutants generated by insertional mutagenesis (see Chapter 74 in this volume). For isolation of temperature-sensitive flagellar mutants using chemical mutagenesis, see Huang *et al.* (1977) and Adams *et al.* (1982). For isolation of inner-arm dynein mutants, see Kamiya (1991). Detailed procedures for genetic analysis, strain maintenance, and culture of *Chlamydomonas* have been described in a comprehensive guidebook by Harris (1989). Wild-type strains and most published mutants can be obtained from the *Chlamydomonas* Genetics Center (Department of Botany, Duke University, Durham, NC).

II. Procedure

A. Medium

TAP medium (Gorman and Levine, 1965) is used.

B. Mutagenesis

1. Culture wild-type *Chlamydomonas reinhardtii* (plus or minus mating type) to a density of 1–2 × 10^6/ml in liquid TAP medium under continuous illumination.

2. Transfer 20 ml of the culture to a sterilized 9-cm Petri dish.

3. Place the dish 30–60 cm below a 10- to 20-W UV lamp. A UV lamp on an ordinary clean bench can be used.

4. Remove the lid and irradiate the cells with UV light for 30–300 seconds while swirling the cell suspension gently by hand (wear a glove to prevent UV exposure of the skin). The irradiation time is chosen by experience such that 1 to 20% of cells will survive. Under our standard conditions with a 15-W UV lamp placed 45 cm above the sample, the time of irradiation is 1–4 minutes.

5. Transfer 1-ml aliquots of the cell suspension to eight test tubes (1 cm diameter × 11 cm high, with a closure to permit gas exchange), containing 3 ml of sterile TAP medium.

6. Keep the test tubes in the dark for 12 hours to prevent photoactivation of DNA repair.

C. Enrichment of Motility-Deficient Cells

1. After the dark incubation, place the test tubes under 12 hours light/12 hours dark and allow the cells to grow.

2. During the second light phase, carefully transfer cells on the bottom of each test tube to another test tube containing 4 ml of fresh medium. The volume of transferred suspension is 50–200 μl. Use a new, sterilized Pasteur pipet for each transfer.

3. When the bottom and the upper part of the culture medium become greenish after a few days, transfer the cells on the bottom to a third test tube, containing 4 ml of medium, and allow them to grow under light/dark conditions.

4. Repeat the above process two or three more times over a period of about 1 week.

5. Transfer the cells growing on the bottom of the last set of test tubes to 0.5–1 ml of medium. Inoculate them on TAP/agar plates (TAP medium containing 1.5% w/v Difco Bacto Agar) so as to obtain single colonies. Keep the agar plates under constant illumination for 3–5 days.

D. Selection of Flagellar Mutants

1. When single colonies are apparent, use sterile toothpicks to transfer the colonies to 96-well culture plates (such as Corning No. 25860), each well containing 200 μl liquid medium. Typically, 24 or 48 colonies are transferred for each test tube. Select colonies with various appearances.

2. Place the 96-well plates under constant illumination for 1 or 2 days and observe each well with an inverted microscope. Cells in most wells probably will display abnormal motility.

3. Examine the flagella of any interesting clones under a dark-field microscope, and select the colonies to be studied further. When more than one clone from one test tube culture appear similar, select only one clone. Inoculate the selected clones on 1.5% agar plates.

4. When mutants of interest are obtained, mate them with wild type to obtain daughter cells of opposite mating types. This step is necessary because the first-generation mutants often carry a second, undesirable mutation. Repeat this step a few times to ensure that all but the most closely linked second mutations are crossed out.

References

Adams, G. M., Huang, B., and Luck, D. J. (1982). Temperature-sensitive, assembly-defective flagella mutants of *Chlamydomonas reinhardtii*. *Genetics* **100,** 579–586.

Dutcher, S. K., and Lux F. G., III. (1989). Genetic interactions of mutations affecting flagella and basal bodies in *Chlamydomonas*. *Cell Motil. Cytoskel.* **14,** 104–117.

Gorman, D. S., and Levine, R. P. (1965). Cytochrome F and plastocyanin: their sequence in the photosynthetic electron transport chain of *Chlamydomonas reinhardtii*. *Proc. Natl. Acad. Sci. U.S.A.* **54,** 1665–1669.

Harris, E. H. (1989). "The *Chlamydomonas* Sourcebook". Academic Press, San Diego.

Horst, C. J., and Witman, G. B. (1993). *ptx1,* a nonphototactic mutant of *Chlamydomonas,* lacks control of flagellar dominance. *J. Cell Biol.* **120,** 733–741.

Huang, B., Rifkin, M. R., and Luck, D. J. L. (1977). Temperature-sensitive mutations affecting flagellar assembly and function in *Chlamydomonas reinhardtii*. *J. Cell. Biol.* **72,** 67–85.

Kamiya, R. (1988). Mutations at twelve independent loci result in absence of outer dynein arms in *Chlamydomonas reinhardtii*. *J. Cell Biol.* **107,** 2253–2258.

Kamiya, R. (1991). Selection of *Chlamydomonas* dynein mutants. *Methods Enzymol.* **196,** 348–354.

Lewin, R. A. (1954). Mutants of *C. moewusii* with impaired motility. *J. Gen. Microbiol.* **11,** 358–363.

Luck, D. J. L. (1984). Genetic and biochemical dissection of the eukaryotic flagellum. *J. Cell Biol.* **98,** 789–794.

CHAPTER 78

Epitope Tagging of Flagellar Proteins

Dennis R. Diener

Department of Biology
Yale University
New Haven, Connecticut 06520

I. Introduction

Antibodies are invaluable for the localization and purification of proteins, but generating an antibody with the desired specificity can be time consuming and problematic, especially if the protein in question is difficult to purify or is a member of a family of closely related proteins. If the gene for the protein has been cloned, however, these difficulties can be circumvented by inserting an oligonucleotide encoding an antigenic determinant into the gene. When the gene is expressed *in vivo* the tagged protein can be immunolocalized or immunoprecipitated with an existing antibody against the appended peptide. Conceptually, this strategy, known as epitope tagging, is similar to fusing a reporter protein (e.g., β-galactosidase) to a protein of interest, but differs in that the tagged protein is identified immunologically by the presence of a single epitope, usually about 10 amino acids long.

Epitope tagging rapidly provides a high-affinity antibody of known specificity that uniquely reacts with the tagged gene product. In addition, cells not expressing the tagged construct provide a type of control not available with the use of traditional antibodies. What follows is a discussion of some uses for epitope tagging and how this technique might be applied to the study of flagella.

II. Uses of Epitope Tagging

One of the most common uses of epitope tagging is to identify a mutagenized protein expressed *in vivo* in the presence of the endogenous wild-type protein. In this application, the antiepitope antibody can differentiate between the tagged protein and an endogenous protein, even if they differ in only one amino acid (and the tag). For example, to study the effect of acetylation on tubulin function, the codon for lysine-40 [the lysine that is normally acetylated (LeDizet and Piperno, 1987)] in the *Chlamydomonas* α_1-tubulin gene was changed to a codon for alanine or arginine. In addition, an oligonucleotide encoding a nine-amino-acid epitope from influenza virus hemagglutinin (Wilson *et al.,* 1984) was inserted into the gene. Following introduction of the mutant gene into *Chlamydomonas,* an antihemagglutinin antibody was used to differentiate the mutant α-tubulin from the endogenous α-tubulin, which continued to be expressed in the transformed cells. Immunofluorescence using the antiepitope antibody showed that the introduced, nonacetylatable tubulin was incorporated into cytoplasmic and flagellar microtubules; otherwise, the cells were indistinguishable from wild type (Kozminski *et al.,* 1993). This approach could be taken to study the effects of mutations generated in any flagellar protein for which the gene has been cloned.

In a similar way, epitope tagging can be used to identify a single protein isotype (Albers and Fuchs, 1987; Chen *et al.,* 1993), even in a background of other isotypes. The distribution of a single tubulin isotype, for example, could be determined by tagging the corresponding gene rather than generating an isotype-specific antibody.

Another application of epitope tagging is to rapidly provide a specific probe for a protein for which the gene has been cloned. Once a cloned gene is tagged and expressed *in vivo,* the existing antiepitope antibody is immediately available to identify, localize, or immunoprecipitate the protein, without the labor, time, and expense required to generate a new antibody.

An extension of this strategy was used to clone proteins that contained nuclear localization signals (Sugano *et al.,* 1992). A cDNA library was cloned downstream of a sequence encoding an epitope; members of the library were transfected, in pools or individually, into a mammalian cell line; and the cells were screened with an antiepitope antibody to look for products that concentrated in the nucleus. A similar approach may be useful to identify cDNAs of proteins that can enter the flagella and/or that bind the axoneme.

Epitope tagging ensures that a high-affinity antibody will be available against the tagged protein, which is especially important when the protein of interest is poorly antigenic or difficult to purify. Antiepitope antibodies typically are suitable to immunoprecipitate or affinity purify the tagged protein. Epitope tagging was used, for example, to affinity purify a RAS-responsive adenylyl cyclase complex from *Saccharomyces cerevisiae,* a complex that had been recalcitrant to purification by conventional methods (Field *et al.,* 1988). In addition, immunoprecipitation of a native protein complex via an epitope tag can provide independent confirmation of results obtained with another antibody, and immunoprecipitation from cells not expressing the tagged protein provides an excellent control.

Epitope tagging also has been used to study the site of tubulin assembly in flagella. *Chlamydomonas* cells expressing epitope-tagged tubulin were fused to wild-type cells having half-length flagella. As the flagella completed regeneration they incorporated the tagged tubulin; both immunofluorescence and immunoelectron microscopy demonstrated that all the newly incorporated, epitope-tagged tubulin assembled at the tip of the flagella (Johnson and Rosenbaum, 1992). In these studies epitope-tagged tubulin was used in much the same way as microinjection of biotinylated tubulin was used to determine the site of assembly of kinetochore microtubules (Mitchison *et al.,* 1986).

III. Practical Considerations

A. Selection of an Epitope

The major considerations in selecting an epitope are the potential effects of the appended peptide on the protein of interest and the availability of a suitable antibody against the epitope. Although the effect of attaching a peptide to a protein is impossible to predict precisely, deleterious steric effects can be minimized by using a small tag. Epitopes can be less than 10 amino acids in length, and so can be inserted repeatedly (Roof *et al.,* 1992) to increase immunoreactivity. Hydrophilic epitopes are more likely to be exposed on the surface of the protein than hydrophobic sequences, and so may be more accessible to the antibody and less disruptive of protein structure. The antiepitope antibody should be of high affinity and should not react with untagged proteins in the cells in which the tagged protein will be expressed.

A variety of epitopes have been used for tagging, perhaps the most popular being those derived from c-*myc* (Munro and Pelham, 1987) and influenza virus hemagglutinin (Wilson *et al.,* 1984). The latter, sometimes referred to as the "HA" or "Flu" tag, was identified in studies of antigenic determinants in the hemagglutinin molecule, and consists of nine amino acids, YPYDVPDYA (Wilson *et al.,* 1984). Both monoclonal and polyclonal antibodies against this epitope are commercially available (Boehringer-Mannheim; Babco, Berkeley, CA).

The Flag epitope, DYKDDDDK (Immunex), was designed to contain the recognition site of the protease enterokinase, which can be used to cleave the tag from the protein (Hopp *et al.,* 1988). Because antibody binding to the FLAG tag is calcium dependent, the protein can be eluted easily and inexpensively from an antibody affinity column with ethylene glycol bis(β-aminoethyl ether)-N,N'-tetraacetic acid (EGTA) (Hopp *et al.,* 1988). Vectors that facilitate the cloning and expression of a FLAG-tagged protein, a vector that allows tagging without knowledge of the reading frame of the gene, the Flag polypeptide, an anti-Flag antibody, and affinity columns are all available commercially (IBI, New Haven, CT).

A technique related to epitope tagging is the insertion of six histidine residues into a protein. Although this sequence is poorly antigenic and so cannot be identified with antibodies, the sequence binds certain metal ions with very high affinity. A Ni^{2+} nitrilotriacetate resin is available (Qiagen, Chatsworth, CA) that can be used for affinity purification of the tagged protein, even in the presence of strong protein denaturants like 8 *M* urea and 6 *M* guanidine hydrochloride (Hochuli, 1990). In addition to isolating proteins expressed in bacteria (Hochuli, 1990), this tag also can be used to purify proteins under native conditions from eukaryotes (Bugge *et al.,* 1992).

B. Placement of the Epitope

The effect of epitope insertion on protein function is impossible to predict, but in the absence of information on functional domains or three-dimensional structure of the target protein, a good place to begin is to append the epitope to either terminus of the protein. Insertions in the terminus of a protein, however, are not always innocuous to protein function (Munro and Pelham, 1984), and insertions in the center of the protein have been successful (Davis and Fink, 1990). To increase the probability that the epitope will be accessible and to minimize its effect on protein structure, the insertion site should be in a hydrophilic region of the protein. In some cases the epitope tag has been attached to the protein via a small hydrophilic peptide spacer to increase the chances that the epitope will be exposed (Munro and Pelham, 1984). Computer programs (e.g., GCG PeptideStructure, Devereux *et al.,* 1984) can be used to predict regions of secondary structure that may be important for function and so should be avoided as sites of insertion. As a complementary approach, insertion of the epitope can purposely be used as a mutagen to assess the importance of specific regions for the function of the target protein (Roof *et al.,* 1992).

Several strategies can be used to introduce the epitope into the desired gene. When convenient, complementary oligonucleotides encoding the epitope can be inserted into restriction sites in the coding region of the gene (e.g., Kozminski *et al.,* 1993). Alternatively, vectors that encode the epitope can be used to append the epitope to the protein (e.g., Flag Expression Vectors, IBI). For more flexibility in the site of insertion, oligonucleotide-directed mutagenesis (Kunkel *et al.,* 1987) or PCR strategies (e.g., Qian *et al.,* 1993) can be used. In designing an oligonucleotide to encode the epitope it may be useful to include a diagnostic restriction site to facilitate identification of insertions and to take into account the codon usage of the organism in which the construct is to be expressed.

C. Choice of Organism

Only organisms in which engineered genes can be introduced and expressed are amenable to epitope tagging. This constraint eliminates many of the multicellular organisms used for the study of flagella. One notable exception is *Drosophila,* in which epitope tagging could be used to study flagellar development and function during spermatogenesis.

One organism in which epitope tagging has been useful in studying flagella is the unicellular alga *Chlamydomonas reinhardtii*. The flagella of *Chlamydomonas* have long been the object of genetic, biochemical, and functional studies, resulting in the characterization of a variety of mutants with defects in constituent proteins of various flagellar structures (Huang, 1986). The genes for several flagellar proteins have been sequenced (for examples, see Curry and Rosenbaum, 1993) and are prime candidates for functional studies that employ epitope tagging. The ease with which engineered genes can be reintroduced into *Chlamydomonas* (Kindle, 1990) makes it the ideal organism for this approach (e.g., Johnson and Rosenbaum, 1992; Kosminski *et al.,* 1993; Diener *et al.,* 1993). The ciliate *Tetrahymena,* another unicellular organism in which motility mutants have been identified (see Chapter 81 in this volume), has the additional advantage that genes can be routinely replaced by homologous recombination (see Chapter 80 in this volume).

IV. Conclusion

Once a gene has been cloned, epitope tagging is a means to generate a specific probe applicable to any standard immunological procedure: immunolocalization, immunoprecipitation, and affinity purification. As more flagellar genes are cloned, this versatile technique should become a valuable tool with which to study the function of flagellar proteins.

Acknowledgments

I thank Keith Kozminski, Mitchell Bernstein, and Joel Rosenbaum for comments on the manuscript. This work was supported by National Institutes of Health Grant GM14642 and National Science Foundation Grant CB45147 to Joel Rosenbaum.

References

Albers, K., and Fuchs, E. (1987). Expression of mutant epidermal keratin cDNAs transfected in simple epithelial and squamous cell carcinoma lines. *J. Cell Biol.* **105,** 791–806.

Bugge, T. H., Pohl, J., Lonnoy, O., and Stunnenberg, H. G. (1992). RXRα, a promiscuous partner of retinoic acid and thyroid hormone receptors. *EMBO J.* **11,** 1409–1418.

Chen, Y.-T., Holcomb, C., and Moore, H.-P. H. (1993). Expression and localization of two low molecular weight GTP-binding proteins, Rab8 and Rab10, by epitope tag. *Proc. Natl. Acad. Sci. U.S.A.* **90,** 6508–6512.

Curry, A. M., and Rosenbaum, J. L. (1993). Flagellar radial spoke: A model molecular genetic system for studying organelle assembly. *Cell Motil. Cytoskel.* **24,** 224–232.

Davis, L. I., and Fink, G. R. (1990). The NUPI gene encodes an essential component of the yeast nuclear pore complex. *Cell* **61,** 965–978.

Devereux, J., Haeberli, P., and Smithies, O. (1984). A comprehensive set of sequence analysis programs for the VAX. *Nucleic Acids Res.* **12,** 387–395.

Diener, D. R., Ang, L. H., and Rosenbaum, J. L. (1993). Assembly of flagellar radial spoke proteins in *Chlamydomonas:* Identification of the axoneme binding domain of radial spoke protein 3. *J. Cell Biol.* **123,** 183–190.

Field, J., Nikawa, J.-I., Broek, D., MacDonald, B., Rodgers, L., Wilson, I. A., Lerner, R. A., and Wigler, M. (1988). Purification of a RAS-responsive adenylyl cyclase complex from *Saccharomyces cerevisiae* by use of an epitope addition method. *Mol. Cell. Biol.* **8,** 2159–2165.

Hochuli, E. (1990). Purification of recombinant proteins with metal chelate adsorbent. *In* "Genetic Engineering, Principles and Methods" (J. K. Setlow, ed.), Vol 12, pp. 87–98. Plenum Press, New York.

Hopp, T. P., Prickett, K. S., Price, V. L., Libby, R. T., March, C. J., Cerretti, D. P., Urdal, D. L., and Conlon, P. J. (1988). A short polypeptide marker sequence useful for recombinant protein identification and purification. *Biotechnology* **6,** 1204–1210.

Huang, B. P.-H. (1986). *Chlamydomonas reinhardtii:* A model system for the genetic analysis of flagellar structure and motility. *Int. Rev. Cytol.* **99,** 181–215.

Johnson, K. A., and Rosenbaum, J. L. (1992). Polarity of flagellar assembly in *Chlamydomonas. J. Cell Biol.* **119,** 1605–1611.

Kindle, K. L. (1990). High-frequency nuclear transformation of *Chlamydomonas reinhardtii. Proc. Natl. Acad. Sci. U.S.A.* **87,** 1228–1232.

Kozminski, K. G., Diener, D. R., and Rosenbaum, J. L. (1993). High level expression of nonacetylatable α-tubulin in *Chlamydomonas reinhardtii. Cell Motil. Cytoskel.* **25,** 158–170.

Kunkel, T. A., Roberts, J. D., and Zakour, R. A. (1987). Rapid and efficient site-specific mutagenesis without phenotypic selection. *Methods Enzymol.* **154,** 367–382.

LeDizet, M., and Piperno, G. (1987). Identification of an acetylation site of *Chlamydomonas* α-tubulin. *Proc. Natl. Acad. Sci. U.S.A.* **84,** 5720–5724.

Mitchison, T., Evans, L., Schulze, E., and Kirschner, M. (1986). Sites of microtubule assembly and disassembly in the mitotic spindle. *Cell* **45,** 515–527.

Munro, S., and Pelham, H. R. B. (1984). Use of peptide tagging to detect proteins expressed from cloned genes: deletion mapping functional domains of *Drosophila* hsp70. *EMBO J.* **3,** 3087–3093.

Munro, S., and Pelham, H. R. B. (1987). A C-terminal signal prevents secretion of luminal ER proteins. *Cell* **48,** 899–907.

Qian, N.-X., Winitz, S., and Johnson, G. L. (1993). Epitope-tagged $G_q\alpha$ subunits: Expression of GTPase-deficient α subunits persistently stimulates phosphatidylinositol-specific phospholipase C but not mitogen-activated protein kinase activity regulated by the M_1 muscarinic acetylcholine receptor. *Proc. Natl. Acad. Sci. U.S.A.* **90,** 4077–4081.

Roof, D. M., Meluh, P. B., and Rose, M. D. (1992). Kinesin-related proteins required for the assembly of the mitotic spindle. *J. Cell Biol.* **118,** 95–108.

Sugano, S., Kim, D. W., Yu, Y.-S., Mizushima-Sugano, J., Yoshitomo, K., Watanabe, S., Suzuki, F., and Yamaguchi, N. (1992). Use of an epitope-tagged cDNA library to isolate cDNAs encoding proteins with nuclear localization potential. *Gene* **120,** 227–233.

Wilson, I. A., Niman, H. L., Houghten, R. A., Cherenson, A. R., Connolly, M. L., and Lerner, R. A. (1984). The structure of an antigenic determinant in a protein. *Cell* **37,** 767–778.

CHAPTER 79

Enrichment of mRNA Encoding Flagellar Proteins

Laura R. Keller

Department of Biological Science
Florida State University
Tallahassee, Florida 32306

I. Introduction

Mechanisms controlling formation of cilia and flagella have been studied extensively in a variety of organisms including sea urchins (Merlino *et al.*, 1978), *Chlamydomonas* (Lefebvre and Rosenbaum, 1986), *Volvox* (Coggin and Kochert, 1980), *Tetrahymena* (Calzone and Gorovsky, 1982), *Naegleria,* (Fulton and Walsh, 1980), *Physarum* (Green and Dove, 1984), and others (Lefebvre and Rosenbaum, 1986). In many cases, growth of cilia or flagella is accompanied by large increases in levels of translatable mRNAs encoding flagellar proteins and correspondingly large increases in synthesis of flagellar polypeptides. Thus, the process of cilium/flagellum formation provides the opportunity to study controls over the expression of a specific large set of structurally diverse yet functionally related genes encoding ciliary or flagellar proteins. It also provides an opportunity to isolate mRNA enriched in transcripts encoding flagellar proteins, thus permitting construction of cDNA libraries enriched for these sequences.

One organism for which the induction of flagellar loss, outgrowth, and gene expression are well characterized is the unicellular biflagellate alga *Chlamydomonas reinhardtii*. In response to certain extracellular stimuli, *Chlamydomonas* cells excise their flagella, induce expression of more than 200 genes encoding flagellar proteins, and assemble a new flagellar pair. The response to stimulation is rapid and transient: flagella shed instantly, stubs of regenerating flagella are visible by ~15 minutes, and nearly full-length flagella reassemble by ~120 minutes after stimulation. During this time, levels of mRNA encoding flagellar protein subunits increase to peak levels by ~45 minutes after stimulation and return to prestimulation levels by ~120 minutes.

This chapter describes a basic protocol for stimulating wild-type *Chlamydomonas* cells by pH shock, collecting samples of fixed cells for flagellar length measurements, and collecting RNA samples enriched for flagellar mRNAs before and 15, 45, 90, and 120 minutes after stimulation. RNA samples prepared in this manner have been used for cDNA library preparation (Schloss *et al.*, 1984; Silflow and Rosenbaum, 1981), Northern (Schloss *et al.*, 1984) and S1 nuclease protection analysis (Cheshire and Keller, 1991; Cheshire *et al.*, 1994), and *in vitro* translation (Lefebvre *et al.*, 1980). At the end of this description, variations in the basic method are discussed.

II. Methods

A. Equipment

Magnetic stirrer
Fluorescent light (desk lamp)
Two 100-ml beakers and stir bars
Vortex Genie
Rotary platform shaker
Six 0.5-ml disposable tubes containing 100 μl of 1% glutaraldehyde
Fifteen baked 15-ml Corex glass centrifuge tubes (RNase-free, 3/RNA sample)
Six 1.5-ml microcentrifuge tubes, RNase-free
One hundred baked 9-in. Pasteur pipets and bulbs
Ten baked 10-ml glass pipets
pH meter
Three baked 125-ml bottles or flasks (for lysis buffer, phenol, and Sevag)
Six 5-ml polypropylene tubes each containing 2 g CsCl

Note: Glassware should be baked in a clean glassware oven at 200–240°C for several hours to overnight.

B. Reagents

Culture of ~10^8 synchronized *Chlamydomonas* cells grown to a density of ~10^5 to 10^6/ml in medium I of Sager and Granick (1953) in alternating cycles of 14 hours light/10 hours dark

Stock solutions of 0.5 *M* acetic acid and 0.5 *M* KOH

Stock solution of 1% glutaraldhyde, prepared in distilled H_2O

Lysis buffer containing 50 m*M* Tris–Cl, pH 8.0, 0.3 *M* NaCl, 5 m*M* ethylene glycol bis(β-aminoethyl ether)-*N,N'*-tetraacetic acid (EGTA), and 2% sodium dodecyl sulfate (SDS), (50 ml)

Lysis buffer (as above) containing 40 μg/ml proteinase K (2 ml/RNA sample)

Neutralized phenol saturated with RNase-free water or TE [10 m*M* Tris–Cl, 1 m*M* ethylenediaminetetraacetic acid (EDTA), pH 8.0], (3 ml/RNA sample)

Sevag (96% chloroform/4% isoamyl alcohol), (3 ml/RNA sample)

2 g CsCl (biotechnology grade) per sample, weighed in polypropylene tubes

CsCl cushion (5.99 *M* CsCl, density = 1.74, refractive index = 1.4030, 1.0098 g/ml in 0.1 *M* EDTA) (Glisin *et al.*, 1974)

Stock solution of 10.5 *M* ammonium acetate

100% ethanol

C. Cell Stimulation

1. Concentrate cells ~10-fold by centrifugation (400*g* for 5 minutes) and gentle resuspension at a concentration of 5×10^6 to 5×10^7 cells/ml in a volume of 60 ml.

2. As some cells lose their flagella during concentration, allow cells to recover from concentration by gently stirring with a magnetic stirrer under illumination for 120 minutes. Ideally, recovery should occur midmorning, so that the experiment can begin near noon. During recovery, perform a trial pH shock stimulation on surplus cells (see step 4 below), to determine the volume of 0.5 *M* acetic acid needed to lower the pH of 50 ml of cell culture to 4.3 and the volume of 0.5 *M* KOH needed to return to the initial pH of 6.8.

3. After recovery, fix a 100-μl aliquot of cells by pipetting them into a tube containing an equal volume of glutaraldehyde, for the later measurement of flagellar lengths in nonstimulated cells. Next, remove 10 ml of culture to a Corex tube. Begin the RNA extraction protocol for this sample, which will serve as the nonstimulated control sample (see Section II,D). *Note:* Be careful not to expose the cell culture to glutaraldehyde or glutaraldehyde fumes, as low concentrations of glutaraldehyde may cause the cells to deflagellate.

4. Stimulate the remainder of the culture by adding the volume of acetic acid (determined in step 2 above) to lower the pH of the culture to 4.3 while the

cells continue to be stirred. Wait 30 seconds, and then add KOH to return the pH to its initial value. Fix a 100-μl aliquot of cells in glutaraldehyde, for the later measurement of flagellar lengths in cells immediately after stimulation.

5. At ~7 minutes after stimulation, remove 10 ml of cells from the culture to a Corex tube, and begin the RNA extraction protocol on this sample. Fifteen minutes after stimulation, fix an aliquot of cells for later measurement of flagellar lengths.

6. At ~8 min prior to each time point of 45, 90, and 120 minutes after pH shock stimulation, remove 10 ml of cells from the culture, and begin the RNA extraction protocol. Exactly 45, 90, and 120 minutes after stimulation, fix an aliquot of cells by pipetting them into one of the tubes containing 100 μl glutaraldehyde.

D. RNA Extraction

1. Eight minutes prior to the desired time after stimulation, transfer a 10-ml aliquot of cells to a 15-ml Corex tube. Spin at 400*g* for 5 minutes. Quickly aspirate the supernatant off the cell pellet. (The 8 minutes allows time for sample collection, centrifugation, and aspiration to be completed before the exact point is reached, and may vary depending on the laboratory centrifuge used.)

2. At the exact time after stimulation, resuspend the cell pellet in 2 ml lysis buffer containing proteinase K. Pipet the suspension vigorously with a Pasteur pipet while adding lysis buffer. Secure the tube on the plateform rotary shaker, and mix at 300 rpm for 5 minutes at room temperature.

3. Add 1.5 ml of phenol and 1.5 ml of Sevag, and mix well by vortexing. Incubate on ice for 10 minutes, keeping the sample emulsified by occasional vortexing.

4. Spin at 8000*g* for 20 minutes.

5. Transfer the aqueous layer to a fresh Corex tube, leaving behind the material at the interface between the aqueous and organic layers. Add 3 ml Sevag. Mix well by vortexing.

6. Spin at 8000*g* for 15 minutes.

7. Transfer the aqueous phase to a polypropylene tube containing 2 g of solid CsCl (1 g/ml of aqueous phase) and mix at room temperature until the CsCl dissolves. Prepare one extra tube of 1 g/ml CsCl in lysis buffer lacking proteinase K to be used as a "standard tube." Adjust the volume of each RNA sample tube by adding lysis buffer lacking proteinase K until it matches the volume of the standard tube. (*Note:* Volume may have been lost from the original sample during organic extraction. This adjustment restores the original sample volume and ensures that all samples are of uniform CsCl density.)

8. Pipet 1.2 ml of CsCl cushion into a 0.5 × 2-in. Beckman polyallomer centrifuge tube for the Beckman SW50.1 rotor. Gently overlay the CsCl cushion

with the RNA sample, taking care to avoid disturbing the cushion/sample interface. Balance the tubes in pairs by adding TE or CsCl solution from the "standard tube," and bring the volume to within 3 mm of the tube top with TE or mineral oil. Spin at 35,000 rpm for 12 hours (Glisin *et al.*, 1974).

9. After centrifugation, remove the tubes from the rotor, and locate the clear RNA pellet at the bottom of each tube. Draw off solution to just above the RNA pellet using a baked Pasteur pipet. Dissolve the RNA pellets on ice in ~0.2 ml TE, and combine with 2.8 ml TE in another fresh Corex tube.

10. Add ammonium acetate to a final concentration of 0.4 *M* (114 μl of a 10.5 *M* stock) and 2–2.5 vol of ethanol (7.5 ml of 100%). Mix well.

11. Precipitate at $-20°C$ overnight or at $-80°C$ for 1 hour. Collect the nucleic acid precipitate by centrifugation at 8000*g* for 20 minutes. Drain, dry, and resuspend the pellet in 300 μl of double-distilled, RNase-free H_2O. Transfer the sample to a 1.5-ml Eppendorf tube.

12. Reprecipitate the nucleic acids by adding ammonium acetate and ethanol, and perform cold incubation as indicated in steps 7 and 8.

13. Collect the precipitated nucleic acids by centrifugation at 10,000 rpm for 15 minutes.

14. Discard the supernatant, wash the pellet with 70% ethanol, and repellet the nucleic acids by centrifugation at 10,000 rpm for 10 minutes. Decant the supernatant, and dry the pellet. Resuspend the pellet on ice in 100 μl of dH_2O.

15. Determine the concentration of nucleic acids by reading the absorbance at 260 nm of a 1:100 dilution of the sample. The protocol as described should yield ~100–200 μg of total nucleic acids.

16. Measure the lengths of flagella on cells fixed before and after deflagellation as described (Cheshire and Keller, 1991).

III. Variations in the Basic Protocol

Isolation of Total Nucleic Acids from Cells. Many protocols for RNA analyses, especially those such as Northern blot and S1 nuclease protection in which formamide is present in the hybridization buffer, can be performed on total nucleic acid samples that contain both RNA and DNA (Cheshire and Keller, 1991). For preparation of total nucleic acid samples, delete steps 7 through 9 of the RNA extraction protocol described above, and precipitate the aqueous phase of the last organic extraction in step 6 with salt and ethanol as in step 10.

For Extraction of RNA from Larger Cell Samples. Increase the volumes of lysis buffer, phenol, and Sevag used for extraction. For example, for more than 5×10^8 cells/sample, use 4 ml lysis buffer containing proteinase K, 3 ml phenol + 3 ml Sevag for RNA extraction step 3, and 6 ml Sevag for RNA extraction step 5, and perform the extraction in 30-ml RNase-free Corex tubes.

For Extraction of RNA from Smaller Cell Samples. Add 10 μg of yeast tRNA to the sample in RNA extraction step 10 as a carrier to coprecipitate the *Chlamydomonas* nucleic acids. Addition of tRNA to samples will affect the determination of nucleic acid concentrations in step 15 of the protocol.

For Samples of Mutant Chlamydomonas Cells That Hatch Poorly. Before recovery from concentration, gently homogenize the concentrated cell cultures at low speed using a Virtis homogenizer to release cells from the mother cell walls. Pellet the cells at 400*g*, decant the supernatant containing cell wall ghosts, resuspend the cell culture in fresh growth medium, and examine a small sample of the culture using the light microscope. Several additional gentle homogenizations may be required to completely release the cells. Extract the RNA sample twice with phenol + Sevag (as in RNA extraction step 3) and once with Sevag (RNA extraction step 5) before precipitation with ethanol. Additional reextractions may be required after ethanol precipitation depending on the initial cell density and success of cell wall removal (Cheshire *et al.,* 1994).

For Enrichment of a Specific Flagellar mRNA. Induction results in stimulation of different kinetic classes of flagellar mRNAs (Schloss *et al.,* 1984). If the kinetic class of the desired RNA is known, time points can be adjusted to collect samples containing the RNA at its abundance peak.

For Stimulation of Flagellar Excision and Gene Expression Using Means Other Than pH Shock. The protocol is adapted easily for stimulation of flagellar excision and mRNA abundance changes using previously described methods such as mechanical shear or addition of the calcium-binding protein antagonists W-7 or TFP (Cheshire and Keller, 1991). Other agents such as mastoparan (Quarmby *et al.,* 1992), dibucaine, and ethanol (Harris, 1989) stimulate flagellar excision, but measurements of flagellar mRNA abundance have not been reported using these agents.

For Stimulation of Ciliary/Flagellar Excision and Gene Induction in Systems Other Than Chlamydomonas. Refer to citations in the Introduction of this chapter to modify the basic protocol for other organisms.

Acknowledgments

I thank J. Schloss, who taught me to isolate *Chlamydomonas* RNA; J. Cheshire and J. Evans, for modifications to the protocol; and the National Science Foundation for support of my research program.

References

Calzone, F. C., and Gorovsky, M. A. (1982). Cilia regeneration in *Tetrahymena. Exp. Cell Res.* **140,** 471–476.

Cheshire, J. L., and Keller, L. R. (1991). Uncoupling of *Chlamydomonas* flagellar gene expression and outgrowth from flagellar excision by manipulation of Ca^{2+}. *J. Cell Biol.* **115,** 1651–1659.

Cheshire, J. L., Evans, J. H., and Keller, L. R. (1994). Ca^{2+} signaling in the *Chlamydomonas* flagellar regeneration system: cellular and molecular responses. *J. Cell Sci.* **107,** 2491–2498.

Coggin, S., and Kochert, G. (1980). Flagellar growth and regeneration in the life cycle of Volvox carteri. *J. Cell Biol.* **87,** 39a.

Fulton, C., and Walsh, C. (1980). Cell differentiation and flagellar elongation in *Naegleria gruberi*. *J. Cell Biol.* **85,** 346–360.

Glisin, V., Crkvenjakov, R., and Byers, C. (1974). Ribonucleic acid isolation by cesium chloride centrifugation. *Biochemistry* **13,** 2633–2637.

Green, L. L., and Dove, W. F. (1984). Tubulin proteins and RNA during the myxamoeba-flagellate transformation of *Physarum*. *Mol. Cell. Biol.* **4,** 1706–1711.

Harris, E. (1989). "The *Chlamydomonas* Sourcebook: A Complete Guide to Biology and Laboratory Use". Academic Press, San Diego.

Lefebvre, P. A., Silflow, C. D., Wieben, E. D., and Rosenbaum, J. L. (1980). Increased levels of mRNAs for tubulin and other flagellar proteins after amputation or shortening of *Chlamydomonas* flagella. *Cell* **20,** 469–477.

Lefebvre, P. A., and Rosenbaum, J. L. (1986). Regulation of the synthesis and assembly of ciliary and flagellar proteins during regeneration. *Annu. Rev. Cell Biol.* **2,** 517–546.

Merlino, G. A., Chamberlain, J. P., and Kleinsmith, L. J. (1978). Effects of the deciliation on tubulin messenger activity in sea urchin embryos. *J. Biol. Chem.* **253,** 7078–7085.

Quarmby, L. M., Yueh, Y. G., Cheshire, J. L., Keller, L. J., Snell, W. J., and Crain, R. C. (1992). Inositol phospholipid metabolism may trigger flagellar excision in *Chlamydomonas reinhardtii*. *J. Cell Biol.* **116,** 737–744.

Sager, R., and Granick, S. (1953). Nutritional studies with *Chlamydomonas reinhardtii*. *Ann. N. Y. Acad. Sci.* **56,** 831–838.

Schloss, J. A., Silflow, C. D., and Rosenbaum, J. L. (1984). mRNA abundance changes during flagellar regeneration in *Chlamydomonas reinhardtii*. *Mol. Cell. Biol.* **105,** 424–434.

Silflow, C. D., and Rosenbaum, J. L. (1981). Multiple α- and β- tubulin genes in *Chlamydomonas* and regulation of tubulin mRNA levels after deflagellation. *Cell* **24,** 81–88.

CHAPTER 80

DNA-Mediated Transformation in *Tetrahymena*

Jacek Gaertig and Martin A. Gorovsky
Department of Biology
University of Rochester
Rochester, New York 14627

I. Introduction

Significant progress has been made in the development of methods for the manipulation of the *Tetrahymena thermophila* genome. *T. thermophila* is the only ciliated or flagellate model currently available in which high-frequency transformation and gene replacement can be performed. *Tetrahymena* can be used to study the function of any cloned gene involved in ciliary function or biogenesis using gene replacement or gene knockout.

Like most ciliates, *Tetrahymena* have two nuclei in each cell (Gorovsky, 1980). The diploid, germinal micronucleus is transcriptionally silent during vegetative growth, divides mitotically, and serves as a genetic repository. The macronucleus is a polyploid somatic nucleus, divides amitotically, is transcrip-

METHODS IN CELL BIOLOGY, VOL. 47

tionally active, and is responsible for the phenotype. *Tetrahymena* grow vegetatively by binary fission. Conjugation, the sexual phase of the ciliate life cycle, is induced by starvation. During each conjugation, cells of different mating types pair and micronuclei undergo meiosis. Three of the four meiotic products break down, and the fourth undergoes a haploid mitosis to produce two gametic nuclei. Following fertilization and two postzygotic nuclear divisions, the new macronucleus is formed as a product of differentiation of a zygotic micronuclear derivative, while the old macronucleus is resorbed (see Bruns, 1986; Orias, 1986, for reviews).

All of the published transformation techniques in *Tetrahymena* introduce genes into the macronucleus (Tondravi and Yao, 1986; Yu and Blackburn, 1989; Yao and Yao, 1989, 1991; Kahn *et al.*, 1993; Gaertig and Gorovsky, 1992). Preliminary studies in collaboration with Peter Brun's laboratory indicate that germ-line transformation techniques for the micronucleus should be available soon. This chapter deals only with transformation techniques for the somatic macronucleus. Genes introduced into the macronucleus are stably maintained either because they become integrated or because they are on origin-containing plasmids and persist there until the next mating when the macronucleus is replaced. As *T. thermophila* can be propagated indefinitely without mating, macronuclear transformation provides a way to obtain stable somatic transformants.

II. Transformation Strategies in *Tetrahymena*

The DNA recombination activity observed to date in the *Tetrahymena* macronucleus is exclusively homologous (Yu *et al.*, 1988; Yao and Yao, 1991; Kahn *et al.*, 1993; Gaertig *et al.*, 1994b), enabling highly predictable targeting of the transforming DNA fragments. Also, unlike another ciliate, *Paramecium* (Gilley *et al.*, 1988), in *Tetrahymena*, DNA introduced into the macronucleus requires a replication origin to be replicated (Gaertig *et al.*, 1994a). Thus, in *Tetrahymena*, genes can be either targeted into homologous locations or expressed on a replicative plasmid.

A. Gene Replacement

Because there are ~45 copies of most genes in the polycopy macronucleus, only partial replacement of all macronuclear copies is initially obtained in a typical gene replacement experiment (Yao and Yao, 1991; Kahn *et al.*, 1993; Gaertig *et al.*, 1994b). During vegetative growth, the macronucleus divides amitotically, and alleles are segregated randomly at each division. This phenomenon is well characterized in *Tetrahymena* and is known as *phenotypic assortment* (Nanney, 1980; Doerder *et al.*, 1992). Thus, all heterozygous clones eventually become homozygous in the macronucleus during vegetative growth. Even alleles initially present in only one of the 45 copies can be expected to become homozygous within 80 generations (Doerder *et al.*, 1992). Complete assortment of an allele is possible only if this allele is not lethal. The assortment rate may be affected if one of the alleles provides a selective advantage. In particular, gene replacement is promoted when the transformed allele provides a strong selective marker.

1. Functional Gene Replacement

Gene replacement can be performed using a drug-resistant version of a *Tetrahymena* gene, such as the cycloheximide-resistant mutation of the ribosomal protein *rpL29* (Yao and Yao, 1991), as shown in Fig. 1. We have used a similar strategy to replace tubulin genes in *Tetrahymena*. There is only a single α-tubulin gene in *T. thermophila* (Callahan *et al.*, 1984; McGrath *et al.*, 1994). There are two β-tubulin genes (BTU1 and BTU2) and both genes encode the same protein (Gaertig *et al.*, 1994b). An analog of the *Chlamydomonas* β-tubulin mutant *col[r]15* (Bolduc *et al.*, 1988; Lee and Huang, 1990) was constructed in the *Tetrahymena* BTU1 gene by replacement of lysine with methionine at position 350. In *Chlamydomonas*, this mutation causes resistance to several microtubule-depolymerizing drugs and hypersensitivity to taxol (Schibler and Huang, 1991). The mutant BTU1 gene transformed *Tetrahymena* by gene replacement and conferred the expected phenotype (Gaertig *et al.*, 1994b). Because the BTU1 mutant gene confers multiple drug resistance and taxol hypersensitivity, mutant cells can be retransformed by introduction of a wild-type BTU1 gene and selection with taxol at a concentration to which wild-type cells are resistant (Gaertig *et al.*, 1994b). Thus, BTU1 gene replacement can be used for both positive and negative selection (Fig. 2).

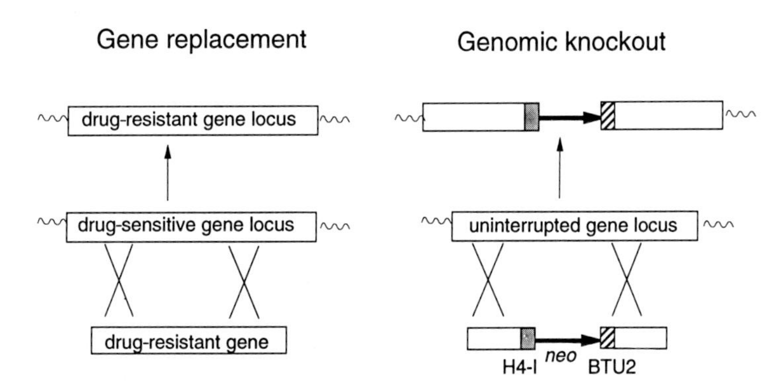

Fig. 1 Schematic representation of transformation techniques based on homologous recombination in *Tetrahymena:* functional gene replacement with a drug-resistant gene (left) and gene knockout using an H4-I*neo*BTU2 cassette (right).

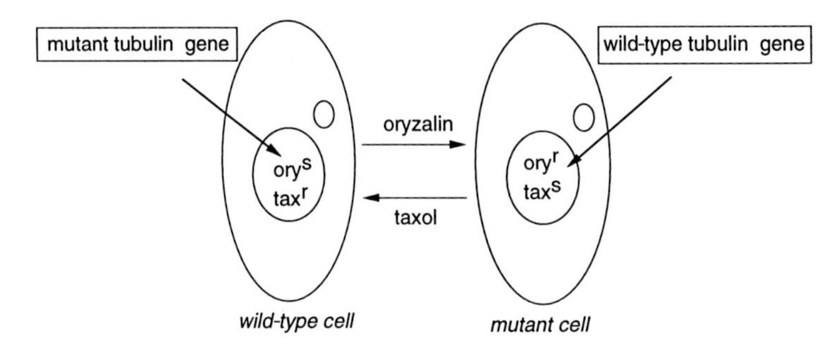

Fig. 2 Functional tubulin gene replacement in *Tetrahymena*. Tubulin gene markers (currently available for α- and β-tubulin subunits) can be used for both positive and negative selection.

Another marker suitable for complete α-tubulin gene replacement in *Tetrahymena* has been developed. A germ-line mutant, E5, was isolated by random chemical mutagenesis and selection for resistance to multiple tubulin-depolymerizing drugs and taxol sensitivity (J. Gaertig, J. Moran, M. Gorovsky, and D. Pennock, unpublished results). This mutation contains a replacement of alanine by threonine at position 65. This mutant strain can be transformed with a wild-type α-tubulin gene and selection for taxol resistance. Thus, like the BTU1 gene, the α-tubulin gene in *Tetrahymena* can be used for positive and negative selection. In the case of the α-tubulin gene, complete replacement of the α-tubulin gene product is obtained because there is only a single α-tubulin gene. Because *T. thermophila* contains two coexpressed β-tubulin genes, replacement of the BTU1 gene results in about half of the total β-tubulin gene expression being produced by the introduced gene. Either of the two β-tubulin genes can be inactivated by a genomic knockout (below). It should be possible to replace the BTU1 gene first and then inactivate the BTU2 gene in a second round of transformation.

These methods of α- and β-tubulin gene replacement in *Tetrahymena* should allow analysis of the *in vivo* functions of tubulin molecular domains such as MAP-binding sites, GTP-binding domains, and sites that are post-translationally modified.

2. Gene Knockout

To determine whether a gene is essential in *Tetrahymena,* one can perform a genomic knockout. Kahn *et al.* (1993) used microinjection-mediated transformation to replace the coding sequence of the H4-I histone gene with that of a neomycin resistance gene. In most transformants, gene replacement occurred, resulting in a gene knockout. In others, integration into the flanking region of the gene occurred. Similar results were obtained when the construct was introduced by electroporation (Gaertig *et al.,* 1994a). Complete macronuclear gene replacement was obtained in the knockout transformants.

A general disruption cassette for genomic knockout, containing the *neo* coding sequence flanked by the H4-I gene promoter and terminator from the BTU2 gene, has been constructed (Fig. 1). The flanking sequences are ~300 bp and the hybrid gene alone does not transform *Tetrahymena* by homologous recombination. To perform a knockout, the hybrid H4-I*neo*BTU2 gene is inserted into the coding sequence of any cloned *Tetrahymena* gene, and the disrupted gene is introduced into the macronucleus where it is targeted to the homologous locus (Fig. 1). Several *Tetrahymena* genes have been knocked out using this method, including both β-tubulin genes (L. Gu, J. Gaertig, M. Gorovsky, unpublished results), three genes encoding histone H2As (X. Liu, M. Gorovsky, unpublished results), and a gene encoding a small RNA induced by starvation or heat shock (P. Funk, R. Hallberg, personal communication). If the disrupted gene is not essential, complete replacement can be achieved by phenotypic assortment promoted by growth in drug (paromomycin)-containing medium.

B. Expression of Genes Using Replicative Plasmids

In *Tetrahymena,* genes can be expressed using replicative plasmids. Two types of vectors, rDNA vectors and plasmids containing rDNA replication origins, are maintained at a high copy number in the macronucleus, resulting in overexpression of cloned genes (Yao and Yao, 1991; Kahn *et al.,* 1993; Gaertig *et al.,* 1994a). The rDNA vectors are introduced into the developing macronucleus during conjugation, when rDNA is excised from the plasmid, converted into a palindrome, and amplified to high copy number characteristic for rDNA. Circular somatic rDNA vectors need two copies of a restriction fragment containing the replication origin to replicate efficiently (Yu and Blackburn, 1989) and transform with high efficiency (Gaertig *et al.,* 1994a). A smaller vector, containing a repeat of the rDNA replication origin and the H4-I*neo*BTU2 gene as a selectable marker, has been constructed (pH4T2), and the *rpL29* gene was expressed when inserted into this vector (Gaertig *et al.,* 1994a).

III. Methods for Introducing Transforming DNA into *Tetrahymena thermophila*

A. Overview

Two methods for the introduction of transforming DNA are macronuclear microinjection (Tondravi and Yao, 1986) and conjugant electrotransformation (CET) (Gaertig and Gorovsky, 1992). Microinjection is usually performed into the macronucleus of vegetatively growing cells but can be done into developing macronuclei of conjugating cells. In the CET method, conjugating cells are electroporated in the presence of the transforming DNA. The optimal timing for transformation (10–11 hours after mixing of complementary mating types) coincides with the period of new macronucleus formation. Using vectors containing origins of replication from the *Tetrahymena* rDNA, transformation rates exceeding 1000 transformants/μg DNA can be obtained. In gene replacement experiments, up to 100 transformants can be obtained in a single experiment. Using CET, cells either can be electroporated with a linear gene replacement/knockout construct and selected directly or can be transformed with one of the high-efficiency replicative vectors, pD5H8 or pH4T2. For cotransformation, electroporated cells are first selected for paromomycin resistance (10–15 generations) conferred by the high-efficiency vector, and subsequently screened for gene replacement by selection for another drug-resistant marker. For direct selection, transformed cells undergo only a few divisions before selection is initiated. The direct selection method is convenient when only a short period of growth is required for expression of the resistance trait and is applicable to experiments using hybrid *neo* genes and paromomycin selection. Cotransformation is the preferred method if the expression of the transformed gene requires a longer period of growth (as for tubulin or *rpL29* genes) to enable cells to replace preexisting drug-sensitive cellular targets. Usually a few percent of the primary transformant pool is cotransformed by gene replacement.

CET does require that two strains of different mating types are fertile, so mutant cells that do not generate conjugation progeny cannot be transformed by this method. CET can only be used for one round of gene replacement because the old macronucleus is discarded and replaced by a new one during each transformation. A reliable method for transformation by macronuclear microinjection is described by Tondravi and Yao (1986). Here we present a general protocol both for CET and for microinjection. For general methods for *Tetrahymena* cell culture, cloning, and genetic analysis, see Orias and Bruns, 1975.

B. Materials

SPP growth medium: 1% proteose peptone, 0.1% yeast extract, 0.2% dextrose, 0.003% sequesterine.

Paromomycin sulfate (Sigma) or Humatin (Parke–Davis): 100 mg/ml in H_2O. Store frozen.

6-Methylpurine: 15 mg/ml in H_2O. Store frozen.

Cycloheximide: 12.5 mg/ml in ethanol. Store frozen.

10 m*M* 4-(2-Hydroxyethyl)-1-piperazineethanesulfonic acid (Hepes), pH 7.5: Autoclave.

10 m*M* Tris, pH 7.5: Autoclave.

100× Antibiotic mix: 1 ml Fungizone (GIBCO No. 600-2595AE), 100 ml; penicillin/streptomycin (GIBCO No. 600-5140PG). Store frozen in aliquots.

50-ml Corning plastic conical centrifuge tubes.

Phenol/chloroform/isoamyl alcohol (25/24/1, v/v/v).

Chloroform/isoamyl alcohol (24/1).

Ethanol.

BTX ECM 600 electroporator (BTX, San Diego, CA), with 0.2-cm cuvettes.

Narashige IM-5B microinjector mounted on an Olympus IMT-2 inverted microscope.

Microinjection needles: Pulled using a Flaming/Brown micropipet puller, Model P-80PC (Sutter Instruments, San Rafael, CA) with the following settings: heat, 470; pull, 150; velocity, 160; time, 80. Glass capillaries are purchased from World Precision Instruments (New Haven, CT, No. TW100F-4).

C. *Tetrahymena* Strains

Two highly fertile strains of different mating types should be used and are available from Dr. P. J. Bruns, Cornell University. One strain should be a functional heterokaryon, like CU428, which is MPR/MPR (homozygous for resistance to 6-methylpurine) in the micronucleus, while the macronuclear phenotype is sensitive to 6-methylpurine. In crosses between CU428 and B2086, parental cells or cells that abort conjugation are sensitive and only conjugation progeny are 6-methylpurine resistant, due to the formation of a new macronucleus derived from the micronucleus. Cells that become 6-methylpurine resistant can be used to monitor the proportion of cells that undergo complete conjugation, which is required for CET.

D. Conjugant Electrotransformation Protocol

1. Grow cells of two different mating types (e.g., CU428, mt VII, and B2086, mt II) at 30°C with shaking. Transfer $\sim 10^5$ cells each morning to 50 ml fresh SPP in 250-ml Erlenmeyer flasks.

2. In the morning, initiate starvation of the day-old culture by washing cells with 50 ml of 10 m*M* Tris, pH 7.5, at 1100*g* for 5 minutes. Resuspend in 50 ml Tris buffer in 250-ml Erlenmeyer flasks. All operations are carried out at room temperature.

3. Starve cells at 30°C. In the afternoon, count the starved cells and adjust to 3×10^5 cells/ml. In the late afternoon or evening of the same day, mix 50 ml each of the two different mating types in a sterile 2-liter flask. Add 0.01th the volume of the antibiotic mix and place on a rotary shaker at 30°C. Shake at 160 rpm. Use a timer to turn the shaker off 10 hours before electroporation. Cells will begin to pair when the shaker stops.

4. Check pairing efficiency 4–6 hours after turning the shaker off. Proceed only if more than 80% of cells are paired.

5. Ten to eleven hours after turning shaker off, transfer cells to 50-ml Corning plastic conical centrifuge tubes and centrifuge 5 minutes at 2000 rpm in a Sorvall GLC centrifuge. Discard the supernatant and resuspend cells in 100 ml of 10 m*M* Hepes, pH 7.5. Repeat centrifugation.

6. Resuspend cells in 1 ml of Hepes buffer at $\sim 3 \times 10^7$ cells/ml.

7. For gene replacement, release the insert from the vector by restriction endonuclease digestion to create homologous ends. Purify the digested DNA by extraction with phenol/chloroform/isoamyl alcohol and chloroform/isoamyl alcohol, and precipitate with ethanol. Resuspend in Hepes electroporation buffer. Use 50 μg of digested DNA per electroporation. To coelectroporate, add 15–18 μg of undigested pD5H8 or pH4T2.

8. Mix 125 μl of cell suspension with the plasmid DNA (e.g., 10–15 μg pD5H8 rDNA processing vector or 50 μg of integrative vector in 125 μl of Hepes buffer), transfer to a 0.2-cm cuvette, and immediately subject to electroporation. Set voltage at 250 V, capacitance at 275 μF, and resistance at 13 ohms. The voltage peak with these conditions is about 225V, pulse length 4 ms, and field strength about 1125 V/cm.

9. Let the cuvette sit at room temperature for 1 minute, then resuspend cells in 20 ml SPP + 1× antibiotic medium (SPPA). Dilute cells according to the protocol shown in Fig. 3 and plate 25 μl per well into 96-well microtiter plates filled with 175 μl/well SPPA. The dilution factor is defined as the number of 96-well plates that would be needed to plate all cells using 25 μl of cell suspension per well.

10. Plating:

a. For high-efficiency transformation vectors (pD5H8, pH4T2, pAU3, and derivatives), make one plate for each dilution.

b. For gene replacement (knockout) experiments using direct selection, plate out all the 8× cells (25 μl/microtiter plate well). Initial dilution with 40 ml medium (16× dilution) improves overall yield but increases the number

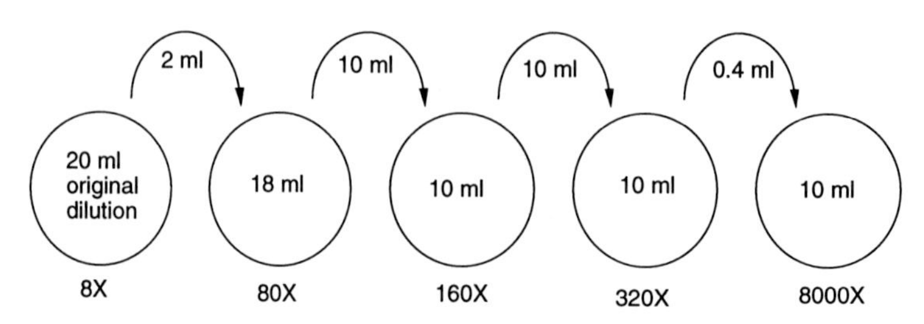

Fig. 3 Scheme for dilution of cells transformed by CET. The dilution factor is defined as the number of 96-well microtiter plates needed to plate the entire volume of cells subjected to electroporation (20 ml) using 25 μl of cell suspension per well.

of plates you must handle. These plates will be directly selected with your replacement marker. Also make one plate at 8000× dilution.

c. For experiments in which a cotransforming vector was included, plate the entire volume of electroporated cells at 8× or 16× dilution. Also make one plate each at the 80×, 160×, 320×, and 8000× dilutions. These plates will be selected first with paromomycin. From the number of primary transformants and the dilution, calculate the total number of primary transformants in the experiment. The primary transformants from more concentrated plates (8–16×) will later be selected for gene replacement.

11. For all three types of experiments, use the 8000× dilution to test overall survival and mating efficiency. After 12–18 hours, add 50 μl medium containing 75 μg/ml 6-methylpurine (if CU428 strain was used) or 75 μg/ml cycloheximide (if CU427 strain was used) to each well. From the number of wells with live cells and the dilution factor, calculate the number of clones in the experiment that survived the electroporation and completed conjugation.

12. After 12–18 hours, add 50 μl/well 600 μg/ml paromomycin (final concn of 120 μg/ml) or cycloheximide (at 75 μg/ml) for *rpL29* vectors. When a vector containing the H4-I*neo* selectable marker is used, an early selection with paromomycin is strongly recommended, because selection efficiency using this marker decreases with increased cell concentration. The transformed clones are apparent after 1–3 days of growth.

13. After 4–5 days, calculate the transformation yield using the Poisson distribution (Orias and Bruns, 1975).

14. In the case of a cotransformation experiment, replicate 4- to 5-day-old primary transformant plates into fresh SPPA medium (150 μl/well) containing 15 μg/ml cycloheximide (*rpL29* gene), 30 μg/ml oryzalin or vinblastine (mutant tubulin gene), or 15 μ*M* taxol (wild-type tubulin gene). Plates containing medium with antitubulin drugs should be protected from excessive light exposure.

15. Critical factors that effect CET transformation efficiency:

a. *Tetrahymena* strains must be highly fertile.

b. Mating should have good synchrony and 80–90% of cells should be paired 4 hours after mixing or turning the shaker off.

c. It is good practice to monitor cell survival rate (the fraction of clones that are conjugation progeny). The survival rate is usually 100–500 × 10^3 conjugation progeny clones.

d. Low transformation yield correlated with low survival rate indicates that mating strains are too old and not fertile, contaminants are present in the plasmid DNA (such as RNA or salt), or cultures are contaminated by fungi or bacteria. Low transformation yield and high survival rate indicate that the optimal timing of transformation was missed or an insufficient amount of transforming DNA was used.

E. Microinjection-Mediated Transformation

This procedure was adapted from Tondravi and Yao (1986).

1. Use strains of *T. thermophila* lacking mucocysts (such as SB255) because mucocyst contents tend to clog the injection needles; however, strains containing mucocycts (e.g., CU428) have been used for microinjection.
2. Grow cells to early log phase ($\sim 1 \times 10^5$ cells/ml) in SPP. Concentrate cells about 10-fold by brief centrifugation.
3. Resuspend the transforming DNA in 1× TE buffer at ~0.25 mg/ml and spin down in a microfuge for 20–30 minutes to sediment any particulate material.
4. Deposit the cell suspension in small droplets under paraffin oil on a microscope slide. Using a dissecting scope and a braking micropipet, remove as much of the growth medium as possible to immobilize cells. Try to leave ~10 cells/drop.
5. Inject cells into the macronuclei using a moderate flow rate from the needle. Try to target the needle into the macronucleus which is visible using the phase contrast 40–60× lens. During injection, a short burst of fluid flow from the needle should be visible inside the cell.
6. After injection of 20–50 cells, add some SPPA into the drops. Using a braking micropipet, isolate cells into small drops of SPPA in a 6 × 8 pattern on a disposable Petri plate. After 3–4 days, transfer cells into a microtiter 96-well plate and grow for 1–2 days. Usually about half of the microinjected cells give viable clones. Transformation rates are usually 1–10% of injected viable clones regardless of the type of vector used.
7. Select transformants by replica plating. Also duplicate plates using fresh nonselective medium so the selection can be repeated after additional growth if necessary.

References

Bolduc, C., Lee, V. D., and Huang, B. (1988). β-tubulin mutants of the unicellular green alga *Chlamydomonas reinhardtii*. *Proc. Natl. Acad. Sci. U.S.A.* **85,** 131–135.

Brunk, C. F., and Navas, P. (1988) Transformation of *Tetrahymena thermophila* by electroporation and parameters effecting cell survival. *Exp. Cell Res.* **174,** 525–532.

Bruns, P. (1986). Genetic organization of Tetrahymena. *In* "The Molecular Biology of Ciliated Protozoa" (J. G. Gall, ed.), pp. 27–44. Academic Press, Orlando.

Callahan, R. C., Shalke, G., and Gorovsky, M. A. (1984). Developmental rearrangements associated with a single type of expressed alpha tubulin in *Tetrahymena*. *Cell* **36,** 442–445.

Doerder, F. P., Deak, J. C., and Lief, J. H. (1992). Rate of phenotypic assortment in *Tetrahymena thermophila*. *Dev. Genet.* **13,** 126–132.

Gaertig, J., and Gorovsky, M. A. (1992). Efficient mass transformation of *Tetrahymena thermophila* by electroporation of conjugants. *Proc. Natl. Acad. Sci. U.S.A.* **89,** 9196–9200.

Gaertig, J., Gu, L., Hai, B., and Gorovsky, M. A. (1994a). Gene replacement and high frequency vector-mediated electrotransformation in *Tetrahymena*. *Nucleic Acids Res.* **22,** 5391–5398.

Gaertig, J., Thatcher, T. H., Gu, L., and Gorovsky, M. A. (1994b). Electroporation-mediated replacement of a positively and negatively selectable β-tubulin gene in *Tetrahymena thermophila*. *Proc. Natl. Acad. Sci. U.S.A.* **91,** 4549–4553.

Gilley, D., Preer, J. R., Jr., Aufterheide, K. J., and Polisky, B. (1988). Autonomous replication and addition of telomere-like sequences to DNA microinjected into *Paramecium tetraurelia* macronuclei. *Mol. Cell. Biol.* **8,** 4765–4772.

Gorovsky, M. A. (1980). Genome organization and reorganization in *Tetrahymena*. *Annu. Rev. Genet.* **14,** 203–239.

Gorovsky, M. A., Yao, M.-C., Keevert, J. B., and Pleger, G. L. (1975). Isolation of micro- and macronuclei of *Tetrahymena pyriformis*. *Methods Cell Biol.* **9,** 311–327.

James, S. W., Silflow, C. D., Stroom, P., and Lefebvre, P. A. (1993). A mutation in the α1-tubulin gene of *Chlamydomonas reinhardtii* confers resistance to anti-microtubule herbicides. *J. Cell Sci.* **106,** 209–218.

Kahn, R. W., Andersen, B. H., and Brunk, C. F. (1993). Transformation of *Tetrahymena thermophila* by microinjection of a foreign gene. *Proc. Natl. Acad. Sci. U.S.A.* **90,** 9295–9299.

Larson, D. D., Blackburn, E. H., Yaeger, P. C., and Orias, E. (1986). Control of rDNA replication in *Tetrahymena* involves a cis-acting upstream repeat of a promoter element. *Cell* **47,** 229–240.

Lee, V. D., and Huang, B. (1990). Missense mutations at lysine 350 in β2-tubulin confer altered sensitivity to microtubule inhibitors in *Chlamydomonas*. *Plant Cell* **2,** 1051–1057.

McGrath, K. E., Yu, S.-M., Heruth, D. P., Kelly, A. A., and Gorovsky, M. A. (1994). Regulation and evolution of the single alpha-tubulin gene of *Tetrahymena thermophila*. *Cell Motil. Cytoskel.* **27,** 272–283.

Nanney, D. L. (1980). "Experimental Ciliatology". John Wiley and Sons, New York.

Orias, E. (1986). Ciliate conjugation. *In* "The Molecular Biology of Ciliated Protozoa" (J. G. Gall, ed.), pp. 45–84. Academic Press, Orlando.

Orias, E., and Bruns, P. J. (1975). Induction and isolation of mutants in *Tetrahymena*. *In* "Methods in Cell Biology" (D. M. Prescott, ed.), pp. 247–282. Academic Press, New York.

Orias, E., Larson, D., Hu, Y.-F., Yu, G.-L., Karttunen, J., Lovlie, A., Haller, B., and Blackburn, E. H. (1988). Replacement of the macronuclear ribosomal RNA genes of a mutant *Tetrahymena* using electroporation. *Gene* **70,** 295–301.

Schibler, M. J., and Huang, B. (1991). The col^R4 and col^R15 b-tubulin mutations in *Chlamydomonas reinhardtii* confer altered sensitivities to microtubule inhibitors and herbicides by enhancing microtubule stability. *J. Cell Biol.* **113,** 605–614.

Spangler, E. A., and Blackburn, E. H. (1985). The nucleotide sequence of the 17S ribosomal RNA of *Tetrahymena thermophila* and the identification of point mutations resulting in resistance to the antibiotics paromomycin and hygromycin. *J. Biol. Chem.* **260,** 6334–6340.

Tondravi, M. M., and Yao, M.-C. (1986). Transformation of *Tetrahymena thermophila* by microinjection of ribosomal RNA genes. *Proc. Natl. Acad. Sci. U.S.A.* **83,** 4369–4373.

Yao, M.-C., and Yao, C.-H. (1989). Accurate processing and amplification of cloned germ line copies of ribosomal DNA injected into developing nuclei of *Tetrahymena thermophila*. *Mol. Cell. Biol.* **9,** 1092–1099.

Yao, M.-C., and Yao, C.-H. (1991). Transformation of *Tetrahymena* to cycloheximide resistance with a ribosomal protein gene through sequence replacement. *Proc. Natl. Acad. Sci. U.S.A.* **88,** 9493–9497.

Yu, G.-L., and Blackburn, E. H. (1989). Transformation of *Tetrahymena thermophila* with a mutated circular ribosomal DNA plasmid vector. *Proc. Natl. Acad. Sci. U.S.A.* **86,** 8487–8491.

Yu, G.-L., Hasson, M., and Blackburn, E. H. (1988). Circular ribosomal DNA plasmids transform *Tetrahymena thermophila* by homologous recombination with endogenous macronuclear ribosomal DNA. *Proc. Natl. Acad. Sci. U.S.A.* **85,** 5151–5155.

CHAPTER 81

Strategies for the Isolation of Ciliary Motility and Assembly Mutants in *Tetrahymena*

David G. Pennock* **and Martin A. Gorovsky**[†]

* Department of Zoology
Miami University
Oxford, Ohio 45056
[†] Department of Biology
University of Rochester
Rochester, New York 14623

I. Introduction

Compared with the extensive genetic analyses of flagellar function and assembly in *Chlamydomonas,* relatively few mutations affecting ciliary motility and/or assembly have been isolated. We have developed a screen for the isolation of mutations affecting ciliary regeneration in *Tetrahymena thermophila* (Pennock *et al.,* 1988b), and the development of transformation and gene replacement techniques in *Tetrahymena* (Gaertig and Gorovsky, 1992; Gaertig *et al.*, 1994) should lead to increased interest in genetic analyses of ciliary function and assembly.

To date, we have isolated and partially characterized three temperature-sensitive mutations affecting ciliary regeneration in *Tetrahymena*. Cells homozygous for the *chp* (cell cycle, heat shock, and phosphorylation defect) mutation fail to initiate ciliogenesis after deciliation and incubation at the restrictive temperature (Pennock *et al.*, 1988a, Huelsman *et al.*, 1992). Thatcher and Gorovsky (1994) showed that *chp* mutants are defective in the heat shock response.

Another group of mutants, those homozygous for the *dcc* (defective in ciliogenesis and cytokinesis) mutation, also fail to recover motility following deciliation and incubation at the restrictive temperature (Pennock *et al.*, 1988b). *dcc* mutants initiate ciliogenesis, but the cilia fail to elongate to full length (Gitz *et al.*, 1993). Interestingly, *dcc* mutants also exhibit a defect in cytokinesis when incubated in growth medium at the restrictive temperature. The mutants initiate cytokinesis, but arrest before it is completed (Pennock *et al.*, 1988b; Gitz *et al.*, 1993). Electron microscopic examination of the division furrow in arrested mutants suggests that a complete contractile ring fails to form (Wilkes and Pennock, unpublished observation).

The *oad* (outer arm deficient) mutation affects assembly of the outer dynein arm into growing cilia. The *oad* mutants grow and divide at the restrictive temperature, but become nonmotile (Attwell *et al.*, 1992). Cilia isolated from nonmotile *oad* mutants lack approximately 90% of their outer dynein arms (Attwell *et al.*, 1992) and approximately 60% of their 22 S dynein (Ludmann *et al.*, 1993; Daigre and Pennock, unpublished observation). These data formally prove that outer dynein arms are composed of 22 S dynein, and they suggest that 22 S dynein may also be located elsewhere in the ciliary axoneme.

Future work with mutations affecting ciliary assembly and function will most likely involve identification of the *dcc* and *oad* gene products and isolation of new mutations affecting ciliary assembly during regeneration and growth. The new mutations will be induced both by chemical mutagenesis *in vivo* as described below and by *in vitro* mutation of cloned genes followed by gene replacement as described by Gaertig and Gorovsky (1992) and Gaertig *et al.* (1994) as more genes encoding ciliary proteins are cloned.

II. Materials

A. Media

Modified Neff's medium: 0.1 m*M* ferric citrate, 2 m*M* KH_2PO_4, 0.05 m*M* $CaCl_2$ dihydrate, 1.0 m*M* $MgSO_4$ heptahydrate, 0.5% glucose, 0.25% yeast extract, 0.25% proteose peptone. Heat water (1/10th of final volume) to boiling. Add ferric citrate and allow it to dissolve. Remove the beaker from the heat and add water to 9/10th final volume. Add the rest of the ingredients in the order listed, allow them to dissolve, and bring to final volume with water. Aliquot into appropriate flasks and autoclave.

Cell maintenance medium: 2% proteose peptone, 0.1% yeast extract, 0.2% glucose. Dissolve all ingredients, bring to volume with water, aliquot approximately 6 ml into capped culture tubes, and autoclave. Cells can be stored for 2–3 weeks between transfers.

Starvation media: (1) 10 m*M* Tris–Cl (pH 7.4); (2) Dryl's—0.5 g sodium citrate, 0.14 g NaH_2PO_4, 0.14 g Na_2HPO_4. Dissolve the salts in 985 ml water and autoclave. Separately dissolve 0.735 g $CaCl_2$ in 50 ml water and autoclave. Add 15 ml of the $CaCl_2$ solution to the sterile salt solution.

Deciliation medium: 10% Ficoll 400 (Sigma), 10 m*M* Na acetate, 10 m*M* $CaCl_2$, 10 m*M* ethylenediaminetetraacetic acid (EDTA). Mix all components, bring to pH 4.2 with concentrated HCl, and bring to final volume with water.

Regeneration medium: 15 m*M* Tris–Cl (pH 7.95), 2 m*M* $CaCl_2$. This solution can be made up as a 10× stock.

B. Drugs

6-Methylpurine (Sigma): 15 mg/ml stock in water. Use at a final concentration of 15 μg/ml.

Cycloheximide (Sigma): 12.5 mg/ml stock in ethanol. Use at a final concentration of 12.5 μg/ml.

Paramomycin (Sigma): 100 mg/ml stock in water. Use at a final concentration of 120μg/ml.

Nitrosoguanidine (Sigma): 2 mg/ml stock in water. Use at a final concentration of 10 μg/ml.

Antibiotic, antimycotic stock (Sigma): 10,000 units penicillin, 10 mg streptomycin, and 25 μg amphotericin B/ml. 1 : 100 for use.

All drug stocks are stored at −20°C.

III. Special Procedures and Equipment

The procedures and equipment for performing genetic analyses on *Tetrahymena* are clearly described in Orias and Bruns (1976).

IV. Strains

We have used three heterokaryon strains (Bruns and Brussard, 1974) of *Tetrahymena* in our work. All three were kindly provided by Dr. Peter Bruns (Cornell University). CU428 [Mpr/Mpr (6mp-s, VII)] is a heterokaryon that becomes resistant to 15 μg/ml 6-methylpurine on completing conjugation.

CU427 [ChxA2/ChxA2 (cy-s, VI)] is a heterokaryon that becomes resistant to 25 μg/ml cycloheximide on completing conjugation. CU438 [Pmr/Pmr (pm-s, IV)] is a heterokaryon that becomes resistant to 120 μg/ml paromomycin on completing conjugation. Nullisomic strains (Altschuler and Bruns, 1984; Bruns and Brussard, 1981; Bruns *et al.,* 1983) are missing a chromosome or part of a chromosome from the micronucleus. Thus, on conjugation with a heterokaryon, the exconjugants are hemizygous for all or part of one chromosome.

V. Bringing Recessive Micronuclear Mutations to Macronuclear Expression

Although the micronucleus is diploid, three ways are commonly used to bring recessive micronuclear mutations to expression in the macronucleus in *Tetrahymena* after mutagenesis and a single round of mating. One method is to induce cytogamy in mating pairs of *Tetrahymena* (Orias and Hamilton, 1979; Orias *et al.,* 1979). To induce cytogamy, mating pairs of heterokaryons are subjected to an osmotic shock during mating. This prevents nuclear exchange and causes the production of whole-genome homozygotes in some fraction of the pairs (Orias *et al.,* 1979). Cytogamonts of the mutagenized strain are selected by treatment with the appropriate drug. One drawback of this method is that heterozygotes (not expressing recessive mutations) are not killed by the drug treatment and can make screening more difficult.

Cole and Bruns (1992) developed a modification of cytogamy called uniparental cytogamy (UPC). In this method, the mutagenized heterokaryon strain of *Tetrahymena* is mated with a ★ strain (star strain). Cytomgamy is induced and cytogamonts are selected by treatment with the appropriate drug. As ★ strains have a defective micronucleus (Bruns, 1986), all drug-resistant progeny are whole-genome homozygotes in which the macronucleus is derived from the mutagenized micronucleus.

A method for targeting mutations to a specific chromosome or part of chromosome has also been developed (Altschuler and Bruns, 1984). A mutagenized heterokaryon strain of *Tetrahymena* is mated with a nullisomic strain (Bruns and Brussard, 1981; Bruns *et al.,* 1983) missing the chromosome or part of the chromosome of interest. Exconjugants selected by treatment with the appropriate drug will be hemizygous for the chromosome missing from the nullisomic, so any recessive mutation on that chromosome will be expressed. This method is especially useful for targeting mutations to a specific gene whose chromosomal location is known.

VI. Chemical Mutagenesis

1. Start overnight cultures of two appropriate strains of different mating types. One strain will be mutagenized. The other strain will be mated with

the mutagenized strain. We have mutagenized both CU428 and CU438 with reasonable success.

2. Count cells and dilute cultures such that the cultures will be in early log phase (approximately 2×10^5 cells/ml) after growth overnight. Fifty milliliters of cells should be enough for most experiments. The generation time of these strains in modified Neff's medium at 28°C is approximately 2.5 hours with a 2.5-hour lag.

3. When cultures have reached 2×10^5 cells/ml, add nitrosoguanidine to one of the cultures to a final concentration of 10 μg/ml. Nitrosoguanidine is a potent mutagen, and appropriate precautions should be taken. One should wear protective clothing, work under a hood, and follow the safety guidelines suggested by the manufacturer.

4. Incubate all cultures with shaking for 3 hours at 28°C.

5. Wash the cells three times with starvation medium. For each wash, pellet the cells in a conical centrifuge tube at 1000*g* for 2–3 minutes. Aspirate the supernatant and resuspend the cells in 50 ml starvation buffer. Adjust the volumes so that all cultures are at the same concentration ($1–2 \times 10^5$ cells/ml). We have starved cells in either 10 m*M* Tris–Cl (pH 7.4) or Dryl's. Cole and Bruns (1992) reported that Dryl's works better than Tris for UPC.

6. Starve the cells overnight at 28°C with shaking.

7. Mate the appropriate strains. We mix 10 ml of each strain in a 10-cm Petri plate and incubate at 28°C.

8. Induce cytogamy or UPC as appropriate: If the mutagenized strain has been mated with a nullisomic strain to target the mutation to a specific chromosome, simply allow the cells to conjugate. If the mutagenized strain has been mated to a ★ strain, induce UPC (Cole and Bruns, 1992). If the mutagenized strain has been mated to another heterokaryon, induce cytogamy (Orias and Hamilton, 1979; Orias *et al.*, 1979).

9. Select for cytogamonts or exconjugants: Dilute the entire mating into 250 ml of modified Neff's containing antibiotic/antimycotic and the appropriate drugs. Incubate the cultures with shaking at 28°C.

10. In 2 to 4 days determine cell concentration, pellet cells as described above, and resuspend in starvation medium at a final concentration of 2×10^5 cells/ml. Keep cells shaking at 28°C.

The cells will not divide in starvation medium but will remain healthy long enough to allow several screens to be performed over the course of 2–3 days.

VII. Screening for Temperature-Sensitive Ciliary Assembly or Motility Mutants

1. Deciliate starved cells essentially as described in Calzone and Gorovsky (1982). Pellet cells as described above and aspirate the supernatant. Leave

enough medium so that cells will be at 5×10^7 cells/ml after the cell pellet is resuspended. Resuspend cells with a disposable transfer pipet. Add cells to 10 vol of deciliation medium in a small flask. Swirl the flask for 30–60 seconds to remove the cilia. Add 5 vol regeneration buffer and mix gently. Check for motile cells. Remove a small aliquot of the culture to incubate at 28°C as a control for proper regeneration. Incubate the rest of the culture at 38–40°C.

2. Allow cells to regenerate for approximately 3 hours. The control cells incubated at 28°C should be motile at this time.

3. Screen for nomotile cells. Approximately 30 minutes before the cells are to be screened, add an equal volume of prewarmed 1 : 250 dilution of India ink to the culture. Place 20 ml of cells into a glass Petri plate and examine with a dissecting microscope using bright-field optics. Isolate nonmotile cells that do not have black vacuoles (oral ciliature is required for feeding) with a braking pipet and deposit the cells into drops of modified Neff's. This is the most tedious part of the procedure. Even after cells have regenerated cilia, they often lie nonmotile on the bottom of the dish. We pipet individual cells into and out of the pipet several times before placing them into drops. Even then many cells will swim happily away and will have to be removed from the drop. We tried several methods of separating motile from nonmotile cells en masse, but the enrichment was never worth the trouble. Invert the plate containing the drops and incubate at 28°C in a moist chamber to allow the cells to recover motility, divide, and grow into clones (2–3 days).

4. Transfer the clones of cells into microtiter plate wells containing 200 ml of modified Neff's and incubate at 28°C.

VIII. Rescreening Putative Regeneration/Motility Mutants

Many of the putative mutants will actually be wild-type cells that slipped through the screen, so the mutants have to be rescreened to identify true mutants. One method is to grow up individual clones and test each for the ability to regain motility at 39°C. We have not been able to work out a satisfactory method to deciliate cells in microtiter plates, so this can be rather tedious if large numbers of putative mutants were isolated during the initial screen. A second method is to screen putative mutants for visible defects during growth at 39°C. This method allows one to pick up mutants that have defects in ciliary assembly, motility, or cell growth and has the advantage that large numbers of mutants can be screened easily in microtiter plates. The disadvantage to this method is that mutations that affect only ciliary regeneration will be missed.

To screen for the ability to regain motility at 39°C following deciliation, 50-ml cultures should be started from each clone. Cells should be starved and deciliated as described above. The deciliated cells should be separated into two cultures which should be incubated at 28 and 39°C, respectively. Score for motility after 2–4 hours.

To screen for motility or cell growth, replica plate cells into fresh Neff's in microtiter plates. Incubate the plates overnight at 39°C and examine the plates for wells with few cells, nonmotile cells, or cells exhibiting abnormal movement.

Once mutant clones have been identified, they should be transferred to standing cultures. Remove 50–100 ml from the appropriate well of the microtiter plate and transfer that to 6 ml of maintenance medium in a capped culture tube. These standing cultures can be stored at room temperature and transferred every 2–3 weeks. It is probably a good idea to freeze aliquots of the different strains at this point in the procedure (Flacks, 1979). This will ensure against unexpected losses and will ensure that one has fertile strains for future genetic analyses.

Mutant strains should be genetically analyzed as described by Orias and Bruns (1976). Briefly, one should determine whether the mutation is dominant or recessive by mating to the parental strain or to another heterokaryon. One should also assign the various mutant strains into complementation groups (Orias and Bruns, 1986; Pennock *et al.,* 1988b).

References

Altschuler, M. I., and Bruns, P. J. (1984). Chromosome-designated mutation selection in *Tetrahymena thermophila. Genetics* **106,** 387–401.

Attwell, G. J., Bricker, C. S, Schwandt, A., Gorovsky, M. A., and Pennock, D. G. (1992). A temperature-sensitive mutation affecting synthesis of outer arm dyneins in *Tetrahymena thermophila. J. Protozool.* **39,** 261–266.

Bruns, P. J. (1986). Genetic Organization of *Tetrahymena. In* "The Molecular Biology of the Ciliated Protozoa" (J. G. Gall, ed.), pp. 27–44. Academic Press, Orlando.

Bruns, P. J., and Brussard, T. B. (1974). Positive selection for mating with functional heterokayons in *Tetrahymena pyriformis. Genetics* **78,** 831–841.

Bruns, P. J., and Brussard, T. E. B. (1981). Nullisomic *Tetrahymena:* eliminating germinal chromosomes. *Science* **213,** 549–551.

Bruns, P. J., Brussard, T. B., and Merriam, E. V. (1983). Nullisomic *Tetrahymena.* II: A set of nullisomic define the germinal chromosomes. *Genetics* **104,** 257–270.

Calzone, F. J., and Gorovsky, M. A. (1982). Cilia regeneration in *Tetrahymena.* A simple reproducible method for producing large numbers of cells. *Exp. Cell Res.* **140,** 471–476.

Cole, E. S., and Bruns, P. J. (1992). Uniparental Cytogamy: A Novel Method for Bringing Micronuclear mutations of *Tetrahymena* into homozygous macronuclear expression with precocious sexual maturity. *Genetics* **132,** 1017–1031.

Flacks, M. (1979). Axenic storage of small volumes of *Tetrahymena* cultures under liquid nitrogen: A miniaturized procedure. *Cryobiology* **16,** 287–291.

Gaertig, J., and Gorovsky, M. A. (1992). Efficient mass transformation of *Tetrahymena thermophila* by electroporation of conjugants. *Proc. Natl. Acad. Sci. U.S.A.* **89,** 9196–9200.

Gaertig, J., Thatcher, T. H., Gu, L., and Gorovsky, M. A. (1994). Electroporation-mediated replacement of a positively and negatively selectable β-tubulin gene in *Tetrahymena thermophila. Proc. Natl. Acad. Sci. U.S.A.* **91,** 4549–4553.

Gitz, D. L., Eells, J. B., and Pennock, D. G. (1993). The *dcc* mutation affects ciliary length in *Tetrahymena thermophila. J. Eukaryotic Microbiol.* **40,** 668–676.

Huelsman, D. A., Gitz, D. L., and Pennock, D. G. (1992). Protein phosphorylation and the regulation of basal body microtubule organizing centers in *Tetrahymena thermophila. Cytobios* **71,** 31–50.

Ludmann, S. A., Schwandt, A., Kong, X., Bricker, C. S., and Pennock, D. G. (1993). Biochemical analysis of a mutant *Tetrahymena* lacking outer dynein arms. *J. Eukaryotic Microbiol.* **40,** 650–660.

Orias, E., and Hamilton, E. P. (1979). Cytogamy: An inducible, alternate pathway of conjugation in *Tetrahymena thermophila*. *Genetics* **91,** 657–671.

Orias, E., and Bruns, P. J. (1976). Induction and isolation of mutants in *Tetrahymena*. *In* "Methods in Cell Biology" (D. M. Prescott ed.), Vol 13, pp. 247–282. Academic Press, New York.

Orias, E., Hamilton, E. P., and Flacks, M. (1979). Osmotic shock prevents nuclear exchange and produces whole-genome homozygotes in conjugating *Tetrahymena*. *Science* **203,** 660–663.

Pennock, D. G., Thatcher, T., and Gorovsky, M. A. (1988a). A temperature-sensitive mutation that affects cilia regeneration and the cell cycle in *Tetrahymena thermophila* is rescued by cytoplasmic exchange. *Mol. Cell. Biol.* **8,** 2681–2689.

Pennock, D. G., Thatcher, T., Bowen, J., Bruns, P. J., and Gorovsky, M. A. (1988b). A conditional mutant having paralyzed cilia and a block in cytokinesis is rescued by cytoplasmic exchange in *Tetrahymena*. *Genetics* **120,** 697–705.

Thatcher, T. H., and Gorovsky, M. A. (1994). A temperature-sensitive cell cycle arrest mutation affecting H1 phosphorylation and nuclear localization of a small heat shock protein in *Tetrahymena thermophila*. *Exp. Cell Res.* **209,** 261–270.

CHAPTER 82

Identification of New Dynein Heavy-Chain Genes by RNA-Directed Polymerase Chain Reaction

David J. Asai and Peggy S. Criswell
Department of Biological Sciences
Purdue University
West Lafayette, Indiana 47907

I. Introduction

A. Multiple Axonemal Dynein Isoforms

Cilia and flagella use multiple isoforms of dynein to produce movement (reviewed in Asai and Brokaw, 1993). Although each inner and outer dynein arm most likely is able to translocate along microtubules, the requirements for the initiation and propagation of bends place different functional demands on different dyneins (reviewed in Brokaw, 1994). In addition to the several differences in the motor activities among dynein isoforms, each isoform faithfully localizes to the appropriate place in the axoneme (Smith and Sale, 1992), suggesting that there are isoform-specific differences in the regions that mediate tethering of the dynein to its site of attachment.

The functional differences among the axonemal dyneins are due, at least in part, to structural differences among the heavy-chain isoforms. Detailed characterization of the various isoforms should reveal the structural bases for functional specialization. To begin the characterization of the dynein heavy-chain family, an RNA-directed polymerase chain reaction (PCR) method was developed in collaboration with Ian Gibbons in his laboratory (Asai *et al.*, 1991). The method has been applied in the characterization of the dynein gene families expressed in sea urchin (Gibbons *et al.*, 1994), *Paramecium tetraurelia* (Asai *et al.*, 1994), and *Chlamydomonas reinhardtii* (Wilkerson *et al.*, 1994). These studies are aimed at answering three general questions: How many different dynein heavy-chain genes are expressed in the organism? How similar/different are these dynein heavy chains to/from one another? What is the function of each of the heavy-chain gene products?

B. The Strategy

The catalytic P-loop has been identified in the complete sequences of several dynein heavy chains and is the absolutely conserved sequence GPAGTGKT which conforms to the GXXXXGKT/S motif found in most ATP-binding proteins (Saraste *et al.*, 1990; Walker *et al.*, 1982). Degenerate oligonucleotide primers were designed to amplify regions of the dynein gene encoding the sequence surrounding the conserved catalytic P-loop (Gibbons *et al.*, 1992). Primers corresponding to four short sequences near the P-loop have been used with good success: (1) S1 corresponds to ITPLTDR located 25 residues upstream of the first glycine in the catalytic P-loop (i.e., located at −25 to −19); (2) S2 corresponds to PAGTGKT (+1 to +7); (3) A1 corresponds to FITMNPG (+99 to +105); and (4) A2 corresponds to CFDEFNR (+50 to +56). The S1 and S2 primers are derived from the DNA sense strand and the A1 and A2 primers are from the antisense strand.

A small amount of mRNA from the cell of interest is reverse transcribed from one of the A primers. The first-strand cDNA is then diluted and a small portion is amplified with the same A primer and an equal concentration of an S primer. The strategy of beginning the reaction with a few cycles that use a relatively high annealing temperature is aimed at increasing the specificity of the initial amplifications; at high annealing temperatures, only a small number of products may be made. Figure 1 shows an example of the amplification of the P-loop encoding fragment of *Paramecium* dynein genes in which different conditions were used.

II. Methods

Recipes for media, solutions, and many of these procedures are from Sambrook *et al.* (1989).

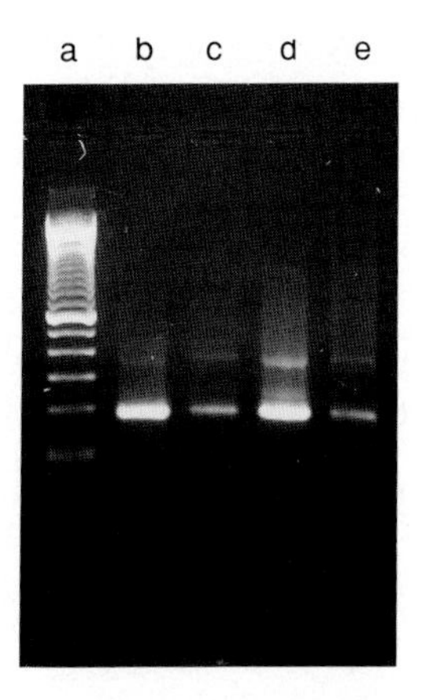

Fig. 1 Amplification of *Paramecium* dynein gene fragments. Approximately 5 μg total RNA was the template for the first-strand cDNA synthesis. Amplification was with the S2 and A2 degenerate primers: ca. 500 pmole of each (lanes b and d), ca. 50 pmole of each (lanes c and e). Lane a: 1 μg of 100-bp interval ladder of DNA. Lanes b and c: amplification by 5 cycles using 50°C annealing followed by 40 cycles of 35°C annealing. Lanes d and e: amplification by 45 cycles using 35°C annealing. Equal amounts (ca. 1/20th of total product) of amplified DNA were loaded in each lane. This is an ethidium bromide-stained 1.5% agarose gel in 0.5X TBE. The ca. 180-bp dynein fragment in lane b was amplified with the best specificity.

A. The Degenerate Primers

A1: 5′-GGGGGAATTCCCIGG[AG]TTCATIGT[ATG]AT[AG]AA-3′

A2: 5′-GGGCGAATTC[TC][AG]TT[AG]AA[TC]TC[AG]TC[AG]AA[AG]CA-3′

S1: 5′-GGGGGAATTCATCACICCI[TC]TIACIGA[TC][AC]G-3′

S2: 5′-CCCCGGATCCTGCTGG[TCAG]AC[TCAG]GG[TCAG]AA[GA]AC-3′

Degeneracies are enclosed in square brackets. The base inosine (I) has been used effectively at positions of fourfold degeneracy. The 5′ ends of primers are modified with (1) a few G's or C's, which help stabilize the ends during amplification, and (2) a restriction enzyme site (*Eco*RI or *Bam*HI), which aids in the cloning of the amplified products. The primers are dissolved in sterile water at a concentration of approximately 100 pmole/ml.

B. Synthesis of the First DNA Strand

1. Ingredients

~0.5 μg of poly(A)+ or 5 μg of total RNA

~200 pmole antisense primer A1 or A2

dNTPs, 10 m*M* each nucleotide, dissolved together in sterile water

Reverse transcriptase (RT, Superscript from BRL), 5× buffer, and 0.1 *M* dithiothreitol (DTT) solution provided by vendor

Diethyl pyrocarbonate (DEPC)-treated sterile water

2. Procedure

1. In a sterile 500-μl Eppendorf tube, mix RNA, primer, and DEPC-treated water to a final volume of 14 μl.
2. Heat at 70°C for 10 minutes.
3. Immediately plunge into liquid nitrogen for 30 seconds.
4. After the tube has returned to room temperature, add 8 μl 5X RT buffer, 4 μl 0.1 *M* DTT, and 3 μl dNTPs.
5. Mix, pulse in microfuge, and warm to 40°C for 2–3 minutes.
6. Add 2 μl RT Superscript (400 units).
7. Allow first-strand synthesis for 1.5 hours at 40°C.
8. Stop the reaction by adding 75 μl of 0.5X TE buffer.

C. PCR Amplification

1. Ingredients

cDNA after first-strand synthesis

Degenerate primers

dNTPs (1.25 m*M* each), dissolved together in sterile water

Taq DNA polymerase

10X Taq reaction buffer (we use buffer with a final Mg^{2+} concentration of 1.5 m*M*)

2. Preparation of the Reactants

1. In a 500-μl Eppendorf tube, mix 45 μl sterile water and 5μl cDNA (from Section II,B,2, step 8).
2. Heat DNA template at 90°C for 10 minutes, then immediately plunge into wet ice.
3. In a separate 500-μl tube, make the master mix: 200 pmole of each primer (e.g., A2 and S2); 16 μl dNTPs; 10 μl 10X buffer (final concentration of 1.5 m*M* Mg^{2+}, note that some vendors supply magnesium separately); sterile water to bring the mixture to a total volume of 50 μl; 5 units of Taq polymerase. Mix well; pulse in a microfuge.
4. Pipet the master mix into the DNA, mix, and pulse in microfuge; final volume should now be 100 μl.
5. Add 3 drops of sterilized mineral oil (Sigma) on top of the solution.

3. Amplification

The precise details of the amplification are empirically determined and may change depending on the concentration of primers and the concentration of dynein mRNA in the initial RNA sample, We have had good luck with the following protocol:

5 cycles: melt at 94°C, 1 minute; anneal at 50°C, 2 minutes; extend at 72°C, 2 minutes

40 cycles: 94°C, 1 minute; 35°C, 2 minutes; 72°C, 2 minutes

1 cycle: 72°C, 10 minutes

4. Recovery and Analysis of the Amplified DNA

1. After amplification, add 150 μl chloroform; mix well. *Caution:* The tube caps tend to leak if they are not pressed down firmly! Centrifuge in a microfuge 10 minutes at top speed.

2. Carefully pipet the aqueous (upper) layer. It is usually slightly opaque. The aqueous layer will form a bubble on top of the chloroform. The best way to remove the 100-μl sample is to penetrate the bubble with the pipet tip while tipping the tube.

3. The amplified DNA may be examined in a 1.5% agarose gel (0.5X TBE).

5. Precipitation of the Amplified DNA

1. To 100 μl of amplified DNA, add 25 μl polyacrylamide (0.125% acrylamide in 0.5 *M* KCl), mix, and add 250 μl cold absolute ethanol.
2. Precipitate overnight at −20°C.
3. Wash the precipitated DNA with cold 70% ethanol.

D. Ligation, Cloning, and Sequencing

1. Digestion of the Amplified DNA

1. Redissolve desiccated precipitated DNA in 20 μl TE and digest with the appropriate enzymes (e.g., with A2 and S2 amplification, *Eco*RI and *Bam*HI)

2. Gel purify the digested DNA in a 1% low-melting agarose gel (be sure to use 1X TAE, not TBE) or by electrophoresing the DNA in a standard agarose gel and then recovering the DNA by the freeze–squeeze method (Thuring *et al.*, 1975)

2. Ligation and Transformation

Ligation into a suitable vector and transformation into an *Escherichia coli* host are performed according to standard procedures (Sambrook *et al.*, 1989).

We use pUC118 and pUC119 vectors and JM101 as the bacterial host. Single-stranded DNA is packaged into M13K07 helper phage and recovered from the supernatants of overnight cultures.

3. Sequencing

We use standard single-strand sequencing employing Sequenase Version 2.0 (USB) following the manufacturer's protocol and label with [^{35}S]dATP.

```
consensus        GPAGTGKTETTKDLGKALG  CVVFNCSE  DYK MGK FKGLAQSGAWGCFDEFNR

urchin β                            IMVY      QM   SC NI    S
urchin 3a                     M     KYV       QM   GL  I        S
urchin 3b                     M  S  KYV       QM   GL  I        S
urchin 3c                     M     KYV       QM   GL  I        S
urchin 4                            LL   T  G GM   AV  I S    C
urchin 5a                           KQ        DL   M   F S        C
urchin 5b                  V      V KL L      GL   M   F S  S   S
urchin 5c                           NYVI V    GL   S   M S
urchin 6                            IQT       QL  MA   FL    S
urchin 7a                         V KQ        GL  IAL  F     SC   S
urchin 7b                         V KQ        GL   A   F      A
urchin 7c                           VQ        GL  LA S F     S
urchin 7d                  C      V KQ ..................................
Paramecium β                      V LPVM      QMGKDS  QI M  S
Paramecium 1                   N    KA Y     SEMN ES  NI     S  C
Paramecium 2                        KQ        SM  IMV  F     SA
Paramecium 3                        KQ        QM  IMV  F     SA
Paramecium 4               V        RQ        GL   A   F     S    S
Paramecium 5               V     NM KFVL      GL   VI  M S  V V
Paramecium 7               V     T  VFV       QHR  D   I    C   L
Chlamydomonas α                SAQ  KSVY     PEM   T  DI     A  S
Chlamydomonas β                     IQ Y      QM   A  HT
Chlamydomonas 1                  SM LL      G GL   A  SI S  V C
Chlamydomonas 4                     KQ        GL  QA   F     S
Chlamydomonas 5                     VN      G NL   F   F S    C
Chlamydomonas 6                     RQ        TLN QD   F     AA
Chlamydomonas 7                     RQ        SL  QA   F     S
Chlamydomonas 8                     MQ        GL  QA   F     S
Chlamydomonas 9                     VN      G NL   F   F S    C
Tetrahymena 1              V     T  VFV  T    QHR  D   I    V
Tetrahymena 2              V        RQ        GL   A E F     S
Tetrahymena 4              V        VQ        GL   A   F     S
Tetrahymena 6                       VQ        GL   T   FS
Tetrahymena 9                    S  RL I      QITAAM N L S  V Q
Rat trachea B                    R LCQ        GL  LA   F     S
Rat trachea F                     V KQ        GL  LAL  F    LS
```

Fig. 2 Partial sequences of several dynein heavy chains presumed to encode axonemal dynein isoforms amplified with the S2 and A2 primers. Only nonconserved differences from the consensus sequences are shown, using the single-letter abbreviations. Conserved substitutions are only E and D, K and R, S and T, G and A, L and I, N and Q, F and Y. The sequences were obtained from the following sources: Gibbons *et al.*, 1994 (sea urchin); Asai *et al.*, 1994 (*Paramecium*); Porter *et al.*, 1992 (*Chlamydomonas*); D. Wilkes, J. Forney, and D. Asai, unpublished (*Tetrahymena*); and S. Ashworth, L. Ostrowski, and D. Asai, unpublished (rat trachea).

E. Analysis of Sequences

When the RNA is from cells that produce cilia, multiple different dynein sequences have been obtained, thus supporting the hypothesis that different dynein functional isoforms are encoded by separate genes. The method described in this chapter has identified numerous dynein genes expressed in several species. Examples of deduced partial sequences of various axonemal dynein heavy chains are presented in Fig. 2. In many cases the sequences have been extended in both directions, but these data are not included in the figure.

Acknowledgments

The method was developed in collaboration with Ian Gibbons and Wen-Jing Y. Tang of the Pacific Biomedical Research Center, University of Hawaii. We thank Ian Gibbons, Grace Tang, Hening Ren, Barbara Gibbons, Mary Porter, and Charles Brokaw for sharing their ideas and results with us. The ciliate dynein work is a collaboration with Jim Forney, Purdue Department of Biochemistry. Our laboratory is supported by grants from the National Science Foundation and National Institutes of Health.

References

Asai, D. J., and Brokaw, C. J. (1993). Dynein heavy chain isoforms and axonemal motility. *Trends Cell Biol.* **3,** 398–402.

Asai, D. J., Beckwith, S. M., Kandl, K. A., Keating, H. H., Tjandra, H., and Forney, J. D. (1994). The dynein genes of *Paramecium tetraurelia:* Sequences adjacent to the catalytic P-loop identify cytoplasmic and axonemal heavy chain isoforms. *J. Cell Sci.* **107,** 839–847.

Asai, D. J., Tang, W. J. Y., Ching, N. S., and Gibbons, I. R. (1991). Cloning and sequencing of the ATP-binding domains of novel isoforms of sea urchin dynein. *J. Cell Biol.* **115,** 369a.

Brokaw, C. J. (1994). The control of flagellar bending: A new agenda based on dynein diversity. *Cell Motil. Cytoskel.* **28,** 199–204.

Gibbons, I. R., Asai, D. J., Tang, W.-J. Y., and Gibbons, B. H. (1992). A cytoplasmic dynein heavy chain in sea urchin embryos. *Biol. Cell* **76,** 303–309.

Gibbons, B. H., Asai, D. J., Tang, W.-J. Y., Hays, T. S., and Gibbons, I. R. (1994). Phylogeny and expression of axonemal and cytoplasmic dynein genes in sea urchins. *Mol. Biol. Cell* **5,** 57–70.

Porter, M., Knott, J., Gardner, L., Farlow, S., Myster, S. and Mansaneres, K. (1992). Characterization of the dynein gene family in *Chlamydomonas reinhardtii. Mol. Biol. Cell* **3,** 161a.

Sambrook, J., Fritsch, E. F., and Maniatis, T. (1989). "Molecular Cloning: A Laboratory Manual" 2nd ed. Cold Spring Harbor Laboratory Press, Cold Spring Harbor, New York.

Saraste, M., Sibbald, P. R., and Wittinghofer, A. (1990). The P-loop—a common motif in ATP- and GTP-binding proteins. *Trends Biochem. Sci.* **15,** 430–434.

Smith, E. F., and Sale, W. S. (1992). Structural and functional reconstitution of inner dynein arms in *Chlamydomonas* flagellar axonemes. *J. Cell Biol.* **117,** 573–581.

Thuring, R. W. J., Sanders, J. P. M., and Borst, P. (1975). A freeze-squeeze method for recovering long DNA from agarose gel. *Anal. Biochem.* **66,** 213–220.

Walker, J. E., Saraste, M., Runswick, M. J., and Gay, N. J. (1982). Distantly related sequences in the α- and β-subunits of ATP synthase, myosin, kinases and other ATP-requiring enzymes and a common nucleotide binding fold. *EMBO J.* **1,** 945–951.

Wilkerson, C. G., King, S. M., and Witman, G. B. (1994). Molecular analysis of the γ heavy chain of *Chlamydomonas* flagellar outer-arm dynein. *J. Cell Sci.* **107,** 497–506.

INDEX

D

N

O

P

Q

R

S

T

VOLUMES IN SERIES

Founding Series Editor
DAVID M. PRESCOTT

Volume 1 (1964)
Methods in Cell Physiology
Edited by David M. Prescott

Volume 2 (1966)
Methods in Cell Physiology
Edited by David M. Prescott

Volume 3 (1968)
Methods in Cell Physiology
Edited by David M. Prescott

Volume 4 (1970)
Methods in Cell Physiology
Edited by David M. Prescott

Volume 5 (1972)
Methods in Cell Physiology
Edited by David M. Prescott

Volume 6 (1973)
Methods in Cell Physiology
Edited by David M. Prescott

Volume 7 (1973)
Methods in Cell Biology
Edited by David M. Prescott

Volume 8 (1974)
Methods in Cell Biology
Edited by David M. Prescott

Volume 9 (1975)
Methods in Cell Biology
Edited by David M. Prescott

Volume 10 (1975)
Methods in Cell Biology
Edited by David M. Prescott

Volume 11 (1975)
Yeast Cells
Edited by David M. Prescott

Volume 12 (1975)
Yeast Cells
Edited by David M. Prescott

Volume 13 (1976)
Methods in Cell Biology
Edited by David M. Prescott

Volume 14 (1976)
Methods in Cell Biology
Edited by David M. Prescott

Volume 15 (1977)
Methods in Cell Biology
Edited by David M. Prescott

Volume 16 (1977)
Chromatin and Chromosomal Protein Research I
Edited by Gary Stein, Janet Stein, and Lewis J. Kleinsmith

Volume 17 (1978)
Chromatin and Chromosomal Protein Research II
Edited by Gary Stein, Janet Stein, and Lewis J. Kleinsmith

Volume 18 (1978)
Chromatin and Chromosomal Protein Research III
Edited by Gary Stein, Janet Stein, and Lewis J. Kleinsmith

Volume 19 (1978)
Chromatin and Chromosomal Protein Research IV
Edited by Gary Stein, Janet Stein, and Lewis J. Kleinsmith

Volume 20 (1978)
Methods in Cell Biology
Edited by David M. Prescott

Advisory Board Chairman
KEITH R. PORTER

Volume 21A (1980)
Normal Human Tissue and Cell Culture, Part A: Respiratory, Cardiovascular, and Integumentary Systems
Edited by Curtis C. Harris, Benjamin F. Trump, and Gary D. Stoner

Volume 21B (1980)
Normal Human Tissue and Cell Culture, Part B: Endocrine, Urogenital, and Gastrointestinal Systems
Edited by Curtis C. Harris, Benjamin F. Trump, and Gary D. Stoner

Volume 22 (1981)
Three-Dimensional Ultrastructure in Biology
Edited by James N. Turner

Volume 23 (1981)
Basic Mechanisms of Cellular Secretion
Edited by Arthur R. Hand and Constance Oliver

Volume 24 (1982)
The Cytoskeleton, Part A: Cytoskeletal Proteins, Isolation and Characterization
Edited by Leslie Wilson

Volume 25 (1982)
The Cytoskeleton, Part B: Biological Systems and *in Vitro* Models
Edited by Leslie Wilson

Volume 26 (1982)
Prenatal Diagnosis: Cell Biological Approaches
Edited by Samuel A. Latt and Gretchen J. Darlington

Series Editor
LESLIE WILSON

Volume 27 (1986)
Echinoderm Gametes and Embryos
Edited by Thomas E. Schroeder

Volume 28 (1987)
***Dictyostelium discoideum:* Molecular Approaches to Cell Biology**
Edited by James A. Spudich

Volume 29 (1989)
Fluorescence Microscopy of Living Cells in Culture, Part A: Fluorescent Analogs, Labeling Cells, and Basic Microscopy
Edited by Yu-Li Wang and D. Lansing Taylor

Volume 30 (1989)
Fluorescence Microscopy of Living Cells in Culture, Part B: Quantitative Fluorescence Microscopy—Imaging and Spectroscopy
Edited by D. Lansing Taylor and Yu-Li Wang

Volume 31 (1989)
Vesicular Transport, Part A
Edited by Alan M. Tartakoff

Volume 32 (1989)
Vesicular Transport, Part B
Edited by Alan M. Tartakoff

Volume 33 (1990)
Flow Cytometry
Edited by Zbigniew Darzynkiewicz and Harry A. Crissman

Volume 34 (1991)
Vectorial Transport of Proteins into and across Membranes
Edited by Alan M. Tartakoff

Selected from Volumes 31, 32, and 34 (1991)
Laboratory Methods for Vesicular and Vectorial Transport
Edited by Alan M. Tartakoff

Volume 35 (1991)
Functional Organization of the Nucleus: A Laboratory Guide
Edited by Barbara A. Hamkalo and Sarah C. R. Elgin

Volume 36 (1991)
***Xenopus laevis:* Practical Uses in Cell and Molecular Biology**
Edited by Brian K. Kay and H. Benjamin Peng

Series Editors
LESLIE WILSON AND PAUL MATSUDAIRA

Volume 37 (1993)
Antibodies in Cell Biology
Edited by David J. Asai

Volume 38 (1993)
Cell Biological Applications of Confocal Microscopy
Edited by Brian Matsumoto

Volume 39 (1993)
Motility Assays for Motor Proteins
Edited by Jonathan M. Scholey

Volume 40 (1994)
A Practical Guide to the Study of Calcium in Living Cells
Edited by Richard Nuccitelli

ISBN 0-12-564148-6

90018

9 780125 641487